# Methods in Ring Theory

# NATO ASI Series

## Advanced Science Institutes Series

*A series presenting the results of activities sponsored by the NATO Science Committee, which aims at the dissemination of advanced scientific and technological knowledge, with a view to strengthening links between scientific communities.*

The series is published by an international board of publishers in conjunction with the NATO Scientific Affairs Division

| | | |
|---|---|---|
| A | Life Sciences | Plenum Publishing Corporation |
| B | Physics | London and New York |
| C | Mathematical and Physical Sciences | D. Reidel Publishing Company Dordrecht, Boston and Lancaster |
| D | Behavioural and Social Sciences | Martinus Nijhoff Publishers |
| E | Engineering and Materials Sciences | The Hague, Boston and Lancaster |
| F | Computer and Systems Sciences | Springer-Verlag |
| G | Ecological Sciences | Berlin, Heidelberg, New York and Tokyo |

Series C: Mathematical and Physical Sciences Vol. 129

# Methods in Ring Theory

edited by

F. van Oystaeyen

Department of Mathematics,
University of Antwerp, Antwerp, Belgium



D. Reidel Publishing Company

Dordrecht / Boston / Lancaster

Published in cooperation with NATO Scientific Affairs Division

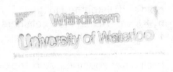

Proceedings of the NATO Advanced Study Institute on
Methods in Ring Theory
Antwerp
August 2-12, 1983

Library of Congress Cataloging in Publication Data

NATO Advanced Study Institute on Methods in Ring Theory
(1983: Antwerp, Belgium)
Methods in ring theory.

(NATO ASI series. Series C, Mathematical and physical sciences, vol, 129)
"Proceedings of the NATO Advanced Study Institute on Methods in Ring Theory, Antwerp, August 2–12, 1983" - T.p. verso.
1. Rings (Algebra)-Congresses. I. Oystaeyen, F. van, 1947- . II. Title. III. Series: NATO advanced study institutes series. Series C, Mathematical and physical sciences, v. 129.
QA247.N37 1983 512′.4 84-9876
ISBN 90-277-1743-5

Published by D. Reidel Publishing Company
P.O. Box 17, 3300 AA Dordrecht, Holland

Sold and distributed in the U.S.A. and Canada
by Kluwer Academic Publishers,
190 Old Derby Street, Hingham, MA 02043, U.S.A.

In all other countries, sold and distributed
by Kluwer Academic Publishers Group,
P.O. Box 322, 3300 AH Dordrecht, Holland

D. Reidel Publishing Company is a member of the Kluwer Academic Publishers Group



Printed in The Netherlands.

TABLE OF CONTENTS


Acknowledgement viii

J. ALAJBEGOVIĆ / R-Prüfer Rings and Approximation Theorems 1

T. ALBU / Certain Artinian Lattices Are Noetherian. Applications to the Relative Hopkins-Levitzki Theorem 37

J. L. BUESO and P. JARA / A Generalization of Semisimple Modules 53

S. CAENEPEEL / Graded Complete and Graded Henselian Rings 67

B. FEIN and M. SCHACHER / Cyclic Classes in Relative Brauer Groups 81

K. R. GOODEARL / Simple Noetherian Non-Matrix Rings 87

R. GORDON / Group-Gradings of Categories 89

W. HABOUSH / Brauer Groups of Homogeneous Spaces I. 111

M. HARADA / Simple Submodules in a Finite Direct Sum of Uniform Modules 145

R. T. HOOBLER / Functors of Graded Rings 161

A. HUDRY / Sur une Classe d'Algèbres Filtrées 171

Y. IWANAGA / Some Special Class of Artin Rings of Finite Type 181

E. JESPERS and P. F. SMITH / Group Rings and Maximal Orders 185

T. KANZAKI / A Note on Infinite Torsion Primes of a Commutative Ring 197

I. KERSTEN and J. MICHALIČEK / Applications of Kummer Theory Without Roots of Unity 201

A. A. KLEIN / The Index of Nility of a Matrix Ring Over a Ring With Bounded Index 207

J. KREMPA and J. MATCZUK / On Algebraic Derivations of Prime Rings 211

L. LE BRUYN / Smooth Maximal Orders in Quaternion Algebras I 231

M. LORENZ / Group Rings and Division Rings 265

L. MAKAR-LIMANOV / On Free Subobjects of Skew Fields 281

W. S. MARTINDALE, 3rd and C. R. MIERS / Herstein's Lie and Jordan Theory Revisited 287

H. MARUBAYASHI / Divisorially Graded Orders in a Simple Artinian Ring 291

K. MASAIKE / $\Delta$-Injective Modules and QF-3 Endomorphism Rings 317

S. MONTGOMERY / Group Actions on Rings: Some Classical Problems 327

B. J. MÜLLER / Links Between Maximal Ideals in Bounded Noetherian Prime Rings of Krull Dimension One 347

I. N. MUSSON / Noetherian Subrings of Quotient Rings 379

C. NĂSTĂSESCU and S. RAIANU / Stability Conditions for Cummutative Rings with Krull Dimension 391

D. S. PASSMAN / Cancellative Group-Graded Rings 403

M. PITTALUGA / The Automorphism Group of a Polynomial Algebra 415

K. W. ROGGENKAMP / Auslander-Reiten Quivers for Some Artinian Torsion Theories and Integral Representations 433

K. W. ROGGENKAMP / Automorphisms and Isomorphisms of Integral Group Rings of Finite Groups 451

H. SATO / Self-Injective Dimension of Serial Rings 455

W. F. SCHELTER / Smooth Affine PI Algebras 483

A. H. SCHOFIELD / Questions on Skew Fields 489

S. K. SEHGAL / Torsion Units in Group Rings 497

J.-P. TIGNOL / On the Length of Decompositions of Central Simple Algebras in Tensor Products of Symbols 505

M. VAN DEN BERGH / A Duality Theorem for Hopf Algebras 517

F. VAN OYSTAEYEN / Note on Central Class Groups of Orders Over Krull Domains 523

A. VERSCHOREN / On the Picard Group of a Quasi-Affine Scheme 541

J. M. ZELMANOWITZ / Duality Theory for Quasi-Injective Modules 551

List of Participants 567

Index 570

LECTURES (one-hour)

2-8-83 M. ARTIN, Algebras of global dimension two.
E. FRIEDLANDER, Foundations of algebraic K-Theory.
H. BASS, Automorphisms of polynomial rings.

3-8-83 M. LORENZ, Group rings and division rings.
S.A. AMITSUR, Generic methods I.
E. FORMANEK, The invariants of nxn matrices.

4-8-83 K. ROGGENKAMP, Isomorphisms and automorphisms of group rings.
L. SMALL, Bergman-Small's theorem revisited.
E. FRIEDLANDER, Topological methods in K-Theory.
H. BASS, The Jacobian conjecture.

5-8-83 S. MONTGOMERY, Group actions on rings; classical problems.
D. PASSMAN, Infinite crossed products.
E. FORMANEK, Noncommutative invariant theory.
M. HARADA, Extending property of simple modules.

8-8-83 S.A. AMITSUR, Generic methods II.
S. MONTGOMERY, Free algebras and generic matrix rings.
K. GOODEARL, Simple Noetherian non-matrix rings.

9-8-83 K. ROGGENKAMP, Auslander-Reiten quivers of some torsion theories.
M. AUSLANDER, Almost split exact sequences and rational singularities.
D. SALTMAN, Brauer-Severi varieties and units in orders.
M. HARADA, Simple submodules in a direct sum of uniform modules.

11-8-83 T. STAFFORD, Stable range and projective modules.
A. VERSCHOREN, Relative invariants of rings.

12-8-83 S.P. SMITH, Sheaves of twisted differential operators on the flag variety.
F. VAN OYSTAEYEN, Clifford systems and generalized crossed products.

ACKNOWLEDGEMENT

We are grateful to the NATO scientific committee for making the Advanced Study Institute "Methods in Ring Theory" possible.

Further financial support towards the organization of this meeting has been supplied to us by : the Belgian Foundation of Scientific Research N.F.W.O., the University of Antwerp U.I.A., the Belgian Friends of the Hebrew University of Jerusalem, the Ministry of Education.

The "Kredietbank" (KB) and "Algemene Spaar en Lijfrente Kas" (ASLK) and also the "Herfurth" travel agency provided some partial assistence in the organization of the meeting and the entertainment program.

Finally the A.S.I. owes all of its success to the cooperation of all lecturers, authors of papers for the proceedings and the hard labour of the local committee. Allow me to single out Alain Verschoren from the latter, not only because he organized the book-exhibit, but also because he promised to help me out with some typing work (if this line does not appear in print, it means that he did not stick to his promise).

# R-PRÜFER RINGS AND APPROXIMATION THEOREMS

Jusuf Alajbegović

Department of Mathematics PMF Sarajevo
Yugoslavia

Abstract

The main result of this paper is a general approximation theorem for pairwise incomparable valuations on a commutative ring R in case that the intersection of the corresponding valuation rings is a R-Prüfer ring. As a corollary we obtain a general approximation theorem for rings with large Jacobson radical. We also get as a corollary a general approximation theorem for pairwise incomparable valuations on a total quotient ring T(A) of the Prüfer ring A, due to M.Arapović [2] .
In the first part of this paper we deduce some results concerning R-Prüfer rings. We show that one form of the Chinese Remainder Theorem holds for a R-Prüfer ring A without Griffin's extra assumption $R \subseteq T(A)$ .

## §0. Preliminary results on the Manis valuation

All rings considered will be commutative and have unity. Suppose A is a subring of a ring R . An element $a \in A$ is said to be R-regular if a lies in U(R) ,



F. van Oystaeyen (ed.), Methods in Ring Theory, 1–36.

the group of units of the ring R . An ideal I of the ring A is called R-regular if $I \cap U(R) \neq \emptyset$ . Also, if B is a subring of the ring R and $A \subseteq B$ , then we shall say B is a R-overring of the ring A .

<u>Definition 0.1.</u> A Manis valuation on a ring R is a mapping $v: R \to \Gamma_\infty$ from R onto a totally ordered additive group $\Gamma$ with a simbol $\infty$ adjoined such that $\Gamma_\infty = \Gamma \cup \{\infty\}$ , where $\infty + \infty = \infty \notin \Gamma$ and for all $\gamma \in \Gamma$ is $\gamma + \infty = \infty + \gamma = \infty > \gamma$ . The mapping v satisfais the following conditions :

i) $v(R) = \Gamma_\infty$ ;

ii) $(\forall x,y \in R) \quad v(xy) = v(x) + v(y)$ ;

iii) $(\forall x,y \in R) \quad v(x+y) \geq \min\{v(x), v(y)\}$ .

<u>Remark 0.2.</u> Using the notations in Def.0.1. it is easy to conclude that $R_v = \{x \in R : v(x) \geq 0\}$ is a subring of the ring R . $R_v$ is called the ring of the valuation v , and $P_v = \{x \in R : v(x) > 0\}$ is a prime ideal of the ring $R_v$ . The infinite ideal of the valuation v is the set $\{x \in R : v(x) = \infty\} = v^{-1}(\infty)$. It is evident that $v^{-1}(\infty)$ is a prime ideal of the ring R contained in $R_v$. A valuation v on a ring R is said to be trivial if $v(R) = \{0, \infty\}$ . It follows easily that there exists a bijection between the set of all trivial valuations of the ring R and the set of all prime ideals of R .

The following result is due to M.E.Manis [8, Prop.1]:

<u>Theorem 0.3.</u> Let A be a subring of a ring R and P a prime ideal of the ring A . Then the following statements are equivalent :

i) If B is a R-overring of A and M a prime ideal of B such that $A \cap M = P$ , then $A = B$ ;

ii) $(\forall x \in R\setminus A)(\exists p \in P)\ xp \in A\setminus P$ ;

iii) There exists a Manis valuation $v:R \to \Gamma_\infty$ of the ring R such that $R_v=R$ and $P_v=P$ .

Remark 0.4. A pair (A,P) having properties i),ii), iii) of Th.0.3. is called a Manis valuation pair for R and if a valuation $v:R\to\Gamma_\infty$ satisfies $R_v=R$ and $P_v=P$ we say that v corresponds to the pair (A,P) .

If v and w are two valuations on a ring R we shall write $w \leq v$ if there exists an epimorphism $f:\Gamma_v \cup\{\infty\} \to \Gamma_w \cup \{\infty\}$ of totally ordered semigroups such that $f(\infty)=\infty$ and $w(x)=f(v(x))$ for all $x\in R$ . We remark that if $w\leq v$ then the infinite ideals of these valuations are equal , i.e. $v^{-1}(\infty)=w^{-1}(\infty)$ .

If $v \not\leq w$ and $w \not\leq v$ , we say that the valuations v and w are incomparable. We say that the valuations v and w are dependent if for some nontrivial valuation $v'$ we have $v' \leq v$ and $v' \leq w$ . Otherwise,we shall say that v and w are independent .

For a valuation $v:R\to\Gamma_\infty$ on a ring R we shall say that an ideal Q of the ring $R_v$ is v-closed if $v(a)\leq v(x)$ for some $a\in Q$ and $x\in R$ implies $x\in Q$ . One can easily deduce that the radical of any v-closed ideal is a prime ideal of the valuation ring.

If A is a subring of a ring R and P a prime ideal of the ring A , then $A_{[P]}=\{x\in R:(\exists s\in A\setminus P)\ xs\in A\}$ is a subring of the ring R and $A\subseteq A_{[P]}$ . Also,the set $[P]A_{[P]}=\{x\in R:(\exists s\in A\setminus P)\ xs\in P\}$ is a prime ideal of the ring $A_{[P]}$ and $A\cap[P]A_{[P]}=P$ . Furthermore,if P is a R-regular prime ideal of the ring A ($\supsetneq$ P) , then $A_{[P]}\subsetneq R$ . It can be shown that $A_{[P]}\subsetneq R$ implies the R-regularity of the ideal P in some special cases,e.g.

if $R=T(A)$ , the total quotient ring of the ring R ,or in the case R has a large Jacobson radical .

In the following proposition we shall list some of the results obtained by M.E.Manis [8,Prop.3] and by M.Griffin [6,Prop.4.] :

Proposition 0.5. Let $v:R\to\Gamma_\infty$ be a valuation on a ring R . Then the following hold :

i) The set of all v-closed ideals of the ring $R_v$ is totally ordered by inclusion ; The prime v-closed ideals of the ring $R_v$ are exactly those prime ideals P of the ring $R_v$ such that

$$v^{-1}(\infty)\subseteq P\subseteq P_v \quad .$$

ii) There exists a bijective inclusion-inverting correspondence $\varphi$ between the set of all prime v-closed ideals P of the ring $R_v$ and the set of all isolated subgroups $\Delta$ of the group $\Gamma$ defined in the following way :

$$\varphi:P\mapsto\Delta_P \quad ; \quad \Delta_P=\{\gamma\in\Gamma:(\forall x\in P)\ \max\{-\gamma,\gamma\}<v(x)\}.$$

On the other hand,to the isolated subgroup $\Delta$ of the group $\Gamma$ corresponds v-closed prime ideal

$$P_\Delta=\{x\in R:(\forall\delta\in\Delta)\ \delta<v(x)\} \quad .$$

iii) If an isolated subgroup $\Delta$ of the group $\Gamma$ corresponds to the prime v-closed ideal P of the ring $R_v$ , i.e. if $\Delta=\Delta_P$ , then the mapping $w=f\circ v$ , where $f|_\Gamma:\Gamma\to\Gamma/\Delta$ is the canonical epimorphism and $f(\infty)=\infty$ , is a valuation on the ring R . Furthermore,we have $R_w=\{x\in R:\ xP\subseteq P\}$ and $P_w=P$ . If $\Delta\subsetneq\Gamma$ , then $R_w=R_{v[P]}$ . Also , $\Delta\subsetneq\Gamma$ if and only if $P\nsubseteq v^{-1}(\infty)$ .

One can easily verify the following lemma :

Lemma 0.6. Let $(A,P)$ and $(A',P')$ be valuation pairs of a ring $R$ . Then the following hold :

i) $A=A' \subsetneq R \Rightarrow P=P'$ ;

ii) $P=P' \Rightarrow A=A'$ ;

iii) If $v$ and $v'$ are the valuations on the ring $R$ corresponding to the pairs $(A,P)$ and $(A',P')$ respectively, then $v' \leq v$ if and only if

$$v^{-1}(\infty) \subseteq P_{v'} \subseteq P_v \text{ and } R_v \subseteq R_{v'} .$$

Proposition 0.7. Let $A$ be a subring of a ring $R$ and $P$ a prime ideal of the ring $A$ such that $(A_{[P]}, [P]A_{[P]})$ is a Manis valuation pair of the ring $R$ . Also, assume that the valuation $v:R \to \Gamma_\infty$ corresponding to the given pair is nontrivial. Then the following is satisfied

$$(\forall x,y \in A)(\{x,y\} \not\subseteq v^{-1}(\infty)) \Rightarrow (\exists a,b \in A)(\{a,b\} \not\subseteq P \wedge \wedge ax=by)$$

Proof - Suppose that $x,y \in A$ and that $\{x,y\} \not\subseteq v^{-1}(\infty)$ and $v(y) \leq v(x)$ . Then $v(y) < \infty$ , and there exists $r \in R$ with $v(r)=-v(y)$ , i.e. $ry \in A_{[P]} \setminus [P]A_{[P]}$ . Further, we have $xr \in A_{[P]}$ since $v(xr)=v(x)-v(y) \geqslant 0$ . Therefore, we have for some $u,v \in A \setminus P$ , $xr \cdot u \in A$ and $yru \cdot v \in A \setminus P$ . So, if we take $a=yruv \in A \setminus P$ and $b=uxvr \in A$ , then $\{a,b\} \not\subseteq P$ and $ax=by$ .

Remark 0.8. i) If the ring $R$ in Proposition 0.7. is equal to the total quotient ring $T(A)$ of the ring $A$ , we get a result of M.D.Larsen [7, Th.10.14., part 1)⇒2)] and M.Griffin [5, Lemma 5 , part 1)⇒3)] . In fact, it suffices to see that in case $R=T(A)$ the core $C(P_v)$ of the ideal $P_v$ is equal to the infinite ideal $v^{-1}(\infty)$.

ii) The next proposition also corresponds to the analogous results of M.Griffin and M.D.Larsen in the case $R=T(A)$ .

Proposition 0.9. Let $A$ be a subring of a ring $R$ and $P$ a prime ideal of the ring $A$ such that $(A_{[P]}, [P]A_{[P]})$ is a Manis valuation pair of the ring R and $A_{[P]} \subsetneq R$ . Then for all ideals $I$ and $J$ of the ring $A$ , at least one of which is R-regular, the following holds :

$$I\,A_P \subseteq J\,A_P \quad \text{or} \quad J\,A_P \subseteq I\,A_P \quad .$$

Proof - Let $I \subsetneq A$ and $J \subsetneq A$ . Further, assume the ideal $I$ is R-regular and $a \in I \cap U(R)$ . Clearly, we have $0 \leq v(a) < \infty$ for a valuation $v$ on the ring $R$ such that $R_v = A_{[P]}$ and $P_v = [P]A_{[P]}$ . We shall first show that $IA_P \nsubseteq JA_P$ implies the existence of an element $a_o \in A$ such that $0 \leq v(a_o) < \infty$ and $a_o s \notin J$ for all $s \in A \setminus P$ . Suppose that $a \in I \cap U(R)$ is not the element we are looking for , i.e. suppose $q \in A \setminus P$ and $aq \in J$ . Then there exists $d \in I$ with $ds \notin J$ for all $s \in A \setminus P$ becouse of the assumption $IA_P \nsubseteq JA_P$ . According to Proposition 0.7. for elements $a$ and $d$ ; $\{a,d\} \nsubseteq v^{-1}(\infty)$ we have $at=dz$ for some $t,z \in A$ and $\{t,z\} \nsubseteq P$ . Therefore, $t \in A \setminus P$ since otherwise $t \in P$ implies $z \in A \setminus P$ and also $zq \in A \setminus P$ . Hence, it follows that $d \cdot zq = dz \cdot q = aq \cdot t \in JA \subseteq J$ , which is a contradiction since $ds \notin J$ for all $s \in A \setminus P$ . Thus we have $t \in A \setminus P$ and $at=dz$ for some $z \in A$ . From $v(t)=0$ and $v(a) < \infty$ we conclude that $v(d) \neq \infty$ , i.e. the following is true :
$d \in I$ ; $0 \leq v(d) < \infty$ ; $(\forall s \in A \setminus P)$ $ds \notin J$ .
So we can take $a_o = d$ .

Now, we can show $JA_P \subseteq IA_P$ . It suffices to show $J \subseteq IA_P$ . Let $b \in J$ and $a_o \in A$ such that $0 \leq v(a_o) < \infty$

and $a_o s \notin J$ for all $s \in A \setminus P$. By Proposition 0.7. we see that $\{b, a_o\} \not\subseteq v^{-1}(\infty)$ which implies $a_o x = by$ for some $x, y \in A$ such that $\{x, y\} \not\subseteq P$. Since $b \in J$, $by \in J$ and therefore $x \in P$, consequently $y \in A \setminus P$. So we now have $b = by/y = a_o x/y$ in $A_P$, i.e. $b \in IA_P$.

Remark 0.10. i) The proof of Proposition 0.9. is given in all its details for the sake of completeness and as an illu stration of how one can avoid the extra hypothesis " $C(P_v) = v^{-1}(\infty)$ " for a nontrivial valuation $v$ on the ring $R$.
ii) If the ideal $P$ in Proposition 0.9. is R-regular prime ideal of the ring $A$ and $P \subsetneq A$, then we must have $A_{[P]} \subsetneq R$. If $P$ is not R-regular and $I$ is R-regular, then $IA_P = A_P$, consequently $JA_P \subseteq IA_P$. In fact, if $a \in I \cap U(R)$ then $a \in A \setminus P$ and we have $1 = a/a \in IA_P$.

## §1. R-Prüfer rings

In the classical case, for domain $A$, we say that $A$ is a Prüfer domain if for every maximal ideal $M$ of $A$ the localization $A_M$ is a valuation domain of the field of fractions $T(A)$ of the domain $A$. M.Griffin [5] introduced the class of Prüfer rings with zero divisors using the notion of the Manis valuation on a commutative ring [8]. More precisely, we say that a commutative ring $A$ is a Prüfer ring if for each maximal ideal $M$ of the ring $A$ the large quotient ring $A_{[M]} = \{x \in T(A) : (\exists s \in A \setminus M) xs \in A\}$ is a Manis valuation ring of the total quotient ring $T(A)$ of the ring $A$. M.Griffin [5] gave different characterizations of Prüfer rings. Latter, in [7], it has been shown that a ring $A$ is a Prüfer ring if and only if for each maximal ideal M of A the pair $(A_{[M]},[M]A_{[M]})$ is a Manis valuation pair of the total quotient ring T(A) of the ring A. Here $[M]A_{[M]}$ denotes the set $\{x\in T(A):(\exists s\in A\setminus M)\ xs\in M\}$. Imitating the methods in [5], M.Griffin in [6] introduced the class of R-Prüfer rings in the following way :

A subring A of a ring R is a R-Prüfer ring (in Griffin's sense) if $A_{[M]}=\{x\in R:(\exists s\in A\setminus M)\ xs\in A\}$ is a Manis valuation ring of R for each maximal ideal M of the ring A.

However, some of the propositions in [6] are incomplete and some of them are incorect as it has been shown by J.Gräter [3,p.285]. Following J.Gräter we shall give the following definition :

<u>Definition 1.1.</u> A subring A of a ring R is a R-Prüfer ring if for each maximal ideal M of A the pair $(A_{[M]},[M]A_{[M]})$ is a Manis valuation pair of the ring R, where $A_{[M]}=\{x\in R:(\exists s\in A\setminus M)\ xs\in A\}$ and $[M]A_{[M]}=\{x\in R:(\exists s\in A\setminus M)\ xs\in M\}$.

<u>Remark 1.2.</u> It is not difficult to conclude that in case R=T(A) the ring A is a R-Prüfer ring if and only if A is a R-Prüfer ring in Griffin's sense. On the other hand, this statement is not true in general as the following example, due to J.Gräter [3], shows :

<u>Example 1.3.</u> Let A=Z[X] be the ring of polynomials with coefficients in the ring of integers Z and $R=Z[X,X^{-1}]$ a subring of the total quotient ring T of the ring A. The set P=X Z[X] is a prime ideal of the ring A and the following hold :

i) (A,P) is a Manis valuation pair of the ring R ;

ii) The ring $A$ is a R-Prüfer ring in Griffin's sense;

iii) For the maximal ideal $M=(2,X)$ of the ring $A$ the pair $(A_{[M]},[M]A_{[M]})$ is not a valuation pair of $R$;

iv) The ideal $P$ of the ring $A$ is R-regular and $P$ is not maximal in $A$.

Remark 1.4. Example 1.3. shows that there exists a nontrivial valuation $v$ on a ring $R$ such that $R$ is a subring of the total quotient ring $T(R_v)$ of the ring $R_v$ and $P_v$ is not a maximal ideal in $R_v$ and on the other hand $R_v$ is a R-Prüfer ring in Griffin's sense. Therefore, Proposition 14. in [6] is not correct.
At the end of this paragraph we shall prove that the positive ideal $P_v$ of a nontrivial valuation $v$ on a ring $R$, such that $R_v$ is a R-Prüfer ring, is a maximal ideal in $R_v$. Also, we shall generalize some of Griffin's results in [6] ommiting the assumption $R\subseteq T(A)$.

Theorem 1.5. If $A$ is a R-Prüfer ring, then the following hold :

i) $I(J\cap L)=IJ\cap IL$ , if $I,J$ and $L$ are ideals in $A$ and either $J$ or $L$ is R-regular ;

ii) $(I+J)(I\cap J)=IJ$ , for all ideals $I,J$ of the ring $A$ if at least one of them is R-regular ;

iii) $I\cap(J+L)=(I\cap J)+(I\cap L)$ , for all ideals $I,J$ and $L$ of the ring $A$ if at least one of them is R-regular.

Proof - The statements i),ii),iii) hold for extensions of ideals in $A_M$ for each maximal ideal $M$ in $A$ according to Proposition 0.9. and Remark 0.10.ii).

Definition 1.6. Let $A$ be a subring of a ring $R$ and $L$ a submodule of the $A$-module $R$. We say $L$ is an

R-fractionary ideal of the ring A if there exists a regular element $r \in R$ such that $rL \subseteq A$ .

An ideal I of the ring A is called R-invertible if there exists an R-fractionary ideal L of the ring A such that IL=A .

<u>Proposition 1.7.</u> Let A be a subring of a ring R . Suppose that $(I+J)(I\cap J)=IJ$ , whenever I and J are ideals in A , of which at least one is R-regular. Then each finitely generated R-regular ideal of the ring A is R-invertible .

<u>Proof</u> - Cf. [7, Theorem 10.18; part 6) ⇒ 1)] .
We remark that for a finitely generated R-regular ideal I and $c \in I \cap U(R)$ we have IL=A if $L=J\cdot Ac^{-1}$ where $J=\{x \in R: xI \subseteq Ac\}$ is an ideal in A .

<u>Theorem 1.8.</u> If A is a R-Prüfer ring, then each finitely generated R-regular ideal of the ring A is R-invertible .

<u>Proof</u> - The proof follows immediately from Theorem 1.5. and Proposition 1.7.

<u>Remark 1.9.</u> Theorem 1.8. generalizes a result of M.Griffin [6, Theorem 7.; part 1) ⇒ 3)] in the case that R is a subring of the total quotient ring T(A) of the ring A .

<u>Proposition 1.10.</u> If A is a R-Prüfer ring and R is a subring of the total quotient ring of the ring A , then each R-overring of the ring A is also a R-Prüfer ring .

<u>Proof</u> - It suffices to observe Proposition 6. in [6].

<u>Definition 1.11.</u> If A is a subring of a ring R we say that Chinese Remainder Theorem holds for A with

respect to R if the following is valid :

For any ideals $M_1,\dots,M_n$ of $A$, of which at most two ideals not R-regular, and for elements $x_1,\dots,x_n \in A$, the system of congruences $x \equiv x_i \pmod{M_i}$ admits a solution $x \in A$ if and only if $x_i \equiv x_j \pmod{M_i+M_j}$ for all $1 \leq i,j \leq n$ .

<u>Proposition 1.12.</u> Suppose $A$ is a subring of a ring $R$ . Then the following statements are equivalent :

i) The Chinese Remainder Theorem holds for $A$ with respect to $R$ ;

ii) For all ideals $L$ , $M$ and $N$ of the ring $A$ , of which at least one is R-regular, we have :

$$L \cap (M+N) = (L \cap M) + (L \cap N) \quad ;$$

iii) For all ideals $L$ , $M$ and $N$ of $A$ , at least one of which is R-regular, we have :

$$L + (M \cap N) = (L+M) \cap (L+N) \quad .$$

<u>Proof</u> - i) ⇒ ii) ⇒ iii) ⇒ i) ;

For ii) ⇒ iii) ⇒ i) see [11] . We shall prove i) ⇒ ii). Let $M$ , $N$ and $L$ are any three ideals of the ring $A$ and of which at most two are not R-regular. It suffices to show that $L \cap (M+N) \subseteq (L \cap M) + (L \cap N)$ . Take an element $a \in L \cap (M+N)$ and consider the system of congruences $x \equiv 0 \pmod{L}$ , $x \equiv 0 \pmod{M}$ , $x \equiv a \pmod{L}$ , $x \equiv a \pmod{N}$ , i.e. $x_1=x_2=0$ and $x_3=x_4=a$ ; $M_1=M_4=L$ and $M_2=M$ , $M_3=N$ . Evidently , $x_i - x_j \in M_i + M_j$ for all $i,j$ in $\{1,2,3,4\}$ . Furthermore, at most two of ideals $M_1, M_2, M_3$ are not R-regular, so there exists an element $x \in A$ such that $x - x_i \in M_i$ , $1 \leq i \leq 3$ . But , $M_4 = M_1$ and we have $x - x_4 = (x-x_1)+(x_1-x_4) \in M_1+M_4 = M_4$ , i.e. $x - x_i \in M_i$ ; $1 \leq i \leq 4$.

Proposition 1.13. If $A$ is a R-Prüfer ring then the Chinese Remainder Theorem holds for $A$ with respect to $R$.

Proof - The proof of the statement follows directly from Theorem 1.4. and Proposition 1.12.

Remark 1.14. Proposition 1.13. generalizes a result of M.Griffin [6,p.425] since we do not assume that $R$ is a subring of $T(A)$.

Proposition 1.15. Let $A$ be an R-Prüfer ring and $P$ a prime ideal of $A$ ; $P \subsetneq A$. Then the pair $(A_{[P]}, [P]A_{[P]})$ is a valuation pair of the ring $R$.

Proof - Let $Q$ denote a maximal ideal of $A$ such that $P \subseteq Q$. Then the pair $(A_{[Q]}, [Q]A_{[Q]})$ is a valuation pair of the ring $R$ and $A_{[Q]} \subseteq A_{[P]}$ holds. If $x \in R \setminus A_{[P]}$, then $x \in R \setminus A_{[Q]}$ and for some $q$ in $[Q]A_{[Q]}$ we have $xq \in A_{[Q]} \setminus [Q]A_{[Q]}$. Therefore, there exists $s \in A \setminus Q$ such that $qs \in Q$ and there exists $t$ in $A \setminus Q$ such that $xq \cdot t \in A \setminus Q$. Consequently, $xqst \in A \setminus Q \subseteq A \setminus P$, while $qst \in A$. Furthermore, we have $qst \in P$ since $x \notin A_{[P]}$. Thus we have $x \cdot qst \in A \setminus P \subseteq$ $\subseteq A_{[P]} \setminus [P]A_{[P]}$ and $qst \in P \subseteq [P]A_{[P]}$.

Proposition 1.16. Suppose $A$ is a R-Prüfer ring and $v : R \to \Gamma_\infty$ is a nontrivial valuation on $R$ such that $A \subseteq R_v$. Then $R_v = A_{[A \cap P_v]}$ $(= \{x \in R : (\exists s \in A \setminus P_v)\ xs \in A\})$.

Proof - The proof is the same as in the case $R=T(A)$.

Proposition 1.17. If $A$ is a R-Prüfer ring and $P_1, \ldots, P_n$ are prime ideals of the ring $A$ satisfying $A = A_{[P_1]} \cap \cdots \cap A_{[P_n]}$, then for a prime ideal $P$ of $A$

such that $P \not\subseteq P_1 \cup \cdots \cup P_n$ we have $A_{[P]}=R$ .

Proof - Cf. [6, Proposition 11.] .

Remark 1.18. Let $A$ be a subring of a ring $R$ . It is not difficult to conclude that if $P$ is a maximal ideal of the ring $A$ and the pair $(A_{[P]}, [P]A_{[P]})$ is a valuation pair of the ring $R$ then the ideal $[P]A_{[P]}$ is maximal in $A_{[P]}$ . Let for some prime ideal $P$ in $A$ the pair $(A,P)$ be a Manis valuation pair for $R$ . The ideal $P$ need not be a maximal ideal in $A$ even in the case $A$ is R-Prüfer ring in Griffin's sense ( cf. Example 1.3.). In place of the incorrect result of M. Griffin [6, Proposition 14.] we shall prove the following theorem :

Theorem 1.19. Let $v:R \rightarrow \Gamma_\infty$ be a nontrivial valuation on a ring $R$ and assume the positive ideal $P_v$ is R-regular. If the ring $R_v$ is a R-Prüfer ring, then the following hold :

i) The ideal $P_v$ is maximal in $R_v$ and each R-regular ideal strictly contained in $R_v$ is contained in $P_v$ ;

ii) Each R-regular ideal of the ring $R_v$ is v-closed.

Proof - For the proof of part i) we shall use an idea of M. Griffin [6, Proposition 14.]. We denote $A=R_v$ and $P=P_v$ .

i) Let $Q$ denote an R-regular ideal of the ring $A$ such that $Q \not\subseteq P$ . Then there exists $a \in Q$ with the property $v(a)=0$ . Suppose $r \in Q \cap U(R)$ and $r^{-1} \notin A$, and $b$ is any element from $A$ . The ideal $I=(r,a,b)$ of the ring $A$ is R-regular, and from the proof of Proposition 1.7. it follows that $(r,a,b)D=A$ if we take $D=B \cdot Ar^{-1}$ and $B=\{x \in R: xI \subseteq rA\}$ . Clearly ,

$r \in B$ implies $1 \in D$. On the other hand, $F=(r,a)D$ is an R-regular ideal of the ring A. If $1 \notin F$, then for some R-regular maximal ideal M of the ring A we have $F \subseteq M$. But $a \in F$ and $v(a)=0$ together with $M \nsubseteq P$ implies, according to Proposition 1.17., that $A_{[M]}=R$ holds. This is a contradiction since $r^{-1}$ is in $R \setminus A_{[M]}$. In fact, if $r^{-1} \in A_{[M]}$ then for some $s \in A \setminus M$ we would have $r^{-1}s \in A$ and thus $s \in rA \subseteq FA \subseteq M$. Therefore $1 \in F$, i.e. $(r,a)D=A$ and the following holds:

$$(r,a)=(r,a)A=(r,a)(r,a,b)D=(r,a,b)(r,a)D=(r,a,b)A=(r,a,b)$$

i.e. $b \in (r,a) \subseteq Q$. Consequently, we have $A=Q$ which is a contradiction.

ii) Let $I \subsetneq A$ denote an R-regular ideal of the ring A. Suppose $x \in A=R_v$, $a \in I$ satisfy $v(x) \geq v(a)$. We must show that $x \in I$.

If $v(x)=\infty$ and $a_o \in I \cap U(R)$, then $xa_o^{-1} \in R_v$ since $0 < v(a_o) < \infty$, and therefore $x \in a_oR_v \subseteq I$. Therefore we can assume that $v(a) \leq v(x) < \infty$. For $Ax \subseteq I$ it suffices to show $(Ax)A_M \subseteq IA_M$ for each maximal ideal M of the ring A. We only need to consider R-regular maximal ideals M of the ring A, since if M is not R-regular then $IA_M=A_M$ (Remark 0.10.ii)).

If M is an R-regular maximal ideal of the ring $A=R_v$ then from i) we conclude $M=P_v=P$ holds, and if $z \in R$ satisfies $v(z)=-v(a)$ then we have $az \in R_v \setminus P_v = A \setminus P$ while $xz \in R_v=A$. Therefore, we obtain

$$x=(axz)/(az)=(a\,xz)/(az) \in IA_P ,$$

since $xz \in A$, $a \in I$ and $az \in A \setminus P$.

<u>Remark 1.20.</u> Let us remark that the proof of Proposition 14. in [6] can not be applied here in the proof of Theorem 1.19., e.g. for the proof of part ii).

Proposition 1.21. Let A be a R-Prüfer ring, R a subring of the total quotient ring T(A) of A and P, Q R-regular prime ideals of A strictly contained in A. Then $A_{[P]} \subseteq A_{[Q]}$ implies $Q \subseteq P$.

Proof - From Proposition 1.10. it follows that $A_{[P]}$ is an R-Prüfer ring and of course $T(A_{[P]})=T(A)$. Since the ideal P is an R-regular ideal strictly contained in A, we have $A_{[P]} \subsetneq R$. Therefore to the pair $(A_{[P]}, [P]A_{[P]})$ corresponds a nontrivial valuation on the ring R (Proposition 1.15.) and the positive ideal of that valuation $[P]A_{[P]}$ is R-regular. According to Theorem 1.19. we have $[Q]A_{[Q]} \cap A_{[P]} \subseteq [P]A_{[P]}$, and $Q \subseteq P$ follows from the fact that $[Q]A_{[Q]} \cap A=Q$ and $[P]A_{[P]} \cap A=P$.

Proposition 1.22. Let A be a R-Prüfer ring and R a subring of the total quotient ring T(A) of the ring A. If $v:R \to \Gamma_\infty$ is a nontrivial valuation on R and $R_v=A$, then for each R-overring B of the ring A the following holds:

B is a R-Prüfer ring and if $B \subsetneq R$ then there exists a nontrivial valuation w on R such that $R_w=B$, $R_w=A_{[P_w]}$ and $P_w$ is a prime v-closed ideal in A.

Proof - Cf. [6, Proposition 13.].

Proposition 1.23. Let A be a R-Prüfer ring and R a subring of the total quotient ring T(A) of the ring A. If v and w are nontrivial valuations on the ring R such that $A \subseteq R_v \cap R_w$ and $A \cap P_v$, $A \cap P_w$ are R-regular ideals of the ring A, then the following statements are equivalent:

i) $v \leq w$; ii) $R_w \subseteq R_v$; iii) $A \cap P_v \subseteq A \cap P_w$.

Proof - i)$\Rightarrow$ii) follows directly from Lemma 0.6.iii). ii)$\Leftrightarrow$iii) follows from Proposition 1.16. and Proposition 1.21. ; iii)$\Rightarrow$i) follows from ii)$\Leftrightarrow$iii) and from Theorem 1.19.ii) .

## § 2. Approximation families of valuations , Compatible families

The class of commutative rings with large Jacobson radical M.Griffin introduced in [6] in the following way:

Definition 2.1. A commutative ring $R$ is said to have large Jacobson radical $J(R)$ if $J(R)\subseteq P$ implies $P$ is maximal in $R$ , for all prime ideals $P$ in $R$ .

Proposition 2.2.([6,Prop.19.]) If $R$ is a commutative ring with unit element $1$ , then the following statements are equivalent :

i) $R$ has a large Jacobson radical ;

ii) $(\forall a\in R)(\exists b\in R)(\forall d\in R)(\forall r\in U(R))\ \{a+rb, 1+dab\}\subseteq$
$\subseteq U(R)$ ;

iii) $(\forall a\in R)(\exists b\in R)\quad ab\in J \ \wedge\ a+b\in U(R)$ .

Remark 2.3. If $R$ is 0-dimensional ring, then $R$ has large Jacobson radical and each regular element in $R$ is invertible in $R$ . Nevertheles, if $R$ has large Jacobson radical then $R$ need not be a total quotient ring (of some ring $A$) as the following easy example shows :

Example 2.4. Let $A=Z[X]$ be the ring of polynomials in $X$ with coefficients in $Z$ the ring of integers , $P=XA$ and $R_i=A_P$ for $1\leq i\leq n<\omega$ . Then the ring

$R=R_1\times\cdots\times R_n$ has large Jacobson radical , R is not a domain , R is not 0-dimensional and there exists an regular element in R which is not invertible in R .

Proposition 2.5. If $v:R\to\Gamma_\infty$ is a nontrivial valuation on a ring R and R has large Jacobson radical, then the positive ideal $P_v$ is R-regular and R is a subring of the total quotient ring $T(R_v)$ of the ring $R_v$ .

Proof - Cf. [6,p.424.] .

Theorem 2.6. Let $v_1,\ldots,v_n$ be nontrivial valuations on a ring R with large Jacobson radical and suppose $A=R_{v_1}\cap\cdots\cap R_{v_n}$ . Then the following hold :

i) $(\forall a\in R)(\exists b\in A\cap U(R))\ ab\in A \ \wedge\ ((\forall i\in\{1,\ldots,n\})\ v_i(a)\geqslant 0 \Rightarrow v_i(b)=0)$ ;

ii) $(\forall i\in\{1,\ldots,n\})(\forall\alpha_i\in\Gamma_{v_i})(\exists a\in A\cap U(R))v_i(a)>\alpha_i$;

iii) A is a R-Prüfer ring and R is a subring of the total quotient ring T(A) of the ring A .

iv) If $R_{v_i}\not\subseteq R_{v_j}$ for all $1\leqslant i\neq j\leqslant n$ , then $A\cap P_{v_i}$ , $1\leqslant i\leqslant n$ , are exactly the R-regular maximal ideals of A .

Proof - i) is proved in [6,Prop.22] . Let us observe that i) implies R is a subring of T(A) . In fact , T(A) is a subring of the ring T(R) since each element $a\in A$ which is regular in A is obviously regular in R. Namely, if a is regular in A but not in R , then there exists $x\in R\setminus A$ such that ax=0 . From i) it follows that there exists $r\in U(R)\cap A$ such that $xr\in A$.

Therefore, we have $a\cdot xr=0$ which implies $x=0$, which is a contradiction. Thus, $R$ and $T(A)$ are both subrings of the ring $T(R)$. Now, for $a\in R$ and $b\in U(R)\cap A$ such that $ab\in A$ from the relation $a=(ab)/b\in T(A)$ we conclude $R\subseteq T(A)$.

ii) Take $i\in\{1,\dots,n\}$ and $0<\alpha_i\in\Gamma_{v_i}$. Let $Q_i$ denote the set $\{x_i\in R_{v_i} : v_i(x_i)\geqslant\alpha_i\}$.

It is easy to see that $r(Q_i)=P_i$ is a $v_i$-closed prime ideal of the ring $R_{v_i}$. Furthermore, $P_i\not\subseteq v_i^{-1}(\infty)$ holds, since otherwise $Q_i\subseteq v_i^{-1}(\infty)$ implies $\alpha_i=\infty$ which is impossible. According to Proposition 1.22. $(R_{v_i[P_i]},P_i)$ is a valuation pair of the ring $R$ and $R_{v_i[P_i]}\subsetneq R$. Therefore the ideal $P_i$ is R-regular (Proposition 2.5.) and obviously $Q_i$ is also R-regular. Take $r_i\in U(R)\cap Q_i$. From i) there exists $r\in U(R)\cap A$ such that $r_ir\in A$ and $v_j(r_i)\geqslant 0$ implies $v_j(r)=0$ for all $1\leqslant j\leqslant n$. Thus, we have $v_i(r)=0$ and $v_i(r_ir)=$ $=v_i(r_i)\geqslant\alpha_i$ and at the same time $r_ir\in U(R)\cap A$. So we conclude that there exists $b\in U(R)\cap A$ such that $v_i(b)\geqslant\alpha_i+\alpha_i>\alpha_i$ and therefore statement ii) follows.

iii) Let $M_i$ denote the set $A\cap P_{v_i}$ for all $1\leqslant i\leqslant n$.

From ii) it follows that the ideals $M_i$ are R-regular prime ideals of the ring $A$. Furthermore i) implies that $R_{v_i}=A_{[M_i]}$. On the other hand, $P_{v_i}\subseteq M_i$ implies $P_{v_i}\subseteq[M_i]A_{[M_i]}$. If $x\in R$ is such that for some $s\in A\setminus M_i\subseteq R_{v_i}\setminus P_{v_i}$ we have $xs\in M_i\subseteq P_{v_i}$, then $x\in P_{v_i}$, i.e. $[M_i]A_{[M_i]}\subseteq P_{v_i}$.

In [6] it has been shown that all R-regular maximal ideals of the ring $A$ are of the form $A\cap P_{v_i}$ for some $1 \leq i \leq n$ . Therefore $(A_{[M]}, [M]A_{[M]})$ is a valuation pair of the ring R for all R-regular maximal ideals M of the ring A . If M is maximal and M is not R-regular ideal of A , then $A_{[M]}=R$ . In fact , if $x \in R \setminus A_{[M]}$ then there exists $r \in U(R) \cap A$ such that $xr \in A$ . Since $x \notin A_{[M]}$ , we must have $r \in M$ , consequently M is R-regular which is a contradiction .

iv) The statement iv) follows immediately from the fact that each maximal R-regular ideal of the ring A has the form $A \cap P_{v_i}$ for some $1 \leq i \leq n$ .

Remark 2.7. i) From Theorem 2.6. it follows that each nontrivial valuation ring A of a ring R , with large Jacobson radical J(R) , is also a R-Prüfer ring and that there exist R-regular elements with arbitrarily large values .

ii) With the same hypotheses as in Theorem 2.6. we conclude that if v , w are valuations on R which are nontrivial and nonnegative on A , then $v \leq w$ if and only if $R_w \subseteq R_v$ (Proposition 1.23.).

iii) The part iii) of Theorem 2.6. motivated M.Griffin to formulate the definition of an approximation family of valuations on a given ring R . We shall give a somewhat different definition that includes the extra hypothesis ii) of Theorem 2.6. We remark that the property ii) in Theorem 2.6. is satisfied in case R=T(A) without the assumption that R has a large Jacobson radical .

Definition 2.8. Let $\Omega$ denote a family of nontrivial valuations on a ring R and let A denote the set $\bigcap_{v \in \Omega} R_v$ . The family $\Omega$ is called an approximation family of valuations on the ring R if the following two conditions are satisfied :

i) A is a R-Prüfer ring and R is a subring of the total quotient ring T(A) of the ring A ;

ii) $(\forall v \in \Omega)(\forall \alpha \in \Gamma_v)(\exists r \in U(R) \cap A)\ v(r) > \alpha$ .

Proposition 2.9. If R has large Jacobson radical, then each finite family of nontrivial valuations on the ring R is an approximation family for R .

Proof - The statement follows immediately from Theorem 2.6.

Remark 2.10. If A is a R-Prüfer ring and R is a subring of the total quotient ring T(A) and v , w are valuations on R which are nonnegative on A , then there exists a valuation $v \wedge w$ on R such that $R_{v \wedge w} = R_v R_w$ and $P_{v \wedge w}$ is both v-closed and a w-closed prime ideal of the ring $R_v$ , resp. $R_w$ (Proposition 1.22.) Furthermore , if $R_v R_w \subsetneq R$ , then $P_{v \wedge w} \not\subseteq v^{-1}(\infty)$ and $P_{v \wedge w} \not\subseteq w^{-1}(\infty)$ . Also , $R_{v \wedge w} = R_{v[P_{v \wedge w}]} = R_{w[P_{v \wedge w}]}$ holds .

In case that for valuations v , w there exist R-regular elements in A with arbitrarily large values, then the v-closed prime ideals of $R_v$ which are not contained in $v^{-1}(\infty)$ are R-regular ; the corresponding statement holds for w-closed prime ideals of $R_w$ . Therefore, in this case, $R_v R_w \subsetneq R$ implies $P_{v \wedge w}$ is the unique R-regular maximal ideal of the ring $R_{v \wedge w}$ .
Let $\Delta_{v,w}$ denote the largest isolated subgroup of $\Gamma_v$ such that $\Delta_{v,w} \cap v(P_{v \wedge w}) = \emptyset$ , and $\Delta_{w,v}$ is the largest

isolated subgroup of $\Gamma_w$ such that $\Delta_{w,v} \cap w(P_{v\wedge w}) = \emptyset$. Further, let $\theta_{v,w}$ denote the canonical epimorphism $\Gamma_v \to \Gamma_v / \Delta_{v,w}$ and let $\theta_{w,v}$ denote the canonical epimorphism $\Gamma_w \to \Gamma_w / \Delta_{w,v}$. The groups $\Gamma_{v\wedge w}$, $\Gamma_v / \Delta_{v,w}$ and $\Gamma_w / \Delta_{w,v}$ can be identified in the following way :

$(\forall x \in R)$ $\theta_{v,w}(v(x)) = \theta_{w,v}(w(x)) = (v\wedge w)(x)$, where $\theta_{v,w}(\infty) = \infty$ and $\theta_{w,v}(\infty) = \infty$, since $R_v R_w \subsetneq R$ implies $v\wedge w$ is a nontrivial valuation and $v\wedge w \lneq v$, $v\wedge w \lneq w$ and therefore $v^{-1}(\infty) = w^{-1}(\infty)$ holds .

In the case that $R_v R_w = R$ let $\Delta_{v,w} = \Gamma_v$ and $\Delta_{w,v} = \Gamma_w$ and $\theta_{v,w}(v(x)) = 0$ if $v(x) \neq \infty$, $\theta_{v,w}(\infty) = \infty$; $\theta_{w,v}(w(x)) = 0$ if $w(x) \neq \infty$, $\theta_{w,v}(\infty) = \infty$.

Now we are in a position to give a definition of an compatible family :

<u>Definition 2.11.</u> Let A be a R-Prüfer ring and R a subring of the total quotient ring T(A) of the ring A . Further, let $\{v_i\}_{i\in I}$ be a family of nontrivial valuations on R which are all nonnegative on A . A family $\{\alpha_i\}_{i\in I} \in \prod_{i\in I} \Gamma_{v_i}$ is called compatible if the following holds :

$$(\forall i,j \in I) \quad i \neq j \Rightarrow \theta_{v_i,v_j}(\alpha_i) = \theta_{v_j,v_i}(\alpha_j) \quad .$$

<u>Remark 2.12.</u> With the same notations as in Def. 2.11. we have :

i) For each $x \in R \setminus \bigcup_{i\in I} v_i^{-1}(\infty)$ the family $\{v_i(x)\}_{i\in I}$ is compatible .

ii) If $R_{v_i} R_{v_j} = R$ for all $i \neq j$ in $I$ , then each family $\{\alpha_i\}_{i\in I} \in \prod_{i\in I} \Gamma_{v_i}$ is compatible .

The next three lemmas can be proved in an almost identical way as for valuations on a field (Cf. [4] , Lemma 1 , Lemma 2. and the proof of Lemma 11.).

<u>Lemma 2.13.</u> Let $\Omega$ be an approximation family of valuations on a ring $R$ . If $w \in \Omega$ and $\alpha \in \Gamma_w \setminus \{0\}$, let $\Delta$ denote the largest isolated subgroup of $\Gamma_w$ not containing $\alpha$ . The mapping $v: R \to (\Gamma_w / \Delta) \cup \{\infty\}$ is defined as follows :

$v(x) = w(x) + \Delta$ if $w(x) \neq \infty$ , and $v(x) = \infty$ if $w(x) = \infty$ .

Then $v$ is a valuation on $R$ and the following holds:

$$\{w' \in \Omega : R_{w'} \subseteq R_w\} = \{w' \in \Omega : \theta_{w,w'}(\alpha) \neq 0\} .$$

Here for $w' = w$ we take $\Delta_{w,w'} = (0)$ and $\theta_{w,w'}(\gamma) = \gamma$ for all $\gamma \in \Gamma_w$ .

<u>Lemma 2.14.</u> Let $\Omega$ be an approximation family of valuations on a ring $R$ . Then the following holds :

$(\forall w_1, w_2 \in \Omega)(\forall \alpha_1 \in \Gamma_{w_1} \setminus \{0\}, \forall \alpha_2 \in \Gamma_{w_2} \setminus \{0\})$

$\Omega_{w_1}(\alpha_1) \cap \Omega_{w_2}(\alpha_2) \neq \emptyset \Rightarrow \Omega_{w_1}(\alpha_1) \subseteq \Omega_{w_2}(\alpha_2) \vee$ $\vee\, \Omega_{w_2}(\alpha_2) \subseteq \Omega_{w_1}(\alpha_1)$ , where $\Omega_{w_i}(\alpha_i)$ is equal to the set $\{w \in \Omega : v_i \leq w\}$ ; $i = 1,2$ ; and $v_1$ , $v_2$ are valuations defined as in Lemma 2.13.

Furthermore, the the following implication is also valid :

$$\theta_{w_1,w_2}(\alpha_1) = \theta_{w_2,w_1}(\alpha_2) = 0 \Rightarrow \Omega_{w_1}(\alpha_1) \cap \Omega_{w_2}(\alpha_2) = \emptyset .$$

Lemma 2.15. Let $\{v_1,\ldots,v_n\}$ , $n \geqslant 3$ , be an approximation family of valuations on a ring R and $0 \leqslant \gamma_i \in \Gamma_{v_i}$ , $1 \leqslant i \leqslant n-1$ be elements such that $(\gamma_1,\ldots,\gamma_{n-1}) \in \Gamma_{v_1} \times \cdots \times \Gamma_{v_{n-1}}$ is compatible family . Then there exists a nonnegative element $\gamma_n \in \Gamma_{v_n}$ such that $(\gamma_1,\ldots,\gamma_{n-1},\gamma_n)$ is a compatible family from $\Gamma_{v_1} \times \cdots \times \Gamma_{v_{n-1}} \times \Gamma_{v_n}$ .

## §3. Approximation theorems

The proof of the Approximation theorems we are going to give use an idea of M.Griffin [4] , for valuations on a field , and some ideas of M.Arapović [2] , for Prüfer - valuations on a total quotient ring . We shall consider a situation more general than that in [2].

In the proof of an approximation theorem in the neighbourhood of zero [2,Theorem 1.] , for pairwise incomparable valuations on a field , it is proved that the set $R \cap (V_1 \setminus M_1) \cap (P_2 \cap \cdots \cap P_n)$ is not empty . The proof of that statement can not be applied to the case of Prüfer - valuations on a quotient ring (Cf. [2]). Therefore we shall first prove the following proposition which throw light on the situation in [2] .

Proposition 3.1. Let $\{v_1,\ldots,v_n\}$ , $n \geqslant 3$ , be an approximation family of valuations on a ring R and let $v_i$ , $v_j$ are incomparable for all $1 \leqslant i \neq j \leqslant n$ . Further, let $(0,\alpha_2,\ldots,\alpha_n) \in \Gamma_{v_1} \times \Gamma_{v_2} \times \cdots \times \Gamma_{v_n}$ be an compatible family , $\alpha_i > 0$ , $2 \leqslant i \leqslant n$ .

If $Q_i = \{y_i \in R_{v_i} : v_i(y_i) \geqslant \alpha_i\}$ and $P_i$ denote the radical $r(Q_i)$ of $Q_i$, $2 \leqslant i \leqslant n$, then the set

$$(R_{v_1} \setminus P_{v_1}) \cap (P_2 \cap \cdots \cap P_n) \cap (R_{v_1} \cap \cdots \cap R_{v_n})$$

is not empty.

<u>Proof</u> - It is obvious that $Q_i$ is an $v_i$-closed ideal of the ring $R_{v_i}$ and $Q_i \not\subseteq v_i^{-1}(\infty)$, $2 \leqslant i \leqslant n$. By Definition 2.8. the ideals $Q_i$ are also R-regular and therefore $P_i$ is an R-regular prime ideal satisfying

$P_i \not\subseteq v_i^{-1}(\infty)$, $P_i \subseteq P_{v_i}$ for $2 \leqslant i \leqslant n$. It follows

that $R_{v_i[P_i]}$ is an R-Prüfer ring and a valuation ring of the ring R whose positive ideal equals $P_i$ (Cf. Proposition 1.22.). Let A denote the set $R_{v_1} \cap \cdots \cap R_{v_n}$ and suppose that the set

$$A \cap (P_2 \cap \cdots \cap P_n) \cap (R_{v_1} \setminus P_{v_1})$$

is empty. If $\bar{P}_i$ denote $A \cap P_i$, $2 \leqslant i \leqslant n$, and $\bar{P}_1$ denote $A \cap P_{v_1}$, then we have $\bar{P}_2 \cap \cdots \cap \bar{P}_n \subseteq \bar{P}_1$ and since $\bar{P}_1, \bar{P}_2, \ldots, \bar{P}_n$ are prime ideals of A it follows that $\bar{P}_i \subseteq \bar{P}_1$ for some $i \in \{2,\ldots,n\}$. By Proposition 1.16. we have $A_{[\bar{P}_1]} = R_{v_1}$. Therefore $\bar{P}_i \subseteq \bar{P}_1$ implies

$R_{v_1} = A_{[\bar{P}_1]} \subseteq R_{v_i[P_i]} \subsetneq R$ and since $R_{v_i} \subseteq R_{v_i[P_i]}$, we conclude that $R_{v_1} R_{v_i} \subseteq R_{v_i[P_i]} \subsetneq R$, i.e.

$R_{v_1 \wedge v_i} = R_{v_1[P_{v_1 \wedge v_i}]} = R_{v_i[P_{v_1 \wedge v_i}]}$ where $P_{v_1 \wedge v_i}$ is an R-regular prime ideal of $R_{v_1}$ and $R_{v_i}$. It follows from Proposition 1.21. that $R_{v_i[P_{v_1 \wedge v_i}]} \subseteq R_{v_i[P_i]}$ implies $P_i \subseteq P_{v_1 \wedge v_i}$.

If $x_i \in R$ satisfies $v_i(x_i)=\alpha_i$, then $x_i \in P_i$, i.e. $\alpha_i \in v_i(P_{v_1 \wedge v_i})$ and since $\Delta_{v_1,v_i} \cap v_i(P_{v_1 \wedge v_i})=\emptyset$, we conclude that $\alpha_i \notin \Delta_{v_1,v_i}$. But, the pair $(0,\alpha_i) \in \in \Gamma_{v_1} \times \Gamma_{v_i}$ is compatible and $\alpha_i$ must lie in $\Delta_{v_1,v_i}$. We have thus obtained a contradiction which shows that the set $A \cap (P_2 \cap \cdots \cap P_n) \cap (R_{v_1} \setminus P_{v_1})$ is not empty.

<u>Theorem 3.2.</u> Let $\{v_1,\ldots,v_n\}$, $n \geq 2$, be an approximation family of pairwise incomparable valuations on a ring $R$ and suppose $(\alpha_1,\ldots,\alpha_n) \in \Gamma_{v_1} \times \cdots \times \Gamma_{v_n}$ is a compatible family. Then there exists an element $x \in R$ such that $v_i(x)=\alpha_i$, for all $1 \leq i \leq n$.

<u>Proof</u> - I) Let us suppose first that $\alpha_1=0$ and $\alpha_i > 0$ for all $2 \leq i \leq n$.

We shall prove that there exists $x \in A=R_{v_1} \cap \cdots \cap R_{v_n}$ such that $v_i(x)>\alpha_i$ for all $2 \leq i \leq n$ and $v_1(x)=0$. By Proposition 3.1. there exists $y \in A$ with the property $v_1(y)=0$ and for each $i \in \{2,\ldots,n\}$ there exists a natural number $m_i$ such that $v_i(y^{m_i})>\alpha_i$, $2 \leq i \leq n$. Therefore, if we take $m=\max\{m_i : 2 \leq i \leq n\}$, then we have $v_1(y^m)=0$ and $v_i(y^m)>\alpha_{i_m}$ for all $2 \leq i \leq n$. It follows that we can take $x=y^m$.

II) Let us suppose now that $\alpha_1=0$ and $\alpha_i \leq 0$ for all $2 \leq i \leq n$. We can take $0<\delta_i \in \bigcap_{j \neq i} \Delta_{v_i,v_j}$, $2 \leq i \leq n$. In fact, there exists $j_o \neq i$ such that $\Delta_{v_i,v_{j_o}} \subseteq \Delta_{v_i,v_j}$ for all $j \neq i$, and for all $2 \leq i \leq n$.

Also, $\Delta_{v_i,v_{j_o}} \neq (0)$, since otherwise we would have

$P_{v_i \wedge v_{j_o}} = P_{v_i}$ and $R_{v_i} = R_{v_i} R_{v_{j_o}}$, i.e. $R_{v_{j_o}} \subseteq R_{v_i}$ and $v_i \leqslant v_{j_o}$ (Proposition 1.23.) which is impossible. The family $(0, \delta_2, \ldots, \delta_n)$ is compatible, and from I) it follows that there exists an $x \in A$ such that

$v_1(x)=0$ and $v_i(x) > \delta_i > 0 \geqslant \alpha_i$ for all $2 \leqslant i \leqslant n$.

III) Let $(\alpha_1, \ldots, \alpha_n) \in \Gamma_{v_1} \times \cdots \times \Gamma_{v_n}$ be any compatible family and

$v_1(x_1) = \alpha_1$. We shall show that for some $a_1 \in A$ the following hold :

$v_1(x_1 a_1) = \alpha_1$ and $v_i(x_1 a_1) > \alpha_i$ for all $2 \leqslant i \leqslant n$.

If there exists $a_1 \in A$ such that $v_1(x_1 a_1) = \alpha_1$ and for all indexes $i \in \{2, \ldots, n\}$ satisfying $v_i(x_1) \neq \infty$ we have $v_i(x_1 a_1) > \alpha_i$, then $v_i(x_1 a_1) > \alpha_i$ also holds for each index $i$ satisfying $v_i(x_1) = \infty$. Let us suppose therefore that $v_i(x_1) \neq \infty$ for all $2 \leqslant i \leqslant n$. Then the family $(v_1(x_1), \ldots, v_n(x_1)) \in \Gamma_{v_1} \times \cdots \times \Gamma_{v_n}$ is compatible, and if $\alpha'_i = \alpha_i - v_i(x_1)$, $1 \leqslant i \leqslant n$, then the family $(\alpha'_1, \ldots, \alpha'_n)$ is also compatible. Certainly, $\alpha'_1 = 0$. If $\alpha'_i \leqslant 0$ holds for all $2 \leqslant i \leqslant n$, then by II) there exists an element $a_1 \in A$ such that $v_1(a_1)=0$ and $v_i(a_1) > 0$ for all $2 \leqslant i \leqslant n$. Therefore, we have :

$v_1(a_1) = \alpha'_1 = \alpha_1 - v_1(x_1)$, i.e. $v_1(a_1 x_1) = \alpha_1$,

while for $2 \leqslant i \leqslant n$ we have

$v_i(a_1) > \alpha'_i = \alpha_i - v_i(x_1)$, i.e. $v_i(a_1 x_1) > \alpha_i$

for all $2 \leqslant i \leqslant n$.

Let us suppose now that there exist $i, j \in \{2, \ldots, n\}$ such that $\alpha'_i \leqslant 0$ and $\alpha'_j > 0$. Let $I$ denote the set $\{1 \leqslant i \leqslant n : \alpha'_i \leqslant 0\}$ and $J$ denote the set $\{2 \leqslant j \leqslant n : \alpha'_j > 0\} \cup \{1\}$. It is clear that the families $\{\alpha'_i\}_{i\in I}$ and $\{\alpha'_j\}_{j\in J}$ are compatible, and from II) and I) we conclude that there exist $a'_1$ , $a''_1 \in A$ such that $v_1(a'_1)=v_1(a''_1)=0$ and

$$v_i(a'_1) > 0 \geq \alpha'_i \quad ; 1\neq i \in I \; ; \; v_j(a''_1) > \alpha'_j \quad , 1\neq j \in J.$$

IV) In the same way as in the proof of the part III) it follows that for each $j \in \{1,\dots,n\}$ we can find an $a_j \in A$ such that the following hold :

$$v_j(a_jx_j)=\alpha_j \quad ; \quad v_i(a_jx_j) > \alpha_i \quad , 1 \leq i\neq j \leq n \quad ,$$

where $x_j \in R$ satisfies $v_j(x_j)=\alpha_j$ for each $1 \leq j \leq n$. Finally, if we take $x=a_1x_1+\dots+a_nx_n \in R$ we see that for all $j \in \{1,\dots,n\}$ we have :

$$v_j(x)=\min\{v_j(a_ix_i) : 1\leq i\leq n\} = \alpha_j \quad .$$

<u>Remark 3.3.</u> - i) Theorem 3.2. is so called Approximation theorem in the neighbourhood of zero . From Theorem 2.6. it follows that each finite family of nontrivial valuations on a ring R with large Jacobson radical is an approximation family for R . Thus, as a corollary of Theorem 3.2. we have that for each finite family of pairwise incomparable nontrivial valuations on a ring R with large Jacobson radical the Approximation theorem in the neighbourhood of zero holds. On the other hand it can be easily seen that the proof of Theorem 3.2. is correct in the case that $R=T(A)$ , where A is a Prüfer ring and $v_1,\dots,v_n$ are valuations on R which are nonnegative on A . In this way we generalize Theorem 4. in [2] since a ring with large Jacobson radical need not be a total quotient ring (cf. Example 2.4.).

ii) It can be shown that an analogue to Theorem 3.2. holds for a strictly wider class of rings than the class of rings with large Jacobson radical (cf. [1]).

Lemma 3.4. If A is a subring of a ring R and $\{M_i\}_{i\in I}$ is the set of all R-regular maximal ideals of the ring A , then for any R-regular ideal Q of the ring A the following holds :

$$Q=A\cap(\bigcap_{i\in I}[Q]A_{[M_i]})=A\cap(\bigcap_{i\in I}QA_{[M_i]})$$

Proof - If $x\in A$ and for all $i\in I$ there exists $s_i\in A\setminus M_i$ such that $xs_i\in Q$ , i.e. $s_i\in(Q:Ax)_A$ , then $(Q:Ax)_A$ is an R-regular ideal of the ring A , since $Q\subsetneq(Q:Ax)_A$ . $(Q:Ax)_A\not\subseteq M_i$ for all $i\in I$ , implies $(Q:Ax)_A=A$ , i.e. $x\in Q$ .

Proposition 3.5. If $\{v_1,\dots,v_n\}$ , $n\geqslant 2$ , is an approximation family of valuations on a ring R , $A=R_{v_1}\cap\cdots\cap R_{v_n}$ , and $v_i$ , $v_j$ are incomparable for all $1\leqslant i\neq j\leqslant n$ , then $A\cap P_{v_i}$ , $1\leqslant i\leqslant n$ , are exactly all R-regular maximal ideals of the ring A .

Proof - The ring A is a R-Prüfer ring and for $M_i=A\cap P_{v_i}$ , $1\leqslant i\leqslant n$ , we have $A=A_{[M_1]}\cap\cdots\cap A_{[M_n]}$ . The ideals $M_i$ , $1\leqslant i\leqslant n$ , are R-regular (Definition 2.8.) and prime ideals of A . It follows from Proposition 1.17. that $A_{[P]}=R$ holds for every proper prime R-regular ideal P of A such that $P\not\subseteq M_i$ for all $1\leqslant i\leqslant n$ . Thus,we see that $[P]A_{[P]}$ is an R-regular ideal of the ring $A_{[P]}=R$ , and therefore $[P]A_{[P]}=R$ . Consequently , P=A which is a contradiction . This means that all R-regular maximal ideals of the ring A are contained in the set $\{M_1,\dots,M_n\}$ . Finally,if some $M_i$ is not maximal , then for some $j\neq i$ we would have $M_i \subseteq M_j$ , and $R_{v_j} \subseteq R_{v_i}$ , i.e. $v_i \leq v_j$ which is impossible .

Now we can prove a General Approximation Theorem generalizing a result of M. Arapović [2] .

<u>Theorem 3.6.</u> Let $\{v_1,\ldots,v_n\}$ , $n \geq 2$ , be an approximation family of pairwise incomparable valuations on a ring $R$ and let $A=R_{v_1}\cap\cdots\cap R_{v_n}$ . Further, let $s_1,\ldots,s_n$ be R-regular elements in $A$ , and $b_1,\ldots,b_n$ any elements in $A$ . If $(\alpha_1,\ldots,\alpha_n)\in \Gamma_{v_1}\times\cdots\times\Gamma_{v_n}$ is a compatible family, then the following implication holds :

$$((\forall i\neq j \in \{1,\ldots,n\})\ v_i(a_is_i^{-1} - a_js_j^{-1}) < \alpha_i \Rightarrow$$
$$\Rightarrow \alpha_i - v_i(a_is_i^{-1} - a_js_j^{-1}) \in \Delta_{v_i,v_j}) \Rightarrow$$
$$\Rightarrow (\exists x\in R)(\forall i \in \{1,\ldots,n\})\ v_i(x-a_is_i^{-1}) = \alpha_i .$$

<u>Proof</u> - I) Let us consider first the case when $b_i = a_is_i^{-1} \in A$ for all $1 \leq i \leq n$ .
We shall prove that there exists $x\in R$ such that $v_i(x-b_i) \geq \alpha_i$ for all $1\leq i\leq n$ if the following implication holds :

$$(\forall i\neq j \in \{1,\ldots,n\})\ v_i(b_i-b_j) < \alpha_i \Rightarrow$$
$$\Rightarrow \alpha_i - v_i(b_i-b_j) \in \Delta_{v_i,v_j} .$$

Let $Q_i$ denote the set $\{b\in A : v_i(b) \geq \alpha_i\}$ , $1\leq i\leq n$ . The ideals $Q_1,\ldots,Q_n$ are R-regular (cf. Definition 2.8.). We are now going to prove that

$b_i - b_j \in Q_i + Q_j$ for all $1 \leq i,j \leq n$.

The ideals $A \cap P_{v_1}, \ldots, A \cap P_{v_n}$ are R-regular maximal ideals of the ring A (Proposition 3.5.) and $R_{v_i} = A_{[A \cap P_{v_i}]}$. $1 \leq i \leq n$, and Lemma 3.4. implies :

$$Q_i + Q_j = \bigcap_{1 \leq k \leq n} (Q_i + Q_j) R_{v_k} .$$

The ideal $(Q_i + Q_j)R_{v_k}$ of the ring $R_{v_k}$ is R-regular since the ideals $Q_1, \ldots, Q_n$ are R-regular and consequently $(Q_i + Q_j)R_{v_k}$ is also $v_k$-closed (Proposition 1.22. and Theorem 1.19.). Therefore, it suffices to show that $v_k(b_i - b_j) \in v_k(Q_i + Q_j)$ for all $1 \leq k \leq n$.

Let $k$ be a fixed element in $\{1, \ldots, n\}$. If $v_i(b_i - b_j) \geq \alpha_i$ or $v_j(b_i - b_j) \geq \alpha_j$, then $b_i - b_j \in Q_i$ or $b_i - b_j \in Q_j$. Let us suppose that $v_i(b_i - b_j) < \alpha_i$ and $v_j(b_i - b_j) < \alpha_j$ hold. Then $0 < \gamma_i = \alpha_i - v_i(b_i - b_j) \in \Delta_{v_i, v_j}$ and $0 < \gamma_j = \alpha_j - v_j(b_i - b_j) \in \Delta_{v_j, v_i}$ hold. Therefore, the pair $(\gamma_i, \gamma_j) \in \Gamma_{v_i} \times \Gamma_{v_j}$ is compatible, in fact $\theta_{v_i, v_j}(\gamma_i) = \theta_{v_j, v_i}(\gamma_j) = 0$. With the notation used in Lemma 2.14. we have :

$$\Omega_{v_i}(\gamma_i) \cap \Omega_{v_j}(\gamma_j) = \emptyset .$$

Consequently, $v_k \notin \Omega_{v_i}(\gamma_i)$ or $v_k \notin \Omega_{v_j}(\gamma_j)$.

If $v_k \notin \Omega_{v_i}(\gamma_i)$, then we must have $k \neq i$. Lemma 2.13. implies $\theta_{v_i, v_k}(\gamma_i) = 0$, i.e. the pair $(\gamma_i, 0) \in \Gamma_{v_i} \times \Gamma_{v_k}$ is compatible. Furthermore,

Lemma 2.15. implies that there exist elements $0 \leq \beta_j \in \Gamma_{v_j}$ , $1 \leq j \leq n$ , such that $\beta_i = \gamma_i$ , $\beta_k = 0$ and the family $(\beta_1, \ldots, \beta_n) \in \Gamma_{v_1} \times \cdots \times \Gamma_{v_n}$ is compatible . Now, by Theorem 3.2. we can conclude that there exists $x \in R$ such that $v_j(x) = \beta_j$ for all $1 \leq j \leq n$ . Of course, then $x \in A$ and $v_i(x) = \gamma_i$ , $v_k(x) = 0$ hold . Thus we have $x(b_i - b_j) \in A$ , $v_i(x(b_i - b_j)) = \alpha_i$ and therefore $x(b_i - b_j) \in Q_i$ . Finally , it follows that the following holds :

$$v_k(b_i - b_j) = v_k(x(b_i - b_j)) \in v_k(Q_i) \subseteq v_k(Q_i + Q_j) \quad .$$

In case that $v_k \notin \Omega_{v_j}(\gamma_j)$ , we can see in the same way that $v_k(b_i - b_j) \in v_k(Q_j) \subseteq v_k(Q_i + Q_j)$ .

Thus, we have shown that $b_i - b_j \in Q_i + Q_j$ for all $1 \leq i \neq j \leq n$ . Now, we can apply the Chinese Remainder Theorem (cf. Proposition 1.13.) to conclude that there exists an element $a \in A$ such that $a - b_i \in Q_i$ for all $1 \leq i \leq n$ . This proves the part I) .

II) Now, we suppose that $a_1, \ldots, a_n$ are arbitrary elements in $A$ and $s_1, \ldots, s_n$ are any R-regular elements in $A$ .

There exist an R-regular element $s \in A$ and elements $c_1, \ldots, c_n \in A$ such that $b_i = c_i s^{-1}$ , $1 \leq i \leq n$ . Of course $v_i(s) \neq \infty$ for all $1 \leq i \leq n$ . The family $(v_1(s), \ldots, v_n(s)) \in \Gamma_{v_1} \times \cdots \times \Gamma_{v_n}$ is compatible and thus the family $(\alpha'_1, \ldots, \alpha'_n)$ , $\alpha'_i = \alpha_i + v_i(s) + \beta_i$ with $0 < \beta_i \in \bigcap_{1 \leq j \neq i \leq n} \Delta_{v_i, v_j}$ , is also compatible .

From the part I) it follows that there exists a $y \in R$ such that $v_i(y - c_i) \geq \alpha'_i$ for all $1 \leq i \leq n$ . In fact , from $v_i(c_i-c_j)<\alpha'_i$ it follows that $v_i(b_i-b_j)<$ $<\alpha_i+\beta_i$ , i.e. $v_i(b_i-b_j)-\alpha_i<\beta_i$ .

If $0\leq v_i(b_i-b_j)-\alpha_i$ , then $v_i(b_i-b_j)-\alpha_i\in\bigcap_{j\neq i}\Delta_{v_i,v_j}\subseteq$ $\subseteq\Delta_{v_i,v_j}$ , and if $v_i(b_i-b_j)<\alpha_i$ , then according to the hypotheses of this theorem,we have $\alpha_i-v_i(b_i-b_j)\in$ $\in\Delta_{v_i,v_j}$ . Therefore ,

$\alpha'_i-v_i(c_i-c_j)=\alpha_i-v_i(b_i-b_j)+\beta_i\in\Delta_{v_i,v_j}$ if $v_i(c_i-c_j)<$ $<\alpha'_i$ holds . Thus,we can apply the part I) for the elements $c_1,\ldots,c_n$ and for the family $(\alpha'_1,\ldots,\alpha'_n)$. Now,we have

$$v_i(ys^{-1}-c_is^{-1})=v_i(ys^{-1}-b_i)\geq\alpha_i+\beta_i\ ,$$

$1\leq i\leq n$ .

Since the family $(\alpha_1,\ldots,\alpha_n)$ is compatible there exists $z\in R$ such that $v_i(z)=\alpha_i$ , $1\leq i\leq n$ (Theorem 3.2.). Now,we can take $x=ys^{-1}+z$ . Thus,we have :

$$v_i(x-b_i)=v_i(ys^{-1}-b_i+z)=\min\{v_i(ys^{-1}-b_i),v_i(z)\}=$$

$=\alpha_i$ for all $1\leq i\leq n$ .

Remark 3.7. - i) If $R$ is a ring with large Jacobson radical,Theorem 2.6. implies that Theorem 3.6. holds for each finite family of pairwise incomparable nontrivial valuations on $R$ and for any elements $b_1,\ldots,b_n\in R$ . In fact,according to Theorem 2.6. each $b_i\in R$ can be written as $b_i=a_is_i^{-1}$ where $a_i,s_i\in A$ and $s_i\in U(R)$ .

ii) Suppose $R=T(A)$ , the total quotient ring of a Prüfer ring $A$ and $v_1,\ldots,v_n$ are pairwise incomparable and nontrivial valuations on $R$ , nonnegative on $A$ and $b_1,\ldots,b_n\in T(A)$ are arbitrarily chosen, then for any compatible family $(\alpha_1,\dots,\alpha_n)$ from $\Gamma_{v_1} \times \cdots \times \Gamma_{v_n}$ the statement of Theorem 3.6. holds .

In fact , the set $\bar{A} = \bigcap_{1\leq i\leq n} R_{v_i}$ is a Prüfer ring and $T(\bar{A})=R$ .
Now, we shall prove that for each $i \in \{1,\dots,n\}$ and for a given $\alpha_i \in \Gamma_{v_i}$ the set

$$\{a\in \bar{A}\cap U(R) : v_i(a) > \alpha_i\}$$

contains at least one element .

Let us suppose that the set $\{a\in A\cap U(R): v_i(a)>\alpha_i\}$ is empty . So for all regular elements $a \in A$ we have $v_i(a) \leq \alpha_i$ . But , $R_{v_i} \subsetneq R$ , and consequently there exists a regular element $s_o \in A$ such that $v_i(s_o) > 0$ (since $r\in R\setminus R_{v_i}$ implies $v_i(r) < 0$ , $r=a_o/s_o$ , $s_o$ regular in $A^1$, $a_o \in A$ , and thus we see $0 \leq v(a_o) < v(s_o)$ ).

Therefore, since there exists $x\in R$ with $x=bt^{-1}$ , where $b,t \in A$ and $t$ regular in $A$ , such that $v_i(x)=-\alpha_i$ , then the following hold :

$$v_i(a) \leq \alpha_i = -v_i(x) = v_i(t) - v_i(b) \leq v_i(t) < v_i(t) + v_i(s_o) = v_i(ts_o)$$

for each regular $a \in A$ . Thus, we have $v_i(ts_o) < v_i(ts_o)$ which is a contradiction . This means, that we have the following :

$$\emptyset \neq \{a\in A\cap U(R) : v_i(a) > \alpha_i\} \subseteq \{a\in \bar{A}\cap U(R) : v_i(a) > \alpha_i\}.$$

Therefore, we have proved that $\{v_1,\dots,v_n\}$ is an approximation family for $R=T(A)$ (cf. Definition 2.8.). Furthermore , if $b_i \in R$ , then $b_i = a_i s_i^{-1}$ with $a_i, s_i \in A$ and $s_i$ regular in $A$ . Therefore $s_i$ is an invertible element in $R=T(A)$ .

As corollaries of Theorem 3.6. we have the following results :

<u>Theorem 3.8.</u> Suppose R is a ring with large Jacobson radical . Then for each finite family of pairwise incomparable and nontrivial valuations on R , for arbitrarily chosen elements $b_1,\dots,b_n \in R$ and for any compatible family $(\alpha_1,\dots,\alpha_n) \in \Gamma_{v_1} \times \cdots \times \Gamma_{v_n}$ , the following implication holds :

$$((\forall i \neq j \in \{1,\dots,n\})\ v_i(b_i - b_j) < \alpha_i \Rightarrow \alpha_i - v_i(b_i - b_j) \in$$
$$\in \Delta_{v_i, v_j}) \Rightarrow (\exists x \in R)(\forall i \in \{1,\dots,n\})\ v_i(x - b_i) = \alpha_i \, .$$

<u>Proof</u> - Cf. Remark 3.7. i) .

<u>Theorem 3.9. (M.Arapović)</u> Suppose R is a total quotient ring T(A) of a Prüfer ring A . If $v_1,\dots$ $\dots,v_n$ are pairwise incomparable and nontrivial valuations on the ring R , each nonnegative on A , then for any elements $b_1,\dots,b_n \in R$ and for any compatible family $(\alpha_1,\dots,\alpha_n) \in \Gamma_{v_1} \times \cdots \times \Gamma_{v_n}$ the implication in Theorem 3.8. holds.

<u>Proof</u> - Cf. Remark 3.7. ii) .

<u>Remark 3.10.</u> - i) Theorem 3.9. generalizes the corresponding classical result of P.Ribenboim [9] for valuations on a field .

ii) J.Gräter gave a form of general approximation theorem [4,Satz 2.5. and Satz 3.5.] . Let us remark (cf. [4 , p.283] ) that Gräter considered the compatible families $(\varepsilon_1,\dots,\varepsilon_n) \in \Gamma_{v_1} \times \cdots \times \Gamma_{v_n}$

where $\varepsilon_i \in \bigcap_{1\leq j\neq i\leq n} \Delta_{v_i,v_j}$ , so $\theta_{v_i,v_j}(\varepsilon_i)=$ $=\theta_{v_j,v_i}(\varepsilon_j)=0$ which is just one special case of the compatibility . Thus,it is easy to conclude,even in the case of valuations on a field,that Theorem 3.6. is more general than Gräter's results in [4] .

ACKNOWLEDGEMENTS

This research was supported by the Republican Council for Scientific Work of Bosna and Hercegovina .

REFERENCES

[1] J.Alajbegović , Approximation theorems for valuations with the inverse property , Commun.Alg. (to appear) .

[2] M.Arapović , Approximation theorems for fields and commutative rings , Glasnik Mat.18(38),(1983), pp.61-66.

[3] J.Gräter , Der allgemeine Approximationssatz für Manisbewertungen , Mh.Math.93 (1982)pp.277-288.

[4] M.Griffin , Rings of Krull type , J.Reine Angew. Math.229 (1968)pp.1-27.

[5] M.Griffin , Prüfer rings with zero divisors , J.Reine Angew.Math. 239/240 (1970)pp.55-67.

[6] M.Griffin , Valuations and Prüfer rings , Canad.J.Math. 26 (1974)pp.412-429.

[7] M.D.Larsen and P.Mc.Carthy , Multiplicative Theory of Ideals , New York and London, Academic Press, 1971.

[8] M.E.Manis , Valuations on a commutative ring , Proc.Amer.Math.Soc. 20 (1969)pp.193-198.

[9] P.Ribenboim , Le théorème d'approximation pour les valuations de Krull , Math. Zeitschr.68(1957)pp.1-18.

[10] P.Ribenboim , Théorie des Groupes Ordonnés , Inst.Mat.Univ.Nacional del sur Bahia blanca 1959.

[11] Zariski,O.,Samuel,P.,Commutative Algebra,Vol.1., Van Nostrand,Princeton,N.J.,1958.

# CERTAIN ARTINIAN LATTICES ARE NOETHERIAN. APPLICATIONS TO THE RELATIVE HOPKINS-LEVITZKI THEOREM

TOMA ALBU
Universitatea Bucureşti
Facultatea de Matematică
Str. Academiei 14
R 70109-Bucharest 1, Romania

The classical Hopkins-Levitzki Theorem states that any right Artinian ring with identity element is right Noetherian. Usually this Theorem is proved by the method of factoring through the nilpotent Jacobson radical of the ring. A proof which avoids the concept of the Jacobson radical was first performed by Shock [1]; he obtains also necessary and sufficient conditions for an Artinian module over a ring not necessarily unitary to be Noetherian.

The relativization of the classical Hopkins-Levitzki Theorem with respect to a Gabriel topology was first proved in the commutative case and conjectured in the noncommutative case by Albu and Năstăsescu [1 ; Théorème 4.7, Problème 4.8]. The noncommutative case of the relative Hopkins-Levitzki Theorem was first proved by Miller and Teply [1]. However, their proof is long and complicated; another module-theoretical proof of this Theorem, also hard, is **available** in Faith [1]. A different way to approach this Theorem is to translate the module-theoretical relative chain conditions occuring in its statement in absolute chain conditions in a suitable Grothendieck category, and to prove thus a general Hopkins-Levitzki Theorem in such a category; this was done by Năstăsescu [1]. Another short proof of this general Hopkins-Levitzki Theorem in a Grothendieck category is also due to Năstăsescu [2], and is somewhat simi-



*F. van Oystaeyen (ed.), Methods in Ring Theory, 37–52.*

lar to the one given by Shock [1] for modules over Artinian rings not necessarily with identity element.

A discussion on the various forms of the Hopkins-Levitzki Theorem and the connection between them may be found in Albu and Năstăsescu [2].

A short noncategorical proof of the relative Hopkins-Levitzki Theorem does not yet exists. The aim of this paper is to give such a proof by placing the Hopkins-Levitzki Theorem in a latticial setting; moreover, we shall obtain even two different proofs of this lattice-theoretical form of the Hopkins-Levitzki Theorem. Our proofs are inspired by some ideas of Shock [1] and Năstăsescu [1],[2], and are based on the concepts of length and Loewy length of an upper continuous and modular lattice of finite length.

## 0. PRELIMINARIES

Throughout this paper L will denote an upper continuous and modular lattice. The least (resp. greatest) element of L will be denoted by 0 (resp. 1). The notation and terminology will follow Stenström [1].

Recall that a non-zero element a of L is an atom if whenever $b \in L$ and $b < a$, then $b = 0$. The lattice L is called semi-atomic if 1 is a join of atoms; L is called semi-Artinian if for every $x \in L$, $x \neq 1$ the sublattice $[x,1]$ of L contains an atom. As in the case of modules, it can be shown (see e.g. Năstăsescu [3]) that if L is a semi-atomic lattice, then L is complemented, and for every $x, y \in L$ with $x \leq y$ the interval $[x,y]$ of L is also a semi-atomic lattice.

The ascending Loewy series of L:

$$s_0(L) < s_1(L) < \dots < s_{\lambda(L)}(L) \qquad (*)$$

is defined inductively: $s_0(L) = 0$, $s_1(L) = So(L)$ where $So(L)$ is the socle of L (i.e. the join of all atoms of L), and if the elements $s_\beta(L)$ of L have been defined for all ordinals $\beta < \alpha$, then we set $s_\alpha(L) = \bigvee_{\beta<\alpha} s_\beta(L)$ if $\alpha$ is a limit ordinal, and $s_\alpha(L) = So([s_\gamma(L),1])$ if $\alpha = \gamma + 1$; $\lambda(L)$ is the least ordinal $\lambda$ such that $s_\lambda(L) = s_{\lambda+1}(L)$. The ordinal $\lambda(L)$ is the Loewy length of L, and it exists always because L is a set. The intervals $[s_\alpha(L), s_{\alpha+1}(L)]$, which are for each $\alpha < \lambda(L)$ semi-atomic lattices, are called the factors of the series (*). As in the case of modules it is easy to show that L is a semi-Artinian lattice if and only if $s_{\lambda(L)}(L) = 1$; moreover, for such a lattice, So(L) is an essential element of L (see e.g. Năstăsescu [3]).

Recall that in the sequel L will always be assumed upper continuous and modular.

0.1. LEMMA Let $x, y \in L$ be such that $x \leq y$. Then

$$s_\alpha([0,x]) = s_\alpha([0,y]) \wedge x$$

for each ordinal $\alpha$.

Proof. The lemma holds trivially if $\alpha = 0$, so assume it holds for each ordinal $\gamma < \alpha$ and proceed by induction. For the sake of brevity denote $s_\beta([0,x]) = x_\beta$ and $s_\beta([0,y]) = y_\beta$ for each ordinal $\beta \leq \alpha$.

If $\alpha$ is a limit ordinal, then

$$x_\alpha = \bigvee_{\beta<\alpha} x_\beta = \bigvee_{\beta<\alpha} (y_\beta \wedge x) = (\bigvee_{\beta<\alpha} y_\beta) \wedge x = y_\alpha \wedge x.$$

If $\alpha = \beta+1$, then

$[x_\beta , y_\alpha \wedge x] = [y_\beta \wedge x , y_\alpha \wedge x] \simeq [y_\beta , y_\beta \vee (y_\alpha \wedge x)] \subseteq [y_\beta , y_\alpha]$.
Since $[y_\beta , y_\alpha]$ is a semi-atomic lattice, so is $[x_\beta , y_\alpha \wedge x]$, and consequently $y_\alpha \wedge x \leq x_\alpha$. On the other hand

$$[y_\beta , y_\beta \vee x_\alpha] \simeq [x_\alpha \wedge y_\beta , x_\alpha] \subseteq [x_\beta , x_\alpha]$$

because $x_\beta \leq y_\beta$ by the induction hypothesis. It follows that $[y_\beta , y_\beta \vee x_\alpha]$ is a semi-atomic lattice, hence $y_\beta \vee x_\alpha \leq y_\alpha$, and so $x_\alpha \leq y_\alpha \wedge x$. ∎

0.2. PROPOSITION Let $(z_i)_{i\in I}$ be a family of elements of L. Then $[0, \bigvee_{i\in I} z_i]$ is a semi-Artinian lattice if and only if $[0,z_i]$ is a semi-Artinian lattice for each $i \in I$, and in this case

$$\lambda([0, \bigvee_{i\in I} z_i]) = \sup_{i\in I} \lambda([0,z_i]).$$

Proof. Suppose that $[0,z_i]$ are all semi-Artinian lattices, and denote $\alpha = \sup_{i\in I} \lambda([0,z_i])$. By Lemma above one has for each $j \in I$

$$z_j = s_\alpha([0,z_j]) \leq s_\alpha([0, \bigvee_{i\in I} z_i]),$$

and so $\bigvee_{j\in I} z_j \leq s_\alpha([0, \bigvee_{i\in I} z_i])$. Hence $\bigvee_{i\in I} z_i = s_\alpha([0, \bigvee_{i\in I} z_i])$, and consequently $[0, \bigvee_{i\in I} z_i]$ is a semi-Artinian lattice and $\lambda([0, \bigvee_{i\in I} z_i]) \leq \alpha$.

Conversely, suppose that $[0, \bigvee_{i\in I} z_i]$ is semi-Artinian, and denote $\lambda([0, \bigvee_{i\in I} z_i]) = \beta$. Then

$$s_\beta([0,z_j]) = z_j \wedge s_\beta([0, \bigvee_{i\in I} z_i]) = (\bigvee_{i\in I} z_i) \wedge z_j = z_j$$

for each $j \in I$, by 0.1. Hence $[0,z_j]$ is semi-Artinian for each $j \in I$ and $\lambda([0,z_j]) \leqslant \beta$. It follows that $\sup_{i\in I} \lambda([0,z_i]) \leqslant \beta$, and the proposition is proved.∎

Recall that a lattice with 0 and 1 is of <u>finite length</u> if there exists a (Jordan-Hölder) composition series between 0 and 1. It is well-known that a modular lattice with 0 and 1 is of finite length if and only if it is both Artinian and Noetherian; the length of such a lattice X will be denoted in the sequel by $\ell(X)$.

The next Lemma is a lattice-theoretical formulation of the Proposition 2 of Shock [1]. For the convenience of the reader we include a proof here.

<u>0.3. LEMMA</u> If A is an Artinian and modular lattice with 1, then there exists an element $a^* \in A$ which is the least element of A such that the sublattice $[a^*, 1]$ of A is of finite length.

<u>Proof</u>. Let $N = \{x \in A \mid [x,1] \text{ is a lattice of finite length}\}$. Since $1 \in N$, $N \neq \emptyset$. If $x_1, x_2 \in N$, then $[x_1 \wedge x_2, x_1] \simeq [x_2, x_1 \vee x_2] \subseteq [x_2, 1]$, hence $[x_1 \wedge x_2, x_1]$ is of finite length, and consequently $[x_1 \wedge x_2, 1]$ is also of finite length because $[x_1, 1]$ is of finite length. It follows that $x_1 \wedge x_2 \in N$. Let $a^*$ be a minimal element of N; if $x \in N$ then $a^* \wedge x \in N$, and so $a^* \wedge x = a^*$ by the minimality of $a^*$, i.e. $a^* \leqslant x$. Hence $a^*$ is the least element of N.∎

If A is a lattice as in the above Lemma, then $\ell([a^*, 1])$ will be called the <u>reduced length</u> of A, and will be denoted in the sequel by $\ell^*(A)$.

As an easy consequence of 0.3 we obtain the following known result which will be used frequently in this paper:

0.4. COROLLARY Let $C$ be a complemented and modular lattice (e.g. $C$ may be any upper continuous, modular and semi-atomic lattice). Then $C$ is Artinian if and only if $C$ is Noetherian.

Proof. Suppose that $C$ is Artinian, and consider the element $c^* \in C$ defined by 0.3. If $c$ is a complement of $c^*$, then $[0,c] = [c^* \wedge c, c] \simeq [c^*, c^* \vee c] = [c^*, 1]$, hence the sublattice $[0,c]$ of $C$ is of finite length. Suppose that $c \neq 1$; since $C$ is Artinian there exists $a \in C$ such that $a$ is an atom of the interval $[c,a]$. It follows that $[0,a]$ is of finite length. If $b$ is a complement of $a$, then $[0,a] \simeq [b,1]$, hence $[b,1]$ is of finite length, and consequently $c^* \leqslant b$. Then $c^* \wedge a \leqslant b \wedge a = 0$, and so

$$a = a \wedge 1 = a \wedge (c \vee c^*) = c \vee (a \wedge c^*) = c \vee 0 = c,$$

a contradiction. Hence $c = 1$, and thus $C$ is Noetherian.

If $C$ is Noetherian, then the opposite lattice $C^{op}$ of $C$ is modular, complemented and Artinian, and then, by the proof above, $C^{op}$ must be Noetherian, i.e. $C$ is Artinian. ∎

0.5. PROPOSITION The following properties of an upper continuous and modular lattice $L$ are equivalent:

(1) $L$ is a lattice of finite length.

(2) $L$ is an Artinian lattice with $\lambda(L)$ finite.

(3) $L$ is a Noetherian and semi-Artinian lattice.

Proof. (1)$\Longrightarrow$(2) is clear.

(2)$\Longrightarrow$(3): Let $n = \lambda(L)$; then the ascending Loewy series of $L$ is

$$0 = s_0(L) < s_1(L) < \dots < s_n(L) = 1.$$

For each $i = 0, \dots, n-1$ $[s_i(L), s_{i+1}(L)]$ is Artinian and semi-atomic, hence Noetherian by 0.4. Consequently $L = [0,1]$ is of finite length.

(3)$\Longrightarrow$(1): $s_{\lambda(L)}(L) = 1$ since $L$ is semi-Artinian, and $\lambda(L)$ is a finite ordinal, say $n$, since $L$ is Noetherian. For each $i = 0, \dots, n-1$ $[s_i(L), s_{i+1}(L)]$ is Noetherian and semi-atomic, hence Artinian by 0.4. Hence $L = [0,1]$ is of finite length. ∎

0.6. REMARK From the proof of the above Proposition it follows that if the lattice $L$ is of finite length, then $\lambda(L) \leq \ell(L)$; clearly $\lambda(L) = \ell(L)$ if and only if for each $i = 0, \dots, \lambda(L)-1$ each $s_{i+1}(L)$ is an atom in the sublattice $[s_i(L), s_{i+1}(L)]$ of $L$. ∎

## 1. MAIN RESULTS

1.1. THEOREM Let $L$ be an Artinian, upper continuous and modular lattice, and let

(*) $x_1 \leq x_2 \leq \dots \leq x_n \leq \dots$

be an ascending chain in $L$ such that the sublattices $[0, x_i]$ of $L$ are Noetherian for all $i \geq 1$. Then, the following two conditions are equivalent:

(1) The chain (*) is stationary.

(2) For each natural number $i \geq 1$ and each $y \in L$ with $y < x_i$ there exists an element $a_{yi} \in L$ such that $a_{yi} \leq x_i$ and $a_{yi} \nleq y$; furthermore, there exists a natural number $t$ such that $\lambda([0, a_{yi}]) < t$ for all $i \geq 1$ and all $y < x_i$.

Proof. (1)$\Longrightarrow$(2): Suppose that $x_n = x_{n+1} = x_{n+2} = \dots$ and denote $k = \lambda([0, x_n])$; then $\lambda([0, x_i]) < k+1$ for all $i \geq 1$. If $y < x_i$, then clearly (2) holds by choosing $a_{yi} = x_i$.

(2)$\Longrightarrow$(1): Suppose that the chain (*) is strictly ascending. Then the sequence $(\lambda([0, x_i]))_{i \geq 1}$ is unbounded, for otherwise, there exists a natural number $m$ such that $\lambda([0,x_i]) \leqslant m$ for all $i \geqslant 1$; then $\lambda([0, \bigvee_{i=1}^{\infty} x_i]) \leqslant m$ by 0.2, and thus, by 0.5 $[0, \bigvee_{i=1}^{\infty} x_i]$ is a Noetherian lattice, a contradiction. Let $k \geqslant 1$ be such that $\lambda([0,x_k]) > t$. Since $[0,x_k]$ is Noetherian, there exists an element $y < x_k$ maximal with the property $\lambda([0,y]) \leqslant t$. By hypothesis, there exists $a = a_{yk}$ such that $a \leqslant x_k$ and $a \nleqslant y$. According to 0.2, $\lambda([0,a \vee y]) = \sup(\lambda([0,a]), \lambda([0,y])) \leqslant t$. But $y < a \vee y < x_k$ since $a \nleqslant y$ and $\lambda([0,x_k]) > t$; this contradicts the maximality of $y$, and consequently, the chain (*) must be stationary. ∎

1.2. COROLLARY Let $L$ be an upper continuous and modular lattice satisfying the following condition:

($\lambda$) For each $y < x$ in $L$ there exists an element $a_{yx} \in L$ such that $a_{yx} \leqslant x$, $a_{yx} \nleqslant y$ and $[0,a_{yx}]$ is semi-Artinian; furthermore, there exists a natural number $t$ such that $\lambda([0,a_{yx}]) < t$ for all $y < x$ in $L$.

If $L$ is Artinian, then $L$ is Noetherian.

Proof. Consider the ascending Loewy series of $L$:

$$0 = s_0(L) < s_1(L) < \ldots\ldots$$

For each $i \geqslant 0$, $[s_i(L), s_{i+1}(L)]$ is semi-atomic and Artinian, hence Noetherian by 0.4; it follows that $[0,s_i(L)]$ is Noetherian for all $i \geqslant 1$. By 1.1, there exists a natural number $n$ such that $s_n(L) = s_{n+1}(L)$. Hence $1 = s_n(L)$ because $L$ is Artinian, and consequently $L = [0,1] = [0,s_n(L)]$ is Noetherian. ∎

1.3. REMARKS (1) The condition ($\lambda$) about an Artinian, upper continuous and modular lattice $L$ is necessary for $L$ to be Noetherian. Indeed, in this case, for each $y < x$ in $L$ choose $a_{yx} = x$; then $\lambda([0,a_{yx}]) \leqslant \ell([0,a_{yx}]) \leqslant \ell(L)$.

(2) The condition $(\lambda)$ may be reformulated as follows, if we denote $I = \{(y,x) \mid x,y \in L,\ y < x\}$ and $a_i = a_{yx}$ for $i=(y,x) \in I$:

$(\lambda)$ There exists a nonempty family $(a_i)_{i \in I}$ of elements of L and a natural number t such that $[0,a_i]$ is semi-Artinian for each $i \in I$, each element of L is a join of a subfamily of $(a_i)_{i \in I}$, and $\lambda([0,a_i]) < t$ for each $i \in I$. ∎

<u>1.4. LEMMA</u> Let L be an upper continuous and modular lattice and $y < x$ elements in L for which there exists $a \in L$ such that $a \leqslant x$, $a \nleqslant y$ and the sublattice $[0,a]$ of L is semi-Artinian. Then, the interval $[y,x]$ of L contains an atom.

<u>Proof</u>. Since L is a modular lattice, it follows that there exists a canonical isomorphism of lattices $[a \wedge y, a] \simeq [y, a \vee y]$. But $a \nleqslant y$, hence $a \wedge y < a$, and so, the interval $[a \wedge y, a]$ contains an atom b, because $[0,a]$ is semi-Artinian. Then, the corresponding element $b \vee y$ by the above isomorphism is clearly an atom of the interval $[y, a \vee y]$, and hence an atom of $[y,x]$. ∎

We are now in a position to give the following

<u>1.5. THEOREM</u> Let L be an upper continuous and modular lattice satisfying the condition $(\lambda)$ from 1.2. Then L is Artinian if and only if L is Noetherian.

<u>Proof</u>. If L is Artinian, then L is Noetherian by 1.2. Conversely, if L is Noetherian, then L is Artinian according to 0.5, since L is semi-Artinian by 1.4. ∎

Recall that if A is an Artinian and modular lattice with 1, we have denoted in the Section 0 by $\ell^*(A)$ the so called reduced length of A; more precisely, $\ell^*(A) = \ell([a^*,1])$, where $a^*$ is the least element of $A$ such that the sublattice $[a^*,1]$ of $A$ is of finite length (see 0.3); if in addition $A$ is upper continuous we can define the <u>reduced Loewy length</u> $\lambda^*(A)$ of $A$ as being $\lambda([a^*,1])$. Clearly $\lambda^*(A) \leq \ell^*(A)$. Note that $a^*$ is also the least element of $A$ such that the sublattice $[a^*,1]$ of $A$ is of finite Loewy length.

We shall consider now to other conditions on a lattice $L$ (upper continuous and modular as always in this paper):

($\lambda^*$) For each $y < x$ in $L$ there exists an element $a_{yx} \in L$ such that $a_{yx} \leq x$, $a_{yx} \nleq y$ and $[0,a_{yx}]$ is Artinian; in addition, there exists a natural number $t$ such that $\lambda^*([0,a_{yx}]) < t$ for all $y < x$ in $L$.

($\ell^*$) For each $y < x$ in $L$ there exists an element $a_{yx} \in L$ such that $a_{yx} \leq x$, $a_{yx} \nleq y$ and $[0,a_{yx}]$ is Artinian; in addition, there exists a natural number $t$ such that $\ell^*([0,a_{yx}]) < t$ for all $y < x$ in $L$.

Clearly, if $L$ satisfies the condition ($\ell^*$), then $L$ satisfies the condition ($\lambda^*$) too. We ignore the other connections between the conditions ($\lambda$), ($\lambda^*$) and ($\ell^*$) on $L$. Note that ($\lambda^*$) and ($\ell^*$) have similar reformulations to the one of ($\lambda$) given in 1.3.

<u>1.6. THEOREM</u> If the upper continuous and modular lattice $L$ satisfies the condition ($\lambda^*$), then $L$ is Artinian if and only if $L$ is Noetherian.

<u>Proof</u>. The proof may be reduced to the proof 1.5 by using the following obvious fact: $\lambda^*([0,a]) = \lambda([0,a])$ for any $a \in A$ such that $[0,a]$ is of finite length.∎

We shall investigate now the condition ($\ell^*$) on $L$.

1.7. THEOREM Let L be an upper continuous and modular lattice satisfying the condition ($\ell^*$) above. Then L is semi-Artinian and has finite Loewy length.

Proof. First of all, L is semiartinian by 1.4. For each natural number n denote by $s_n$ the term $s_n(L)$ of the ascending Loewy series of L, and suppose that $s_n \neq s_{n+1}$ for all natural numbers n.

Let $x \in L$ be such that $x \leq s_k$ for some natural number k and $[0,x]$ is Artinian. Then $s_k([0,x]) = s_k([0,1]) \wedge x = s_k \wedge x = x$ by 0.1, hence $\lambda([0,x]) \leq k$, and then, by 0.5, $[0,x]$ is of finite length.

If now $n \geq 1$ is a natural number and x is an element of L such that $x \leq s_n$, $x \nleq s_{n-1}$ and $[0,x]$ is Artinian, then we shall prove inductively that $\ell([0,x]) \geq n$. If $n = 1$, then $x \neq 0$, and so $\ell([0,x]) \geq 1$. Let $x \in L$ **be** such that $x \leq s_{n+1}$, $x \nleq s_n$ and $[0,x]$ is Artinian. Denote $z = x \wedge s_n$ and $y = z \vee s_{n-1}$. Then $s_{n-1} \leq y \leq s_n$ and $y = (x \wedge s_n) \vee s_{n-1} = (x \vee s_{n-1}) \wedge s_n$. But $x \nleq s_{n-1}$, hence $s_{n-1} < s_{n-1} \vee x$, and consequently $(x \vee s_{n-1}) \wedge s_n \neq s_{n-1}$ because the socle $s_n$ of the semi-Artinian lattice $[s_{n-1}, 1]$ is an essential element of the lattice $[s_{n-1}, 1]$. Thus $y \neq s_{n-1}$ and therefore $z \nleq s_{n-1}$; it follows that $z \wedge s_{n-1} < z$. By condition ($\ell^*$), there exists $a \in L$ such that $[0,a]$ is Artinian, $a \leq z$ and $a \nleq z \wedge s_{n-1}$. Then $a \leq z \leq s_n$, $a \nleq s_{n-1}$ and $a \leq z < x$, hence $\ell([0,a]) < \ell([0,x])$. On the other hand, by the induction hypothesis $\ell([0,a]) \geq n$, and consequently $\ell([0,x]) \geq n+1$.

Since he have assumed that $s_n \neq s_{n+1}$ for all natural numbers n, it follows that for each $n \geq 1$ there exists $a_n \in L$ such that $a_n \leq s_n$, $a_n \nleq s_{n-1}$, $[0,a_n]$ **is Artinian** and $\ell^*([0,a_n]) < t$. Then $[0,a_n]$ is of finite length and $\ell([0,a_n]) \geq n$. On the other hand, $n \leq \ell([0,a_n]) = \ell^*([0,a_n]) < t$ for all $n \geq 1$, a contradiction. This completes the proof.■

1.8. COROLLARY If the upper continuous and modular lattice L satisfies the condition $(\ell^*)$, then L is Artinian if and only if L is Noetherian.

Proof. Apply 1.7 and 0.5.∎

1.9. REMARKS (1) An other proof of 1.8 can be obtained from 1.6 since L satisfies clearly the condition $(\lambda^*)$ too.

(2) The condition $(\ell^*)$ about an Artinian, upper continuous and modular lattice L is necessary for L to be Noetherian: see 1.3.∎

## 2. APPLICATIONS

Let $\mathcal{C}$ be a Grothendieck category, i.e. an abelian category with exact direct limits and with a generator, and let X be an object of $\mathcal{C}$. $\mathcal{L}(X)$ will denote the lattice of all subobjects of X. It is well-known that $\mathcal{L}(X)$ is a modular and upper continuous lattice (see e.g. Stenström [1]). If U and M are objects of $\mathcal{C}$ then M is said to be U-generated if there exists an epimorphism $U^{(I)} \longrightarrow M$ for some set I, or equivalently, if whenever M' is a subobject of M, M' ≠ M, there exists $f \in Hom_{\mathcal{C}}(U,M)$ such that $Im(f) \nleq M'$. M is said to be strongly U-generated if each subobject of M is U-generated.

2.1. THEOREM (Năstăsescu [1],[2]) Let $\mathcal{C}$ be a Grothendieck category and U an Artinian object of $\mathcal{C}$. If M is an Artinian object of $\mathcal{C}$ which is strongly U-generated, then M is Noetherian.

Proof. By 1.8 it will suffice to check that the lattice L = $= \mathcal{L}(M)$ satisfies the condition $(\ell^*)$. Let $X, Y \in L$ be such that $Y < X$. Since X is U-generated there exists $f \in Hom_{\mathcal{C}}(U,X)$ such that $A = Im(f) \nleq Y$. But $A \simeq U/Ker(f)$, hence the lattice $\mathcal{L}(A) = [0,A]$ is isomorphic to the interval $[Ker(f),U]$ of $\mathcal{L}(U)$. Note also that $A \leqslant X$ and $[0,A]$ is Artinian because $U$ is an Artinian object of $\mathcal{C}$. Thus $\ell^*([0,A]) = \ell^*([Ker(f),U]) \leqslant \ell^*([0,U]) = \ell^*(\mathcal{L}(U))$, and so $L = \mathcal{L}(M)$ satisfies the condition $(\ell^*)$. Let us mention that according to 1.7, **any** strongly U-generated object of $\mathcal{C}$ is a Loewy object having finite Loewy length. ∎

Our next aim is to apply 1.8 to get a simple noncategorical proof of the relative Hopkins-Levitzki Theorem. For this, we shall recall briefly some basic definitions, notations and properties concerning the lattice of F-saturated submodules of a module.

Let $R$ be an associative, unitary and nonzero ring, and Mod-R the category of unitary right R-module. If $M$ is a right R-module, then $\mathcal{L}(M)$ will denote the lattice of all submodules of $M$. Let $F$ be a right Gabriel topology on $R$, $(\mathcal{T},\mathcal{F})$ the corresponding hereditary torsion theory on Mod-R, and $t$ the torsion radical associated to $(\mathcal{T},\mathcal{F})$. If $M \in$ Mod-R, we shall use the following notation

$$C_F(M) = \{N \in \mathcal{L}(M) \mid M/N \in \mathcal{F}\}.$$

If $P \in \mathcal{L}(M)$, then $\widetilde{P}$ will denote the F-saturation of $P$ in $M$, i.e. $\widetilde{P}/P = t(M/P)$; note that $P \in C_F(M)$ if and only if $P = \widetilde{P}$, i.e. $P$ is F-saturated. If $(N_i)_{i\in I}$ is a family of elements of $C_F(M)$, then $\bigvee_{i\in I} N_i = \widetilde{\sum_{i\in I} N_i}$ and $\bigwedge_{i\in I} N_i = \bigcap_{i\in I} N_i$ are elements of $C_F(M)$. Moreover, $C_F(M)$ is an upper continuous and modular lattice with respect to the partial ordering given by "$\subseteq$" (inclusion) and with respect to the operations "$\vee$" and "$\wedge$". $C_F(M)$ is called the lattice of all F-saturated submodules of $M$ and is sometimes denoted also by $Sat_F(M)$.

Let us mention the following properties of the lattice $C_F(M)$;

(1) If $N \in \mathcal{L}(M)$ and $N \in \mathcal{T}$, then the lattices $C_F(M)$ and $C_F(M/N)$ are canonical isomorphic; in particular $C_F(M) \simeq C_F(M/t(M))$.

(2) If $N \in \mathcal{L}(M)$ and $M/N \in \mathcal{T}$, then the lattices $C_F(M)$ and $C_F(N)$ are canonical isomorphic; in particular $C_F(N) \simeq C_F(\widetilde{N})$.

(3) If $M \in \mathcal{F}$ and $N \in C_F(M)$, then $C_F(N) = [0,N]$ and $C_F(M/N) \simeq [N,M]$, where the intervals are considered in the lattice $C_F(M)$.

(4) If $M$ and $M'$ are isomorphic R-modules, then the lattices $C_F(M)$ and $C_F(M')$ are isomorphic. ∎

For all these summarized facts on the lattices $C_F(M)$ the reader is referred to Stenström [1] or Albu and Năstăsescu [2].

Recall that $M \in$ Mod-R is said to be <u>F-Noetherian</u> (resp. <u>F-Artinian</u>) if $C_F(M)$ is a Noetherian (resp. Artinian) lattice. R is said to be F-Noetherian (resp. F-Artinian) if the R-module $R_R$ is F-Noetherian (resp. F-Artinian).

<u>2.2. THEOREM</u> (Miller and Teply [1]) Let F be a right Gabriel topology on the ring R such that R is F-Artinian. Then, a right R-module M is F-Artinian if and only if M is F-Noetherian.

<u>Proof</u>. By the property (1) above, $C_F(M) \simeq C_F(M/t(M))$, hence we can suppose that $M \in \mathcal{F}$. According to 1.8 it will suffice to check that the lattice $C_F(M)$ satisfies the condition $(\ell^*)$. Let $Y < X$ be elements in $C_F(M)$. Then, there exists $x \in X \setminus Y$, and denote $B = xR$, $I = \mathrm{Ann}_R(x)$, $A_{YX} = A = \widetilde{B}$. Clearly $A \in C_F(M)$, $A \leqslant X$, and $A \nleqslant Y$. Since $R/I \simeq B \leqslant M$, it follows that $R/I \in \mathcal{F}$, and so $I \in C_F(R)$. By the properties (2), (3), (4) above one gets:

$$[I,R] \simeq C_F(R/I) \simeq C_F(B) \simeq C_F(A) = [0,A],$$

where the interval $[I,R]$ is considered in $C_F(R)$ and the interval $[0,A]$ in $C_F(M)$. Since $C_F(R)$ is an Artinian lattice, it follows that $[0,A]$ is an Artinian lattice, and then

$$\ell^*([0,A]) = \ell^*([I,R]) \leqslant \ell^*(C_F(R)).$$

Thus $C_F(R)$ satisfies the condition $(\ell^*)$. ∎

2.3. REMARK When the proofs of 1.6 and 1.8 are carried out on the particular lattice $C_F(M)$, F being a right Gabriel topology on R such that R is F-Artinian, one gets two different short module-theoretical proofs of the relative Hopkins-Levitzki Theorem, quoted in Faith [1] as the Teply-Miller Theorem. ∎

The next result has been proved by Năstăsescu and Raianu [1] by using the notion of quotient category. We shall present below a much shorter latticial proof. The terminology involved in all that follows can be found in Năstăsescu and Van Oystaeyen [1].

2.4. THEOREM (Năstăsescu and Raianu [1]) Let G be a group, $R = \bigoplus_{\sigma \in G} R_\sigma$ a graded ring of type G, and F a graded right Gabriel topology on R such that R is gr F-Artinian. Then, a graded right R-module M is gr F-Artinian if and only if M is gr F-Noetherian.

Proof. By definition, M is gr F-Artinian (resp. gr F-Noetherian) if the lattice $C_F^g(M) = \{N \in \mathcal{L}_g(M) \mid M/N \in \mathcal{F}\}$ is Artinian (resp. Noetherian), where $\mathcal{L}_g(M)$ is the lattice of all graded submodules of M and $(\mathcal{T},\mathcal{F})$ is the hereditary rigid torsion theory defined by F. Let us preserve the notations from the proof of 2.2; this proof can be adapted to the graded case as follows. The element $x \in X \setminus Y$ can be choosed homogeneous, say of degree $\tau$. Then there exists an isomorphism of graded R-modules $R(\sigma)/I \simeq B$, where $\tau^{-1} = \sigma$ and $R(\sigma)$ is the $\sigma$-suspension of R. On the other hand, since the torsion theory $(\mathcal{T},\mathcal{F})$ is rigid, the correspondence $J \longmapsto J(\sigma)$ yields an isomorphism of lattices $C_F^g(R) \simeq C_F^g(R(\sigma))$. Consequently $\ell^*([0,A]) = \ell^*([I,R(\sigma)]) \leqslant \ell^*(C_F^g(R(\sigma))) = \ell^*(C_F^g(R))$. ∎

2.5. REMARK Applying 2.4 to the particular case $F = \{R_R\}$ one gets another proof, which avoids the concept of the Jacobson graded radical, of the graded version of the Hopkins-Levitzki Theorem (see Năstăsescu and Van Oystaeyen [1]): any right gr-Artinian ring is right gr-Noetherian.■

REFERENCES

T. ALBU and C. NĂSTĂSESCU

[1] Décompositions primaires dans les catégories de Grothendieck commutatives, II, J. Reine Angew. Math. 282 (1976), 172-185.

[2] Relative Finiteness in Module Theory, Lecture Notes in Pure and Applied Mathematics (Marcel Dekker, Inc., New York and Basel 1983) (to appear).

C. FAITH

[1] Injective Modules and Injective Quotient Rings, Lecture Notes in Pure and Applied Mathematics 72 (Marcel Dekker, Inc., New York and Basel 1982).

R. W. MILLER and M. L. TEPLY

[1] The descending chain condition relative to a torsion theory, Pacific J. Math. 83 (1979), 207-220.

C. NĂSTĂSESCU

[1] Conditions de finitude pour les modules, Rev. Roumaine Math. Pures Appl. 24 (1979), 745 - 758.

[2] Théorème de Hopkins pour les catégories de Grothendieck, in "Ring Theory", Proceedings of the 1980 Antwerp Conference, Lecture Notes in Mathematics 825 (Springer-Verlag, Berlin Heidelberg New York 1980).

[3] Teoria Dimensiunii în Algebra Necomutativă (Editura Academiei, București 1983).

C. NĂSTĂSESCU and Ș. RAIANU

[1] Finiteness conditions for graded modules (gr- $\Sigma(\Delta)$-injective modules) (in preparation).

C. NĂSTĂSESCU and F. VAN OYSTAEYEN

[1] Graded Ring Theory, North-Holland Mathematical Library 28 (North-Holland Publishing Company, Amsterdam New York Oxford 1982).

R. C. SHOCK

[1] Certain Artinian rings are Noetherian, Canad. J. Math. 24 (1972), 553-556.

B. STENSTRÖM

[1] Rings of Quotients, Grundlehren der mathematischen Wissenschaften 217 (Springer-Verlag, Berlin Heidelberg New York 1975).

# A GENERALIZATION OF SEMISIMPLE MODULES

José L. Bueso and Pascual Jara

Departamento de Algebra. Universidad de Granada. España.

ABSTRACT. This communication generalizes the concepts of the socle of a module and of the semisimple module, by replacing simples with $\tau$-critical modules for a hereditary torsion theory $\tau$, showing that $\tau$ is strongly semiprime if and only if $\tau = \chi(M)$ for a $\tau$-semicritical module. As a further application, properties of the endomorphism ring of the $\tau$-injective hulls of these modules are related to its structure.

## PRELIMINARIES.

Throughout the following, a ring is always an associative ring with identity and R denotes such a ring. The category of all (unital) right R-modules is denoted by Mod-R. Unless otherwise specified, all modules are objects of Mod-R.

If $\tau$ is a hereditary torsion theory, we denote the class of all $\tau$-torsion right R-modules by $T_\tau$, and the class of all $\tau$-torsionfree right R-modules by $F_\tau$. The idempotent filter of $\tau$ is denoted by $L_\tau$ , and the torsion radical by $T_\tau$. For information on torsion theories, the reader is referred to [6], [11], and [13].

## 1. $\tau$-CLOSURE OF A SUBMODULE.

For each submodule N of an R-module M, we denote

$$Cl_\tau^M(N) = \{x \in M \mid (N:x) \in L_\tau\}.$$

It is called the $\tau$-closure of N in M. We say that N is $\tau$-dense ($\tau$-closed) in M if $M = Cl_\tau^M(N)$ ($N = Cl_\tau^M(N)$). It is clear that



*F. van Oystaeyen (ed.), Methods in Ring Theory, 53–65.*

$T_\tau(M/N) = Cl_\tau^M(N)/N$.

We denote the set of all $\tau$-closed submodules of M by $C_\tau(M)$. Note that $N \in C_\tau(M)$ if and only if M/N is $\tau$-torsionfree. By the set-theoretical inclusion, the set $C_\tau(M)$ is a complete, modular and upper-continuous lattice, and hence pseudo-complemented, with the greatest element M and the smallest element $T_\tau(M)$. The join and the meet of a family $\{N_\alpha \mid \alpha \in A\}$ of elements in $C_\tau(M)$ are:

$$\vee\{N_\alpha \mid \alpha\in A\} = Cl_\tau^M(\Sigma\{N_\alpha \mid \alpha\in A\}) \quad \text{and}$$
$$\wedge\{N_\alpha \mid \alpha\in A\} = \cap\, \{N_\alpha \mid \alpha\in A\}.$$

1.1. Remarks.

Also the following facts on the lattice $C_\tau(M)$ are useful:

1. A submodule N of M is $\tau$-closed if and only if $C_\tau(N)$ is a subset of $C_\tau(M)$. Moreover, if H is a submodule of N, then

$$Cl_\tau^N(H) = Cl_\tau^M(H).$$

2. A submodule N of M is $\tau$-dense if and only if the lattice $C_\tau(N)$ is isomorphic to the lattice $C_\tau(M)$, by the correspondence $H \longmapsto Cl_\tau^M(H)$. Moreover, we have that $Cl_\tau^N(H) = Cl_\tau^M(H) \cap N$.

3. If $f:M \longrightarrow M'$ is an onto R-homomorphism with Ker f $\tau$-torsion, then $C_\tau(M)$ is isomorphic to $C_\tau(M')$, by the correspondence $H \longmapsto f(H)$, moreover, we have that $f(Cl_\tau^M(H)) = Cl_\tau^{M'}(f(H))$, for every submodule H of M.

4. For every family $\{N_\alpha \mid \alpha\in A\}$ of submodules of M, we have:

$$Cl_\tau^M(\cap\{Cl_\tau^M(N_\alpha) \mid \alpha\in A\}) = \cap\{Cl_\tau^M(N_\alpha) \mid \alpha\in A\} \quad \text{and}$$
$$Cl_\tau^M(\Sigma\{N_\alpha \mid \alpha\in A\}) = Cl_\tau^M(\Sigma\{Cl_\tau^M(N_\alpha) \mid \alpha\in A\}).$$

1.2. Lemma.

Let M be a $\tau$-torsionfree R-module, then:

1. Every pseudo-complement submodule of M is $\tau$-closed.
2. Each $\tau$-dense submodule of M is essential.

1.3. Proposition.

Let M be a $\tau$-torsionfree R-module, then the following statements are equivalent:

1. $C_\tau(M)$ is a complemented lattice.
2. Every $\tau$-closed submodule of M is a pseudo-complement.
3. Each essential submodule of M is $\tau$-dense.

## 2. $\tau$-CRITICAL MODULES.

A right R-module M is said to be $\tau$-critical if $M \neq 0$, is $\tau$-torsionfree and each nonzero submodule is $\tau$-dense.

**2.1. Proposition.**

The following conditions are equivalent for a nonzero R-module M:

1. M is $\tau$-critical.
2. $C_\tau(M) = \{0, M\}$.

**2.2. Lemma.**

Let M be a $\tau$-torsionfree R-module. If K is a $\tau$-critical submodule of M, then its $\tau$-closure is $\tau$-critical.

**2.3. Lemma.**

If K is a $\tau$-critical submodule of M and $N \in C_\tau(M)$, then $N \cap K$ is either zero or K.

**2.4. Lemma.**

Let $f: M \longrightarrow M'$ be an R-homomorphism. If K is a $\tau$-critical submodule, then either f(K) is $\tau$-torsion or f(K) is $\tau$-critical.

## 3. $\tau$-SEMICRITICAL MODULES.

Throughout the following, M is an R-module $\tau$-torsionfree. We adopt the following convention: the empty family of $\tau$-critical submodules of M is independent and its direct sum is zero.

**3.1. Lemma.**

Let M be a $\tau$-torsionfree R-module such that $M = Cl_\tau^M(\Sigma\{K_\alpha | \alpha\varepsilon A\})$ for a family of $\tau$-critical submodules. If N is a $\tau$-closed submodule of M, then there is a subset B of A such that:

1. The family $\{K_\beta | \beta\varepsilon B\}$ is independent.
2. $M = Cl_\tau^M(N \oplus (\oplus\{K_\beta | \beta\varepsilon B\}))$.

*Proof*: If $N = M$, we must have $B = \emptyset$. Assume that $N \neq M$. Let $\Sigma = \{B | B \subseteq A, \{K_\beta | \beta\varepsilon B\}$ is an independent family and $N \cap (\Sigma\{K_\beta | \beta\varepsilon B\}) = 0\}$. As $\emptyset \in \Sigma$, then $\Sigma \neq \emptyset$. $\Sigma$ is ordered by inclusion. Let $B_1 \subseteq B_2 \subseteq \ldots \subseteq B_n \subseteq \ldots$ a totally ordered subset in $\Sigma$. Then $B = \cup\{B_n | n \in \mathbb{N}\}$ is an upper bound of this chain in $\Sigma$. By Zorn's lemma, there is a maximal element B in $\Sigma$. Moreover B satisfies (2).

**3.2. Corollary.**

Let M be a $\tau$-torsionfree R-module such that $M = Cl_\tau^M(\Sigma\{K_\alpha | \alpha\varepsilon A\})$ for a family of $\tau$-critical submodules, then for some $B \subseteq A$, we have:

$$M = Cl_\tau^M(\oplus\{K_\beta | \beta\varepsilon B\}).$$

3.3. Corollary.

If M is a $\tau$-torsionfree R-module and $M = Cl_\tau^M(\Sigma\{K_\alpha \mid \alpha\varepsilon A\})$ for a family of $\tau$-critical submodules of M, then $C_\tau(M)$ is a complemented lattice.

Let M be a $\tau$-torsionfree R-module, we shall denote $S_\tau(M)$ the $\tau$-socle of M,i.e.,

$S_\tau(M) = Cl_\tau^M(\Sigma\{K \mid K$ is a $\tau$-critical submodule of $M\})$ or

$S_\tau(M) = 0$ if M has no $\tau$-critical submodules.

3.4. Lemma.

Let $f:M \longrightarrow M'$ be an R-homomorphism of $\tau$-torsionfree R-modules, then $f(S_\tau(M)) \leq S_\tau(M')$.

Proof: If $S_\tau(M) = 0$, then it is clear. If $S_\tau(M) \neq 0$, then;

$$\begin{aligned} f(S_\tau(M)) &= f(Cl_\tau^M(\Sigma\{K \mid K \text{ is a } \tau\text{-critical submodule of } M\})) \\ &\leq Cl_\tau^{M'}(f(\Sigma\{K \mid K \text{ is a } \tau\text{-critical submodule of } M\})) \\ &= Cl_\tau^{M'}(\Sigma\{f(K) \mid K \text{ is a } \tau\text{-critical submodule of } M\}) \\ &\leq Cl_\tau^{M'}(\Sigma\{K' \mid K' \text{ is a } \tau\text{-critical submodule of } M'\}) \\ &= S_\tau(M'). \end{aligned}$$

3.5. Lemma.

Let M be a $\tau$-torsionfree R-module. For every $\tau$-critical submodule K of M, and for every H submodule of K, we have that $Cl_\tau^M(H) = Cl_\tau^M(K)$.

3.6. Lemma.

Let M be a $\tau$-torsionfree R-module. For every $\tau$-dense submodule N of M, we have that $Cl_\tau^M(S_\tau(N)) = S_\tau(M)$.

Proof: By Lemma 3.4, we have that $S_\tau(N) \leq S_\tau(M)$. It is clear that every $\tau$-critical submodule of N is an intersection of a $\tau$-critical submodule of M and N. Then we have the following identities:

$$\begin{aligned} S_\tau(M) \cap N &= Cl_\tau^M(\Sigma\{K \mid K \leq M \text{ is } \tau\text{-critical}\}) \cap N \\ &= Cl_\tau^M(\Sigma\{Cl_\tau^M(K) \mid K \leq M \text{ is } \tau\text{-critical}\}) \cap N \\ &= Cl_\tau^M(\Sigma\{Cl_\tau^M(K \cap N) \mid K \leq M \text{ is } \tau\text{-critical}\}) \cap N \\ &= Cl_\tau^M(\Sigma\{K \cap N \mid K \leq M \text{ is } \tau\text{-critical}\}) \cap N \\ &= Cl_\tau^N(\Sigma\{K \cap N \mid K \leq M \text{ is } \tau\text{-critical}\}) \\ &= S_\tau(N). \end{aligned}$$

Therefore, by Remark 2, we have that $Cl_\tau^M(S_\tau(N)) = S_\tau(M)$.

3.7. Theorem.
The following statements are equivalent for every $\tau$-torsionfree R-module M:

1. $M = S_\tau(M)$.
2. For each $\tau$-closed submodule N of M, we have $N = S_\tau(N)$.
3. For each $\tau$-closed submodule N of M, we have $M/N = S_\tau(M/N)$.
4. M has a $\tau$-dense direct sum of $\tau$-critical submodules.
5. For each $\tau$-closed submodule N of M there is an independent family $\{K_\beta \mid \beta \in B\}$ of $\tau$-critical submodules of M such that $M = Cl_\tau^M(N \oplus (\oplus\{K_\beta \mid \beta \in B\}))$.

Proof: 5 ⟹ 4, 4 ⟹ 1, 2 ⟹ 1, and 3 ⟹ 1 are obvious.
1 ⟹ 5 is a consequence of 3.1.
5 ⟹ 3. Let N be a $\tau$-closed submodule of M. If N = M, then $S_\tau(0) = 0$. If $N \neq M$, then there is a non empty independent family $\{K_\beta \mid \beta \in B\}$ of $\tau$-critical submodules, such that $M = Cl_\tau^M(N \oplus (\oplus\{K_\beta \mid \beta \in B\}))$. We call $L = Cl_\tau^M(\oplus\{K_\beta \mid \beta \in B\})$. Then $L \cap N = 0$ and L + N is $\tau$-dense in M. Since $(L + N)/N \cong L/(L \cap N) = L$, then $S_\tau((L + N)/N) = (L + N)/N$. So, by Lemma 3.6, $S_\tau(M/N) = M/N$.
5 ⟹ 2. Let N be a $\tau$-closed submodule of M. There is an independent family $\{K_\beta \mid \beta \in B\}$ of $\tau$-critical submodules of M such that $M = Cl_\tau^M(N \oplus (\oplus\{K_\beta \mid \beta \in B\}))$. We call $L = Cl_\tau^M(\oplus\{K_\beta \mid \beta \in B\})$. Then $L \cap N = 0$ and L + N is $\tau$-dense in M. Since (3 ⟺ 5), then $S_\tau(M/L) = M/L$. We have that $N = N/(L \cap N) \cong (L + N)/L$, and by Lemma 3.6, we have $S_\tau(M/L) = Cl_\tau^{M/L}(S_\tau((L + N)/L))$, then by Remark 2 $S_\tau((L + N)/L) = (L + N)/L$, therefore $S_\tau(N) = N$.

We call a $\tau$-torsionfree R-module $\tau$-semicritical if it satisfies the conditions of the theorem.

3.8. Lemma.
For every $\tau$-torsionfree R-module M, we have $S_\tau(S_\tau(M)) = S_\tau(M)$.

3.9. Lemma.
If N is a submodule of a $\tau$-torsionfree R-module, then $S_\tau(N) = S_\tau(M) \cap N$.

3.10. Corollary.
Every submodule and every $\tau$-torsionfree quotient of a $\tau$-semicritical R-module is a $\tau$-semicritical R-module.

3.11. Corollary.
The following conditions are equivalent for a $\tau$-torsionfree R-module M:

1. M is τ-semicritical.

2. a) Every τ-closed submodule of M contains a τ-critical submodule.

b) $C_\tau(M)$ is a complemented lattice.

3. For every τ-closed submodule N of M, N ≠ M, there is a τ-critical submodule K of M such that $N \cap K = 0$.

3.12. Theorem.
The following conditions are equivalent for a τ-torsionfree ring R:

1. R is τ-semicritical as a right R-module.

2. Every τ-torsionfree R-module is τ-semicritical.

Proof: We only need to show that 1 ⟹ 2. As R is a τ-semicritical R-module, then there is an independent family $\{K_\alpha \mid \alpha \in A\}$ of τ-critical ideals of R, such that $R = Cl_\tau^R(\oplus\{K_\alpha \mid \alpha\in A\})$. For every τ-closed submodule N of a τ-torsionfree R-module M, and for every element $x \in M\setminus N$, we have that $(N:x) \notin L_\tau$. If for each $\alpha \in A$ we have $x K_\alpha \leq N$, then $x(\Sigma\{K_\alpha \mid \alpha\in A\}) \leq N$, and since $\Sigma\{K_\alpha \mid \alpha\in A\}$ is τ-dense, then $(N:x) \in L_\tau$, which is a contradiction. Therefore, there is $\alpha \in A$ such that $x K_\alpha \not\leq N$. It is clear that $x K_\alpha$ is τ-critical and $x K_\alpha \cap N = 0$. It follows from Corollary 3.11 that M is τ-semicritical.

A ring R is called τ-semicritical if it satisfies the equivalent conditions of the theorem.

4. STRONGLY SEMIPRIME TORSION THEORY.

4.1. Proposition.
The following conditions are equivalent for a nonzero R-module M:

1. M is χ(M)-semicritical.

2. There is a τ ε R-prop with M τ-semicritical.

Proof: 1 ⟹ 2 is obvious.
2 ⟹ 1. If M is a τ-torsionfree R-module, for some τ ε R-prop, then $\tau \leq \chi(M)$. Thus if there is an independent family of τ-critical submodules of M, $\{K_\alpha \mid \alpha\in A\}$ such that $M = Cl_\tau^M(\oplus\{K_\alpha \mid \alpha\in A\})$, then as $M = Cl_{\chi}^M(\oplus\{K_\alpha \mid \alpha\in A\})$ and every $K_\alpha$ is χ(M)-critical, we have that M is χ(M)-semicritical.

4.2. Lemma.
Let τ ε R-prop and let M be a τ-semicritical R-module, then M contains an essential sum of uniform submodules.

We will say that $\tau \in$ R-prop is strongly semiprime if and only if $\tau = \wedge U$, for some $U \subseteq \{\chi(M) \mid M$ is a $\tau$-critical R-module$\}$.

4.3. Theorem.
The following conditions are equivalent for $\tau \in$ R-prop:

1. There is a $\tau$-semicritical R-module M such that $\tau = \chi(M)$.
2. $\tau$ is strongly semiprime.

Proof: 1 ⟹ 2. By hypothesis, $\tau = \chi(M)$ for a $\tau$-semicritical R-module M. By definition, there is a family $\{K_\alpha \mid \alpha \in A\}$ of $\tau$-critical submodules of M such that $M = Cl_\tau^M(\oplus\{K_\alpha \mid \alpha\in A\})$. Then $\oplus\{K_\alpha \mid \alpha\in A\}$ is essential in M, and by [6; Proposition 8.6]:

$$\tau = \chi(M) = \chi(\oplus\{K_\alpha \mid \alpha\in A\}) = \wedge\{\chi(K_\alpha) \mid \alpha\in A\}.$$

2 ⟹ 1. By hypothesis $\tau = \wedge\{\chi(K_\alpha) \mid \alpha\in A\}$ for a family $\{K_\alpha \mid \alpha\in A\}$ of $\tau$-critical R-modules. Let $E = E(\oplus\{K_\alpha \mid \alpha\in A\})$, and let $M = Cl_\tau^E(\oplus\{K_\alpha \mid \alpha\in A\})$, then M is $\tau$-semicritical and essential in E, and so

$$\chi(M) = \chi(E) = \chi(\oplus\{K_\alpha \mid \alpha\in A\}) = \wedge\{\chi(K_\alpha) \mid \alpha\in A\} = \tau.$$

4.4. Corollary.
The following conditions are equivalent for a ring R:

1. R is a right seminoetherian ring.
2. Every $\tau \in$ R-prop has a $\tau$-semicritical right R-module.

## 5. THE CANONICAL DECOMPOSITION OF THE $\tau$-SOCLE.

Two $\tau$-critical R-modules H and K will be said to be related - written as $H \sim K$ - if $E_\tau(H) \cong E_\tau(K)$, or equivalently, if they have nonzero isomorphic submodules. Let $\Omega$ denote the set of the equivalence class modulo the equivalence relation "$\sim$" of related $\tau$-critical submodules of the fixen module M.

For every $\omega \in \Omega$, we may define a $\omega$-socle relative to the torsion theory $\tau$, as:

$$S_{\tau\omega}(M) = Cl_\tau^M(\Sigma\{K \mid K \leq M \text{ and } K \in \omega\}) \quad \text{or}$$
$$S_{\tau\omega}(M) = 0 \quad \text{if M has no } \tau\text{-critical submodules in } \omega.$$

Let $\Omega'$ be the analogue of $\Omega$ for the $\tau$-injective hull of M, $E_\tau(M)$. Since $\omega \longmapsto \omega' = \{H' \mid H' \leq E_\tau(M)$, and $E_\tau(H') \cong E_\tau(H)$ for some $H \in \omega\}$ is a bijection from $\Omega$ to $\Omega'$, the same index set $\Omega$ will be used for M and $E_\tau(M)$.

5.1. Lemma.
Let M be a $\tau$-torsionfree R-module, and N a submodule of M. If $\{K_\alpha \mid \alpha\in A\}$ is an independient family of submodules of M such that $N \cap (\oplus\{K_\alpha \mid \alpha\in A\}) \neq 0$, then there is $\alpha \in A$ such that $K_\alpha$ contains a

nonzero submodule isomorphic to a submodule of N.
Proof: See [5; Proposition 2].

5.2. Lemma.
For every $\tau$-torsionfree R-module M there is a maximal independent family $\{K_\alpha | \alpha\varepsilon A\}$ of $\tau$-critical submodules of M such that

$$S_\tau(M) = Cl_\tau^M(\oplus\{K_\alpha | \alpha\varepsilon A\}).$$

Moreover, for every $\omega \varepsilon \Omega$, if $A_\omega = \{\alpha | \alpha\varepsilon A, \text{ and } K_\alpha \varepsilon\omega\}$, then $\{K_\alpha | \alpha\varepsilon A_\omega\}$ is a maximal independent family of $\tau$-critical submodules of M in $\omega$, such that:

$$S_{\tau\omega}(M) = Cl_\tau^M(\oplus\{K_\alpha | \alpha\varepsilon A_\omega\}).$$

Proof: The first affirmation is an immediate consequence of 3.2. Now assume that H is a $\tau$-critical submodule of M such that $H \varepsilon \omega$, then $H \cap (\oplus\{K_\alpha | \alpha\varepsilon A\}) \neq 0$, so there is a minimal subset A' of A such that $H \cap (\oplus\{K_\alpha | \alpha\varepsilon A'\}) \neq 0$. We call $\pi_\alpha : \oplus\{K_\alpha | \alpha\varepsilon A'\} \longrightarrow K_\alpha$ the canonical projection and $\pi_\alpha' : H \cap (\oplus\{K_\alpha | \alpha\varepsilon A'\}) \longrightarrow K_\alpha$ its restriction. We will show that $\pi_\alpha'$ is a monomorphism for every $\alpha \varepsilon A'$. If there is $\alpha' \varepsilon A'$ such that $\pi_{\alpha'}'$ is not a monomorphism, then $\pi_{\alpha'}' = 0$, so $H \cap (\oplus\{K_\alpha | \alpha' \neq \alpha\varepsilon A'\}) \neq 0$, which is a contradiction. Then $H \sim K_\alpha$ for every $\alpha \varepsilon A'$, and thus we have $A' \subseteq A_\omega$. So $H \cap (\oplus\{K_\alpha | \alpha\varepsilon A_\omega\}) \neq 0$ and $A_\omega$ is maximal and independent. The rest is now clear.

5.3. Lemma.
Let M be a $\tau$-torsionfree R-module such that $S_\tau(M) = Cl_\tau^M(\oplus\{K_\alpha | \alpha\varepsilon A\})$ For each $\omega \varepsilon \Omega$, we define $M_\omega = \oplus\{K_\alpha | \alpha\varepsilon A_\omega\}$, then $\{M_\omega | \omega\varepsilon\Omega\}$ is an independent family of submodules of M.
Proof: Assume that $\omega_1$ and $\omega_2$ are elements of $\Omega$, then $M_{\omega_1} \cap M_{\omega_2} = 0$. In fact, if $0 \neq L \leq M_{\omega_1} \cap M_{\omega_2}$, then we have that $L \cap (\oplus\{K_\alpha | \alpha\varepsilon A_{\omega_1}\})$ is nonzero, by Lemma 5.1, L contains a nonzero submodule $H \varepsilon \omega_1$. Also we have that $H \cap (\oplus\{K_\alpha | \alpha\varepsilon A_{\omega_2}\}) \neq 0$, then H contains a nonzero submodule in $\omega_2$, which is a contradiction. Let $\omega_1$, $\omega_2$ and $\omega_3$ be elements of $\Omega$, and assume that $M_{\omega_1} \cap (M_{\omega_2} \oplus M_{\omega_3}) \neq 0$, then there is a nonzero submodule H in $\omega_2$ or $\omega_3$, and H by Lemma 5.1,is also in $\omega_1$, which is a contradiction. As a consequence, we have the result for every finite family of elements in $\Omega$. Assume now that $\omega \varepsilon \Omega$ and $M_\omega \cap (\Sigma\{M_{\omega'} | \omega \neq \omega'\varepsilon\Omega\}) \neq 0$, then there is a finite family C of $\Omega$ such that $M_\omega \cap (\Sigma\{M_{\omega'} | \omega \neq \omega'\varepsilon C\}) \neq 0$, which is a contradiction.

5.4. Corollary.
Let M be a $\tau$-torsionfree R-module, then $S_\tau(M) = Cl_\tau^M(\oplus\{S_{\tau\omega}(M) | \omega\varepsilon\Omega\})$.

5.5. Lemma.
Let M be a $\tau$-torsionfree R-module, and $\{K_\alpha | \alpha\in A\}$ and $\{H_\beta | \beta\in B\}$ be two maximal independent families of $\tau$-critical submodules of M in $\omega$, such that

$$S_\tau(M) = Cl^M_\tau(\oplus\{K_\alpha | \alpha\in A\}) = Cl^M_\tau(\oplus\{H_\beta | \beta\in B\}),$$

then card(A) = card(B).
Proof: It is similar to [7; Lemme 1.2.2].

5.6. Lemma.
Let M be a $\tau$-torsionfree R-module and $\{K_\alpha | \alpha\in A\}$ and $\{H_\beta | \beta\in B\}$ be two maximal families of $\tau$-critical submodules of M, such that

$$S_\tau(M) = Cl^M_\tau(\oplus\{K_\alpha | \alpha\in A\}) = Cl^M_\tau(\oplus\{H_\beta | \beta\in B\}),$$

then there is a bijection $\sigma: A \longrightarrow B$ and an R-isomorphism $f: \oplus\{E_\tau(K_\alpha) | \alpha\in A\} \longrightarrow \oplus\{E_\tau(H_\beta) | \beta\in B\}$.
Proof: By 5.4, $S_\tau(M) = Cl^M_\tau(\oplus\{S_{\tau\omega}(M) | \omega\in\Omega\})$ and, by 5.2, $S_{\tau\omega}(M) = Cl^M_\tau(\oplus\{K_\alpha | \alpha\in A_\omega\}) = Cl^M_\tau(\oplus\{H_\beta | \beta\in B_\omega\})$. Moreover, by 5.5, $card(A_\omega) = card(B_\omega)$, and so there is a bijection $\sigma: A \longrightarrow B$, and as $K_\alpha \sim H_{\sigma(\alpha)}$, then there is an R-isomorphism $f_\alpha: E_\tau(K_\alpha) \longrightarrow E_\tau(H_{\sigma(\alpha)})$, and so there is an R-isomorphism $f: \oplus\{E_\tau(K_\alpha) | \alpha\in A\} \longrightarrow \oplus\{E_\tau(H_\beta) | \beta\in B\}$.

5.7. Lemma.
Let M be a $\tau$-torsionfree R-module, then

$$Hom_R(S_{\tau\omega}(M), S_{\tau\omega'}(M)) = 0 \quad \text{for} \quad \omega \neq \omega' \in \Omega.$$

Proof: Let $f \in Hom_R(S_{\tau\omega}(M), S_{\tau\omega'}(M))$ for $\omega \neq \omega' \in \Omega$, then

$$f(Cl^M_\tau(\oplus\{K_\alpha | \alpha\in A_\omega\}) \leq Cl^M_\tau(\oplus\{f(K_\alpha) | \alpha\in A_\omega\}) = Cl^M_\tau(0) = 0.$$

5.8. Lemma.
Let M be a $\tau$-torsionfree R-module, then $S_{\tau\omega}(E_\tau(M)) \leq E_\tau(S_{\tau\omega}(M))$.
Proof: If K is a $\tau$-critical submodule of $E_\tau(M)$ such that $K \in \omega$, then $K \cap M$ is a $\tau$-critical submodule of M, so $K \cap M \in \omega$, and thus $E_\tau(K \cap M) = E_\tau(K)$, hence $K \leq E_\tau(K \cap M) \leq E_\tau(S_{\tau\omega}(M))$.

5.9. Lemma.
Let M be a $\tau$-torsionfree R-module. If $S_\tau(M) = Cl^M_\tau(\oplus\{K_\alpha | \alpha\in A\})$ for an independent family of $\tau$-critical submodules of M, then

$$S_{\tau\omega}(E_\tau(M)) = Cl^{E_\tau(M)}_\tau(\oplus\{E_\tau(K_\alpha) | \alpha\in A_\omega\}) \quad \text{and}$$

$$S_\tau(E_\tau(M)) = Cl^{E_\tau(M)}_\tau(\oplus\{E_\tau(K_\alpha) | \alpha\in A\}).$$

Proof: If $K_\alpha \in \omega$, then $E_\tau(K_\alpha) \in \omega$ and so $Cl^{E_\tau(M)}_\tau(\oplus\{E_\tau(K_\alpha) | \alpha\in A_\omega\})$ is a submodule of $S_{\tau\omega}(E_\tau(M))$. Let H be a $\tau$-critical submodule of $E_\tau(M)$ such that $H \in \omega$, then $0 \neq H \cap M \in \omega$, and so:

$$H \cap M \leq S_\tau(M) = Cl^M_\tau(\oplus\{K_\alpha | \alpha\in A\setminus A_\omega\}) \oplus Cl^M_\tau(\oplus\{K_\alpha | \alpha\in A_\omega\}).$$

We call $L = H \cap M \cap ((\oplus\{K_\alpha | \alpha \in A \setminus A_\omega\}) \oplus (\oplus\{K_\alpha | \alpha \in A_\omega\}))$, if $L \cap (\oplus\{K_\alpha | \alpha \in A_\omega\}) = H \cap (\oplus\{K_\alpha | \alpha \in A_\omega\})$ is zero, then

$$L \cong \frac{L \oplus (\oplus\{K_\alpha | \alpha \in A_\omega\})}{\oplus\{K_\alpha | \alpha \in A_\omega\}} \leq \frac{\oplus\{K_\alpha | \alpha \in A\}}{\oplus\{K_\alpha | \alpha \in A_\omega\}} \cong \oplus\{K_\alpha | \alpha \in A \setminus A_\omega\},$$

which is a contradiction. Then for every $0 \neq x \in H \cap (\oplus\{K_\alpha | \alpha \in A_\omega\})$, there is a finite family $K_1, \ldots, K_n \in \omega$ such that $x \in H \cap (K_1 + \ldots + K_n)$, so $E_\tau(xR) \leq \oplus\{E_\tau(K_i) | 1 \leq i \leq n\}$ and we have $H \leq E_\tau(H) = E_\tau(xR) \leq \oplus\{E_\tau(K_i) | 1 \leq i \leq n\} \leq \oplus\{E_\tau(K_\alpha) | \alpha \in A_\omega\}$ and so $S_{\tau\omega}(E_\tau(M)) \leq Cl_\tau^{E_\tau(M)}(\oplus\{E_\tau(K_\alpha) | \alpha \in A_\omega\})$. The rest is now clear.

An R-module M will be said to be $\tau$-quasi-injective if for each $\tau$-dense submodule N of M, every R-homomorphism $f: N \longrightarrow M$ can be extended to an R-endomorphism of M, or equivalently, if M is stable under the endomorphisms of $E_\tau(M)$.

5.10. Lemma.
Let M be a $\tau$-torsionfree R-module. If M is $\tau$-quasi-injective and $C_\tau(M)$ is a complemented lattice, then M is quasi-injective.
Proof: See [7; Proposition 1.1.6] .

5.11. Proposition.
Let K be a $\tau$-critical R-module. Then $E_\tau(K)$ is quasi-injective and $End(E_\tau(K))$ is a division ring.
Proof: $E_\tau(K)$ is $\tau$-quasi-injective and $\tau$-critical, so it is quasi-injective. Since K is $\tau$-critical, then K is uniform and E(K) is indecomposable, and so End(E(K))/J(End(E(K))) is a division ring. If we define $\rho: End(E(K))/J(End(E(K))) \longrightarrow End(E_\tau(K))$ by $\rho(f + J(End(E(K)))) = f_{E_\tau(K)}$, then $\rho$ is clearly an isomorphism of rings.

5.12. Lemma.
Let M be a $\tau$-torsionfree R-module and N a $\tau$-quasi-injective $\tau$-dense submodule of $E_\tau(M)$, then $End(N) \cong End(E_\tau(M))$.
Proof: We define the map $\delta: End(N) \longrightarrow End(E_\tau(M))$ by the diagram

$$\begin{array}{ccccc} 0 & \longrightarrow & N & \longrightarrow & E_\tau(M) \\ & & f \downarrow & & \downarrow \delta(f) \\ 0 & \longrightarrow & N & \longrightarrow & E_\tau(M) \end{array}$$

Then it is clear that $\delta$ is an isomorphism of rings.

5.13. Lemma.
Let M be a $\tau$-torsionfree R-module. If M is $\tau$-semicritical, then $E_\tau(M)$ is also $\tau$-semicritical.

Proof: Since M is $\tau$-dense in $E_\tau(M)$, then the lattices $C_\tau(M)$ and $C_\tau(E_\tau(M))$ are isomorphic, and so $E_\tau(M)$ is $\tau$-semicritical.

5.14. Theorem.
Let M be a $\tau$-torsionfree R-module. If M is $\tau$-semicritical, then

$$\mathrm{End}(E_\tau(M)) \cong \Pi\{\mathrm{End}(S_{\tau\omega}(E_\tau(M)))\mid \omega\varepsilon\Omega\}.$$

Proof: By 5.13, $E_\tau(M)$ is $\tau$-semicritical, by 5.4, we have that $\oplus\{S_{\tau\omega}(E_\tau(M))\mid \omega\varepsilon\Omega\}$ is $\tau$-dense in $E_\tau(M)$. As $S_{\tau\omega}(E_\tau(M)) \;\varepsilon\; C_\tau(E_\tau(M))$, then $S_{\tau\omega}(E_\tau(M))$ is $\tau$-injective, and so $\oplus\{S_{\tau\omega}(E_\tau(M))\mid\omega\varepsilon\Omega\}$ is $\tau$-quasi-injective and thus we have:

$$\mathrm{End}(E_\tau(M)) \cong \mathrm{End}(\oplus\{S_{\tau\omega}(E_\tau(M))\mid\omega\varepsilon\Omega\}) \cong \Pi\{\mathrm{End}(S_{\tau\omega}(E_\tau(M)))\mid\omega\varepsilon\Omega\}.$$

5.15. Theorem.
Let M be a $\tau$-torsionfree R-module. If M is $\tau$-semicritical, then $\mathrm{End}(S_{\tau\omega}(E_\tau(M)))$ is isomorphic to a ring of endomorphisms on a left vector space over a division ring.

Proof: Let $\{K_\alpha\mid\alpha\varepsilon A\}$ be a maximal independent family of $\tau$-critical submodules of M, such that $S_\tau(M) = Cl_\tau^M(\oplus\{K_\alpha\mid\alpha\varepsilon A\})$. Then

$$S_{\tau\omega}(E_\tau(M)) \leq E_\tau(S_{\tau\omega}(M)) = S_{\tau\omega}(E_\tau(S_{\tau\omega}(M))).$$

Moreover, $E_\tau(S_{\tau\omega}(M)) \leq E_\tau(M)$, so we have $E_\tau(S_{\tau\omega}(M)) = S_{\tau\omega}(E_\tau(M))$, and then, since $S_{\tau\omega}(E_\tau(M)) = Cl_\tau^{E_\tau(M)}(\oplus\{E_\tau(K_\alpha)\mid\alpha\varepsilon A_\omega\})$, we have that $\oplus\{E_\tau(K_\alpha)\mid\alpha\varepsilon A_\omega\})$ is $\tau$-quasi-injective and $\tau$-dense in $E_\tau(S_{\tau\omega}(M))$, and, by 5.12, $\mathrm{End}(S_{\tau\omega}(E_\tau(M)) \cong \mathrm{End}(\oplus\{E_\tau(K_\alpha)\mid\alpha\varepsilon A_\omega\})$.
We will study the structure of this last ring. For this purpose, let E denote an R-module isomorphic to $E_\tau(K_\alpha)$, and let $\sigma_\alpha : E \longrightarrow E_\tau(K_\alpha)$ the mentioned isomorphism. We define the following R-homomorphism

$$E \xrightarrow{\sigma_\alpha} E_\tau(K_\alpha) \xrightarrow{\iota_\alpha} \underset{\alpha}{\oplus} E_\tau(K_\alpha) \xrightarrow{f} \underset{\alpha}{\oplus} E_\tau(K_\alpha) \xrightarrow{\pi_\beta} E_\tau(K_\beta) \xrightarrow{\sigma_\beta^{-1}} E,$$

where $\iota_\alpha$ and $\pi_\beta$ are the canonical R-homomorphisms. For almost every index $\beta$, this R-homomorphism is zero. We consider then a left vector space V over the division ring $S' = \mathrm{End}(E)$, with dimension $\mathrm{card}(A_\omega)$. If $\{e_\alpha\mid\alpha\varepsilon A_\omega\}$ is a base of V, we define $f' : V \longrightarrow V$ by

$$(e_\alpha)f' = \sum_{\beta\varepsilon A_\omega} (\sigma_\beta^{-1}\pi_\beta f \iota_\alpha \sigma_\alpha) e_\beta .$$

The map defined by $f \longmapsto f'$ from $\mathrm{End}(\oplus\{E_\tau(K_\alpha)\mid\alpha\varepsilon A_\omega\})$ to $\mathrm{End}(V)$ is an isomorphism of rings.

5.16. Corollary.
Let M be a $\tau$-torsionfree R-module. If M is $\tau$-semicritical, then $\mathrm{End}(E_\tau(M))$ is isomorphic to a direct product of rings of endomorphisms on vector spaces.

## 6. EXAMPLES.

6.1. Trivial torsion theory. An R-module is defined to be $\xi(0)$-torsion if and only if it is zero. Hence every R-module is $\xi(0)$-torsionfree. Thus an R-module is $\xi(0)$-critical if and only if it is simple and an R-module is $\xi(0)$-semicritical if and only if it is semisimple or zero. See [11; § 11].

6.2. Goldie's torsion theory. An R-module is nonsingular if and only if it is $\tau_G$-torsionfree. An R-module is therefore $\tau_G$-critical precisely when it is uniform and nonsingular, and an R-module is $\tau_G$-semicritical if and only if it contains an essential sum of uniform nonsingular submodules or it is zero. See [4].

6.3. Finitely $\tau$-semicritical R-modules. We say that an R-module M is finitely $\tau$-semicritical if it is $\tau$-semicritical and $S_\tau(M)$ is the sum of a finite family of $\tau$-critical submodules of M. In [9], Lau has defined a $\tau$-semicritical R-module as an R-module such that zero is a finite intersection of $\tau$-cocritical submodules of M. By [9; Proposition 3.6] it is clear that M is $\tau$-semicritical in the sense of Lau if and only if it is finitely $\tau$-semicritical.

## REFERENCES

[1] Bueso, J. L. and Jara P. "A generalization of semicritical modules". 1.983, to appear.

[2] Bueso, J. L. and Jara P. "Loewy series relative to a torsion theory" 1.983, to appear.

[3] Chew, K. L. "Closure operations in the study of rings of quotients". Bull. Math. Soc. Nanyang. pp. 1 - 20. 1.965.

[4] Dauns, J. "Sums of uniform modules". 1.982. Lecture Notes in Math., 951. pp. 68 - 87. Springer-Verlag.

[5] Fort, J. "Sommes directs de sous-modules coirréductibles d'un module". Math. Z. 103, 363 - 388. 1.967.

[6] Golan, J. "Localization of noncommutative rings". Marcel-Dekker. 1.975.

[7] Hudry, A. "Applications de la theorie de la localization aux anneaux et aux modules". Thése (3ème cycle) Univ. Lyon I. 1.971.

[8] Jara, P. "Teorias de torsion: zócalo y radical". Tesis Doctoral. Univ. de Granada. 1.983.

[9] Lau, W. G. "Torsion theoretic generalizations of semisimple modules". Ph. D. Thesis. Univ. Wisconsin-Milwaukee. 1.980.

[10] Luedeman, J. K. "Σ-quasi-injective modules and Utumi's Σ-quotient ring", Bull. Soc. Math. Belg., 22, 368 - 375. 1.970.

[11] Nastasescu, C. "Inele, module, categorii". Ed. Academiei. 1.976.

[12] Raynaud, J. "Localizations et anneaux semi-noetheriens". Publ. Dép. Math. Lyon. 8-3, 77 - 112. 1.971.

[13] Stenström, B. "Rings of quotients". Springer-Verlag. 1.975.

[14] Teply, M. L. "Torsionfree modules and the semicritical socle series" Private notes. 1.982.

# GRADED COMPLETE AND GRADED HENSELIAN RINGS

S. CAENEPEEL

Free University of Brussels, VUB, Belgium

If one considers a two-sided graded ideal I of a Z-graded ring R, one can construct the completion $\hat{R}$ of R with respect to the I-adic valuation. As $\hat{R}$ is not a graded ring in general, we are forced to introcuce the *graded completion* $\hat{R}^g$ of R. It is shown that $\hat{R}^g$ can be described by means of a non-Archimedean uniformity, which is not necessarily metrizable. Also relations with the completeness properties of $R_o$ are studied. In the last section, we study the notion of gr-Henselian rings. It turns out that a gr-local ring R is gr-Henselian if and only if $R_o$ is Henselian. Also a graded version of Hensel's lemma is presented.

## 1. Notations

All graded rings concerned are $\mathbb{Z}$-graded rings. For full detail concerning graded ring theory, we refer to [6]. For a graded ring R and a graded R-module M, $R_i$ and $M_i$ will denote the set of homogeneous elements of degree i in R and M. h(R) is the set of all homogeneous elements in R. For an element a of degree i, we denote $\deg_R a = i$.



*F. van Oystaeyen (ed.), Methods in Ring Theory, 67–80.*

A gr-maximal ideal m of a commutative ring is a graded ideal which is maximal in the set of graded ideals. A gr-local ring is a graded ring containing a unique gr-maximal ideal m. Note that a gr-local ring is not necessary local. R is gr-local if every element a in h(R-m) is invertible in R. A graded field is a graded ring in which every homogeneous element is invertible. It is either a field - with trivial gradation - or of the form $k[T,T^{-1}]$, where k is a field.

## 2. Gr-complete rings

In this section, R is a graded ring, and I is a two-sided graded ideal of R. I defines the I-adic valuation on R (cf. e.g. [5]).

$$v(x) = \max\{n \in \mathbb{N} : x \in I^n\} \quad \text{for } x \neq 0$$
$$v(0) = +\infty$$

Choose a fixed number $\alpha \in ]0,1[$ and put $d(x,y) = \alpha^{v(x-y)}$. Then R,d becomes an ultrametric space, which is Hausdorff if and only if $\cap_{n\in N} I^n = (0)$. This condition is fulfilled e.g. when R is commutative and Noetherian. We shall always assume that it holds. The zero-dimensional topology $\mathcal{T}$ - I generated by d is called the I-adic topology on R. The $I_o$-adic valuation on $R_o$ will be denoted by w. R is called I-adically complete if it is complete for the ultrametric d. If R is not complete, one can construct the completion $\hat{R}$, which is not a graded ring in general (e.g. take $R = k[X]$, $I = (X)$, then $\hat{R} = k[[X]]$).

We therefore introduce the notion of gr-complete rings. For each n in $\mathbb{N}_o$, $I^n$ is a graded ideal, so $I^n = \oplus_{r\in Z}(I^n)_r$. For each $r \in Z$, let $v_r$ be the restriction of v to $R_r$. Hence, for $x_r \neq 0$ in $R_r$, $v_r(x_r) = \max\{n \in \mathbb{N} : x_r \in (I^n)_r\}$, and for $x \in R$, $v(x) = \min\{v_r(x_r) : r \in \mathbb{Z}\}$. Also note that in general $v_o(x) \neq w(x)$.

2.1. Definition. The graded ring R is said to be *gr-I-complete* if for every r in $\mathbb{Z}$, $R_r$ is complete for the valuation $v_r$.

If R is not gr-I-complete, we can consider the completion $\hat{R}_r$ of $R_r$ with respect to $v_r$. Clearly $v_r$ extends to $\hat{R}_r$. Also, $\hat{R}_o$ is a ring, and $\hat{R}_r$ is an $\hat{R}_o$-module. Before introducing the graded completion, we establish a lemma.

2.2. Lemma. With notations as above, we have

a) $\hat{R} = \{(x_r)_r \in \Pi_{r\in Z}\hat{R} : \lim_{|r|\to\infty} v_r(x_r) = +\infty\}$

b) $\hat{R}_r\hat{R}_s \subset R_{s+t}$

Proof. Let $(x^{(n)})_n$ be a v-Cauchy sequence, then for each r in $\mathbb{Z}$, $(x_r^{(n)})_n$ is a $v_r$-Cauchy sequence. So putting $x_r = v_r\text{-}\lim x_r^{(n)}$ in $\hat{R}_r$, we get easily $x = (x_r)_r = v\text{-}\lim x^{(n)} \in \Pi\,\hat{R}_r$. Hence for all $M > 0$, there exists N in $\mathbb{N}$ such that for all $n \geq N$ : $v(x^{(n)}-x) > M$. As $x^{(N)} \in R$, we have $s > 0$ such that $x_r^{(N)} = 0$ for $|r| \geq s$. But then for each $|r| \geq s$ : $v_r(x_r) = v_r(x_r^{(N)}-x_r) > v(x^{(N)} - x) > M$, so $\lim v_r(x_r) = +\infty$.

Conversely, take $x = (x_r)_r$ in $\Pi\,\hat{R}_r$ such that $\lim_{|r|\to\infty} v_r(x_r) = +\infty$. for each $r \in \mathbb{Z}$, then we have a sequence $(x_r^{(n)})_n$ in $R_r$ such that $v_r\text{-}\lim x_r^{(n)} = x_r$. So for all $M > 0$ and r in $\mathbb{Z}$, we have $N_r(M)$ such that for all $n \geq N_r(M)$ : $v_r(x_r^{(n)} - x_r) > M$.

Also there exists $R(M) > 0$ such that for all $|r| > R(M)$ : $v_r(x_r) > M$.

Now, let $N(M) = \max\{N_r(M) : |r| \leq R(M)\}$, and define $y^{(n)}$ in R as follows :

$$y_r^{(n)} = x_r^{(N(n))} \quad \text{for } |r| \leq R(n)$$
$$y_r^{(n)} = 0 \quad \text{for } |r| > R(n)$$

It is then easily checked that $v\text{-}\lim y^{(n)} = x$. The proof of the second part of the lemma is obvious.

It follows from the preceding lemma that $\hat{R}$ is a filtered ring. We then define the *graded completion* $\hat{R}^g$ as the graded ring

associated to $\hat{R}$ (cf.[6], 1.4). It is clear that we have that $\hat{R}^g = \oplus_{r\in\mathbb{Z}} \hat{R}_r$. In the sequel we shall show that $\hat{R}^g$ can also be defined by means of a non-Archimedean uniformity.
Let F be the set of function from Z to the open interval ]0,1]. For each $\sigma$ in F, x in R, we define

$$v_\sigma(x) = \min \{v_r(x_r)\sigma(r) : r \text{ in } Z\}.$$

Then, fixing $\alpha$ in ]0,1[, let $d_\sigma(x,y) = \alpha^{v_\sigma(x-y)}$. $d_\sigma$ determines a new ultrametric on R. Let $U$-I-gr be the uniformity generated by $P = \{d_\sigma : \sigma \in F\}$. For details about uniformities, we refer to [4]. It is easily seen (cf. [8], p.34) that $U$-I-gr is a non-Archimedean uniformity.

<u>2.3 Theorem</u>. The completion of R with respect to the $U$-I-gr-uniformity is equal to the graded completion $\hat{R}^g$ of R.

Before giving the proof of the theorem, we first prove a lemma.

<u>2.4. Lemma</u>. $(x^{(n)})_n$ is a $U$-I-gr-Cauchy sequence in R if and only if the following conditions are fulfilled :
a) for each r in $\mathbb{Z}$, $(x_r^{(n)})_n$ is a $v_r$-Cauchy sequence in $R_r$
b) there exists a finite part J of $\mathbb{Z}$ such that for all n in $\mathbb{N}$, $x^{(n)} \in \oplus_{r\in J} R_r$.

<u>Proof</u>. Suppose a) and b) true. Choose $M > 0$, $\sigma : \mathbb{Z} \to ]0,1]$. Then from a) we get that for all r in J, there is $N_r$ such that for all n, $m > N_r$ : $v_r(x_r^{(n)} - x_r^{(m)}) > M/\sigma(r)$.
Set $N = \max \{N_r : r \in J\}$. Then for all n, m > N , for all r in $\mathbb{Z}$, $\sigma(r)v_r(x_r^{(n)} - v_r^{(m)}) > M$. If $r \notin J$, this is trivial.
Conversely, let $(x^{(n)})_n$ be a $U$-I-gr-Cauchy sequence. Then a) follows immediately. Suppose b) is false. Put $J = \{r\in \mathbb{Z} : x_r^{(n)} \neq 0$ for some n}, and assume $\sup J = +\infty$. The case $\inf J = -\infty$ is

treated in a similar way. For $x \in R \subset \Pi \hat{R}_r$ we introduce the following notation : $g(x) = \sup \{r \in \mathbb{Z} : x_r \neq 0\}$ . Up to replacing $(x^{(n)})$ by a subsequence, we can suppose that $g(x^{(n)}) = m_n$, with $m_1 < m_2 < m_3 < \ldots$

We define $\sigma$ in F as follows :

$$\sigma(m_j) = 1/v_j(x^{(j)}_{m_j}) \text{ if } v_j(x^{(j)}_{m_j}) \neq 0$$

$$\sigma(n) = 0 \qquad \text{in all other cases}$$

Then for all $j < k$, $v_\sigma(x^{(k)} - x^{(j)}) \leqslant \sigma(m_k)v_k(x^{(k)}_{m_k}) \leqslant 1$, so it follows that $(x^{(n)})_n$ is not a $\mathcal{U}$-I-gr-Cauchy sequence.

<u>Proof of Theorem 2.3</u>. From the preceding results, we know that $\oplus_{r\in Z} \hat{R}_r \subset \bar{R}$, the completion of R with respect to $\mathcal{U}$-I-gr, and that no $\mathcal{U}$-I-gr-Cauchy sequence converges to an element of $\Pi\hat{R}_r - \oplus \hat{R}_r$. Since in general, $\mathcal{U}$-I-gr does not have a countable base (cf. e.g. [4], p.72), we still have to convince ourselves that no Cauchy *net* converges to $x \in \Pi\hat{R}_r - \oplus\hat{R}_r$.

Suppose that $(x^{(\alpha)})_{\alpha\in X}$ is such a Cauchy net. We assume $g(x) = +\infty$ (as in lemma 2.4), and $x^{(\alpha)} \to x$ in the $\mathcal{U}$-I-gr-topology.

Recall that each cofinal subnet $(x^{(\beta)})_{\beta\in Y}$ also converges to x. A subnet is called cofinal if for all $\alpha\in X$, there exists $\beta\in Y$ such that $\beta \geqslant \alpha$. Define $X_i$ as follows ($i = 1, 2,\ldots$) :

$$X_1 = \{\alpha : g(x^{(\alpha)}) \geqslant 1\}$$

$$X_{n+1} = X_n - \{\alpha : g(x^{(\alpha)}) = n+1,\ g(x^{(\beta)}) \leqslant n \text{ for some } \beta \geqslant \alpha\}$$

Let $Y = \cap X_n$. Then $(x^{(\beta)})_{\beta\in Y}$ has the property that for all $\alpha$, $\beta$ in Y, $\alpha \leqslant \beta$ implies $g(x^{(\alpha)}) \leqslant g(x^{(\beta)})$.

We also have that $g(x^{(\alpha)}) \leqslant n$ and $\alpha \in X_n$ imply $\alpha \in Y$.

Next, for all n in N, there is $m \geqslant n$, $\alpha \in X_m$ such that $g(x^{(\alpha)})=m$. Indeed, otherwise we have for all $\alpha$ : $g(x^{(\alpha^m)}) \geqslant n$ implies that there exists $\beta \geqslant \alpha$ : $g(x^{(\beta)}) < n$. Then $x^{(\alpha)} \to x$ implies that $g(x) \leqslant n$ which is a contradiction.

This implies that we are able to take $x^{(\alpha_1)}, x^{(\alpha_2)}, \ldots$ such that $g(x^{(\alpha_i)}) = m_i$, $x^{(\alpha_i)} \in X_{m_i}$, $m_1 < m_2 < \ldots$

It follows that $(x^{(\beta)})_{\beta\in Y}$ is a cofinal subnet. Indeed, take $\alpha \in X - Y$, and suppose $g(x^{(\alpha)}) = n$, then there is a $\beta \in X_{n-1}$ such that $g(x^{(\beta)}) \leqslant n-1$ and $\beta \geqslant \alpha$. But then $\beta \in Y$.
Finally, $(x^{(\alpha_i)})_{i\in N}$ is a cofinal subsequence of $(x^{(\beta)})_{\beta\in Y}$, for if it were not, then there would exist $\beta \in Y$ such that $\beta > \alpha_i$ for each $i \in N$, so $g(x^{(\beta)}) > g(x^{(\alpha_i)}) = m_i$ so $g(x^{(\beta)}) = +\infty$, which is a contradiction.
Now $(x^{(\alpha_i)})_{i\in N}$ is a sequence converging to x, so from lemma 2.4 it follows that $g(x) < \infty$, establishing the result.

<u>2.5. Proposition</u>. If for each N in $\mathbb{N}$, there exists n, with $|n| > N$ such that for all M in $\mathbb{N}$, there is $x \in R_n$ such that $v(x) > M$, then $\mathcal{U}$-I-gr is not metrizable.

<u>Proof</u>. If $\mathcal{U}$-I-gr were metrizable, then it would have a countable base, hence one would be able to generate it by a countable subset $P = \{d_\sigma : \sigma \in F\}$(cf. [4], P. 186). Let $\mathcal{U}$ be the uniformity generated by some $\{d_{\sigma_i} : i \in N\} \subset P$ ; we shall construct a $\mathcal{U}$-Cauchy sequence which is not a $\mathcal{U}$-I-gr-Cauchy sequence.
For each $r \in \mathbb{Z}$, define $p(r) = \exp \sup \{1/\sigma_i(r) : i \leqslant r\}$ .
Then we have that $\lim_{r\to\infty} \sigma_i(r)p(r) = +\infty$. We assumed that $\lim_{i\to\infty} \sigma_i(r) = o$, because, in other situations, $d_{\sigma_i}$ becomes equivalent to d and the statement is trivial.
Using induction, one can construct a sequence $(x_i)_i$ in h(R) such that $\deg x_i = n_i$, with $|n_1| < |n_2| < \ldots$ and $v_{n_i}(x_i) > p(n_i)$.
Start e.g. with $n_o = 0$ ; then, putting $N = |n_{i-1}|$, we get $|n_i| > |n_{i-1}|$ such that there exist $x_i$ with $\deg x_i = n_i$ and $v_{n_i}(x_i) > p(n_i)$. Let $y_n = \Sigma_{i=1}^n x_i$. Then $(y_n)_n$ is not a $\mathcal{U}$-I-gr-Cauchy sequence (from lemma 2.4), but it is a $\mathcal{U}$-Cauchy sequence, as we have for each i in $\mathbb{N}$, $t > s$ : $v_{\sigma_i}(y_t - y_s) = v_{\sigma_i}(\Sigma_{j=s+1}^t x_j)$
$= \min_{s<j\leqslant t} \sigma_i(n_j)v_{n_j}(x_j) > \min_{s<j\leqslant t} \sigma_i(n_j)p(x_{n_j}) \to +\infty$
as s, $t \to +\infty$.

<u>2.6. Example</u>. Let $R = \mathbb{Z}[X]$, deg $X = 1$, $I = (p, X)$ ; Then $\mathcal{U}$-$I$-gr is not metrizable. $\hat{R}^g = \mathbb{Z}_p[X]$, $\hat{R} = \mathbb{Z}_p[[X]]$.

<u>2.7. Example</u>. Let $R = k[X]$, deg $X = 1$, $I = (X)$, where $k$ is a field. Then $\hat{R}^g = k[X]$, $\hat{R} = k[[X]]$. Note that in this case $\mathcal{U}$-$I$-gr is metrizable : we have that $v_r(X^r) = r$ ; put $\sigma(0) = 1$ and $\sigma(r) = 1/r$ for $r \neq 0$. Then $v_\sigma$ generates $\mathcal{U}$-$I$-gr ($v_\sigma$ generates the discrete topology, as $v_\sigma(x, y) < 1$ if $x \neq y$).

<u>2.8. Algebraic characterization of gr-completeness.</u>
Using the fact that $R$ is gr-$I$-complete if and only if every $R_r$ is complete for $v_r$, we have easily that this is again equivalent to the fact that, for each $r$ in $\mathbb{Z}$, the map $\nu_r\colon R_r \to \lim R_r/(I^n)_r$, induced by the commutative diagram

$$\begin{array}{ccc} & R_r & \\ {}^{\pi_{n+1}}\swarrow & & \searrow^{\pi_n} \\ R_r/(I^{n+1})_r & \xrightarrow{f_i^r} & R_r/(I^n)_r \end{array}$$

is a bijection. Injectivity is equivalent to $\cap I^n = (0)$, surjectivity to one of the three following facts :

S1 Given $x_i$ in $R_r/(I^i)_r$, for each $i$ in $\mathbb{N}$, with $f_i^r(x_{i+1}) = x_i$, there exists $x$ in $R$ such that $\pi_i(x) = x_i$ for each $i$.

S2 Given $a_1, a_2, \ldots$ in $R_r$ such that $a_{i+1} - a_i \in (I^i)_r$, there exists $a \in R_r$ such that $a - a_i \in (I^i)_r$, for each $i$.

S3 Given $E \subset Z$ finite, and $a_1, a_2, \ldots \in \oplus_{r \in E} R_r$ such that $a_{i+1} - a_i \in I^i$ for all $i \in N$, there exists $a \in R$ such that $a - a_i \in I^i$ for all $i \in N$.

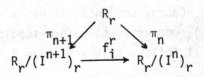

## 3. Relations with the completeness of $R_o$.

Suppose R gr-I-complete. In some applications, it is interesting to know whether the part of degree 0, $R_o$ is $I_o$-complete (cf. [3]). In the following proposition, we show that this is true in general, if R is commutative.

3.1. Proposition. If a graded ring R is gr-I-complete, then $R_o$ is $I_o$-complete (we suppose $\cap I^n = (0)$).

Proof. We use the notations of sec. 2. Since $(I_o)^n \subset (I^n)_o$, we have $w(x) \leqslant v_o(x)$, for all x in $R_o$ (but not necessarily the converse). Now consider a w-Cauchy sequence $(x_n)_n$ in $R_o$. Then $(x_n)_n$ has a limit x in the w-completion $S_o$ of $R_o$. Let $S = S_o \otimes R$, $J = S_o \otimes I$, the tensor products taken over $R_o$. v extends to a valuation on S, and $v_o$ extends to a valuation on $S_o$. So $S_o$, $v_o$ becomes a Hausdorffspace.
As $(x_n)_n$ is also a $v_o$-Cauchy sequence, it has a $v_o$-limit in $R_o$, say y. But since $(x_n)_n$ converges to x in w-topology, it converges to x in w-topology. It follows that x = y, so $R_o = S_o$.

3.2. Note. It is not necessary true that the restriction of the $\mathcal{U}$-I-gr-uniformity to $R_o$ is equal to the $I_o$-uniformity.
As an example, let $R = k[Y, X_1, X_2, \ldots]$, with deg $X_i = i$, deg $Y = -1$ and $I = (Y, X_1, X_2, \ldots)$. Then $w(Y^{m-1}X_{m-1}) = 1$, $v_o(Y^{m-1}X_{m-1}) = 2m$. So $Y^{m-1}X_{m-1} \to 0$ in $\mathcal{U}$-I-gr-topology, but not in $I_o$-adic topology.
As was shown in [3], the statement is true if R contains a homogeneous unity with degree different from zero in its center.

3.3 Note. The converse of prop. 3.1 does not hold in general. Let $S = S_o \otimes R$, where $S_o$ is the w-completion of $R_o$, and R as in 3.2. Then $(\Sigma_{m=1}^{n} Y^{m-1}X_{m-1})_n$ is a Cauchy sequence in $U$-I-gr having no limit.

3.4. Note. $v_o$-completeness of $R_o$ does not imply gr-completeness of R : let $S = S' \otimes R$, where R and $R_o$ are again as in 3.2, and S' the $v_o$-completion of $R_o$. Then $(\Sigma_{m=1}^{n} Y^{m-1}X_{m-1})_n$ is a $U$-I-gr-Cauchy sequence having no limit.

## 4. Gr-Henselian rings.

In this section, R will be a commutative gr-local ring.
It is well known that complete local rings satisfy Hensel's lemma (cf. e.g. [5], sec.30), i.e. they are Henselian rings. An advantage of this notion is that it can also be applied to the non-Noetherian case. In this section we discuss the notion of gr-Henselian rings.

4.1. Definition. A commutative gr-local ring R is called *gr-Henselian* if every commutative graded R-algebra is gr-decomposed, i.e. if it is the direct sum of gr-local rings.

This definition is inspired by the one given in [7]. Let us recall some basic facts on gr-Artinian rings. For the proofs, the reader may consult the corresponding ungraded theorems, e.g. in [1], [2]. A *gr-Artinian* ring R is a graded ring satisfying the descending chain condition on graded ideals. In a gr-Artinian ring R, each graded prime ideal is gr-maximal, hence the nilradical N(R) equals the graded Jacobson radical $J^g(R)$ and the Jacobson radical J(R). Then R is graded isomorphic (i.e. the connecting isomorphism having degree zero) to a unique finite direct sum of gr-Artinian gr-local rings, in fact $R \cong \oplus Q^g_{n_i}(R)$, where the $n_i$ are the graded prime ideals of R. If in a gr-local Noetherian ring R, with gr-maximal ideal m, $m^n = (0)$ for some n, then R is gr-Artinian. Now using the obvious fact that every finite graded algebra over a graded field is gr-Artinian, we have

<u>4.2. Proposition</u>. If R is a gr-local ring with gr-maximal ideal m, then every finite graded R-algebra B is gr-semilocal, and the gr-maximal ideals of B are just the graded primes $\{n_i : i \in I\}$ lying above m. The natural mapping $\bar{B} \to \oplus Q^g_{n_i}(B)$ is a graded isomorphism (of course, $\bar{B} = B \otimes R/m$).
Furthermore, the following conditions are equivalent :

1) B is gr-decomposed
2) $B \to \oplus Q^g_{n_i}(B)$ is a graded isomorphism
3) The gr-decomposition of $\bar{B}$ can be lifted to a gr-decomposition of B

<u>Proof</u>. An easy adaption of [7], ch. I, prop. 1, 2, 3.

<u>4.3. Corollary</u>. A graded field is gr-Henselian.

Next, we want to describe gr-Henselian rings using the lifting property for idempotents.

<u>4.4. Lemma</u>. The idempotents of a commutative graded ring R are homogeneous of degree zero.

<u>Proof</u>. Suppose first that R is a reduced gr-local ring. Then if e is a non-homogeneous idempotent, the component of highest or lowest degree is a nilpotent element, contradicting the hypothesis. We therefore have that 0 and 1 are the only idempotents in R. Next, assume that R is a general gr-local ring, let m be the gr-maximal ideal of R. $\bar{R} = R/m$ is a graded field, so if e is an

idempotent in R, $\bar{e} = 0$ or 1 in $\bar{R}$. First suppose $\bar{e} = 0$, so $e \in m$. Let N be the nilradical of R. From the reduced case, we know that e mod N = 0 in R/N, hence $e \in N$. But then e is a nilpotent idempotent, so e = 0. If $\bar{e} = 1$ in $\bar{R}$, then $1 - \bar{e} = 0$, so e = 1. In general, suppose e is an idempotent of R, and let $e_o$ be the part of degree zero in the homogeneous decomposition. For every graded prime ideal p, $e-e_o$ maps to 0 in $Q_p^g(R) = S^{-1}R$, where $S = (R - p) \cap h(R)$. Hence $e - e_o = 0$.

<u>4.5. Proposition</u>. Let B be a finite graded algebra over the gr-local ring R. Then the natural mapping $\pi$ : Idemp(B) $\to$ Idemp($\bar{B}$) is injective. B is gr-decomposed if $\pi$ is bijective.

<u>Proof</u>. Let e, e' be idempotents (of degree zero) such that $\bar{e} = \bar{e}'$, i.e. $x = e - e' \in mB$. Then $x^3 = (e-e')^3 = e^3 - 3e^2e' + 3ee'^2 - e'^3 = e - e' = x$, so $x(1 - x^2) = 0$. As $x \in mB \subset J^g(B)$, $x^2 \in J^g(B)$, so $1 - x^2$ is invertible, implying $x^2 = 0$. Hence x = 0.
For $1 \leqslant i \leqslant n$, let $\bar{e}_i = (0,\ldots,1,\ldots,0)$ $\bar{B} = \oplus Q^g_{n_i}(B)$. Each idemptent of $\bar{B}$ is a sum of some of the $\bar{e}_i$, and $\bar{e}_i$ can be lifted to $e_i$ in $B_o$ if and only if $Q^g_{n_i}(B)$ is a direct factor of B.

Next, we give other characterizations of gr-local rings. As it happens, R is gr-Henselian if and only if $R_o$ is Henselian. If we compare this to the situation encountered in sec. 3, this is a remarkable fact. It is due to lemma 4.4.

4.6. Theorem. Let R be a gr-local ring with gr-maximal ideal m, and $\bar{R} = R/m$. Then the following conditions are equivalent :

1) R is gr-Henselian ;
2) Each graded free R-algebra is gr-decomposed ;
3) For each monic homogeneous polynomial $P \in R[X, \deg X = t]$, $R[X]/(P)$ is gr-decomposed (for each t in N) ;
4) For each monic polynomial $P \in R_o[X]$, $R[X]/(P)$ is gr-decomposed ;
5) For each $t \in N$, and for each monic homogeneous polynomial $P \in R[X, \deg X = t]$, with $\bar{P} = \bar{Q}\,\bar{S}$, for some monic homogeneous polynomials $\bar{Q}, \bar{S} \in \bar{R}[X, \deg X = t]$, with $(\bar{Q}, \bar{S}) = 1$, we have $P = Q\,S$ for some monic homogeneous polynomials lifting $\bar{Q}, \bar{S}$ ;
6) For each monic $P \in R_o[X]$, with $\bar{P} = \bar{Q}\,\bar{S}$ for some monic $\bar{Q}, \bar{S} \in \bar{R}_o[X]$, and $(\bar{Q}, \bar{S}) = 1$, we have $P = Q\,S$ for some monic homogeneous polynomials lifting $\bar{Q}, \bar{S}$ ;
7) $R_o$ is Henselian ;
8) Each monic $P \in R_o[X]$ such that the image $\bar{P} \in \bar{R}_o[X]$ has a simple root $\bar{a}$ in $\bar{R}_o$, has a root in $R_o$ lifting $\bar{a}$ ;
9) Each monic homogeneous polynomial $P \in R[X, \deg X = t]$ $(t \in N)$, such that $\bar{P}$ has a simple root $\bar{a}$ in $h(\bar{R})$ has a root in $h(R)$ lifting $\bar{a}$.

Proof. $1 \Rightarrow 2 \Rightarrow 3 \Rightarrow 4$ and $5 \Rightarrow 6$ are clear. The implications $3 \Rightarrow 5$, $4 \Rightarrow 6$, $6 \Rightarrow 4$, $4 \Rightarrow 1$ are just easy modifications of the proof of the classical theorem given by Raynaud, [7], ch. I, prop. 5. $6 \Rightarrow 7$ also follows from this theorem, and $7 \Leftrightarrow 8$ follows from [7], ch. VII, prop. 3. $9 \Rightarrow 8$ is clear, and $5 \Rightarrow 9$ follows from the observation that we have $\bar{P} = (X - \bar{a})\bar{Q}$ with $(X - \bar{a}, \bar{Q}) = 1$.

We close this note giving Hensel's lemma for gr-local rings.

4.7. Proposition. Every gr-complete gr-local ring is gr-Henselian

Proof. If B is graded free and finitely generated, then B is gr-mB-complete. Indeed, it is easy to see that $\cap_n (mB)^n = (0)$, and one can check that condition S3 in 2.8 holds, by chosing a set of homogeneous generators of B as a graded R-module, and reducing the statement S3 to a statement about the coefficients of the elements expressed in the chosen generators.
Therefore, we have, for all $r \in Z$, $B_r \cong \varprojlim_n B_r/(m^n B)_r$. Consider the ring $R/m^n$. Since the gr-maximal ideal $m/m^n$ is a nilideal, $\mathrm{idemp}\,((B/m^n B)/(mB/m^n B)) = \mathrm{idemp}(B/mB) \cong \mathrm{idemp}(B/m^n B)$. Hence $R/m^n$ is gr-Henselian, as R/m is gr-Henselian. It follows that the finitely generated graded $R/m^n$-algebra $B/m^n B$ is gr-decomposed, so, letting $\{n_i : i \in I\}$ be the gr-maximal ideals in B lying above m, we get $B/m^n B \cong \oplus Q^g_{n_i}(B/m^n B) \cong \oplus Q^g_{n_i}(B)/m^n Q^g_{n_i}(B)$.
Now $(B/m^n B)_r = (B_r + m^n B)/m^n B \cong B_r/(m^n B)_r$, hence $(B/m^n B)_r \cong \oplus (Q^g_{n_i}(B)/m^n Q^g_{n_i}(B))_r$, and, taking direct limits, $B_r \cong \oplus (Q^g_{n_i}(B))_r$, finishing the proof.

If a gr-local ring R is not gr-Henselian, one can define the notion of gr-Henselisation $\tilde{R}^g$ of R. In the vein of [7], we call *gr-Henselisation* of R each pair $(\tilde{R}^g, i)$, where $\tilde{R}^g$ is a gr-local gr-Henselian ring, and $i: R \to \tilde{R}^g$ is a gr-local homomorphism of degree zero, such that for each gr-Henselian ring S and each gr-local homomorphism $u : R \to S$, there exists a unique gr-local homomorphism $\tilde{u} : \tilde{R}^g \to S$ such that $u = \tilde{u} \circ i$. From the equivalence $1 \Leftrightarrow 7$ in 4.6, we immediately have

4.8. Proposition. Let $(\tilde{R}_o, i_o)$ be a Henselisation of $R_o$, then $(\tilde{R}^g = \tilde{R}_o \otimes R, i_o \otimes 1)$ is a gr-Henselisation of R (tensor products being taken over $R_o$). The gr-Henselisation is unique up to graded isomorphism.

References.

[1] Atiyah, M.F., McDonald, S. J., *Introduction to commutative algebra*, Addison-Wesley, Reading, Mass. 1969.

[2] Bourbaki, N., *Algèbre Commutative*, Hermann, Paris 1965.

[3] Caenepeel, S., Van Oystaeyen, F., *Crossed Products over Graded Local Rings*, in *Brauer Groups in Ring Theory and Algebraic Geometry*, pp. 25-42, Lect. Notes in Math. 917, Springer Verlag, Berlin, 1982.

[4] Kelley, J., *General Topology*, Van Nostrand, New York, 1955.

[5] Nagata, M., *Local Rings*, Interscience Tracts in Pure and Apllied Math. 13, John Wiley and Sons, New York 1962.

[6] Nastacescu, C., Van Oystaeyen, F., *Graded and Filtered Rings and Modules*, Lect. Notes in Math. 758, Springer Verlag, Berlin, 1979.

[7] Raynaud, M., *Anneaux Locaux Henséliens*, Lect. Notes in Math. 169, Springer Verlag, Berlin 1970.

[8] Van Rooij, A.C.M., *Non-Archimedean Functional Analysis*, Marcel Dekker, New York 1978.

# CYCLIC CLASSES IN RELATIVE BRAUER GROUPS

Burton Fein

Oregon State University, Corvallis, Oregon

Murray Schacher

University of California, Los Angeles, California

Let K be a finitely generated infinite field and let L be a finite algebraic extension of K, $L \neq K$. As we showed in (4), one consequence of the classification of the finite simple groups is that the relative Brauer group $B(L/K)$ must be infinite under these conditions. On the other hand, the fundamental result of Merkurjev-Suslin (6) shows that if K continues appropriate roots of unity then every element of $B(L/K)$ can be expressed as a sum of classes in $B(K)$ represented by cyclic algebras. This raises the following natural question: must $B(L/K)$ contain infinitely many cyclic classes? We show in this note that the arguments of [4] can be modified to yield this stronger result.

We begin by fixing some notation and terminology. Suppose L is a finite separable extension of K, $L \neq K$, and E is a Galois closure of L over K. Let $G = Gal(E/K)$, $H = Gal(E/L)$ and let $\Omega$ be the set of left cosets of H in G. We view G as acting on $\Omega$ by left translation. By (3), Theorem 1, there exists a prime p and a p-element $\sigma \in G$ such that $\sigma$ acts without fixed points on $\Omega$. In this situation we will say that p and $\sigma$ are <u>special</u> for L/K. The proof of (4), Theorem, shows that if p is special for L/K and K is a finitely generated infinite field, then $B(L/K)_p$ is infinite. Here



*F. van Oystaeyen (ed.), Methods in Ring Theory, 81–86.*

${}_pB(L/K)$ denotes the subgroup of $B(L/K)$ consisting of those elements of order 1 or p. Our main result generalizes this.

Theorem Let K be a finitely generated infinite field and let $L_1,\ldots,L_m$ be finite non-trivial separable extensions of K having Galois closures $E_1,\ldots,E_m$ respectively in some fixed algebraic closure of K. Assume that $\{E_i\}, i=1,\ldots,m$ is linearly disjoint and that there is a prime p which is special for each $L_i/K, i=1,\ldots,m$.

Then ${}_pB(L_1/K)\cap {}_pB(L_2/K)\cap\cdots\cap {}_pB(L_m/K)$ is infinite.

This theorem has the following striking consequence.

Corollary. Let K be a finitely generated infinite field and let L be a finite algebraic extension of K, $L \neq K$. Then $B(L/K)$ contains infinitely many cyclic classes.

Proof of the Corollary. If L/K is purely inseparable, the result follows from the proof of Case 2 on page 39 of (4). We may clearly assume, then that L/K is separable. By (3), Theorem 1, there is a prime p which is special for L/K. Set $L_1 = L$ and let $L_2$ be a cyclic extension of K of degree p with $L_2$ linearly disjoint from the Galois closure of L over K. By the Theorem, ${}_pB(L_1/K)\cap {}_pB(L_2/K)$ is infinite. But since $L_2/K$ is cyclic, every element of ${}_pB(L_2/K)$ is a cyclic class, proving the Corollary.

We turn next to the proof of the Theorem. Since the arguement is quite similar to the arguments appearing in (3) and (4), we will only sketch the proof; the reader should note, however, that the sequence of steps we will follow yields a simpler proof of the main result of (4) than the one appearing in that paper.

Let $K, L_1,\ldots,L_m, E_1,\ldots,E_m$, and p be as in the statement of the Theorem and let $\sigma_i \in (\mathrm{Gal}(E_i/K), i = 1,\ldots,m$, be special for p. Set $G_i = \mathrm{Gal}(E_i/K)$, $H_i = \mathrm{Gal}(E_i/L_i)$, $E = E_1 \ldots E_m$, and $G = \mathrm{Gal}(E/K)$. Because of our linear disjointness assumption, $G \cong G_1\times\ldots\times G_m$; we identify G with this direct product. Let

$\sigma = (\sigma_1,\sigma_2,\ldots,\sigma_m) \in G$, let $M = E^{\langle\sigma\rangle}$, and let $p^k = [E:M] = |\langle\sigma\rangle|$. If $T$ is a finite extension of $F$, we denote by $\mathrm{Res}_{T/F}$ and $\mathrm{Cor}_{T/F}$ the restriction and corestriction maps of cohomology theory; we refer to (1), pages 254-257, for the relevant properties of these maps.

The first half of the proof of the Theorem consists of showing that if $[A] \in B(E/M)$, then for each $i = 1,\ldots,m$, $\mathrm{Res}_{L_i/K}\ \mathrm{Cor}_{M/K}\ p^{k-1}[A] = 0$. This is proved by a purely cohomological argument similar to that appearing in (4), page 39, and uses the fact that $\sigma_i$ is special for $L_i/K$. Fix $i$ and let $L = L_i$. For $\tau \in G$, let $c_\tau$ denote the conjugation isomorphism between $B(E/M)$ and $B(E/\tau(M))$. Let $N_\tau$ denote the fixed field of $\tau\langle\sigma\rangle\tau^{-1}(G_1 \times\ldots\times G_{i-1} \times H_i \times G_{i+1}\times\ldots\times G_m)$. By (1), Proposition 9.1, page 257, $\mathrm{Res}_{L/K}\ \mathrm{Cor}_{M/K}\ p^{k-1}[A]$ is a sum of terms each of the form $\mathrm{Cor}_{L/N_\tau}\ \mathrm{Res}_{N_\tau/\tau(M)}\ c_\tau\ (p^{k-1}[A])$. To show that each of these terms is $0$ it suffices, because $[E:M] = p^k$, to show that $p\,|\,[N_\tau : \tau(M)]$ for each $\tau \in G$. But if not then $N_\tau = \tau(M)$ for some $\tau \in G$ and this would imply that $\tau\sigma\tau^{-1} \subset G_1\times\ldots\times G_{i-1} \times H_i \times\ldots\times G_m$. But $\sigma = (\sigma_1,\ldots,\sigma_m)$ so $\tau\sigma_i\tau^{-1} \in H_i$, contradicting the fact that $\sigma_i$ is special for $L_i/K$.

We have shown that if $[A] \in B(E/M)$, then $\mathrm{Cor}_{M/K}\ p^{k-1}[A] \in {}_pB(L_i/K)$ for each $i = 1,\ldots,m$. By assumption,

$K$ contains a global subfield $F_0$; we denote the transcendence degree of $K$ over $F_0$ by $n$. We will prove by induction on $n$ that there exists an infinite set $\{[A_j]\}$, $j = 1,2,\ldots,$ of elements of order $p^k$ in $B(E/M)$ such that $\{\mathrm{Res}_{M/K}\ \mathrm{Cor}_{M/K}\ p^{k-1}[A_j]\}$, $j = 1,2,\ldots,$ is an infinite set of independent elements of order $p$ in $B(M)$. Since this implies that $\{\mathrm{Cor}_{M/K}\ p^{k-1}[A_j]\}$, $j = 1,2,\ldots,$ is an infinite set in ${}_pB(L_1/K)\cap\ldots\cap{}_pB(L_m/K)$, this suffices to prove the Theorem.

Suppose first that $n = 0$. Then $K$ is a global field. By the Tchebotarev density theorem (7), Theorem 12, page 289, there is an infinite set $\pi_0, \pi_1, \ldots,$ of primes of $M$ lying over distinct primes $\delta_0, \delta_1, \ldots,$ of $K$ such that $\sigma$ is the Frobenius automorphism of $E/K$ at $\delta_i$. We produce the desired $[A_j]$ and verify their properties using standard arguments involving Hasse invariants; we refer the reader to (2), Chapter 7 for the relevant material. We define $[A_i]$, $i > 0$, to be the element of $B(M)$ having Hasse invariants $1/p^k$ and $-1/p^k$ at $\pi_0$ and $\pi_i$ and invariants $0$ at all other primes of $M$. Arguing, for example, as on page 54 of (3), it is not difficult to see that the $\{[A_j]\}$, $j = 1,2,\ldots,$ have the desired properties.

Now suppose $n > 0$. We argue as on pages 54-56 of (3). We first express $K$ as a finite separable extension of $F(t)$ where $t$ is transcendental over $F$. Using the Hilbert Irreducibility Theorem (5), Theorem 2, page 155, we show that there is a $c \in F$ such that the discrete rank one valuation $\phi$ of $F(t)$ trivial on $F$ and having $t - c$ as uniformizing parameter is unramified and inertial from $F(t)$ to $E$. For $V \subset E$, we let $\overline{V}$ denote the residue class field of $V$ under $\phi$. Our inductive hypothesis applied to $\overline{M}$ gives the existence of an infinite set

$\{[\bar{A}_j]\}$, $j = 1,2,\ldots$ of elements of order $p^k$ in $B(\bar{E}/\bar{M})$ such that $\{\mathrm{Res}_{\bar{M}/\bar{K}}\ \mathrm{Cor}_{\bar{M}/\bar{K}}\ p^{k-1}[\bar{A}_j]\}$ is an infinite set of independent elements of order $p$ in $B(\bar{M})$. Each $[\bar{A}_j]$ is a cyclic class $[(\bar{E}/\bar{M},\bar{\sigma},\bar{b}_j)]$. Lift each $\bar{b}_j$ to $b_j \in M$ and form $A_j = (E/M,\sigma,b_j)$. Suppose that

$\{\mathrm{Res}_{M/K}\ \mathrm{Cor}_{M/K}\ p^{k-1}[A_j]\}$, $j = 1,2,\ldots$ is not a set of independent elements of $B(M)$. Then there are values of $j$, without loss of generality $j = 1,\ldots,r$, integers $n_1,\ldots,n_r$ prime to $p$, and $\tau_1,\ldots,\tau_r \in G$ such that $(E/M,\sigma,t_1(b)^{n_1}\ldots t_r(b_r)^{n_r}$ is split. But then $\tau_1(b_1)^{n_1}\ldots\ \tau_r(b_r)^{n_r}$ is a norm from $E$ to $M$ so $\tau_1(\bar{b}_1)^{n_1}\ldots\bar{\tau}_r(\bar{b}_r)^{n_r}$ is a norm from $\bar{E}$ to $\bar{M}$. This contradicts the independence of $\{\mathrm{Res}_{\bar{M}/\bar{K}}\ \mathrm{Cor}_{\bar{M}/\bar{K}}\ p^{k-1}[\bar{A}_j]\}$, $j = 1,\ldots,m$, and completes the proof of the Theorem.

We remark that it is not known in the context of the introduction, whether ${}_pB(L/K)$ infinite implies there is a $\sigma \in G = \mathrm{Gal}(E/K)$ with $p$ and $\sigma$ special for $L/K$.

## References


1. Cartan, H. and Eilenberg, So., Homological Algebra, Princeton 1956.
2. Deuring, M., Algebra, Berlin-Heidelberg - New York, 1968.
3. Fein, B., Kantor, W.M., and Schacher, M., Relative Brauer groups. II. J. reine angew, Math. 328 (1981), 39-57.
4. Fein, B. and Schacher, M. Relative Brauer groups. III, J. reine angew.

5. Lang, S., Diophantine Geometry, New York 1962.
6. Merkurjev, A.S. and Suslin, A.A., K-cohology of Severi-Brauer varieties and norm residue momomorphisms, Izv, Akad. Nauk. SSSR 46 (1982), 1011-1046.
7. Weil, A., Basic Number Theory, Berlin-Heidelberg-New York 1974.

# SIMPLE NOETHERIAN NON-MATRIX RINGS

K. R. Goodearl

Department of Mathematics, University of Utah, Salt Lake City, Utah 84112, U.S.A.

For each integer $n \geqslant 2$, simple noetherian hereditary rings $R$ are constructed such that $R$ has uniform (Goldie) dimension $n$ but $R$ is not isomorphic to a matrix ring over an integral domain. That $R$ is not a matrix ring is verified K-theoretically, using Quillen's machinery [3] to compute $K_0(R)$. In fact, for these examples $K_0(R)$ is free abelian of rank $n$.

As this material is to be published in [1], we only give a brief sketch of it here.

Choose a field $L$ of characteristic zero which contains a primitive $n$-th root of unity, $\xi$, and let $S$ denote the Weyl algebra $A_1(L)$. We may express $S$ as the $L$-algebra with generators $x$ and $\theta$ subject to the relation $\theta x - x\theta = 1$. There is an $L$-algebra automorphism $\alpha$ of $S$ such that $x^{\alpha} = \xi x$ and $\theta^{\alpha} = \xi^{-1}\theta$. The group $G$ generated by $\alpha$ has order $n$, and for our example we take the skew group ring $R = S * G$.

It follows from standard arguments that $R$ is a simple noetherian hereditary ring with uniform dimension at most $n$. On the other hand, $R$ contains the group algebra $L[G]$, which is isomorphic to the algebra $L^n$ because $\xi \in L$. Hence, $R$ contains a set of $n$ nonzero pairwise orthogonal idempotents, and thus the uniform dimension of $R$ is exactly $n$.

As $L[G] \cong L^n$, there is an isomorphism of $K_0(L[G])$ onto $\mathbb{Z}^n$ sending $[L[G]]$ to $(1,1,\ldots,1)$. On the other hand, by [3] the inclusion map $L[G] \to R$ induces an isomorphism of $K_0(L[G])$ onto $K_0(R)$. Consequently, there exists an isomorphism of $K_0(R)$ onto $\mathbb{Z}^n$ sending $[R]$ to $(1,1,\ldots,1)$.



*F. van Oystaeyen (ed.), Methods in Ring Theory, 87–88.*

Now in $K_o(R)$ , the element $[R]$ is not divisible by any integer $k \geqslant 2$ . Applying an observation of Hart [2], it follows that R cannot be isomorphic to a $k \times k$ matrix ring (over any ring) for any integer $k \geqslant 2$ . In particular, R is not isomorphic to a matrix ring over an integral domain.

## REFERENCES

1. Goodearl, K. R., "Simple noetherian rings not isomorphic to matrix rings over domains," Communications in Algebra (to appear).

2. Hart, R., "Invertible $2 \times 2$ matrices over skew polynomial rings," in Ring Theory Antwerp 1980 (F. van Oystaeyen, Ed.), Lecture Notes in Mathematics, No. 825, Springer-Verlag, Berlin, 1980, pp. 59-62.

3. Quillen, D., "Higher algebraic K-theory: I," in Algebraic K-theory I (H. Bass, Ed.), Lecture Notes in Mathematics, No. 341, Springer-Verlag, Berlin, 1973, pp. 85-147.

# GROUP-GRADINGS OF CATEGORIES

Robert Gordon †

Temple University
Philadelphia, Pennsylvania 19122
USA

A general theory of group-gradings of a category is presented which subsumes, for a given group G, internal gradings of modules over a G-graded ring, external gradings of modules over a G-graded ring, and gradings of coverings of a category with fundamental group G.

## INTRODUCTION

In this paper we use the same notation and conventions as in [1]. In particular, G is a fixed group, $|\mathbb{X}|$ stands for the object class of a category $\mathbb{X}$ with morphism class $\mathbb{X}$, and when we write, symbolically, $* \in \mathbb{X}$, we mean that $*$ is either an object or a morphism of $\mathbb{X}$.

The aim of the paper is to extend the theory of pregradings of categories developed in [1] to a theory of gradings of categories. The reader will recall that given a left G-category $\mathbb{A}$ together with an endofunctor $E$ on $\mathbb{A}$ such that $EE^g = E^gE$ for all $g \in G$, an $E$-pregrading of a category $\mathbb{B}$ in $\mathbb{A}$ is a functor T: $\mathbb{A} \to \mathbb{B}$ having a right adjoint H such that 1) for some adjunction $\lambda$: $1_{\mathbb{A}} \to HT$, ($\lambda$; HT) is an $E$-stable coproduct system in $\mathbb{A}$ (see Section III) and 2) corestriction of H to the $E$-stabilizer, $\mathrm{Stab}_E\,\mathbb{A}$, of $\mathbb{A}$ is an isomorphism (see Section I). We use the term "pregrading" since, as one sees from the definition, a given object or morphism of $\mathbb{A}$ need not have a G-gradation in any sense of the term. Basically, an $E$-grading of $\mathbb{B}$ in $\mathbb{A}$ is an $E$-pregrading of $\mathbb{B}$ in $\mathbb{A}$ together with a choice,

† Partially supported by the National Science Foundation.



F. van Oystaeyen (ed.), Methods in Ring Theory, 89–110.

for each object and each morphism of $\mathbb{A}$, of a G-gradation which preserves the G-action, kills $E$, reflects any structure inherent in the gradation, and is compatible with the pregrading. We call such a choice an $E$-component structure for $\mathbb{A}$.

More formally, an $E$-component structure for $\mathbb{A}$ consists of two ingredients. One ingredient is a faithful, $E$-limit creating G-functor $\sigma$ with $\sigma E = \sigma$ from $\mathbb{A}$ to a product category of the form $\mathbb{P} = \prod_G \mathbb{C}$. (We refer to $\sigma$ as an $E$-component functor for $\mathbb{A}$ in $\mathbb{C}$.) The other ingredient is a stable coproduct system in $\mathbb{P}$. The compatibility condition is that, in the above notation, HT$E$ is assumed to be the $E$-stable coproduct functor on $\mathbb{A}$ induced by $\sigma$ via reflection of the given stable coproduct system in $\mathbb{P}$. (We say that the $E$-stable coproduct system in $\mathbb{A}$ which corresponds to HT$E$ is $\sigma$-admissible with respect to the stable system in $\mathbb{P}$.)

Section I deals with the intricacies of creation of $E$-limits, this being a variant of MacLane's notion [4] of creation of limits. Section II studies the influence of the existence of an $E$-component functor for $\mathbb{A}$ on the structure of $\mathbb{A}$ and $\mathrm{Stab}_E\,\mathbb{A}$. In Section III, we study admissibility--for example, given an $E$-component functor $\sigma$ for $\mathbb{A}$ in $\mathbb{C}$, we show that, modulo an equivalence relation, there is a canonical one-to-one correspondence between $\sigma$-admissible, $E$-stable coproduct systems in $\mathbb{A}$ and choices, for each family of G-indexed objects of $\mathbb{C}$, of a coproduct for the family. Gradings are defined in Section IV and their basic properties exploited. One consequence of the major result of the paper--Theorem 4.6--is that, just as for pregradings (see [1, §VI]), given a left G-category $\mathbb{A}$ and a suitable endofunctor $E$ on $\mathbb{A}$, there is a bijection between a certain prespecified class of $E$-gradings on $\mathbb{A}$ and the class of $1_{\mathbb{A}'}$-gradings on some left G-category $\mathbb{A}'$ canonically related to $\mathbb{A}$.

We refer the reader to [3] for applications of the theory presented here to external gradings of modules over G-graded rings and to gradings of coverings of a category with fundamental group G (see [2]).

## 1. CREATION OF LIMITS

We start by recalling the definition of the category A from [1, §I]. Objects of A are ordered pairs $(\mathbb{A},E)$, where $\mathbb{A}$ is a left G-category and $E$ is an endofunctor on $\mathbb{A}$ such that $EE^g = E^gE$ for every $g \in G$. A morphism $F: (\mathbb{A},E) \to (\mathbb{A}',E')$ in A is a functor $F: \mathbb{A} \to \mathbb{A}'$ such that $FE^g = E'^gF$ for every $g \in G$. As in [1], we denote the subcategory $\{* \in \mathbb{A} | E(*) = *\}$ of $\mathbb{A}$ by $\mathbb{A}_E$ and the G-subcategory $\{* \in \mathbb{A} | E({}^g*) = E(*)\ \forall\ g \in G\}$ of $\mathbb{A}$ by $\overline{\mathrm{Stab}}_E\,\mathbb{A}$. We put $\mathrm{Stab}_E\,\mathbb{A} = \mathbb{A}_E \cap \overline{\mathrm{Stab}}_E\,\mathbb{A}$.

Throughout the paper, $\Gamma$ will stand for a small category. If $\mathbb{X}$ is a category, we freely identify constant functors $\Gamma \to \mathbb{X}$ with objects of $\mathbb{X}$. Likewise, we identify natural transformations between constant functors $\Gamma \to \mathbb{X}$ with morphisms in $\mathbb{X}$.

Consider a morphism $S: (\mathbb{A}, \mathcal{E}) \to (\mathbb{A}', 1_{\mathbb{A}'})$ in A. Thus, S is a G-functor $\mathbb{A} \to \mathbb{A}'$ such that $S\mathcal{E} = S$. We say that S <u>creates left $\mathcal{E}$-limits</u> (cf. MacLane [4, p. 108]) if given a functor $F: \Gamma \to \mathbb{A}$ such that $\theta: A' \to SF$ is a left limit of SF in $\mathbb{A}'$, there is a unique natural transformation $\eta: A \to F$ with $A \in |\mathbb{A}_{\mathcal{E}}|$ such that $S\eta = \theta$ and, moreover, the natural transformation $\eta$ is a left limit of F in $\mathbb{A}$. Creation of right $\mathcal{E}$-limits is defined dually. If S creates left and right $\mathcal{E}$-limits, we say that S <u>creates $\mathcal{E}$-limits</u>.

One important merit of this notion is that it respects the relation ~ on $|A|$ studied in [1, §VI]. We recall that $(\mathbb{A}, \mathcal{E}) \sim (\mathbb{A}', \mathcal{E}')$ with factorization (R,L) if

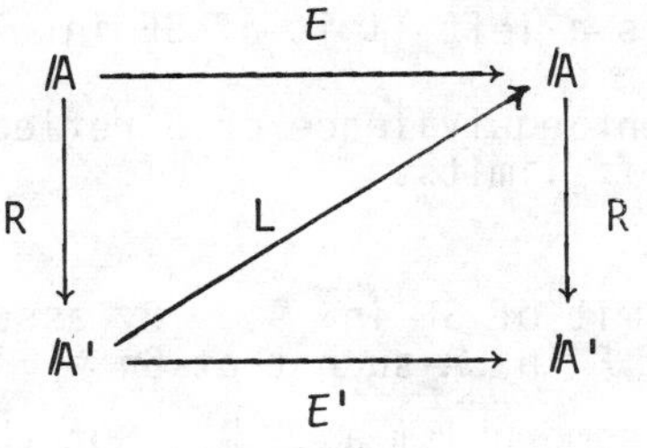


is a commutative diagram in A.

THEOREM 1.1. Suppose that $(\mathbb{A}, \mathcal{E}) \sim (\mathbb{A}', \mathcal{E}')$ with factorization (R,L), where $\mathcal{E}$ is an equivalence and $\mathcal{E}'$ is an idempotent equivalence. Let $(\mathbb{X}, 1_{\mathbb{X}}) \in |A|$ and let $S: (\mathbb{A}, \mathcal{E}) \to (\mathbb{X}, 1_{\mathbb{X}})$ be a morphism in A. If S creates $\mathcal{E}$-limits, then the morphism $S' = SL: (\mathbb{A}', \mathcal{E}') \to (\mathbb{X}, 1_{\mathbb{X}})$ creates $\mathcal{E}'$-limits.

<u>Proof</u>. We show that, if S creates left $\mathcal{E}$-limits, then S' creates left $\mathcal{E}'$-limits. We let $\tau': \mathcal{E}' \to 1_{\mathbb{A}'}$ be the isomorphism such that $\mathcal{E}'\tau' = 1_{\mathcal{E}'}$ given by [1, Corollary 3.6].

Let $F: \Gamma \to \mathbb{A}'$ be a functor and let $\theta$ be a left limit of $S'F$ in $\mathbb{X}$. Since S creates left $\mathcal{E}$-limits and $S(LF) = S'F$, there is a unique natural transformation $\delta: A \to LF$ with $A \in |\mathbb{A}_{\mathcal{E}}|$ such that $S\delta = \theta$; and $\delta$ is a left limit of LF in $\mathbb{A}$.

Let $\eta: RA \to F$ be the natural transformation defined by $\eta = (\tau' F)(R\delta)$. Since, plainly, R is an equivalence, $R\delta$ is a left limit of $E'F$ in $\mathbb{A}'$. Thus, since $\tau' F$ is an isomorphism, $\eta$ is a left limit of F in $\mathbb{A}'$. Furthermore, since $R \in A$, $RA \in |\mathbb{A}'_{E'}|$; and hence $S'\eta = (S'\tau'F)(S'R\delta) = (S'E'\tau'F)(SE\delta) = 1_{S'F}(S\delta) = \theta$.

Let $A' \in |\mathbb{A}'_{E'}|$ and let $\eta': A' \to F$ be a natural transformation such that $S'\eta' = \theta$. But then, $S(L\eta') = S'\eta' = \theta$ and $LA' \in |\mathbb{A}_E|$, and, consequently, $L\eta' = \delta$. In particular, $LA' = A$. In addition, $S(L\eta) = (SL\tau'F)(SLR\delta) = (S'\tau'F)(S'R\delta) = \theta$. Thus, since $L\eta: LRA \to LF$ and $LRA = EA = A$, $L\eta = \delta$. But, L is faithful, and $RA = RLA' = E'A' = A'$. Therefore, $\eta' = \eta$. □

We require two lemmas.

LEMMA 1.2. Let $S: (\mathbb{A}, E) \to (\mathbb{A}', 1_{\mathbb{A}'})$ be a left $E$-limit creating morphism in A.
(i) If $F: \Gamma \to \mathbb{A}$ is a functor, $\theta$ is a left limit of F in $\mathbb{A}$, and SF has a limit in $\mathbb{A}'$, then $S\theta$ is a left limit of SF in $\mathbb{A}'$ (cf. [4, p. 113, Theorem 2]).
(ii) If either $E$ is an idempotent equivalence or S reflects isomorphisms, then S reflects left limits.

<u>Proof</u>. (i) Let $\theta'$ be a left limit of SF in $\mathbb{A}'$. By assumption, there is a left limit, say, $\eta$ of F in $\mathbb{A}$ such that $S\eta = \theta'$. In particular, $\theta a = \eta$ for some isomorphism $a \in \mathbb{A}$. But then, $S\theta = \theta'(Sa)^{-1}$.
(ii) Let $F: \Gamma \to \mathbb{A}$ be a functor, let $A \in |\mathbb{A}|$, and let $\theta: A \to F$ be a natural transformation such that $S\theta$ is a left limit of SF in $\mathbb{A}'$. Then, by assumption, there is a unique natural transformation $\eta: B \to F$ with $B \in |\mathbb{A}_E|$ such that $S\eta = S\theta$; and $\eta$ is a left limit of F. In particular, $\eta a = \theta$ for some $a \in (A,B)_{\mathbb{A}}$. But then, $(S\theta)(Sa) = S\theta$, and this implies that $Sa = 1_{SA}$. Thus, if S reflects isomorphisms, a is an isomorphism; and so $\theta$ is a left limit of F.

Now, suppose that $E$ is an idempotent equivalence, and consider the natural transformation $\delta = \theta\tau_A: EA \to F$, where $\tau: E \to 1_{\mathbb{A}}$ is the isomorphism such that $E\tau = 1_E$ ([1, Corollary 3.6]). Inasmuch as $EA \in |\mathbb{A}_E|$ and $S\delta = (S\theta)(S\tau_A) = (S\theta)(SE\tau_A) = (S\theta)1_{SA} = S\theta$, $\delta = \eta$. Thus, since $\tau_A$ is an isomorphism, $\theta$ is a left limit of F. □

Naturally, this result works equally well for right limits.

LEMMA 1.3. Let $\mathbb{X}$ and $\mathbb{Y}$ be categories, let $U: \mathbb{X} \to \mathbb{Y}$ be a functor having a left adjoint V, and let Y be an object of $\mathbb{Y}$.
(i) If Y is a generator and U is faithful, then VY is a generator.
(ii) If Y is projective and U preserves epimorphisms, then VY is projective.
(iii) If the functor $(Y, \ )_{\mathbb{Y}}$ preserves coproducts and U preserves coproducts, then the functor $(VY, \ )_{\mathbb{X}}$ preserves coproducts.

Proof. We have $(VY,X)_{\mathbb{X}} \simeq (Y,UX)_{\mathbb{Y}}$ naturally in X for each $X \in |\mathbb{X}|$.
(i) Since Y is a generator and U is faithful, the functor $(Y,U(\ ))_{\mathbb{Y}} : \mathbb{X} \to$ SETS is faithful. Thus, the functor $(VY, \ )_{\mathbb{X}}$ is faithful.
(ii) The hypothesis implies that the functor $(Y,U(\ ))_{\mathbb{Y}}$ preserves epimorphisms. Hence, $(VY, \ )_{\mathbb{X}}$ preserves epimorphisms.
(iii) Let $\amalg X_i$ be a coproduct in $\mathbb{X}$. The result follows by analysis of the sequence of canonical isomorphisms

$$(VY,\amalg X_i)_{\mathbb{X}} \simeq (Y,U(\amalg X_i))_{\mathbb{Y}} \simeq (Y,\amalg(UX_i))_{\mathbb{Y}}$$
$$\simeq \amalg(Y,UX_i)_{\mathbb{Y}} \simeq \amalg(VY,X_i)_{\mathbb{X}}. \quad \square$$

By a Grothendieck category we mean a right complete abelian category with exact direct limits and a generator.

PROPOSITION 1.4. Let $(\mathbb{A},E)$ and $(\mathbb{A}',1_{\mathbb{A}'})$ be objects of A, suppose that $S: (\mathbb{A},E) \to (\mathbb{A}',1_{\mathbb{A}'})$ creates $E$-limits, and suppose that either S reflects isomorphisms or else that $E$ is an idempotent equivalence.
(i) If $\mathbb{A}'$ is abelian, then $\mathbb{A}$ is abelian.
(ii) If S is faithful and $\mathbb{A}'$ is Grothendieck, then $\mathbb{A}$ is Grothendieck if and only if S has a left adjoint. Moreover, if this is the case and $\mathbb{A}'$ has a (small) projective generator, then $\mathbb{A}$ has a (small) projective generator.

Proof. (i) First, we note that since $\mathbb{A}'$ has a zero object and S creates $E$-limits, $\mathbb{A}$ has a zero object 0 and S(0) = 0.
Let $\alpha: A \to B$ be a monomorphism in $\mathbb{A}$. Since S creates right $E$-limits and morphisms in $\mathbb{A}'$ have cokernels, it follows that $\alpha$ has a cokernel, say, $\beta: B \to C$. In particular, we have this

pushout diagram:

$$\begin{array}{ccc} A & \longrightarrow & 0 \\ {\scriptstyle\alpha}\downarrow & & \downarrow \\ B & \underset{\beta}{\longrightarrow} & C \end{array} \qquad (1)$$

If we can show that this diagram is a pullback, then $\alpha = \ker \beta = \ker(\operatorname{coker} \alpha)$. But, by Lemma 1.2(i),

$$\begin{array}{ccc} SA & \longrightarrow & 0 \\ {\scriptstyle S(\alpha)}\downarrow & & \downarrow \\ SB & \underset{S(\beta)}{\longrightarrow} & SC \end{array} \qquad (2)$$

is a pushout diagram in $\mathbb{A}'$. Now, inasmuch as $\alpha$ is a monomorphism, $\ker\alpha = 0$. Thus, by the result just cited, $\ker S(\alpha) = S(\ker\alpha) = S(0) = 0$. Therefore, since $\mathbb{A}'$ is abelian, $S(\alpha)$ is a monomorphism and the diagram (2) is a pullback. By Lemma 1.2 (ii), the diagram (1) is a pullback too.

Dually, if $\alpha$ is an epimorphism, then $\alpha$ is the cokernel of its kernel. But, since S preserves $\mathcal{E}$-limits and $\mathbb{A}'$ is finitely complete, $\mathbb{A}$ is finitely complete.

(ii) Suppose that $\mathbb{A}$ is Grothendieck. Then, the Popescu-Gabriel Theorem implies that $\mathbb{A}$ is a left complete, locally small category with an (injective) cogenerator. By the same token, $\mathbb{A}'$ is left complete and, so, by Lemma 1.2(i), S preserves left limits. Thus, the fact that S has a left adjoint is a known consequence of the Adjoint Functor Theorem.

Conversely, suppose that S has a left adjoint. Thus, $\mathbb{A}$ has a generator, by Lemma 1.3. Also, $\mathbb{A}$ is abelian, by (i), and right complete, since $\mathbb{A}'$ is right complete and S creates right $\mathcal{E}$-limits. Next, let $A \in |\mathbb{A}|$, let $\{A_i\}$ be a direct family of subobjects of $\mathbb{A}$, and let $u: \varinjlim A_i \to A$ be the morphism such that the diagram

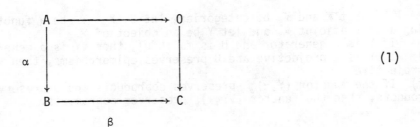

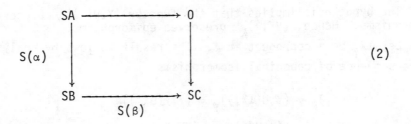

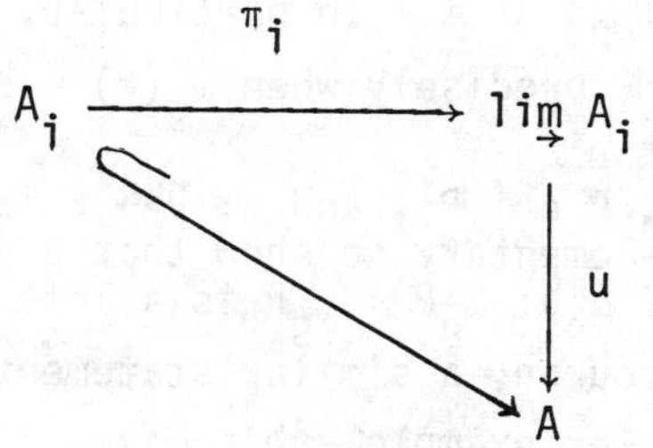


commutes for all i, where the $\pi_i$ are the injections for the direct limit. Then, by Lemma 1.2(i), $\{SA_i\}$ is a direct family of subobjects of SA and the diagram

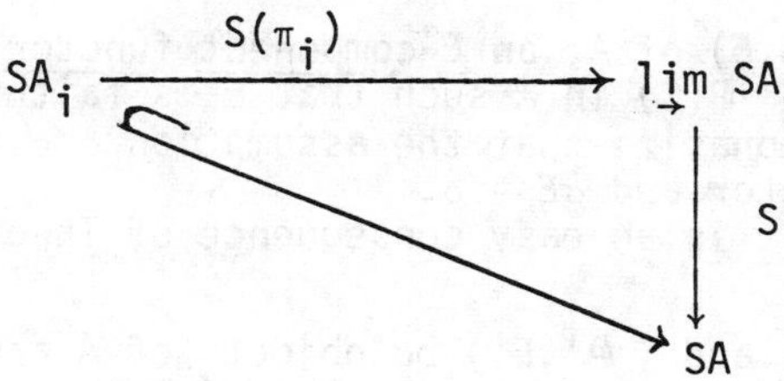


commutes for all i. But then, since $\mathbb{A}'$ is Grothendieck, S(u) is a monomorphism and, so, since S is faithful, u is a monomorphism. It follows that $\mathbb{A}$ has exact direct limits. Consequently, $\mathbb{A}$ is Grothendieck.

The rest is an immediate consequence of Lemma 1.3 since, by Lemma 1.2(i), S preserves epimorphisms and coproducts. □

Concerning the hypothesis of this result, we remark that, of course, if $\mathbb{A}$ is abelian and S is faithful, then S reflects isomorphisms.

## II. COMPONENT FUNCTORS

For the rest of the paper we fix a category $\mathbb{C}$ with zero object 0, and we denote the product category $\prod_G \mathbb{C}$ by $\mathbb{P} = \mathbb{P}(\mathbb{C})$. If $* \in \mathbb{P}$ and $g \in G$, we let $\omega_g(*)$ be the $g^{th}$ component of $*$. Thus, each $\omega_g$ is a functor $\mathbb{P} \to \mathbb{C}$. We make $\mathbb{P}$ into a left G-category by defining, for each $h \in G$, $\omega_g({}^h*) = \omega_{gh}(*)$ for all g.

We regard $\mathbb{P}$ as an object $(\mathbb{P}, 1_{\mathbb{P}})$ of $A$. In particular, note that, given $* \in \mathbb{P}$, $* \in \text{Stab}_{1_{\mathbb{P}}} \mathbb{P}$ precisely when $\omega_g(*) = \omega_1(*)$ for all g.

If $F: \Gamma \to \mathbb{P}$ is a functor, $P \in |\mathbb{P}|$, and $\theta: P \to F$ is a natural transformation, it is elementary to show that $\theta$ is a left limit of F in $\mathbb{P}$ if and only if $\omega_g\theta: \omega_g P \to \omega_g F$ is a left limit of $\omega_g F$ in $\mathbb{C}$ for every $g \in G$. Of course, a similar statement can be made concerning right limits. For example, this makes it plain that the inclusion functor Stab $\mathbb{P} \to \mathbb{P}$ preserves limits.

Now, if $C \in |\mathbb{C}|$ is a (projective) generator, then the object $\hat{C}$ of $\mathbb{P}$ defined by $\omega_g \hat{C} = C$ for all g is a (projective) generator. Thus, since $\mathbb{C} \simeq$ Stab $\mathbb{P}$, it follows that

LEMMA 2.1. If $\mathbb{C}$ is abelian or Grothendieck, then $\mathbb{P}$ is, respectively, abelian or Grothendieck.

Given an object $(\mathbb{A}, E)$ of $A$, an <u>$E$-component functor for $\mathbb{A}$ in $\mathbb{C}$</u> is a functor $\sigma: \mathbb{A} \to \mathbb{P}(\mathbb{C})$ in $A$ such that $\sigma$ is faithful and creates $E$-limits. We emphasize that the assumption $\sigma \in A$ just means that $\sigma$ is a G-functor and $\hat{\sigma}E = \sigma$.

The following result is an easy consequence of Theorem 1.1.

THEOREM 2.2. Let $(\mathbb{A}, E)$ and $(\mathbb{A}', E')$ be objects of $A$ and assume that $E$ and $E'$ are idempotent equivalences. If $(\mathbb{A}, E) \sim (\mathbb{A}', E')$ with factorization (R,L), then the map $\sigma \mapsto \sigma L$, from $E$-component functors for $\mathbb{A}$ to $E'$-component functors for $\mathbb{A}'$, is a bijection with inverse $\sigma' \mapsto \sigma' R$.

We postpone further study of component functors until after the next, fundamental result concerning the $E$-stabilizer of $\mathbb{A}$.

PROPOSITION 2.3. The inclusion functor $\text{Stab}_E \mathbb{A} \to \mathbb{A}$ reflects limits.

<u>Proof</u>. We denote inclusion Stab $\mathbb{A} \hookrightarrow \mathbb{A}$ by I. Let $F: \Gamma \to$ Stab $\mathbb{A}$, let $B \in |\text{Stab } \mathbb{A}|$, and let $\theta: B \to F$ be a natural transformation such that $\theta$ is a left limit of IF in $\mathbb{A}$. Let $\theta': B' \to F$ be a natural transformation, where $B' \in |\text{Stab } \mathbb{A}|$. Then, there is a unique morphism $a: B' \to B$ in $\mathbb{A}$ such that $\theta' = \theta a$. But then, if $g \in G$, $\theta' = E^g\theta' = (E^g\theta)(E^g a) = \theta(E^g a)$; and this implies that $E^g a = a$, since $E^g a \in (B', B)_{\mathbb{A}}$. Thus, $a \in$ Stab $\mathbb{A}$. Consequently, $\theta$ is left limit of F in Stab $\mathbb{A}$. □

If E preserves limits, one can show that inclusion $\overline{\text{Stab}_E} \mathbb{A} \hookrightarrow \mathbb{A}$ reflects limits.

By [1, Lemma 1.3], if $\sigma: \mathbb{A} \to \mathbb{P}$ is an $E$-component functor for $\mathbb{A}$, then restriction of $\sigma$ to $\mathrm{Stab}_E \mathbb{A}$ can be viewed as a functor $\tilde{\sigma}: \mathrm{Stab}_E \mathbb{A} \to \mathrm{Stab}\ \mathbb{P}$. We have

LEMMA 2.4. The functor $\tilde{\sigma}: \mathrm{Stab}_E \mathbb{A} \to \mathrm{Stab}\ \mathbb{P}$ is faithful and creates limits.

Proof. Inasmuch as $\sigma$ is faithful, $\tilde{\sigma}$ is faithful. Let $F: \Gamma \to \mathrm{Stab}\,\mathbb{A}$ be a functor and let $\theta$ be a left limit of $\tilde{\sigma}F$ in $\mathrm{Stab}\ \mathbb{P}$. Then, since, as we observed above, the inclusion functor $\mathrm{Stab}\ \mathbb{P} \to \mathbb{P}$ preserves left limits, $\theta$ is a left limit of $\sigma F$ in $\mathbb{P}$. Thus, since $\sigma$ creates left $E$-limits, there is a unique natural transformation $\eta: A \to F$ with $A \in |\mathbb{A}_E|$ such that $\sigma\eta = \theta$; and $\eta$ is a left limit of F. But, for each $g \in G$, $E(E^gA) = E^gEA = E^gA$ and $\sigma(E^g\eta) = \sigma^g\eta$ ${}^g\sigma\eta = {}^g\theta = \theta$. Thus, $E^g\eta = \eta$ and, consequently, $\eta$ is a natural transformation in $\mathrm{Stab}\ \mathbb{A}$; that is, $\eta_v \in \mathrm{Stab}\ \mathbb{A}$ for each $v \in |\Gamma|$. Moreover, $\tilde{\sigma}\eta = \theta$.

Let $\eta': A' \to F$, $A' \in |\mathrm{Stab}\ \mathbb{A}|$, be a natural transformation such that $\tilde{\sigma}\eta' = \theta$. Then, since $\sigma\eta' = \tilde{\sigma}\eta' = \theta$ and $A' \in |\mathbb{A}_E|$, $\eta' = \eta$. But, by Proposition 2.3, $\eta$ is a left limit of F in $\mathrm{Stab}\ \mathbb{A}$. Thus, $\tilde{\sigma}$ creates left limits. Similarly, $\tilde{\sigma}$ creates right limits. □

THEOREM 2.5. Let $(\mathbb{A}, E)$ be an object of $A$ having a component functor $\sigma$ in $\mathbb{C}$.
(i) If $\mathbb{C}$ is (finitely) complete, then the inclusion functor $\mathrm{Stab}_E \mathbb{A} \to \mathbb{A}$ preserves (finite) limits.
(ii) If $E$ is an idempotent equivalence and $\mathbb{C}$ is abelian, then $\mathrm{Stab}_E \mathbb{A}$ and $\mathbb{A}$ are abelian.
(iii) If $\sigma$ has a left adjoint, $E$ is an idempotent equivalence, and $\mathbb{C}$ is Grothendieck, then $\mathrm{Stab}_E \mathbb{A}$ and $\mathbb{A}$ are Grothendieck.

Proof.

(i) Let $F: \Gamma \to \mathrm{Stab}\ \mathbb{A}$ and let $\theta: B \to F$ be a left limit of F in $\mathrm{Stab}\ \mathbb{A}$. Using Lemmas 2.4 and 1.2(i), the fact that inclusion $\mathrm{Stab}\ \mathbb{P} \hookrightarrow \mathbb{P} = \mathbb{P}(\mathbb{C})$ preserves left limits shows that $\sigma\theta$ is a left limit of $\sigma F$ in $\mathbb{P}$. Thus, since $\sigma$ creates left $E$-limits and $B \in |\mathbb{A}_E|$, it follows that $\theta$ is a left limit of F in $\mathbb{A}$.
(ii) Since $\mathrm{Stab}\ \mathbb{P} \simeq \mathbb{C}$, $\mathbb{A}$ is abelian by Lemma 2.1 and Proposition 1.4; and $\mathrm{Stab}\ \mathbb{A}$ is abelian by Lemma 2.4 and Proposition 1.4 (or by (i) in conjunction with Proposition 2.3).

(iii) By Lemma 2.1 and Proposition 1.4, $\mathbb{A}$ is Grothendieck. By Lemma 2.4 and the proof of Proposition 1.4, Stab $\mathbb{A}$ is a right complete abelian category with exact direct limits. Now, inasmuch as $\mathbb{C}$ is right complete, in particular, $\mathbb{C}$ has G-indexed coproducts. Since $\sigma$ creates $E$-limits, this can be seen -- in Lemma 3.1 and Theorem 3.4 -- to imply that $\mathbb{A}$ has an $E$-stable coproduct functor. But then, by [1, Theorem 5.2], inclusion Stab $\mathbb{A} \hookrightarrow \mathbb{A}$ has a left adjoint. Thus, since we already know $\mathbb{A}$ has a generator, Stab $\mathbb{A}$ has a generator, by Lemma 1.3. □

We mention that in (ii) of this result, the assumption that $E$ is an idempotent equivalence is not needed for $\mathrm{Stab}_E\, \mathbb{A}$ to be abelian.

COROLLARY 2.6. In the notation of the Theorem, suppose that $E$ is an idempotent equivalence, $\sigma$ has a left adjoint, and $\mathbb{C}$ is a module category. Then $\mathbb{A}$ is a Grothendieck category with a projective generator, and $\mathrm{Stab}_E\, \mathbb{A}$ is equivalent to a module category.

<u>Proof</u>. Let $\tilde{\sigma}$: $\mathrm{Stab}_E\, \mathbb{A} \to \mathrm{Stab}\ \mathbb{P}$ be the functor induced by $\sigma$. Since we know that $\tilde{\sigma}$ creates limits and that $\mathrm{Stab}_E\, \mathbb{A}$ is Grothendieck, the rest follows as in Proposition 1.4. □

## III. ADMISSIBILITY

Given $(\mathbb{A},E) \in |A|$, we recall that an endofunctor $\mu$ on $\mathbb{A}$ is called $E$-stable if $\mathrm{im}\mu \subseteq \mathrm{Stab}_E\, \mathbb{A}$. We defined, in [1, §IV], an $E$-stable coproduct system in $\mathbb{A}$ to be an ordered pair $(\lambda;\mu)$, where $\mu$ is an $E$-stable endofunctor on $\mathbb{A}$ and $\lambda$ is a natural transformation $1_{\mathbb{A}} \to \mu$ such that, for each $A \in |\mathbb{A}|$, the G-indexed family $\{E({}^g\lambda_A)\}$ is a coproduct. Thereto, if an endofunctor $\mu$ is such that $(\lambda;\mu)$ is an $E$-stable coproduct system for some $\lambda$, then we call $\mu$ an $E$-stable coproduct functor.

In this section, we study the $E$-stable coproduct systems in $\mathbb{A}$ which are "admissible", in the sense of a given $E$-component functor for $\mathbb{A}$ in $\mathbb{C}$, with respect to some $1_{\mathbb{P}}$-stable coproduct system in $\mathbb{P} = \mathbb{P}(\mathbb{C})$. We start by determining the stable coproduct systems in $\mathbb{P}$. (We will always omit the prefix "$1_{\mathbb{P}}$".)

LEMMA 3.1. There is a canonical one-to-one correspondence between stable coproduct systems in $\mathbb{P}(\mathbb{C})$ and choices, for each G-indexed family of objects of $\mathbb{C}$, of a coproduct for the family.

Proof. We identify G-indexed families of objects of $\mathbb{C}$ with objects of $\mathbb{P}$. Let $(\pi;\rho)$ be a coproduct system in $\mathbb{P}$. Then, if $P \in |\mathbb{P}|$, the family

$$\{{}^h\pi_P: {}^hP \to \rho P\}_{h \in G}$$

is a coproduct in $\mathbb{P}$. Thus, since $\rho P \in |\mathrm{Stab}\ \mathbb{P}|$, the family

$$\{\omega_g({}^h\pi_P): \omega_g({}^hP) \to \omega_1(\rho P)\}_{h \in G}$$

is a coproduct in $\mathbb{C}$ for all $g \in G$. So the family

$$\{\omega_{gh}(\pi_P): \omega_{gh}(P) \to \omega_1(\rho P)\}_{h \in G}$$

is a coproduct for all $g \in G$ too. In particular, taking $g = 1$, we get that $\{\omega_h(\pi_P)\}_{h \in G}$ is a coproduct for the family $P = (\omega_g P)$ of G-indexed objects of $\mathbb{C}$.

Conversely, for each G-indexed family $P = (P_g)$ of objects of $\mathbb{C}$, choose a coproduct

$$\{\nu_h(P): P_h \to \nu P\}_{h \in G}$$

in $\mathbb{C}$. Let $\rho P$ be the constant sequence $(\nu P) \in |\mathbb{P}|$ and put $\pi_P = (\nu_g(P)) \in \mathbb{P}$. Since

$$\{\nu_{gh}(P): P_{gh} \to \nu P\}_{h \in G}$$

is a coproduct for all $g$,

$$\{\omega_{gh}(\pi_P): \omega_{gh}(P) \to \omega_g(\rho P)\}_{h \in G}$$

is a coproduct for all $g$. Thus,

$$\{\omega_g({}^h\pi_P): \omega_g({}^hP) \to \omega_g(\rho P)\}_{h \in G}$$

is a coproduct for all g. This implies that the family $\{{}^h\pi_P : {}^hP \to \rho P\}_{h \in G}$ is a coproduct in $\mathbb{P}$. It follows, by the proof of [1, Lemma 4.2], or directly, that there is a unique way of making $\rho$ into an endofunctor on $\mathbb{P}$ such that $(\pi;\rho)$ is a coproduct system in $\mathbb{P}$. □

Let $\sigma$ be an $E$-component functor for $\mathbb{A}$ in $\mathbb{C}$, where $(\mathbb{A},E) \in |A|$. We call an $E$-stable coproduct system $(\lambda;\mu)$ in $\mathbb{A}$ *$\sigma$-admissible* if there is a stable coproduct system $(\pi;\rho)$ in $\mathbb{P} = \mathbb{P}(\mathbb{C})$ such that $\sigma\lambda = \pi\sigma$; we say that $(\lambda;\mu)$ is *$\sigma$-admissible with respect to $(\pi;\rho)$*.

LEMMA 3.2 Retaining the above notation, if $(\lambda;\mu)$ is $\sigma$-admissible with respect to $(\pi;\rho)$, then $\sigma\mu = \rho\sigma$.

*Proof.* Clearly, $\sigma\mu = \rho\sigma$ on objects. To show that the same is true of morphisms, let $\alpha \in (A,A')_{\mathbb{A}}$. Then,

$$\begin{array}{ccc} E^gA & \xrightarrow{E({}^g\lambda_A)} & \mu A \\ {\scriptstyle E({}^g\alpha)}\downarrow & & \downarrow{\scriptstyle \mu(\alpha)} \\ E^gA' & \xrightarrow[E({}^g\lambda_{A'})]{} & \mu A' \end{array}$$

is a commutative diagram in $\mathbb{A}$, for every $g \in G$. Applying $\sigma$, we get that

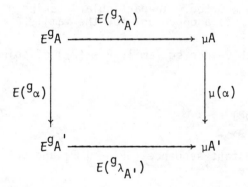

$$\begin{array}{ccc} {}^{g}\sigma A & \xrightarrow{\;{}^{g}\pi_{\sigma A}\;} & \rho(\sigma A) \\ {\scriptstyle {}^{g}\sigma(\alpha)}\downarrow & & \downarrow{\scriptstyle \sigma(\mu(\alpha))} \\ {}^{g}\sigma A' & \xrightarrow[\;{}^{g}\pi_{\sigma A'}\;]{} & \rho(\sigma A') \end{array}$$

is a commutative diagram in $\mathbb{P}$, for every $g \in G$. Thus, $\sigma(\mu(\alpha)) = \rho(\sigma(\alpha))$, by [1, Lemma 4.1]. $\square$

The next result enables one to identify a given $\mathcal{E}$-stable coproduct system as being $\sigma$-admissible.

PROPOSITION 3.3. If $\mathbb{C}$ has G-indexed coproducts and $\sigma$ is an $\mathcal{E}$-component functor for $\mathbb{A}$ in $\mathbb{C}$, then an $\mathcal{E}$-stable coproduct system $(\lambda;\mu)$ in $\mathbb{A}$ is $\sigma$-admissible if and only if for each pair of objects A, A' of $\mathbb{A}$, $\sigma(\lambda_A) = \sigma(\lambda_{A'})$ whenever $\sigma A = \sigma A'$.

Proof. $\Rightarrow$ Trivial.
$\Leftarrow$ Let $A \in |\mathbb{A}|$. It follows, by Lemma 1.2(i), that $\{{}^{g}\sigma(\lambda_A): {}^{g}\sigma A \to \sigma(\mu A)\}_{g \in G}$ is a coproduct in $\mathbb{P}(\mathbb{C})$. Define $\rho(\sigma A) \in |\mathbb{P}|$ to be $\sigma(\mu A)$ and define $\pi_{\sigma A} \in (\sigma A, \rho(\sigma A))_{\mathbb{P}}$ to be $\sigma(\lambda_A)$. These definitions make sense, by assumption. But, since $\mathbb{C}$ has G-indexed coproducts, we can use Lemma 3.1 to make $(\pi;\rho)$ into a stable coproduct system in $\mathbb{P}$. But then, $(\lambda;\mu)$ is $\sigma$-admissible with respect to $(\pi;\rho)$, by construction. $\square$

We define stable coproduct systems $(\pi;\rho)$ and $(\pi';\rho')$ in $\mathbb{P}$ to be <u>$\sigma$-equivalent</u> if $\pi\sigma = \pi'\sigma$. We denote the equivalence class of $(\pi;\rho)$ by $<\pi,\rho>$.

By [1, Lemma 1.3], if $(\mathbb{A},\mathcal{E}) \in |\mathcal{A}|$, then $\mathcal{E}$ induces an endofunctor on $\mathrm{Stab}_{\mathcal{E}}\,\mathbb{A}$ via restriction.

THEOREM 3.4. Let $\sigma$ be a component functor in $\mathbb{C}$ for an object $(\mathbb{A},\mathcal{E})$ of $\mathcal{A}$. If the endofunctor on $\mathrm{Stab}_{\mathcal{E}}\,\mathbb{A}$ induced by $\mathcal{E}$ is idempotent and faithful, then there is a canonical bijection between $\sigma$-admissible $\mathcal{E}$-stable coproduct systems in $\mathbb{A}$ and equivalence classes of $\sigma$-equivalent stable coproduct systems in $\mathbb{P}(\mathbb{C})$.

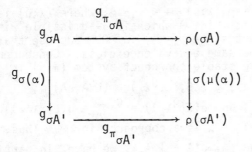

Proof. Consider the map $(\lambda;\mu) \mapsto <\pi,\rho>$, where $(\lambda;\mu)$ is an $E$-stable coproduct system in $\mathbb{A}$ which is $\sigma$-admissible with respect to the stable coproduct system $(\pi;\rho)$ in $\mathbb{P}$. Suppose that $(\lambda;\mu)$ and $(\lambda';\mu')$ map to the same equivalence class. Consequently, $\sigma\lambda = \pi\sigma = \sigma\lambda'$ for some stable coproduct system $(\pi;\rho)$ in $\mathbb{P}$. Let $A \in |\mathbb{A}|$. Then, for all $g \in G$, $E({}^g\lambda_A) \in (E^gA,\mu A)_{\mathbb{A}}$, $E({}^g\lambda'_A) \in (E^gA,\mu'A)_{\mathbb{A}}$, and $\sigma(E({}^g\lambda_A)) = {}^g\pi_{\sigma A} = \sigma(E({}^g\lambda'_A))$. But, the family $\{\sigma(E({}^g\lambda_A))\}_{g \in G}$ is a coproduct in $\mathbb{P}$. Thus, since $\sigma$ creates $E$-limits and $\mu A$, $\mu'A \in |\mathbb{A}_E|$, we have, in particular, that $E(\lambda_A) = E(\lambda'_A)$. This implies that $\mu A = \mu'A$. But then, since $\sigma$ is faithful, $\lambda_A = \lambda'_A$.

This argument shows that the map $(\lambda;\mu) \mapsto <\pi,\rho>$ is one-to-one. To show that it is onto, let an equivalence class $<\pi,\rho>$ be given, and let $A \in |\mathbb{A}|$. Since the family

$$\{{}^g\pi_{\sigma A}: {}^g\sigma A \to \rho\sigma A\}_{g \in G}$$

is a coproduct in $\mathbb{P}$ and $\sigma$ creates $E$-limits, there is a unique family

$$\{\mu_g(A): {}^gA \to \mu A\}_{g \in G}$$

of morphisms in $\mathbb{A}$ with $E\mu A = \mu A$ such that $\sigma(\mu_g(A)) = {}^g\pi_{\sigma A}$ for every $g \in G$ and, moreover, the family $\{\mu_g(A)\}_{g \in G}$ is a coproduct in $\mathbb{A}$. In particular, note that the $\mu_g(A)$ depend only on the equivalence class of $(\pi;\rho)$.

Let $g \in G$. Now, $E(\mu_g(A)) \in (E^gA,\mu A)_{\mathbb{A}}$ and $\sigma(E(\mu_g(A)) = {}^g\pi_{\sigma A}$. But, $E({}^g\mu_1(A)) \in (E^gA,E^g\mu A)_{\mathbb{A}}$, $E(E^g\mu A) = E^gE\mu A = E^g\mu A$, and $\sigma(E({}^g\mu_1(A))) = {}^g\sigma(\mu_1(A)) = {}^g\pi_{\sigma A}$. Thus, since $\sigma$ creates E-limits, $E(\mu_g(A)) = E({}^g\mu_1(A))$, and $\{E(\mu_g(A))\}_{g \in G}$ is a coproduct. So, $E^g\mu A = \mu A$ and, if we set $\lambda_A = \mu_1(A)$, then, inasmuch as $E({}^g\lambda_A) = E(\mu_g(A))$, $\{E({}^g\lambda_A)\}_{g \in G}$ is a coproduct.

We must show how to define $\mu$ on morphisms so that $\mu$ is an $E$-stable endofunctor. For this, let $\alpha \in (A,A')_{\mathbb{A}}$ and let $\mu(\alpha)$ be the morphism in $\mathbb{A}$ such that the diagram

$$\begin{array}{ccc} E^{g}A & \xrightarrow{E({}^{g}\lambda_A)} & \mu A \\ \Big\downarrow{\scriptstyle E({}^{g}\alpha)} & & \Big\downarrow{\scriptstyle \mu(\alpha)} \\ E^{g}A' & \xrightarrow[E({}^{g}\lambda_{A'})]{} & \mu A' \end{array}$$

commutes for all g. Then, $\mu(\alpha)$ is well-defined, since we know that the families $\{E({}^{g}\lambda_A)\}$ and $\{E({}^{g}\lambda_{A'})\}$ are coproducts. Thereto, the $\lambda_A$ define a natural transformation $\lambda\colon 1_{\mathbb{A}} \to \mu$. But, since $\sigma(E^2({}^{g}\lambda_A)) = {}^{g}\pi_{\sigma A}$ for each g, it follows that the family $\{E^2({}^{g}\lambda_A)\}$ is a coproduct too. Thus, by the proof of [1, Lemma 4.2], $(\lambda;\mu)$ is an $E$-stable coproduct system in $\mathbb{A}$. Furthermore, $(\lambda;\mu)$ is admissible with respect to $(\pi;\rho)$, by construction. □

Combining this with Lemma 3.1, we get

COROLLARY 3.5. With the notation and assumptions of Theorem 3.4, $\mathbb{A}$ has $\sigma$-admissible $E$-stable coproduct systems if and only if $\mathbb{C}$ has G-indexed coproducts. Moreover, given a stable coproduct system $(\pi;\rho)$ in $\mathbb{P}$, there is a unique $E$-stable coproduct system in $\mathbb{A}$ which is $\sigma$-admissible with respect to $(\pi;\rho)$.

We end the section with the following observation.

PROPOSITION 3.6. If $\sigma$ is an $E$-component functor for $\mathbb{A}$, $E$ is either idempotent or faithful, and $(\lambda;\mu)$ is a $\sigma$-admissible, $E$-stable coproduct system in $\mathbb{A}$, then $\mu E = \mu$.

Proof. Suppose that $(\lambda;\mu)$ is $\sigma$-admissible with respect to $(\pi;\rho)$. Let $A \in |\mathbb{A}|$ and $g \in G$. We have $E({}^{g}\lambda_{EA}) \in (E^{g}EA,\mu EA)_{\mathbb{A}}$ and $EE({}^{g}\lambda_A) \in (EE^{g}A,\mu A)_{\mathbb{A}}$. Also, $\sigma(E^{g}\lambda_{EA})) = {}^{g}\pi_{\sigma EA} = {}^{g}\pi_{\sigma A}$ and $\sigma(EE({}^{g}\lambda_A)) = {}^{g}\pi_{\sigma A}$. Thus, since $E^{g}EA = EE^{g}A$, it follows that $E({}^{g}\lambda_{EA}) = EE({}^{g}\lambda_A)$ and, hence, that $\mu EA = \mu A$.

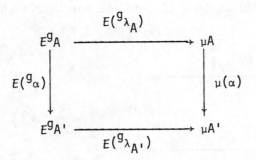

Now, if $\alpha \in (A,A')_{\mathbb{A}}$, then $\mu(\alpha)$ is the morphism such that the diagram

$$\begin{array}{ccc} \mathcal{E}^gA & \xrightarrow{\mathcal{E}({}^g\lambda_A)} & \mu A \\ {\scriptstyle \mathcal{E}({}^g\alpha)}\big\downarrow & & \big\downarrow{\scriptstyle \mu(\alpha)} \\ \mathcal{E}^gA' & \xrightarrow[\mathcal{E}({}^g\lambda_{A'})]{} & \mu A' \end{array} \qquad (1)$$

commutes for all g. But also, after some manipulation, we see that the diagram

$$\begin{array}{ccc} \mathcal{E}\mathcal{E}^gA & \xrightarrow{\mathcal{E}\mathcal{E}({}^g\lambda_A)} & \mu A \\ {\scriptstyle \mathcal{E}\mathcal{E}({}^g\alpha)}\big\downarrow & & \big\downarrow{\scriptstyle \mu(\mathcal{E}(\alpha))} \\ \mathcal{E}\mathcal{E}^gA' & \xrightarrow[\mathcal{E}\mathcal{E}({}^g\lambda_{A'})]{} & \mu A' \end{array}$$

commutes for all g. Thus, since $\mathcal{E}$ is either idempotent or faithful, the diagram (1) commutes with $\mu(\mathcal{E}(\alpha))$ in place of $\mu(\alpha)$. In other words, $\mu(\mathcal{E}(\alpha)) = \mu(\alpha)$. □

## IV. GRADINGS

Let $(\mathbb{A},\mathcal{E}) \in |\mathcal{A}|$ and let $\mathbb{B} \in |\mathcal{B}|$, where $\mathcal{B}$ is the full subcategory of $\mathcal{A}$ whose objects have the form $(\mathbb{X},1_{\mathbb{X}})$ with $\mathbb{X}$ a trivial left G-category. In [1, §II], we defined an $\mathcal{E}$-pregrading of $\mathbb{B}$ in $\mathbb{A}$ to be a functor $T\colon \mathbb{A} \to \mathbb{B}$ having a right adjoint H such that H is an embedding with $\mathrm{im}H = \mathrm{Stab}_{\mathcal{E}}\,\mathbb{A}$ and such that, for some adjunction $\lambda\colon 1_{\mathbb{A}} \to HT$, $(\lambda;HT)$ is an $\mathcal{E}$-stable coproduct system in $\mathbb{A}$. We said that a functor $H\colon \mathbb{B} \to \mathbb{A}$ was an $\mathcal{E}$-cograding of $\mathbb{B}$ in $\mathbb{A}$ if H was an embedding, $\mathrm{im}H = \mathrm{Stab}_{\mathcal{E}}\,\mathbb{A}$, and some left adjoint of H was an $\mathcal{E}$-pregrading of $\mathbb{B}$ in $\mathbb{A}$. Given functors

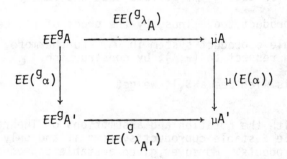

T: $\mathbb{A} \to \mathbb{B}$ and H: $\mathbb{B} \to \mathbb{A}$ we proved, in [1, Corollary 5.4], that, provided $E$ is, for instance, an idempotent equivalence, T is an $E$-pregrading if and only if (T,H') is a universal $E$-pair for some H'; and H is an $E$-cograding if and only if (T',H) is a universal $E$-pair for some T'. We remind the reader that an $E$-pair (T,H) = ${}_{\mathbb{B}}$(T,H) on $\mathbb{A}$, that is, a pair of functors T: $\mathbb{A} \to \mathbb{B}$ and H: $\mathbb{B} \to \mathbb{A}$ such that H $\varepsilon$ $A$ and HT is an $E$-stable coproduct functor, is universal if for any $E$-pair ${}_{\mathbb{B}'}$(T',H') on $\mathbb{A}$ there is a unique functor U: $\mathbb{B}' \to \mathbb{B}$ such that HU = H' and UT' $\simeq$ T.

An _$E$-component structure_, $\zeta = \zeta(\sigma,\langle \pi,\rho \rangle)$, _for $\mathbb{A}$ in $\mathbb{C}$_ consists of an $E$-component functor $\sigma$ for $\mathbb{A}$ in $\mathbb{C}$ together with the $\sigma$-equivalence class of a stable coproduct system $(\pi;\rho)$ in $\mathbb{P}(\mathbb{C})$. In this terminology, by Lemma 3.1, we have

_PROPOSITION_ 4.1 If there is an $E$-component functor for $\mathbb{A}$ in $\mathbb{C}$, then there is an $E$-component structure for $\mathbb{A}$ in $\mathbb{C}$ if and only if $\mathbb{C}$ has G-indexed coproducts.

We point out that, by the proof of Theorem 3.4, if $\zeta = \zeta(\sigma,\langle \pi,\rho \rangle)$ is an $E$-component structure for $\mathbb{A}$ in $\mathbb{C}$, then there is at most one $E$-stable coproduct system $(\lambda;\mu)$ in $\mathbb{A}$ such that $(\lambda;\mu)$ is $\sigma$-admissible with respect to $(\pi;\rho)$ (and, hence, with respect to every stable coproduct system $(\pi';\rho')$ in $\mathbb{P}(\mathbb{C})$ which is $\sigma$-equivalent to $(\pi;\rho)$): We refer to $\mu$ as the _$\zeta$-compatible, $E$-stable coproduct functor on $\mathbb{A}$_. We stress that, by Theorem 3.4, _a sufficient condition for $\mu$ to exist is that the endofunctor on $\mathrm{Stab}_E\,\mathbb{A}$_ induced by $E$ is faithful and idempotent.

Given $\mathbb{B}$ $\varepsilon$ $|\mathcal{B}|$, we define an _$E$-grading of $\mathbb{B}$ in $\mathbb{A}$_ to be an $E$-component structure $\zeta$ for $\mathbb{A}$ together with a functor T: $\mathbb{A} \to \mathbb{B}$ which has a right adjoint H such that 1) H is an $E$-cograding and 2) the $\zeta$-compatible, $E$-stable coproduct functor on $\mathbb{A}$ exists and is equal to HT$E$. We say that T is an _$E$-grading relative to $\zeta$_. We point out that, by the proof of Proposition 3.6, T must satisfy (T$E$)$E$ = T$E$. Also, we hasten to mention that, by [1, Lemma 2.2], any functor which is a grading relative to some component structure is a pregrading.

Next, we investigate the existence and uniqueness of gradings.

PROPOSITION 4.2. Let T: $\mathbb{A} \to \mathbb{B}$ be an $E$-grading relative to an $E$-component structure $\zeta$ for $\mathbb{A}$. A functor T': $\mathbb{A} \to \mathbb{B}'$, where $\mathbb{B}'$ $\varepsilon$ $|\mathcal{B}|$, is an $E$-grading relative to $\zeta$ precisely when there exists an isomorphism V: $\mathbb{B}' \to \mathbb{B}$ such that VT'$E$ = T$E$ and VT' $\simeq$ T.

Proof. Let H be a right adjoint of T such that H is a cograding and $HTE$ is the $\zeta$-compatible, stable coproduct functor on $\mathbb{A}$.
$\Leftarrow$ We have $H'T'E = HTE$, where $H' = HV$ and $V: \mathbb{B}' \to \mathbb{B}$ is an isomorphism. In particular, by [1, Corollary 2.4], H' is a cograding of $\mathbb{B}'$ in $\mathbb{A}$. But, since H is a right adjoint of T, it is readily checked that H' is a right adjoint of $V^{-1}T \simeq T'$.
$\Rightarrow$ Let H' be a cograding of $\mathbb{B}'$ in $\mathbb{A}$ such that H' is a right adjoint of T' and $H'T'E = HTE$. By [1, Corollary 2.4], there is an isomorphism $V: \mathbb{B}' \to \mathbb{B}$ such that $H' = HV$. Consequently, since H is an embedding, $VT'E = TE$. But, inasmuch as we just saw that H' is a right adjoint of $V^{-1}T$, $V^{-1}T \simeq T'$. □

In this result, obviously, if $E$ is an equivalence, then $VT' \simeq T$ is implied by $VT'E = TE$.

THEOREM 4.3. Let $(\mathbb{A},E)$ be an object of $\mathcal{A}$, let $\mathbb{B}$ be an object of $\mathcal{B}$ isomorphic to $\mathrm{Stab}_E\,\mathbb{A}$, and let $\zeta$ be an $E$-component structure for $\mathbb{A}$. If $E$ is faithful and induces an idempotent equivalence on $\mathrm{Stab}_E\,\mathbb{A}$, then there is an $E$-grading $T: \mathbb{A} \to \mathbb{B}$ relative to $\zeta$. In fact, T can be chosen so that $TE = T$. (Cf. [1, Corollary 6.9].)

Proof. This is an easy consequence of Theorem 3.4, Proposition 3.6, and [1, Theorem 5.2]. □

If $\zeta$ is an $E$-component structure for $\mathbb{A}$, we call an $E$-pair $(T,H)$ on $\mathbb{A}$ $\zeta$-compatible if HT is the $\zeta$-compatible, $E$-stable coproduct functor on $\mathbb{A}$. We remark that if $E$ is faithful and induces an idempotent equivalence on $\mathrm{Stab}_E\,\mathbb{A}$, then a $\zeta$-compatible $E$-pair on $\mathbb{A}$ is a universal $E$-pair if and only if it is universal amongst $\zeta$-compatible $E$-pairs on $\mathbb{A}$--this follows by [1, Theorem 5.2] and [1, Lemma 5.1]. We get the following characterization of $E$-gradings.

THEOREM 4.4. Let $(\mathbb{A},E) \in |\mathcal{A}|$ such that $E$ is an idempotent equivalence and let $\zeta$ be an $E$-component structure for $\mathbb{A}$. The following are equivalent properties of a functor $T: \mathbb{A} \to \mathbb{B}$, where $\mathbb{B} \in |\mathcal{B}|$.

(i) T is an $E$-grading relative to $\zeta$.

(ii) T is an $E$-pregrading such that, for some $E$-cograding H of $\mathbb{B}$ in $\mathbb{A}$, $HTE$ is the $\zeta$-compatible, $E$-stable coproduct functor on $\mathbb{A}$.

(iii) $HTE$ is the $\zeta$-compatible, $E$-stable coproduct functor on $\mathbb{A}$ for some embedding $H: \mathbb{B} \to \mathbb{A}$ such that $\mathrm{im}H = \mathrm{Stab}_E\,\mathbb{A}$.

(iv) $(TE,H)$ is a $\zeta$-compatible, universal $E$-pair for some functor H: $\mathbb{B} \to \mathbb{A}$.

We showed, in [1, Corollary 5.6], that if $E$ is an idempotent equivalence and the action of G on $\mathbb{A}$ is fixed point free, then the existence of $E$-pregradings $\mathbb{A} \to \mathbb{B}$ implies the existence of $E$-pregradings $\mathbb{A} \to \mathbb{B}$ which are G-functors. Also, as we commented in [1, §II], the fixed point free assumption is necessary for the validity of this result. We do not know, and doubt, whether the result for gradings analogous to [1, Corollary 5.6] is valid. We have, however,

THEOREM 4.5. Let $\zeta$ be an $E$-component structure for $\mathbb{A}$ in $\mathbb{C}$, where $(\mathbb{A},E) \in |A|$ and $E$ is an idempotent equivalence. Suppose that the action of G on $\mathbb{A}$ is fixed point free and that there is at most one member of $|\mathrm{im}E|$ in each G-orbit. If $\mathbb{B} \in |B|$ and $\mathbb{B} \simeq \mathrm{Stab}_E \mathbb{A}$, then, relative to $\zeta$, there is an $E$-grading $\mathbb{A} \to \mathbb{B}$ which is a G-functor.

Proof. Let $\zeta = \zeta(\sigma, <\pi,\rho>)$, where $\sigma$ is an $E$-component functor for $\mathbb{A}$ in $\mathbb{C}$ and $(\pi;\rho)$ is a stable coproduct system in $\mathbb{P}(\mathbb{C})$. Let $(\lambda;\mu)$ be the $E$-stable coproduct system in $\mathbb{A}$ which is $\sigma$-admissible with respect to $(\pi;\rho)$.

Now, by assumption, we can find a representative class $R$, say, of objects of $\mathbb{A}$ re the action of G such that $R \supseteq |\mathrm{im}E|$. Then, by the proof of [1, Corollary 4.5], there is an $E$-stable coproduct system $(\delta;\nu)$ in $\mathbb{A}$ such that $\nu$ is a G-functor and $E(\delta_{g_R}) = E({}^g\lambda_R)$ for every $R \in R$ and every $g \in G$.

Set $\delta' = (\delta E)\tau^{-1}$, where, using [1, Corollary 3.6], $\tau: E \to 1_{\mathbb{A}}$ is the unique natural isomorphism such that $E\tau = 1_E$. Since, for $g \in G$, $E^g\delta' = E^g\delta E$, it is plain that $(\delta';\nu E)$ is an $E$-stable coproduct system. But, since $\sigma = \sigma E, \sigma\delta' = (\sigma\delta E)(\sigma\tau^{-1}) = (\sigma\delta E)1_\sigma = \sigma\delta E$. But then, since $R \supseteq |\mathrm{im}E|$, $\sigma(\delta'_A) = \sigma(\delta_{EA}) = \sigma(E(\delta_{EA}))^\sigma = \sigma(E(\lambda_{EA})) = \sigma(\lambda_{EA}) = \pi_{\sigma EA} = \pi_{\sigma A}$ for every $A \in |\mathbb{A}|$. Thus, $\nu E = \mu$.

By [1, Theorem 5.2] and [1, Corollary 5.4(i)], there exists a universal $E$-pair ${}_{\mathbb{B}}(T,H)$ on $\mathbb{A}$ such that $HT = \nu$. Inasmuch as H is an embedding, T is a G-functor. Furthermore, since $HTE = \mu$, T is an $E$-grading relative to $\zeta$, by the preceding theorem. □

In the context of this result, we point out a surprising phenomenon; namely, if T: $\mathbb{A} \to \mathbb{B}$ is an $E$-grading and $TE = T$, then, except in certain degenerate cases, T is not a G-functor.

For, otherwise, T would be an $E$-pregrading which is a morphism in $A$ whereas, in [1, §II], we observed this would be possible only when $G = 1$ or $\mathbb{A} \approx \{0\}$. Additionally, we mention that the assumptions of Theorem 4.5 hold, for a canonical choice of $E$ and $\mathbb{C}$, when $\mathbb{A}$ is the category of internally G-graded left modules over a G-graded ring.

Next, concerning gradings and factorizations, we have

THEOREM 4.6. Let $(\mathbb{A},E)$ and $(\mathbb{A}',E')$ be objects of $A$ such that $(\mathbb{A},E) \sim (\mathbb{A}',E')$. If $E$ and $E'$ are idempotent equivalences, then there is a bijection between $E$-gradings $(\zeta,T)$ on $\mathbb{A}$ such that $TE = T$ and $E'$-gradings $(\zeta',T')$ on $\mathbb{A}'$ such that $T'E' = T'$.

Proof. Let $\zeta = \zeta(\sigma,<\pi,\rho>)$ be an $E$-component structure for $\mathbb{A}$ and, using [1, Theorem 5.2], let $(T,H)$ be a universal $E$-pair on $\mathbb{A}$ such that, say, $(\lambda;HT)$ is the $E$-stable coproduct system in $\mathbb{A}$ which is $\sigma$-admissible with respect to $(\pi;\rho)$. By [1, Corollary 6.8], Theorem 4.4, Proposition 3.6, and Theorem 2.2 it suffices, where $T' = TL$ and $H' = RH$, to find an $E'$-stable coproduct system $(\lambda';T'H')$ in $\mathbb{A}'$ which is $\sigma' = \sigma L$-admissible with respect to $(\pi;\rho)$. For this let $\tau': E' \to 1_{\mathbb{A}'}$ be the isomorphism such that $E'\tau' = 1_{E'}$ and put $\lambda' = (R\lambda L)\tau'^{-1}$. Since $E'^g\lambda' = E'^g R\lambda L = RE^g\lambda L$ whenever $g \in G$, it follows that $(\lambda';T'H')$ is an $E'$-stable coproduct system. But also, $\sigma'\lambda' = \sigma' R\lambda L = \sigma LR\lambda L = \sigma\lambda L = \pi\sigma L = \pi\sigma'$. Thus, $(\lambda';T'H')$ is $\sigma'$-admissible with respect to $(\pi;\rho)$. □

Let $\zeta = \zeta(\sigma,<\pi,\rho>)$ be an $E$-component structure for $\mathbb{A}$ in $\mathbb{C}$ and let $(\zeta,T)$ be an $E$-grading of $\mathbb{B}$ in $\mathbb{A}$. By Lemma 3.2, we have a commutative diagram of the form

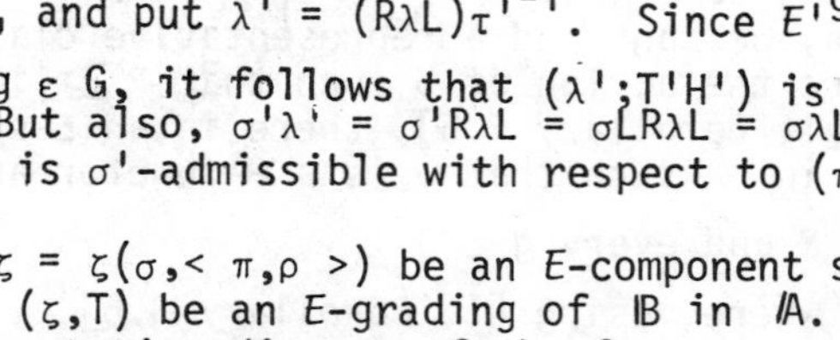

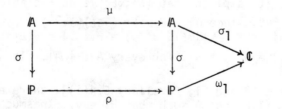

,

where $\mu$ is the $\zeta$-compatible, $E$-stable coproduct functor on $\mathbb{A}$ and $\mathbb{P} = \mathbb{P}(\mathbb{C})$. Thereto, if H is a right adjoint of T such that H is an $E$-cograding with $HTE = \mu$, then, by the proof of Proposition 3.6, we get this commutative diagram:

$$\begin{array}{ccc} \mathbb{A} & & \\ {\scriptstyle TE}\downarrow & \searrow^{\Sigma} & \\ \mathbb{B} & \xrightarrow[\sigma_1 H]{} & \mathbb{C} \end{array},$$

where $\Sigma = \sigma_1\mu$. Note that $\Sigma = (\omega_1\rho)\sigma$.

Now, by Lemma 3.1, $\omega_1\rho$: $\mathbb{P} \to \mathbb{C}$ is the functor defined in the obvious manner by a certain choice of a coproduct for each G-indexed family of objects of $\mathbb{C}$. We remind the reader that, if $\mathbb{C}$ is Grothendieck, it is well-known that $\omega_1\rho$ preserves monomorphisms.

PROPOSITION 4.7. In the above notation, the functor $\sigma_1 H$ is faithful and creates limits.

Proof. Since $\omega_g(\sigma H) = \sigma_1 H$ for all $g \in G$, it is clear that the diagram

$$\begin{array}{ccc} \mathrm{Stab}_{E}\,\mathbb{A} & \xrightarrow{\tilde{\sigma}} & \mathrm{Stab}\ \mathbb{P} \\ {\scriptstyle \tilde{H}}\uparrow & & \uparrow{\scriptstyle W} \\ \mathbb{B} & \xrightarrow[\sigma_1 H]{} & \mathbb{C} \end{array}$$

commutes, where $\tilde{\sigma}$ is induced by $\sigma$, $\tilde{H}$ is induced by H, and W is the obvious isomorphism. But $\tilde{H}$ is an isomorphism and, by Lemma 2.4, $\tilde{\sigma}$ is faithful and creates limits. □

We should mention that, in this result, the assertion that $\sigma_1 H$ is faithful is redundant.

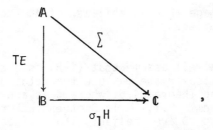

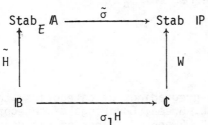

THEOREM 4.8. Let T: $\mathbb{A} \to \mathbb{B}$ be an $E$-grading relative to an $E$-component structure $\zeta = \zeta(\sigma, <\pi, \rho>)$ for $\mathbb{A}$ in some Grothendieck category; and assume that $E$ is an idempotent equivalence. Then the categories $\mathbb{A}$ and $\mathbb{B}$ are abelian and T is an exact functor. If, moreover, the functor $\omega_1\sigma H$ has a left adjoint for some

embedding H: $\mathbb{B} \to \mathbb{A}$ with image the $E$-stabilizer of $\mathbb{A}$, then $\mathbb{A}$ and $\mathbb{B}$ are Grothendieck categories.

Proof. Plainly, we may as well assume that (T,H) is a $\zeta$-compatible, universal $E$-pair. In particular, we may use the notation established in the discussion following Theorem 4.6, where $\mathbb{C}$ is a Grothendieck category. Note that, by Lemma 2.1, $\mathbb{P}$ is Grothendieck too.

By Theorem 2.5, $\mathbb{A}$ and $\mathbb{B}$ are abelian. Thus, since T is a left adjoint, T is right exact. Now, it follows, by Lemma 1.2(i), that $\sigma$ preserves monomorphisms and, as we pointed out above, $\omega_1\rho$ preserves monomorphisms. Thus, $\sum$ preserves monomorphisms. Consequently, the diagram in (1) shows that $TE$ preserves monomorphisms, since we know that $\sigma_1 H$ is faithful. Inasmuch as $T \simeq TE$, we get that T is exact.

Finally, assume that $\sigma_1 H = \omega_1 \sigma H$ has a left adjoint. Then, by Proposition 4.7 and the proof of Proposition 1.4, $\mathbb{B}$ is Grothendieck. Also, since $\sigma$ creates $E$-limits, $\mathbb{A}$ is left complete. But then, since $\mathbb{P}$ is locally small, $\mathbb{A}$ is abelian, and $\sigma$ is faithful, it follows that $\mathbb{A}$ is locally small. Furthermore, since $\mathbb{B}$ has a cogenerator, $\mathbb{A}$ has a cogenerator, by the dual of Lemma 1.3(i) and [1, Proposition 2.1]. Thus, by Lemma 1.2(i) and the Adjoint Functor Theorem, $\sigma$ has a left adjoint. Therefore, $\mathbb{A}$ is Grothendieck, by Theorem 2.5. □

REFERENCES

[1]. R. Gordon, Group-pregradings of categories, to appear.
[2]. R. Gordon and E.L. Green, Coverings of categories, to appear.
[3]. R. Gordon and E.L. Green, Gradings of categories, to appear.
[4]. S. MacLane, "Categories for the Working Mathematician," Graduate Texts in Mathematics 5, Springer-Verlag, New York/Heidelberg/Berlin, 1971.

BRAUER GROUPS OF HOMOGENEOUS SPACES, I

W. Haboush

State University of New York at Albany

INTRODUCTION

This work originates with my desire to find an algebraic homogeneous space, with respect to an affine algebraic group over $\mathbb{C}$ the field of complex numbers which is not a rational algebraic variety. The question in fact is open, but I believe that a non-rational homogeneous space will be found. A quick perusal of the facts involved shows that the chief candidate for such a homogeneous space would be one of the form G/K for G and K both semi-simple over $\mathbb{C}$ and K not contained in any parabolic subgroup of G. In fact this last condition is not essential, but by considering Levi factors one sees that the essential part of the problem lies here.

There are only a few methods which have been successfully employed to establish the non-rationality of a variety. The first is the use of intermediate Jacobeans which seems to be quite unwieldy in the present setting. The method of Manin and Itzkovitch involves showing that the Cremona group of the variety is too small. This method may possibly be applicable in the present situation, but it does not seem suited to my own mathematical temperament. The other method would be the computation of a "Mumford-Artin invariant" of some sort. It has been observed, perhaps by Serre, that if X is complete, smooth and non-singular over $\mathbb{C}$, then if there is non-zero torsion in $H^3(X,\mathbb{Z})$, then X cannot be rational. This comes to the same thing as showing that the Brauer group of X is not zero.

In [1], Artin and Mumford constructed a non-rational threefold as follows. They started with a configuration consisting of two smooth cubics intersecting transversally and a conic doubly tangent three times to each cubic, all in $\mathbb{P}^2(\mathbb{C})$. Then they construct a quaternion algebra on the complement of the cubics which extends to an algebra ramified precisely along the two cubics.



F. van Oystaeyen (ed.), Methods in Ring Theory, 111–144.

After blowing up the points of tangency of the conic to the cubics in a certain order, they are able to analyze the splitting behavior of the algebra on this conic, essentially, and to use the analysis to show that the Brauer Severi variety of the ramified algebra is non-rational, by producing a non-zero element of its Brauer group.

This suggests a certain approach to our rationality problem. One begins by observing that the generic Azumaya algebras studied by Amitsur, Artin, Procesi, Saltman and others is an algebra central simple over the function field of a homogeneous space. Other "generic" algebras occur in a similar fashion. Hence one could hope to find an appropriate compactification of this space with an algebra satisfying conditions related perhaps to ramification behavior. One might hope then to use a strategy like that of the Artin-Mumford paper [1], to establish non-rationality provided that the compactification can be sufficiently well understood. This has motivated much of the work in this paper and a subsequent one as well.

The paper is organized as follows. Section one is largely an expository review of such topics as the results of Grothendieck [7], the theory of G-structures on sheaves and some basics concerning principal homogeneous spaces. In section 2, the G-equivariant Brauer group of a variety is introduced. In addition certain objects called Azumaya representations are introduced and shown to be the objects which induce equivariant Azumaya algebras. They are used to construct a "representation theoretic Brauer Group" of a group G which is nothing more than the group of algebraic central extensions of G by $G_m$ under Baer multiplication.

In section 3 the induction functor taking Azumaya representations to the equivariant Brauer group is studied. It contains a proof that the equivariant Brauer group of G/K is isomrophic to the group of Azumaya representations.

Section 4 is devoted to an exposition of a result of Iverson computing the Brauer group of a semi-simple group.

Section 5 is devoted to the proof of an important result. Namely, it is shown that if G is semi-simple and $K \subset G$ is closed and connected, then the map from the G-equivariant Brauer group of G/K to the Brauer group of G/K given by "forgetting G-structures" is injective.

Section 6 is a study of the Brauer-Severi varieties of induced Azumaya algebras. They are shown to be "generic splitting varieties" in the scheme theoretic sense. I found it quite surprising to note how canonical many of the constructions involved were.

Section 7 if given over to the study of a certain generic Azumaya algebra. While much of it represents a reinterpretation of known results concerning the algebra generated by $n^2$ generic n×n matrices in the language of algebraic groups, I believe that the functoriality of the generic Azumaya Algebra in section 7 is discussed with a previously unachieved precision. In addition the Brauer Severi variety of the generic Azumaya algebra is shown to be itself a homogeneous space.

My interest in this subject was piqued by very enlightening discussions with David Saltman and Claudio Procesi. Pierre DeLigne also made several useful suggestions. In addition I would like to thank Professors Michael Artin and Jacques Tits for encouragement in this work.

## PRELIMINARIES

In this section we shall fix notation and we shall review certain standard facts from the theory of Brauer groups and the theory of affine algebraic groups.

First we recall the results of Grothendieck [1] and [7]. There, non-abelian cohomology in the étale topology with coefficients in certain group schemes is utilized. In particular, the pointed cohomology sets, $H^1_{\text{ét}}(\underline{\underline{X}},Gl(n,k))$, $H^1_{\text{ét}}(\underline{\underline{X}},Pgl(n,k))$ and $H^1_{\text{ét}}(\underline{\underline{X}},Sl(n,k))$ are introduced and demonstrated to classify respectively, vector bundles of rank n on $\underline{\underline{X}}$, Azumaya algebras of rank $n^2$ central over $\mathcal{O}_{\underline{\underline{X}}}$, and pairs consisting of a vector bundle of rank n, $F$, together with a fixed isomorphism $\varphi:\Lambda^n F\to \mathcal{O}_{\underline{\underline{X}}}$, from the highest exterior power of $F$ to $\mathcal{O}_{\underline{\underline{X}}}$. (These last objects, the pairs $(F,\varphi)$, shall be referred to as "bundles of determinant one".)

The major device in this context is the exact sequence associated to a short exact sequence of sheaves of groups in the étale topology:

$$1 \to C \to G \to \overline{G} \to 1,$$

where C is central in G. Then there is a sequence of pointed sets and maps,

$$\begin{aligned} 1 &\to H^0_{\text{ét}}(\underline{\underline{X}},C) \to H^0_{\text{ét}}(\underline{\underline{X}},G) \to H^0_{\text{ét}}(\underline{\underline{X}},\overline{G}) \\ &\to H^1_{\text{ét}}(\underline{\underline{X}},C) \to H^1_{\text{ét}}(\underline{\underline{X}},G) \to H^1_{\text{ét}}(\underline{\underline{X}},\overline{G}) \to H^2_{\text{ét}}(\underline{\underline{X}},C) \end{aligned} \quad (1)$$

which is exact in the following sense. First of all since C is central the $H^i_{ét}(\underline{\underline{X}},C)$ are all abelian groups, while the sets $H^0(\underline{\underline{X}},G)$ and $H^0(\underline{\underline{X}},\overline{G})$ are equal to $G(\underline{\underline{X}})$ and $\overline{G}(\underline{\underline{X}})$ respectively, and hence are groups and so the first four terms in the sequence form an exact sequence in the conventional sense. Moreover $H^1_{ét}(\underline{\underline{X}},C)$ operates on $H^1_{ét}(\underline{\underline{X}},G)$. Then the continued exactness means the following:

i) $H^1_{ét}(\underline{\underline{X}},C)/\mathrm{im}(H^0_{ét}(\underline{\underline{X}},G)$ is $H^1_{ét}(\underline{\underline{X}},C)$-

<u>isomorphic to its image in</u> $H^1_{ét}(\underline{\underline{X}},G)$, <u>and this image is the inverse image of the distinguished point in</u> $H^1_{ét}(\underline{\underline{X}},\overline{G})$; (2)

ii) <u>two elements of</u> $H^1_{ét}(\underline{\underline{X}},G)$ <u>map to the same element of</u> $H^1_{ét}(\underline{\underline{X}},\overline{G})$ <u>if and only if they lie in the same</u> $H^1_{ét}(\underline{\underline{X}},C)$ <u>orbit in</u> $H^1_{ét}(\underline{\underline{X}},G)$.

Now let, as usual, $G_{m,\underline{\underline{X}}}$ and $\mu_n$ denote the multiplicative group and the group of n-th roots of 1. Then consider the exact sequences

$$\text{i)}\quad 1 \to G_{m,\underline{\underline{X}}} \to G\ell(n,\underline{\underline{X}}) \to Pg\ell(n,\underline{\underline{X}}) \to 1$$
$$\text{ii)}\quad 1 \to \mu_n \to S\ell(n,\underline{\underline{X}}) \to Pg\ell(n,\underline{\underline{X}}) \to 1. \qquad (3)$$

Now, recall that $H^1_{ét}(\underline{\underline{X}},G_{\underline{\underline{X}}})$ is the Picard group of $\underline{\underline{X}}$ while $H^1_{ét}(\underline{\underline{X}},\mu_n)$ is the elements of order n in Pic $\underline{\underline{X}}$. Then we derive exact sequences:

$$\text{i)}\quad \ldots \to H^1_{ét}(\underline{\underline{X}},G_{m,\underline{\underline{X}}}) \xrightarrow{j_1} H^1_{ét}(\underline{\underline{X}},G\ell(n,\underline{\underline{X}})) \xrightarrow{\pi_1}$$
$$H^1_{ét}(\underline{\underline{X}},Pg\ell(n,\underline{\underline{X}})) \xrightarrow{\delta} H^2_{ét}(\underline{\underline{X}},G_{m\underline{\underline{X}}})$$
$$\text{ii)}\quad \ldots \to H^1_{ét}(\underline{\underline{X}},\mu_n) \to H^1_{ét}(\underline{\underline{X}},S\ell(n,\underline{\underline{X}})) \xrightarrow{\overline{\pi}_1} \qquad (4)$$
$$H^1_{ét}(\underline{\underline{X}},Pg\ell(n,\underline{\underline{X}})) \xrightarrow{\overline{\delta}} H^2_{ét}(\underline{\underline{X}},\mu_n).$$

The maps $j_1$ and $\overline{j}_1$ assign to the line bundle $L$ the direct sum $L^{\oplus n}$. For $L$ an n-torsion bundle there is also a natural isomorphism $\varphi:\Lambda^n(L^{\oplus n})\to\mathcal{O}_{\underline{\underline{X}}}$. The maps $\pi_1,\overline{\pi}_1$ assign to the bundle $F$ the sheaf of Azumaya algebras $\underline{\mathrm{End}}_{\mathcal{O}_{\underline{\underline{X}}}}(F)$. In both cases the operation of $H^1$ on the middle term is given by tensoring with the corresponding line bundle. This sheaf of Azumaya algebras is Brauer equivalent to 0 if and only if it is identically a trivial algebra $\underline{\underline{\mathrm{End}}}_{\mathcal{O}_{\underline{\underline{X}}}}(F)$. Moreover $F$ may be taken to be of determinant one. Two bundles $F_1$ and $F_2$ give the same endomorphism ring if and only if there is a line bundle $L$ so that $F_1 = L \otimes_{\mathcal{O}_{\underline{\underline{X}}}} F_2$ and moreover when the $F_i$ are of determinant one $L$ must be of order n in $\mathrm{Pic}(\underline{\underline{X}})$. These results are all to be found in [7].

The second fundamental notion in our discussion will be the idea of a homogeneous bundle. Let G be an affine algebraic group over the field k and let $\underline{\underline{X}}$ be a k-scheme with a left G action, $\lambda : G \times_k \underline{\underline{X}} \to \underline{\underline{X}}$. Let $\mu : G \times_k G \to G$ be the multiplication on G. Then consider the maps, $\lambda$ and $p_2$ from $G \times_k \underline{\underline{X}}$ to $\underline{\underline{X}}$, where $p_2$ is projection on the second factor, and the three maps, $\mu \times id$, $id \times \lambda$ and $p_{2,3}$, from $G \times_k G \times_k \underline{\underline{X}}$ to $G \times_k \underline{\underline{X}}$. Then, since $\lambda$ is an action the following hold:

$$\begin{aligned} p_2 \circ \mu \times id_{\underline{\underline{X}}} &= p_2 \circ p_{23} \\ \lambda \circ (id \times \lambda) &= \lambda \circ (\mu \times id_{\underline{\underline{X}}}) \qquad (5) \\ p_2 \circ (id_G \times \lambda) &= \lambda \circ p_{23} \ . \end{aligned}$$

Let $F$ be a quasi-coherent sheaf on $\underline{\underline{X}}$ . Then we recall that a G-structure on $F$ is an isomorphism:

$$\Phi : p_2^* F \to \lambda^* F$$

which satisfies the identity.

$$(id_G \times \lambda)^*(\Phi) \circ p_{23}^*(\Phi) = (\mu \times id)^*(\Phi). \qquad (6)$$

This requires some explanation and it is, I hope, contained in the diagram:

$$\begin{array}{ccc} & (p_2 \circ p_{23})^*(F) = (p_2 \circ \mu \times id))^*(F) & \\ \swarrow p_{23}^*(\Phi) & & \searrow (\mu \times id)^*(\Phi) \\ (\lambda \circ p_{23})^*(F) & & (\lambda \circ (\mu \times id))^*(F) \\ \| & & \| \\ (p_2 \circ (id \times \lambda))^*(F) & \xrightarrow{(id \times \lambda)^*(\Phi)} & (\lambda \circ (id \times \lambda))^*(F) \end{array}$$

More intuitively $\Phi$ can be interpreted as an algebraic family $\{\varphi_g\}_{g \in G(k)}$ where $\varphi$ is an isomorphism from $F$ to $\lambda^*_g F$ where $\lambda_g$ is left translation by g on $\underline{\underline{X}}$ . Then (5) may be interpreted as an algebraicization of the diagram

$$\begin{array}{ccc} \mathcal{F} & \xrightarrow{\varphi_h} & \lambda^*_h(\mathcal{F}) \\ & {\searrow}^{\varphi_{gh}} \quad {\swarrow}_{\lambda^*_h(\varphi_g)} & \\ & \lambda^*_{gh}(\mathcal{F}) & \end{array} \tag{7}$$

That is, pointwise (5) becomes

$$\lambda^*_h(\varphi_g)\circ\varphi_h = \varphi_{gh} \text{ for all } g,h.$$

When G is a linear algebraic group and $K \subset G$ is a k-closed subgroup, a sheaf on G/K with a G-structure is called a <u>homogeneous sheaf</u> and, if it is coherent, a <u>homogeneous bundle</u>. Let $e_k \in G/K$ denote the closed k-point corresponding to the class of e in G/K. If $\mathcal{F}$ is a homogeneous bundle, let $\mathcal{V}(G) = \mathcal{F}_{e_K}$, the fibre of $\mathcal{F}$ over $e_K$. Then $\mathcal{V}(\mathcal{F})$ is a finite dimensional k-space and since $x\cdot e_K = e_K$ for all $x \in K$, $\varphi_x$ maps $\mathcal{V}(\mathcal{F})$ to $\mathcal{V}(\mathcal{F})$. Thus $\mathcal{V}(\mathcal{F})$ is a finite dimensional representation of K. Moreover if V is a finite dimensional representation of K one may define a homogeneous sheaf on G/K as follows.

Let $\pi: G \to G/K$ be the natural projection. Then the sheaf $\mathcal{O}_G \otimes V$ admits a right K-structure corresponding to the diagonal action induced by a right translation on $\mathcal{O}_G$ and the representation on V. (The definition of a right K-structure is left to the reader.) Hence $\pi_*(\mathcal{O}_G \otimes_k V)$ is a sheaf on G/K with an $\mathcal{O}_{G/K}$-linear K-representation. Set $I_{G/K}(V) = (\pi_*(\mathcal{O}_G \otimes_k V))^K$, the subsheaf of K invariants in $\pi_*(\mathcal{O}_G \otimes_k V)$. Then it is elementary to check that the functors $\mathcal{V}$ and $I_{G/K}$ are inverse to one another, and hence give an isomorphism of the appropriate categories.

When $\mathcal{F}$ is a sheaf of $\mathcal{O}_{G/K}$-algebras and $\Phi$, the G-structure, is an algebra isomorphism $\mathcal{F}$ will be called a homogeneous sheaf of $\mathcal{O}_{G/K}$-algebras. Then $\mathcal{V}(\mathcal{F})$ is a k-algebra and K acts by algebra automorphisms. Similarly, if V is a k-algebra and K acts by automorphisms, $I_{G/K}(V)$ is a homogeneous sheaf of $\mathcal{O}_{G/K}$-algebras. Similar considerations may be extended to such structure as homogenous sheaves of representations of a homogeneous sheaf of algebras, etc.

Finally let $\underline{\underline{X}}$ and $\underline{\underline{Y}}$ be k-varieties and let $q:\underline{\underline{X}}\to\underline{\underline{Y}}$ be a k-morphism. Let G and K denote two linear algebraic groups over k and let $''G\times K\times\underline{\underline{X}}\to\underline{\underline{X}}$ be an action. Write $\lambda_1 = \lambda|G\times e\times\underline{\underline{X}}$, and

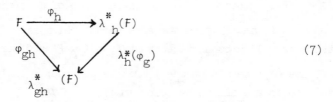

$\lambda_2 = \lambda|e\times K\times \underline{\underline{X}}$ and regard the $\lambda_i$ as separate commuting actions. Then recall the:

1.7 DEFINITION: Let $\gamma:K\times\underline{\underline{X}}\to\underline{\underline{X}}$ be an action of K on $\underline{\underline{X}}$ and let $q:\underline{\underline{X}}\to\underline{\underline{Y}}$ be a morphism. Then we shall say that $\underline{\underline{X}}$ is a principal homogeneous space over $\underline{\underline{Y}}$ with respect to $\gamma$ if and only if

i) q is faithfully flat affine and a geometric quotient mapping which splits in the étale topology.

ii) The map

$$(\gamma,\mathrm{id}):K\times\underline{\underline{X}} \to \underline{\underline{X}}\times_{\underline{\underline{Y}}}\underline{\underline{X}}$$

is a k isomorphism.

1.8 DEFINITION: Let $q:\underline{\underline{X}}\to\underline{\underline{Y}}$ be a morphism and let $\lambda:G\times K\times\underline{\underline{X}}\to\underline{\underline{X}}$ and $\bar{\lambda}:G\times\underline{\underline{Y}}\to\underline{\underline{Y}}$ be actions. Then say that $\underline{\underline{X}}$ is a K-principal G-space over $\underline{\underline{Y}}$ if and only if:

i) $\underline{\underline{X}}$ is a principal homogeneous K-space over $\underline{\underline{Y}}$ with respect to $\lambda_2$.

ii) $q:\underline{\underline{X}}\to\underline{\underline{Y}}$ splits in the étale topology.

Note that in general, if H is a group and N a closed normal subgroup, then whenever H acts on $\underline{\underline{X}}$ and the quotient of $\underline{\underline{X}}$ with respect to N exists, and is equal to say $\underline{\underline{Y}}$ with quotient morphism $q:\underline{\underline{X}}\to\underline{\underline{Y}}$, then there is a naturally induced action of H/N on $\underline{\underline{Y}}$. If this action is viewed as an H action, then q becomes H-equivariant. These considerations apply when $\underline{\underline{X}}$ is a K-principal homogeneous G-space over $\underline{\underline{Y}}$, with H = G×K, N = e×K. In particular there is a unique natural G action on $\underline{\underline{Y}}$, with respect to which q is G-equivariant. Whenever this situation occurs we shall refer to this action as the natural G-action or the induced G-action.

Further if $\rho:K\to G\ell(V)$ is a representation of K, then $\mathcal{O}_{\underline{\underline{X}}}\otimes V$ admits a natural K structure with respect to $\lambda_1$. Hence $q_*(\mathcal{O}_{\underline{\underline{X}}}\otimes_k V)$ admits a K-representation and so $q_*(\mathcal{O}_{\underline{\underline{X}}}\otimes_k V)^K$, the subsheaf of invariants, is a quasi-coherent sheaf of $\mathcal{O}_{\underline{\underline{Y}}}$-modules. In fact by flat descent $q_*(\mathcal{O}_{\underline{\underline{X}}}\otimes_k V)^K$ will be locally free of rank equal to the rank of V over k. [5].

1.9 DEFINITION: Let $\underline{\underline{X}}$ be a K-principal G-space over $\underline{\underline{Y}}$ with quotient morphism $q: \underline{\underline{X}}\to\underline{\underline{Y}}$ as in 1.8. Then if $\rho: K\to G\ell_k(V)$ is a representation, $\mathcal{O}_{\underline{\underline{X}}/K}(V)$ will denote the sheaf $q_*(\mathcal{O}_{\underline{\underline{X}}}\otimes_k V)^K$. It will be referred to as the sheaf on $\underline{\underline{Y}}$ principally induced by $\rho:K\to G\ell_k(V)$.

Evidently, $q_{\underline{\underline{X}}/K}$ is a functor. That is there is a natural map $q_{\underline{\underline{X}}/K}(\beta)$ whenever $\beta:V\to V'$ is a G-morphism. The definition is clear.

In addition $q_{\underline{\underline{X}}/K}$ commutes with certain functorial constructions. The most conspicuous among these are summarized here without proof.

1.10 PROPOSITION: *Let $\rho: K \to G\ell_k(V)$ and $\gamma: K \to G\ell_k(W)$ be two representations. Let $q: \underline{\underline{X}} \to \underline{\underline{Y}}$ be a quotient morphism making $\underline{\underline{X}}$ into a K principal G-space over $\underline{\underline{Y}}$. Then*

i) $q_{\underline{\underline{X}}/K}(V \otimes_k W) \simeq q_{\underline{\underline{X}}/K}(V) \otimes_{\mathcal{O}_{\underline{\underline{X}}}} q_{\underline{\underline{X}}/K}(W)$.

ii) *If V is finite dimensional,*
$q_{\underline{\underline{X}}/K}(\mathrm{Hom}_k(V,W)) = \underline{\mathrm{Hom}}_{\mathcal{O}_{\underline{\underline{X}}}}(q_{\underline{\underline{X}}/K}(V), q_{\underline{\underline{X}}/K}(W)$.

iii) $q_{\underline{\underline{X}}/K}(S^q_k(V)) \simeq S^q_{\mathcal{O}_{\underline{\underline{Y}}}}(q_{\underline{\underline{X}}/K}(V))$

iv) $q_{\underline{\underline{X}}/K}(\Lambda^q_k(V)) = \Lambda^q_{\mathcal{O}_{\underline{\underline{Y}}}}(q_{\underline{\underline{X}}/K}(V))$

v) $q_{\underline{\underline{X}}/K}$ *is exact.*

When $\underline{\underline{X}} = G$ is a group K is a closed subgroup and G and K operate by left and right translation respectively, then G becomes a principal G-space over G/K. In this case we shall revert to the traditional terminology and notation of the induced bundle. In particular, if $\pi: G \to G/K$ is the natural map we shall write $I_{G/K}(V)$ for $\pi_{G/K}(V)$.

2. THREE GROUP VALUED FUNCTORS: In this section we introduce three groups closely related to the Brauer group, the third of them almost entirely for heuristic purposes. In this section k is an algebraically closed field.

2.1 DEFINITION: *Let G be a linear algebraic group over k and let $\underline{\underline{X}}$ be a variety with a left G-action $\lambda: G \times_k \underline{\underline{X}} \to X$. A sheaf of Azumaya algebras over $\mathcal{O}_X$, $\mathcal{A}$ will be said to admit a G-structure as a sheaf of algebras if and only if $\mathcal{A}$ admits a G-structure $\Phi: p_2^*\mathcal{A} \to \lambda^*\mathcal{A}$ which is an isomorphism of algebras. In this case the pair $(\mathcal{A}, \Phi)$ will be called a G-sheaf of Azumaya algebras over $\mathcal{O}_{\underline{\underline{X}}}$.*

In the language of diagram 1, (6), definition 2.1 forces the isomorphisms $\varphi_g$ to be algebra isomorphism. There are now natural notions of triviality and equivalence for G-sheaves of Azumaya algebras.

2.2 DEFINITION: *A G-sheaf of Azumaya algebras $\mathcal{A}$ over $\mathcal{O}_{\underline{X}}$ is said to be G-trivial if and only if there is a locally free sheaf of $\mathcal{O}_{\underline{\underline{X}}}$ modules $\mathcal{F}$ and a G-structure $\psi: p_2^*\mathcal{F} \to \lambda^*\mathcal{F}$, so that $\mathcal{A}$ is G-isomorphic* to $\underline{\underline{\mathrm{End}}}_{\mathcal{O}_{\underline{\underline{X}}}} F$ *under its induced G-structure.*

2.3 DEFINITION: *Let $A$ and $B$ be two G sheaves of Azumaya algebras. Then $A$ will be said to be G-equivalent to $B$ if and only if there are G-trivial G-sheaves of Azumaya algebras $C$ and $D$ so that $A \otimes_{\mathcal{O}_{\underline{\underline{X}}}} C$ is G-isomorphic to $B \otimes_{\mathcal{O}_{\underline{\underline{X}}}} D$.*

Now it is evident that G-isomorphism classes of G-Azumaya algebras from a monoid under tensor product. If the pair $(A,\Phi)$ is a G-sheaf of Azumaya algebras, then the pair $(A^{\circ},\Phi)$ is likewise. ($A^{\circ}$ is the opposite algebra.) Clearly, then $A \otimes_{\mathcal{O}_{\underline{\underline{X}}}} A^{\circ}$ is naturally a G-sheaf of algebras and the isomorphism:

$$A \otimes_{\mathcal{O}_{\underline{\underline{X}}}} A^{\circ} \simeq \underline{\underline{\mathrm{End}}}_{\mathcal{O}_{\underline{\underline{X}}}}(A)$$

is a G isomorphism. It follows that G-equivalence classes under tensor product form, as in the usual case, a group. [2,7]

2.4 DEFINITION: *Let $\underline{\underline{X}}$ be a scheme with a left G-action, $\lambda: G \times_k \underline{\underline{X}} \to \underline{\underline{X}}$. Let $\mathrm{Br}_G(\underline{\underline{X}})$ denote the set of G equivalence classes of G-Azumaya algebras over $\mathcal{O}_{\underline{\underline{X}}}$ under the product induced by tensoring algebras. Then $\mathrm{Br}_G(\underline{\underline{X}})$ is an abelian group called the G-equivariant Brauer group of $\underline{\underline{X}}$.*

Now G-Azumaya algebras are by definition Azumaya algebras and G equivalent pairs of G-Azumaya algebras are evidently equivalent. Hence the map which assigns to a class in $\mathrm{Br}_G(\underline{\underline{X}})$ the class in $\mathrm{Br}(\underline{\underline{X}})$ of a representative of the G class is a well-defined homomorphism of groups.

2.5 DEFINITION: *Let $\underline{\underline{X}}, G, \lambda$ be as in 2.3. Then the natural map from $\mathrm{Br}_G(\underline{\underline{X}})$ to $\mathrm{Br}(\underline{\underline{X}})$ given by sending the class of the pair $(A,\Phi)$ in $\mathrm{Br}_G(\underline{\underline{X}})$ to the class of $A$ in $\mathrm{Br}(\underline{\underline{X}})$ will be denoted $j_{\underline{\underline{X}}/G}$.*

We now proceed to the second of our three group valued functors. As usual $M_{n,k}$ denotes the algebra of n×n matrices over k. We recall that by the Skolem-Noether theorem, $\mathrm{Aut}_k(M_{n,k}) = \mathrm{Pg}\ell(n,k)$, where $\mathrm{Pg}\ell(n,k)$ acts on $M_{n,k}$ by conjugation.

2.6 DEFINITION: *Let G be a linear algebraic group over k. An Azumaya representation of G of degree n is a morphism of algebraic groups over k,* $\rho: G \to \mathrm{Aut}_k(M_{n,k}) = Pg\ell(n,k)$.

As with linear representations, we shall often write down the representation algebra $M_{n,k}$ in place of the morphism $\rho$, when there is no danger of confusion. We may speak of the tensor product of two Azumaya representations. It is evidently again an Azumaya representation under the diagonal action. There are natural notions of triviality and equivalence:

2.7 DEFINITION: *Let* $\rho: G \to Pg\ell(n,k)$ *be an Azumaya representation of degree n. Then* $\rho$ *will be called trivial if and only if there is a homomorphism* $\tilde{\rho}: G \to G\ell(n,k)$ *so that the diagram,*

$$\begin{array}{ccc} & \overset{\tilde{\rho}}{\nearrow} & G\ell(n,k) \\ G & & \downarrow \pi \\ & \underset{\rho}{\searrow} & Pg\ell(n,k) \end{array}$$

*commutes. Here* $\pi$ *is the natural map.*

2.8. DEFINITION: *Let G be a linear algebraic group and let* $\rho: G \to Pg\ell(n,k)$ *and* $\rho': G \to Pg\ell(n',k)$ *be Azumaya representations. Then* $\rho$ *and* $\rho'$ *will be called equivalent if and only if there are trivial Azumaya representations,* $\alpha: G \to Pg\ell(r,k)$, $\alpha': G \to Pg\ell(r',k)$ *so that* $\rho \otimes_k \alpha$ *is isomorphic to* $\rho' \otimes_k \alpha'$.

Now if $\rho: G \to Pg\ell(n,k)$ is an Azumaya representation, we may regard it as a map $\rho: G \to \mathrm{Aut}_k(M_{n,k})$. Then, if $M^\circ_{n,k}$ is the opposite algebra, the representing transformations $\rho(g)$ also act as algebra automorphisms. But transpose $A \to {}^tA$ defines an isomorphism $\tau: M^\circ_{n,k} \quad M_{n,k}$. Thus, the representation of G on $M_{n,k}$ may be transported via $\tau$ to another Azumaya representation of degree n which we call the opposite representation. If $g \in G(k)$ and $\rho(g)(A) = uAu^{-1}$ for all $A \in M_{n,k}$, the opposite representation is given by $\rho^\circ(g)(A) = {}^t(u^{-1})A\,{}^tu$.

2.8. LEMMA: *Let* $\rho: G \to Pg\,(n,k)$ *be any Azumaya representation of G, and let* $\rho^\circ$ *be the opposite representation. Then* $\rho \otimes_k \rho^\circ$ *is a trivial Azumaya representation of G.*

PROOF: First observe that $Pg\ell(n,k) = \mathrm{Aut}(M_{n,k})$ is a closed subgroup of the general linear group of $M_{n,k}$, namely $G\ell(n^2,k)$. Hence the composition of $\rho$ with this inclusion gives a map

$$\rho_1: G \to G\ell(n^2,k),$$

where $\rho_1(g)$ is $\rho(g)$ regarded as only a linear map from $M_{n,k}$ to $M_{n,k}$.

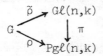

For any $g \in G(k)$, there is at least one $\alpha_g \in G\ell(n,k)$ so that

$$\rho(g)(A) = \alpha_g A \alpha_g^{-1}.$$

Then recall that $\rho^\circ(g)$ is given by

$$\rho^\circ(g)(A) = {}^t\alpha_g^{-1}\, A\, {}^t\alpha_g.$$

Now $M_{n,k} \otimes M_{n,k}$ is naturally isomorphic to $M_{n^2,k} = \mathrm{Hom}_k(M_{n,k}, M_{n,k})$ under the map $\gamma(U\otimes V)(A) = UA\,{}^tV$.

Now

$$[(\rho\otimes\rho^\circ)(g)\gamma(U\otimes V)]\ (A)$$

$$=\gamma(\alpha_g U\alpha_g^{-1} \otimes {}^t\alpha_g^{-1}\ V\,{}^t\alpha_g)(A)$$

$$=\alpha_g U\alpha_g^{-1}\ A\ \alpha_g\ {}^tV\alpha_g^{-1}$$

$$=(\rho_1(g)\circ\gamma(U\otimes V)\circ\rho_1(g)^{-1})(A).$$

That is, $\gamma(\rho\otimes\rho^\circ\ (g)(U\otimes V)) = \rho_1(g)\gamma(U\otimes V)\rho_1(g)^{-1}$.

Hence $\gamma$ gives an isomorphism from $\rho\otimes\rho^\circ$ to the representation on $M_{n^2,k}$, conjugation by $\rho_1(g)$. Since the natural map $G\ell(n^2,k) \to Pg\ell(n^2,k)$ is the map which sends a matrix A to conjugation by A, this implies the result.

The result of all this is that the Azumaya representations of G behave like Azumaya algebras.

2.9 DEFINITION: Let RBr(G/k) denote the set of equivalence classes of Azumaya representations of G over k. If $\xi,\eta \in RBr(G/k)$ and $\xi$ is the class of $\rho$ and $\eta$ is the class of $\rho'$, let $\xi\cdot\eta$ denote the class of $\rho\otimes\rho'$. Then RBr(G/k) is a group called the representation theoretic Brauer group of G over k.

That RBr(G/k) is a group is proven in just the same way as one proves that Brauer equivalence classes form a group under tensor product, and so it follows from 2.8.

To interpret RBr(G/k) clearly, we introduce our third functor and show that it is equal to RBr(G/k).

2.10 DEFINITION: An algebraic central extension of G by $G_{m,k}$ is an exact sequence of algebraic groups over k and morphisms of algebraic groups:

$$1 \to G_{m,k} \xrightarrow{j} E \xrightarrow{p} G \to 1,$$

such that $j(G_{m,k})$ is in the center of E.

As usual, the extension may be donoted only by the pair (E,p). We say that (E,p) and E',p' are isomorphic if and only if there is an algebraic isomorphism $\varphi: E \to E'$ such that $p' \circ \varphi = p$, and such that p is an isomorphism on ker(p). The Baer product of two extensions may be constructed algebraically to yield a third algebraic extension and so isomorphism classes of algebraic central extensions of G by $G_{m,k}$ under Baer product form a group. We write E#E' for the Baer product of E and E'.

2.11 DEFINITION: Let $E_{al}(G;G_{m,k})$ denote the group of algebraic central extensions of G by $G_{m,k}$ under Baer product.

Now we wish to describe the relation between $E_{al}(G;G_{m,k})$ and RBr(G/k). We begin by considering the monoid of isomorphism classes of Azumaya representations of G. Denote it Az(G/k). If the class of $\rho$ is in Az(G/k) it may be regarded as a k-morphism of groups:

$$\rho: \ G \to Pg\ell(n,k)$$

for some n. Let $\pi_n$: $G\ell(n,k) \to Pg\ (n,k)$ be the natural map and define an extension of G by:

$$E_\rho = G \times_{Pg\ell(n,k)} G\ell(n,k). \tag{12}$$

Since the two morphisms involved are group morphisms $E_\rho$ is a group and since ker $\pi_n = G_{m,k}$, the kernel of the projection on the first factor $p_\rho: E_\rho \to G$, is just $G_{m,k}$. Hence the pair $(E_\rho, p_\rho)$ is an algebraic central extension of G by $G_{m,k}$. If $\rho$ is trivial, there is a map $\tilde{\rho}: G \to G\ell(n,k)$ such that $\pi_n \circ \tilde{\rho} = \rho$. Thus, the fibre product $(id_G, \tilde{\rho})$ is a map from G into $E_\rho$ such that $p_\rho\ (id_G, \tilde{\rho}) = id_G$. That is, when $\rho$ is trivial $E_\rho$ is the trivial extension.

2.13 LEMMA: Let $\rho$ and $\gamma$ be two Azumaya representations of G over k. Then there is a natural isomorphism:

$$E_{\rho \otimes_k \gamma} \simeq E_\rho \ \#\ E_\gamma.$$

PROOF: We recall the definition of E # E'. If (E,p) and (E',p') are two algebraic extensions, set U equal to $E \times_G E'$. Let $j: G_{mk} \to E$, and $j': G_{m,k} \to E'$ be the two kernels. Then $p \circ j = \bar{e} = p' \circ j'$ where $\bar{e}$ is the trivial map, and so using $p \circ j$ and $p' \circ j'$ as structure maps, $G_{m,k} \times_k G_{m,k}$ and $G_{m,k} \times_G G_{m,k}$ are the same group. Put K = ker(m) where m: $G_{m,k} \times G_{m,k} \to G_{mk}$ is multiplication $\bar{U} = U/K$ and let q be the induced projection of $\bar{U}$ on G. Then $\bar{U}$ is a central extension of G by $G_{m,k}$ with projection q and $(\bar{U}, q) = E \# E'$.

Now suppose $\rho$ and $\gamma$ are Azumaya representations of degree n and r respectively. Let $\tau_{n,r}$: $G\ell(n,k) \times G\ell(r,k) \to G\ell(nr,k)$ be the morphism $\tau_{n,r}(u,v) = u \otimes v$. Then there is a commutative diagram:

$$
\begin{array}{ccccc}
1 & & 1 & & 1 \\
| & & \downarrow & & \downarrow \\
1 \to K & \longrightarrow & G_{m,k} \times G_{m,k} & \xrightarrow{m} & G_{mk} \\
| & & \downarrow & & \downarrow \\
1 \to K & \longrightarrow & G\ell(n,k) \times_k G\ell(r,k) & \xrightarrow{\tau_{m,r}} & G\ell(nr,k) \\
| & & \downarrow \pi_n \times \pi_r & & \downarrow \pi_{nr} \\
1 & \longrightarrow & Pg\ell(n,k) \times_k Pg\ell(r,k) & \xrightarrow{\overline{\tau}_{m,r}} & Pg\ell(nr,k)
\end{array}
$$

One readily verifies that $\rho\otimes\gamma = \overline{\tau}_{n,r}\circ\rho\times_k\gamma$ and that, in consequence, $\tau_{n,r}\,(E_\rho \times_k E_\gamma) = E_{\rho\otimes\gamma}$. The result follows immediately.

2.15 LEMMA: *Let* $(E,p)$ *be an algebraic central extension of* $G$ *by* $G_{m,k}$. *Then there is an Azumaya representation* $\rho: G \to Pg\ell(n,k)$ *and a natural isomorphism* $\varphi: E \xrightarrow{\sim} E_\rho$.

PROOF: Suppose that there is a faithful finite dimensional representation $\tilde{\rho}: E \to G\ell(U)$ such that $\tilde{\rho}^{-1}(G^c_{m,k}) = \ker(p)$, where $G^c_{m,k}$ denotes the scalars in $G\ell(U)$. Then, if $p_u: G\ell(U) \to Pg\ell(U)$ is the natural map, $\ker p_u \circ \tilde{\rho} = \operatorname{Ker} p$. Hence $\tilde{\rho}$ induces a map $\rho: G = E/G_{m,k} \to Pg\ell(U)$, and a natural isomorphism $E \to G \times_{Pg\ell(U)} G\ell(U)$. This is clearly an Azumaya representation, $\rho$, such that $E \simeq E_\rho$. Thus we proceed to the construction of $\tilde{\rho}$.

Let $\overline{q}_1: G \to G\ell(V_1)$ be a faithful representation. Replacing $V_1$ by a direct sum of $V_1$ and a trivial representation if necessary, we may assume that $\overline{q}_1^{-1}(G^c_{m,k}) = e$. Let $q_1 = p\circ\overline{q}_1$. Then $q_1(\ker p) = e$.

Notice that the restriction homomorphism from $k[E]$, the ring of regular functions on $E$, to the ring $k\,[G_{m,k}]$, the functions on the kernel of $p$, is both surjective and $G_{m,k}$ equivariant for right translation by $G_{m,k}$. Thus there is a function $u \in k[E]$ such that $u(xa) = au(x)$ for all $a \in G_{m,k}$. Let $V_2$ be the $k$ span of the set of all right translations of $u$. Then since $G_{m,k}$ is central $V_2$ is a representation of $E$ under right translation and it is evident that the restriction of this representation to $G_{m,k}$ is just ordinary scalar multiplication. Denote it $q_2$. Let $\rho = q_1\otimes q_2$, and let $U = V_1\otimes V_2$.

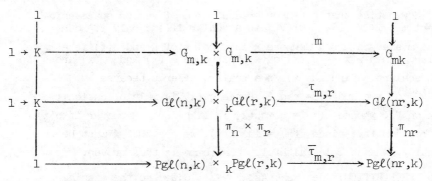

Now notice that a tensor product $\omega\otimes\overline{\omega}$ of vector space endomorphisms, $\omega\in G\ell(V_1)$, $\overline{\omega}\in G\ell(V_2)$ is a scalar if and only if both $\omega$ and $\overline{\omega}$ are scalars. Suppose x is a point of E and $\rho(x)$ is a scalar. Then $\rho(x) = q_1(x)\otimes q_2(x)$ and hence $q_1(x)$ and $q_2(x)$ are both scalars. Since $q_1(x)$ is a scalar, $x\in \mathrm{Ker}\ p$ because $\overline{q}_1{}^{-1}(G^c_{m,k}) = e$ and $q_1 = p\circ\overline{q}_1$. Hence $q_2(x)$ is multiplication by x on $V_2$, whence $\rho(x) = q_1\otimes q_2(x)$ is likewise. Moreover if $\rho(x)$ is the identity, since $\rho(x) = \mathrm{id}\otimes q_2(x)$, we see that x must be the identity, whence $\rho$ is faithful. The Lemma is thus proven.

2.16 PROPOSITION: Let $\varepsilon_\circ: Az_{G/k} \to E_a(G;G_m)$ be the mapping which sends the class of $\rho$ to the class of $E_\rho$. Then

i) $\varepsilon_\circ$ is a morphism of monoids,

ii) The kernel of $\varepsilon_\circ$ is exactly the set of isomorphism classes of trivial Azumaya representations,

iii) $\varepsilon_\circ$ induces a natural isomorphism $\tilde{\varepsilon}_{G/k}: RBr_{G/k} \to E_{a\underset{\sim}{\ell}}(G;G_m)$.

PROOF: That $\varepsilon_\circ$ is a morphism is just Lemma 2.13, while ii) is a consequence of the remarks immediately preceding that lemma. The existence of a natural injection, $\tilde{\varepsilon}_{G/k}$, is an elementary consequence of i) and ii) and the surjectivity of $\tilde{\varepsilon}_{G/k}$ follows immediately from 2.15. Q.E.D.

Observe that when G is semi-simple $E_{a\underset{\sim}{\ell}}(G;G_{m,k}) = \mathrm{Hom}(\mu_G,G_{m,k})$ where $\mu_G$ is the kernel of the natural map $\tilde{G} \to G$ and $\tilde{G}$ is the simply connected cover of G.

3. INDUCED ALGEBRAS: Let G and K be linear algebraic groups over k and let $\underline{\underline{X}}$ be a K-principal G-space over $\underline{\underline{Y}}$ with respect to the quotient mapping q: $\underline{\underline{X}}\to\underline{\underline{Y}}$. Write $\lambda: G\times K\times\underline{\underline{X}}\to\underline{\underline{X}}$, $\lambda_1$: $G\times X\to X$, $\lambda_2$: $K\times\underline{\underline{X}}\to\underline{\underline{X}}$ $\overline{\lambda}: G\times\underline{\underline{Y}}\to\underline{\underline{Y}}$ for the actions involved.

Let $\rho$: $K\to \mathrm{Aut}(M_m(k))$ be an Azumaya representation of K over k. Let $m: M_n\otimes_k M_n\to M_n$ be the multiplication and note that it is a K-map. By 1.10, $q_{\underline{\underline{X}}/K}(M_n\otimes_k M_n)\simeq q_{\underline{\underline{X}}/K}(M_n)\otimes_{\mathcal{O}_{\underline{\underline{X}}}} q_{\underline{\underline{X}}/K}(M_n)$ and hence $q_{\underline{\underline{X}}/K}(m) = \overline{m}$, maps $q_{\underline{\underline{X}}/K}(M_n)\otimes_{\mathcal{O}_{\underline{\underline{X}}}} q_{\underline{\underline{X}}/K}(M_n)$ to $q_{\underline{\underline{X}}/K}(M_n)$. Since m is associative, the functoriality of $q_{\underline{\underline{X}}/K}$ assures that $q_{\underline{\underline{X}}/K}(M_n)$ will be an associative algebra with unit. Moreover, since $q^*(q_{\underline{\underline{X}}/K}(M_n)) \simeq \mathcal{O}_{\underline{\underline{X}}} \otimes_k M$, and the isomorphism is an algebra morphism, the faithful flatness of $\underline{\underline{X}}$ over $\underline{\underline{Y}}$ forces $q_{\underline{\underline{X}}/K}(M_n)$ to be a sheaf of Azumaya algebras on $\underline{\underline{Y}}$.

3.1 DEFINITION: Let $\underline{\underline{X}}$ be a K-principal G-space over $\underline{\underline{Y}}$ with respect to the quotient morphism q and let $\rho$: K-$\mathrm{Aut}_k(M_n)$ be an Azumaya representation of K. Then $q_{\underline{\underline{X}}/K}(\rho)$ is called the Azumaya algebra principally induced by $\rho$ on $\underline{\underline{Y}}$ and it is written $A_{\underline{\underline{X}}/K}(\rho)$.

Now, 1.9 may be applied to $A_{\underline{\underline{X}}/K}(\rho)$. The consequence is:

3.2 PROPOSITION: Let X be a K-principal G-space over $\underline{\underline{Y}}$ with respect to q: $\underline{\underline{X}}\to\underline{\underline{Y}}$. Let $\rho$, $\gamma$ be two Azumaya representations of K over k. Then:

i) If $\rho$ is trivial, then $A_{\underline{\underline{X}}/K}(\rho)$ is G-trivial as a G-sheaf of Azumaya algebras over $\mathcal{O}_{\underline{\underline{X}}}$

ii) $A_{\underline{\underline{X}}/K}(\rho\otimes_k\gamma) = A_{\underline{\underline{X}}/K}(\rho)\otimes_{\mathcal{O}_{\underline{\underline{X}}}} A_{\underline{\underline{X}}/K}(\gamma)$

PROOF: First we prove i). Suppose $\rho$: $K\to Pg\ell(n,k)$ is trivial. Then there is a representation $\tilde{\rho}$: $K\to G\ell(n,k)$ lifting $\rho$. Let K act on the n dimensional vector space V via $\tilde{\rho}$ and on $M_{n,k}$ $\mathrm{End}_k(V)$ via $\rho$. Then the isomorphism is K-equivariant. On the other hand $q_{\underline{\underline{X}}/K}(\mathrm{Hom}_k(V,V)) \simeq \underline{\underline{\mathrm{Hom}}}_{\mathcal{O}_{\underline{\underline{X}}}}(q_{\underline{\underline{X}}/K}(V), q_{\underline{\underline{X}}/K}(V))$ by 1.10. Since $q_{\underline{\underline{X}}/K}$ is a functor we find that $A_{\underline{\underline{X}}/K}(\rho) \simeq q_{\underline{\underline{X}}/K}(M_{n,k}) \simeq \underline{\underline{\mathrm{End}}}_{\mathcal{O}_{\underline{\underline{X}}}}(q_{\underline{\underline{X}}/K}(V))$ and the morphisms preserve G-structures. This proves i). The second assertion is a straightforward consequence of 1.10, i), the only issue in question being whether the isomorphism is an algebra isomorphism. This however follows from the corresponding fact for matrix algebras. Q.E.D.

3.3 PROPOSITION: Let $\underline{\underline{X}}$, $\underline{\underline{Y}}$, G,K be as in 3.2, and let $\xi$ be a class in $\mathrm{RB}(K_{/k})$. Let $\rho$ $K\to Pg\ell(n,k)$ be a representative of $\xi$ and let $\alpha$ be the class of $A_{\underline{\underline{X}}/K}(\rho)$ in $\mathrm{Br}_G(\underline{\underline{Y}})$. Then $\alpha$ is independent of the choice of $\rho$.

PROOF: Let $\rho$, $\rho'$ be two representatives for $\xi$. Then there are trivial Azumaya representations $\gamma$ and $\gamma'$ so that $\rho\otimes\gamma \simeq \rho'\otimes\gamma'$. In particular $A_{\underline{\underline{X}}/K}(\rho)\otimes_{\mathcal{O}_{\underline{\underline{Y}}}} A_{\underline{\underline{X}}/K}(\gamma) \simeq A_{\underline{\underline{X}}/K}(\rho')\otimes_{\mathcal{O}_{\underline{\underline{Y}}}} A_{\underline{\underline{X}}/K}(\gamma')$ as G-sheaves of Azumaya algebras and $A_{\underline{\underline{X}}/K}(\gamma)$ and $A_{\underline{\underline{X}}/K}(\gamma')$ are both G-trivial.

Hence the classes of $A_{\underline{\underline{X}}/K}(\rho)$ and $A_{\underline{\underline{X}}/K}(\rho')$ represent the same class in $Br_G(\underline{\underline{Y}})$. Q.E.D.

3.4 DEFINITION: Let G and K be algebraic groups over k and let $\underline{\underline{X}}$ be a K-principal G-space over $\underline{\underline{Y}}$ with respect to the quotient morphism q: $\underline{\underline{X}}\to\underline{\underline{Y}}$. For, any $\xi \in RBr(K/_k)$ let $i_{\underline{\underline{X}}/K}(\xi)$ be the element of $Br_G(\underline{\underline{Y}}/_K)$ obtained by choosing a representative $\xi_o$ of $\xi$ and sending $\xi$ to the class of $A_{\underline{\underline{X}}/K}(\xi_o)$. Then $i_{\underline{\underline{X}}/K}$ is called the induction morphism.

3.5 PROPOSITION: Let G be a linear algebraic k-group and K a closed k-subgroup. Regard $\underline{\underline{X}}$ = G as a K principal G space with respect to $\pi$: $G\to {}^{G}/K$. Then the induction morphism $i_{G/K}$ :RBr(K/k) $\to$ $Br_G({}^{G}/K)$ is an isomorphism.

PROOF: Consider the full category of Azumaya representations of K over k, which we write $\underline{\underline{A}}$, and the category of G sheaves of Azumaya algebras on ${}^{G}/K$ which we denote $\underline{\underline{B}}$. Then $\pi_{G/K} = I_{G/K}$ is a tensor product preserving functor from $\underline{\underline{A}}$ to $\underline{\underline{B}}$. On the other hand there is the functor $V:\underline{\underline{B}} \to \underline{\underline{A}}$ (cf. §1). Since k is algebraically closed for any $M \in ob(\underline{\underline{B}})$ $V(M)$ is simply a matrix algebra admitting an Azumaya representation and so it is in $\underline{\underline{A}}$. Thus the two categories are isomorphic. That $i_{G/K}$ is an isomorphism follows immediately.

4. SEMI-SIMPLE GROUPS; PRELIMINARIES: The object of this section is to give a proof of 4.1 below. This result has been established quite well by Iverson ([9]). This proof is given mainly for completeness, and because it is relatively compact. It is only a slight variation on Iverson.

4.1 THEOREM: Let G be a semi-simple simply connected linear algebraic group over $\mathbb{C}$. Then the Brauer group of G is $(0)$.

We prove this in steps. First note that since G is affine and regular, by a recent result of Hoobler and Gaber (Hoobler [8] or Gaber [6]) the Brauer group of G is equal to its cohomological Brauer group (i.e. the torsion subgroup of $H^2_{ét}(G,G_m)$). Hence once and for all we drop the distinction between the two objects. Write $Br_n(G)$ for the subgroup of Br(G) consisting of elements of exponent n.

(2) $Br_n(G)$ <u>is functorially isomorphic to</u> $H^2_{ét}(G,\mu_n)$ <u>where</u> $\mu_n$ <u>denotes the n'th roots of unity.</u>
This is nothing more than the Kummer theory in Grothendieck [7]. Consider the short exact sequence,

$$1 \to \mu_n \to G_m \overset{\varphi_n}{\to} G_m \to 1$$

where $\varphi_n$ is raising to the n'th power. The associated exact sequence of étale cohomology groups contains the sequence

$$\to H^1_{ét}(G,G_m) \to H^2_{ét}(G,\mu_n) \to H^2_{ét}(G,G_m) \overset{\tilde{\varphi}_n}{\to} H^2(G,G_m)$$

But $\tilde{\varphi}_n$ is again multiplication by n and $H^1_{ét}(G,G_m) = \text{Pic}(G) = (0)$ because G is simply connected.

Write $H^q_s(G,A)$ for the q'th topological (singular) cohomology group of G with coefficients in A. Then the following is just an application of the Artin approximation theorem and requires no proof.

(3) $Br_n(G)$ <u>is isomorphic to</u> $H^2_s(G,\mathbb{Z})$.
Now let $K \subset G$ be a maximal compact subgroup of G. Then, by the Iwasawa decomposition,

$$G = KAN$$

where N is a maximal unipotent subgroup and A is a positive real torus, i.e., a product of groups $\mathbb{R}^*_+$, the multiplicative group of the positive reals. As AN is contractible, G can be contracted to K whence

(4) $Br_n(G)$ <u>is isomorphic to</u> $H^2_s(K,\mathbb{Z}/_{n\mathbb{Z}})$.
Now recall what is classically known about $H^q_s(K,\mathbb{Z})$. Since G is simply connected, K is topologically simply connected and so $H^1_s(K,\mathbb{Z})=0$.

Cartan proved that $\Pi_2(K)$ and hence $H^2_s(K,\mathbb{Z})$ is a direct sum of copies of $\mathbb{Z}$, one for each simple component of K. Moreover, if $\alpha_n : \mathbb{Z} \to \mathbb{Z}$ is multiplication by n, so is the corresponding cohomology map. [4,3]

Now the exact sequence,

$$0 \to \mathbb{Z} \to \mathbb{Z} \to \mathbb{Z}/_{n\mathbb{Z}} \to 0$$

gives the following at the level of 2 cohomology:

$$H^2_s(K,\mathbb{Z}) \to H^2_s(K,\mathbb{Z}) \to H^3_s(K,\mathbb{Z}) \to H^3_s(K,\mathbb{Z}).$$

Since $H^2_s(K,\mathbb{Z}) = (0)$, $H^2_s(K,\mathbb{Z}/_{n\mathbb{Z}})$ is the n-torsion in $H^3_s(K,\mathbb{Z})$. This latter is free, hence

(5) $Br_n(G)$ is trivial.

Since (5) is true for each n, the theorem is established.

5. HOMOGENEOUS SPACES: Our purpose in this section is to prove the following:

5.1 THEOREM: *Let G be a semi-simple simply connected linear algebraic group over $\mathbb{C}$. Let $K \subset G$ be a closed connected $\mathbb{C}$-subgroup. Then the natural map*

$$j_{G/K/G} : Br_G(G/K) \to Br(G/K)$$

*is injective.*

We begin with several Lemmas.

5.2 LEMMA: *Let G and K be as in 5.1. Then $\mathrm{Pic}(G/K) \cong X(K)$ the group of algebraic characters of K in $G_m$.*

PROOF: Define a map $i$: $X(K) \to \mathrm{Pic}(G/K)$ as follows. Let $\pi : G \to G/K$ be the natural map and let $i(\lambda) = L_\lambda$, where $L_\lambda(U) = \{f \in \mathcal{O}_G(\pi^{-1}(U)) : f(gx) = x^\lambda \cdot f(g)\}$. Then $L_\lambda$ is just the sheaf induced by the one dimensional module associated to $\lambda$. It is clear that $i$ is a morphism.

Conversely, let $D \subset G/K$ be an effective irreducible divisor of degree one. Then $\pi^{-1}(D)$ is a right K-stable divisorial set in G. Since G is simply connected, $\mathbb{C}[G]$ is a U.F.D. and so, if J is the ideal defining $\pi^{-1}(D)$, $J = f_D\, \mathbb{C}[G]$ for some $f_D \in \mathbb{C}[G]$. Further $r_x(f_D) \in J$ and $r_x(f_D)/f_D$ is readily seen to be a unit and hence, by the theorem of Rosenlicht, a constant since G has no characters. Hence $x \to r_x(f_D)/f_D$ is a certain uniquely defined character $\lambda_D$

on K. It is clear that this maps the group of (Weil) divisors to X(K) and two linearly equivalent effective divisors, which after all differ by multiplication by a rational right K-invariant function on G, must give rise to the same character. This gives a map $j$:Pic G/K → X(K).

If D ∈ Div(G/K) is effective with $j(D) = \lambda$, clearly the map $\alpha: L_\lambda(U) \to O_D(U)$ defined by $\alpha(g) = g/f_D$, is an isomorphism. Hence $i \circ j = \mathrm{id}$. The opposite identity is equally clear. Q.E.D.

5.3 LEMMA: <u>If $F$ and $G$ are two vector bundles of determinant one</u> on G/K, <u>and</u> $\underline{\mathrm{End}}_{O_{G/K}}(F) \simeq \underline{\mathrm{End}}_{O_{G/K}}(G)$, <u>then</u> $F \simeq G$.

PROOF: Let $\varphi: \Lambda^n_{O_{G/K}} F \simeq O_{G/K}$ and $\psi: \Lambda^n_{O_{G/K}} G \simeq O_{G/K}$ be the isomorphism associated respectively to $F$ and $G$. By the theorem of Grothendieck, (§1, (3) above) there is an isomorphism $F \simeq G \otimes L$. Since $\Lambda^n F \simeq \Lambda^n G \simeq O_{G/K}$, $L^{\otimes n} \simeq O_{G/K}$, and so $L$ is a torsion element of Pic G/K. But Pic(G/K) ≃ X(K), and so, since K is connected, Pic(G/K) is torsion free, whence $L$ is trivial.

Q.E.D.

We now proceed to the proof. Suppose $\xi \in \mathrm{Br}_G(G/K)$ and $j_{G/K/G}(\xi) = 0$. Let $A$ be a G-sheaf of Azumaya algebras of rank $n^2$ representing $\xi$. Since $j_{G/K/G}(\xi) = 0$, we find that there is a locally free module of rank n, say $F$, admitting an $A$-representation so that $F$ is a sheaf of $A$-irreducible modules. Write $\lambda: G\times G/K \to G/K$ and $p_2$: $G\times G/K \to G/K$ for the action and the projection respectively and let $\Psi: p_2^*A \to \lambda^*A$ be the G-structure on $A$. Then $\Psi$ is an algebra isomorphism.

If $F$ admits a G-structure compatible with $\Psi$ then $A$ is trivial in $\mathrm{Br}_G(G/K)$. We shall demonstrate this to be the case.

First of all by Grothendieck [7] (        ) we may assume that $F$ is a bundle of determinant one. Now consider the two bundles on G×G/K, $\lambda^*F$ and $p_2^*F$ . Then $\lambda^*F$ is an irreducible $\lambda^*A$-module, whence, by transport of structure via $\Psi$, it is an irreducible $p_2^*A$ module. Hence $p_2^*A \simeq \mathrm{End}(p_2^*F)$ on the one hand while it may be made isomorphic via $\Psi$ to $\mathrm{End}(\lambda^*F)$. Since $F$ and both of its pull-backs are bundles of determinant one and since G×G/K is the same as $G\times G/_{e\times K}$, we may apply 5.3 to conclude that there is an isomorphism $\Phi_o: p_2^*F \to \lambda^*F$ so that the map from $\underline{\underline{\mathrm{End}}}(p_2^*F)$ to $\underline{\underline{\mathrm{End}}}(\lambda^*F)$ is the induced map. That is let $\varphi: \to \underline{\underline{\mathrm{End}}}\ F$ be the trivializing map with respect to G/K. Then $\Phi_o$ satisfies:

$$\begin{array}{ccc}
\underline{\underline{\mathrm{End}}}\ p_2^*F & \xrightarrow{\mathrm{Hom}(\Phi_\circ^{-1},\Phi_\circ)} & \underline{\underline{\mathrm{End}}}\ \lambda^*F \\
\Big\downarrow{\scriptstyle p_2^*\varphi} & & \Big\downarrow{\scriptstyle \lambda^*\varphi} \\
p_2^*A & \xrightarrow{\quad\Psi\quad} & \lambda^*A
\end{array} \qquad (4)$$

Hence $p_2^*A \simeq \underline{\underline{\mathrm{End}}}\ (p_2^*F)$ on the one hand while it is isomorphic via $\Psi$ to $\underline{\underline{\mathrm{End}}}\ \lambda^*F$. Since $F$ and hence both $p_2^*F$ and $\lambda^*F$ are determinant one, the conditions of Lemma 5.3 hold and so there is a unique $p_2^*A$ isomorphism $\Phi{:}p_2^*F \to \lambda^*F$.

Now if a map is to be a G-structure it must satisfy §1, (5), and initially $\Phi_\circ$ does not. However if $\Phi_\circ$ is replaced by $u\Phi_\circ$ for u a unit (4) is unaffected. Hence consider the two morphisms of sheaves on $G{\times}G{\times}G/K$:

$$\tilde{\Phi}_1 = (id_G{\times}\lambda)^*(\Phi_\circ)\circ p_{23}^*\Phi_\circ)$$

and

$$\tilde{\Phi}_2 = (\mu{\times}id)^*(\Phi_\circ).$$

When $\alpha{:}\,A \to B$ is a map of sheaves of algebras and $\beta{:}\,F \to G$ is a map of sheaves of modules and $F$ $(G)$ is an $A$-module ($B$ module respectively), then when we say that $\alpha$ and $\beta$ are compatible we mean that

$$\begin{array}{ccc}
A\otimes F & \xrightarrow{\alpha\otimes\beta} & B\otimes G \\
\Big\downarrow & & \Big\downarrow \\
F & \xrightarrow{\quad\beta\quad} & G
\end{array}$$

commutes. If $\beta$ is an isomorphism this is the same as saying that

$$\begin{array}{ccc}
A & \xrightarrow{\quad\alpha\quad} & B \\
\Big\downarrow & & \Big\downarrow \\
\underline{\underline{\mathrm{End}}}\ F & \xrightarrow{\mathrm{Hom}(\beta^{-1},\beta)} & \underline{\underline{\mathrm{End}}}\ G
\end{array}$$

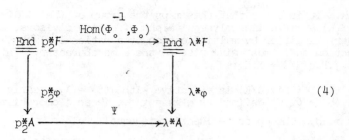

commutes. Now $\Phi_\circ$ is compatible with $\Psi$ in this sense and so $\tilde{\Phi}_1$ is with $(id{\times}\lambda)^*(\Psi)\ p_2^*(\Psi)$ as $\tilde{\Phi}_2$ is with $(\mu{\times}id)^*(\Psi)$. But $\Psi$, being a G-structure, these two expressions are equal and so $\tilde{\Phi}_1$ and $\tilde{\Phi}_2$ are both compatible with $(\mu{\times}id)^*\Psi$. Now $(\mu{\times}id)^*p_2^*F$ and $(\mu{\times}id)^*\lambda^*F$ are irreducible modules over $(\mu{\times}id)^*p_2^*A$ and $(\mu{\times}id)^*\lambda^*A$ respectively and so any two compatible maps between them must differ by a unit. Thus $u\tilde{\Phi}_1 = \tilde{\Phi}_2$. First note that unit on $G{\times}G{\times}G/K$ is actually a unit on $G{\times}G{\times}G$ and by the theorem of Rosenlicht every unit on a group is a character. However the group is semi-simple whence u is a non-zero constant. Thus we may consider $\Phi = u\Phi_o$. Again since u is constant, it passes unaltered through the pull-back symbols $(id_G\times\lambda)^*$ and $(\mu\times id)^*$. Hence $(id_G\times\lambda)^*(\Phi)\circ p^*_{23}(\Phi) = u^2\tilde{\Phi}_1 = u\tilde{\Phi}_2 = (\mu\times id)^*(\Phi)$. That is $\Phi$ satisfies §1, (5) and moreover since it is a simple multiple of $\Phi_o$ it remains compatible with $\Psi$. Thus $F$ admits a G-structure so that $A$ is G isomorphic to End $F$ and so 5.1 is proven.

6. BRAUER-SEVERI VARIETIES: In this section G and K are linear algebraic groups over $\mathbb{C}$. We begin by considering the $Pg\ell(n,\mathbb{C})$ action on $\mathbb{P}(V)$ the space of lines in the natural n-dimensional representation of $S\ell(n,\mathbb{C})$. Then $\mathbb{P}(V)$ is Proj $S_{\mathbb{C}}(V^*)$ where $S_{\mathbb{C}}$ denotes the symmetric algebra over $\mathbb{C}$ and $V^*$ is the contragredient representation. Of course, the $Pg\ell(n,\mathbb{C})$ action cannot be realized as a $Pg\ell(n,\mathbb{C})$ action on $S_{\mathbb{C}}(V^*)$, but merely as an $S\ell(n,\mathbb{C})$ action. However the center of $S\ell(n,\mathbb{C})$, which we write $\mu_n$, acts trivially on the symmetric powers $S^{rn}_{\mathbb{C}}(V^*)$ and so $Pg\ell(n,\mathbb{C})$ acts as a group of graded automorphisms on the algebra $B_{n,\mathbb{C}} = \coprod_{r=0}^{\infty} S^{rn}_{\mathbb{C}}(V^*)$. Moreover Proj $B_{n,\mathbb{C}} = \mathbb{P}(V)$. We write $\mathcal{O}_{\mathbb{P}}$ for the structure sheaf $\mathcal{O}_{\mathbb{P}(V)}$.

Let $F_{n,\mathbb{C}} = \coprod_{r=0}^{\infty} S^{rn+1}_{\mathbb{C}}(V^*)\otimes V$ where $S^{rn+1}_{\mathbb{C}}(V^*)\otimes_{\mathbb{C}} V$ is regarded as being of degree rn. Then $F_{n,\mathbb{C}}$ is a natural $B_{n,\mathbb{C}}$-module. Letting $S\ell(n,\mathbb{C})$ act diagonally on $S^{rn+1}_{\mathbb{C}}(V^*)\otimes_{\mathbb{C}} V$ we observe that $\mu_n$ acts trivially and so $Pg\ell(n,\mathbb{C})$ acts as a group of graded automorphisms on $F_{n,\mathbb{C}}$. The action is compatible with the $B_{n,\mathbb{C}}$ module structure on $F_{n,\mathbb{C}}$ in the sense that $r^1_g(fm) = r^o_g(f)r^1_g(m)$ where $r^o_g$ and $r^1_g$ are the appropriate representing transformations.

Observe that $\text{Proj}(B_{n,\mathbb{C}}) = \mathbb{P}(V)$ and that the module induced by $F_{n,\mathbb{C}}$ is just $\mathcal{O}(1)\otimes_{\mathbb{C}} V$. The module of global sections, $\Gamma(\mathbb{P}(V), \mathcal{O}(1)\otimes V)$, is canonically isomorphic to $V^*\otimes V$. Moreover, the $Pg\ell(n,\mathbb{C})$ actions described above determine the canonical action on $\mathbb{P}(V)$ as well as a $Pg\ell(n,\mathbb{C})$-structure on $\mathcal{O}(1)\otimes V = \tilde{F}_{n,\mathbb{C}}$. We shall always write $\tilde{F}_{n,\mathbb{C}}$ to denote $\mathcal{O}(1)\otimes V$ with this natural $Pg\ell(n,\mathbb{C})$ structure. The $Pg\ell(n,\mathbb{C})$ structure on $\tilde{F}_n$ induces a representation of $Pg\ell(n,\mathbb{C})$ in $\Gamma(\mathbb{P}(V), \tilde{F}_{n,\mathbb{C}}) = V^*\otimes V$. Let $\Psi: V^*\otimes V\to \text{End}(V)$ be the natural isomorphism. Then $\Psi$ is a $Pg\ell(n,\mathbb{C})$-isomorphism from $V^*\otimes V$ under the induced action to $M_{n,\mathbb{C}}$ under the natural (conjugating) representation. Put $\omega_{V_{\mathbb{C}}} = \Psi^{-1}(id_V)$. Then, up to a scalar,

$\omega_{V_{\mathbb{C}}}$ is the unique nowhere vanishing $Pg\ell(n,\mathbb{C})$ invariant section of $\mathcal{O}(1)\otimes_{\mathbb{C}}V$. Furthermore, if $\sigma\in V^*\otimes V$ and $\alpha\in M_{n,\mathbb{C}}$ then $\psi(\mathrm{id}\otimes\alpha(\sigma)) = \alpha\circ\psi(\sigma)$. This last identity admits the following interpretation. The sheaf of matrix algebras $M_n(\mathcal{O}_{\mathbb{P}}) = \mathcal{O}_{\mathbb{P}}\otimes M_{n,\mathbb{C}}$ operates naturally on $\tilde{F}_{n,\mathbb{C}}$ by letting $\mathcal{O}_{\mathbb{P}}$ act as scalars on $\mathcal{O}$ (1) and letting $M_{n,\mathbb{C}}$ operate on V by its natural representation. Further $M_n(\mathcal{O}_{\mathbb{P}}) = \mathcal{O}_{\mathbb{P}} \otimes M_{n,\mathbb{C}}$ admits a diagonal $Pg\ell(n,\mathbb{C})$-structure which we refer to as the natural structure (i.e. $Pg\ell(n,\mathbb{C})$ operates on $M_{n,\mathbb{C}}$ via conjugation.). Then the representation of $M_n(\mathcal{O}_{\mathbb{P}})$ on $\tilde{F}_{n,\mathbb{C}}$ is $Pg\ell(n,\mathbb{C})$-equivariant.

6.1 THEOREM: Let $\rho$: $K\to Pg\ell(n,\mathbb{C})$ be an Azumaya representation of K, and let $\underline{\underline{X}}$ be a G-principal K space over $\underline{\underline{Y}}$ with respect to the quotient morphism q: $\underline{\underline{X}}\to\underline{\underline{Y}}$ (c.f. 1.8). Let V be the natural representation of $S\ell(n,\mathbb{C})$. Then

i) The fibre coproduct $\underline{\underline{X}} \times^K \mathbb{P}(V)$ exists.

ii) Let $\gamma$: $X\times^K\mathbb{P}(V)\to\underline{\underline{Y}}$ be the natural projection. Then there is a canonical vector bundle $F_{\underline{\underline{X}}/K}(\rho)$ on $\underline{\underline{X}}\times^K\mathbb{P}(V)$, a representation of $\gamma^*A_{\underline{\underline{X}}/K}(\rho)$ on $F_{X/K}(\rho)$ and a surjective left $\gamma^*A_{\underline{\underline{X}}_K}(\rho)$-module map $\omega_{\underline{\underline{X}}/K}(\rho)$: $\gamma^*A_{\underline{\underline{X}}/K}(\rho) \to F_{\underline{\underline{X}}/K}(\rho) \to 0$.

PROOF: The natural representation of $Pg\ell(n,\mathbb{C})$ on $S^{nq}_{\mathbb{C}}(V^*)$ may be composed with $\rho$ to give representations, $\rho_{nq}$, of K on $S^{nq}_{\mathbb{C}}(V^*)$. Hence the sheaves $\mathcal{O}_{\underline{\underline{X}}}\otimes_{\mathbb{C}}S^{nq}_{\mathbb{C}}(V^*)$ admit a diagonal K structure and so one may consider the sheaf of graded $\mathcal{O}_{\underline{\underline{Y}}}$-algebras

$\coprod_{q=0}^{\infty} q_{\underline{\underline{X}}/K}(\mathcal{O}_{\underline{\underline{X}}}\otimes S^{nq}_{\mathbb{C}}(V^*)) = \mathcal{B}_{\underline{\underline{X}}/K}(\rho)$. Then, clearly,

$$\mathrm{Proj}(\mathcal{B}_{\underline{\underline{X}}/K}(\rho)) = \underline{\underline{X}} \times^K \mathbb{P}(V).$$

Similarly $\mathcal{O}_{\underline{\underline{X}}} \otimes S^{nq+1}_{\mathbb{C}}(V^*) \otimes V$ admits a K-structure and so

$\coprod_{q=0}^{\infty} q_{\underline{\underline{X}}/K}(\mathcal{O}_{\underline{\underline{X}}} \otimes S^{nq+1}_{\mathbb{C}}(V^*) \otimes V)= F_{X/K}$ is a sheaf of graded modules over $\mathcal{B}_{\underline{\underline{X}}/K}(\rho)$. Let $F_{X/K}(\rho)$ denote the coherent sheaf of modules on $\underline{\underline{\text{Proj}}}\,(\mathcal{B}_{\underline{\underline{X}}/K}(\rho))$ associated to $F_{\underline{\underline{X}}/K}$.

The section $\omega_{V/\mathbb{C}} \in S'(V^*)\otimes V$ is K-invariant and nonvanishing on $\mathbb{P}(V)$. Hence it is a K-invariant section in $\mathcal{O}_{\underline{\underline{X}}}\otimes S'_{\mathbb{C}}(V^*)\otimes V$ whence it gives a nowhere vanishing section in $F_{\underline{\underline{X}}/K}(\rho)$. Then just define $\omega_{\underline{\underline{X}}/K}(\rho)$ locally by

$$\omega_{\underline{\underline{X}}/K}(\rho)(\alpha) = \alpha\cdot\omega_{V,\mathbb{C}}\big|_U$$

for all $\alpha\in\gamma^*(A_{\underline{\underline{X}}/K}(\rho))(U)$. Q.E.D.

By the above it is clear that if a morphism $f:\underline{\underline{Z}}\to\underline{\underline{Y}}$ lifts to a morphism $\tilde{f}:\underline{\underline{Z}}\to\underline{\underline{X}}\times^K\mathbb{P}(V)$ so that $f = \gamma\circ\tilde{f}$, then $f^*A_{\underline{\underline{X}}/K}(\rho)\simeq\text{End}\,\tilde{f}^*F_{\underline{\underline{X}}/K}(\rho)$. Moreover $\tilde{f}^*(\omega_{\underline{\underline{X}}/K}(\rho))$ is a surjective $f^*(A_{\underline{\underline{X}}/K}(\rho))$-module map, whose kernel is hence a global sheaf of maximal left ideals in $f^*(A_{\underline{\underline{X}}/K}(\rho))$. What is more interesting is the existence of a converse.

6.2 THEOREM: <u>Let</u>, p, <u>X</u>, <u>Y</u> <u>and</u> q <u>denote what they did in</u> 6.1. <u>Let</u> $f:\underline{\underline{Z}}\to\underline{\underline{Y}}$ <u>be a morphism of schemes. Suppose that</u> $f^*(A_{\underline{\underline{X}}/K}(\rho))\simeq$ $\underline{\underline{\text{End}}}_{\mathcal{O}_{\underline{\underline{Z}}}}(F)$ <u>and that there is a surjective left</u> $f^*(A_{\underline{\underline{X}}/K}(\rho))$ <u>module map</u> $\omega_F:f^*(A_{\underline{\underline{X}}/K}(\rho))\to F$. <u>Then there is a morphism</u> $\tilde{f}:\underline{\underline{Z}}\to\underline{\underline{X}}\times^K\mathbb{P}(V)$ <u>such that</u> $\gamma\circ\tilde{f}=f$ <u>and a canonical</u> $f^*(A_{\underline{\underline{X}}/K}(\rho))$ <u>isomorphism</u> $\psi:F\to\tilde{f}^*(F_{\underline{\underline{X}}/K}(\rho))$ <u>so that</u> $\psi\circ\omega_F = \tilde{f}^*(\omega_{\underline{\underline{X}}/K}(\rho))$.

PROOF: Let $\underline{\underline{Z}}$ and $f:\underline{\underline{Z}}\to\underline{\underline{Y}}$ be as in the hypotheses. Let $\tilde{\underline{\underline{Z}}}=\underline{\underline{X}}\times_{\underline{\underline{Y}}}\underline{\underline{Z}}$ and write $q_{\underline{\underline{Z}}}:\tilde{\underline{\underline{Z}}}\to\underline{\underline{Z}}$ for the projection. Then the action of K on $\underline{\underline{X}}$ over $\underline{\underline{Y}}$ base extends to an action of K on $\tilde{\underline{\underline{Z}}}$ over $\underline{\underline{Z}}$. Further, $\tilde{\underline{\underline{Z}}}$ is a principal homogeneous space over $\underline{\underline{Z}}$ with respect to K.

Let $\tilde{K}=E_\rho$ be the central extension of K determined by the Azumaya representation $\rho$ as in §2, (12). Then $\tilde{K}$ operates on $\underline{\underline{X}}$ and on $\underline{\underline{Z}}$ and, while these varieties are not principal homogeneous with respect to $\tilde{K}$, we may nonetheless write $\tilde{\underline{\underline{Z}}}/_{\tilde{K}}\simeq\underline{\underline{Z}}$ and $\underline{\underline{X}}/_{\tilde{K}}=\underline{\underline{Y}}$ and $q_{\underline{\underline{Z}}}$ and q are the quotient morphisms.

Consider the cartesian square:

$$\begin{array}{ccc} \tilde{Z} & \xrightarrow{f_1} & X \\ \downarrow & & \downarrow \\ Z & \xrightarrow{f} & Y \end{array} \tag{3}$$

Then $f_1$ is $\tilde{K}$ equivariant, and $q^*A_{\underline{\underline{X}}/K}(\rho) = M_n(\mathcal{O}_{\underline{\underline{X}}})$ has a $\tilde{K}$ equivariant representation on $\mathcal{O}_{\underline{\underline{X}}} \otimes V$. Now $q^*_{\underline{\underline{Z}}}(f^*A_{\underline{\underline{X}}/K}(\rho)) = f_1^*\, q^*A_{\underline{\underline{X}}/K}(\rho) = M_n(\mathcal{O}_{\underline{\underline{\tilde{Z}}}})$. Moreover $q^*_{\underline{\underline{Z}}}(f^*A_{\underline{\underline{X}}/K}(\rho))$ admits a $\tilde{K}$ equivariant representation on $q^*_{\underline{\underline{Z}}}F$, where $q^*_{\underline{\underline{Z}}}F$ has its natural K (and hence $\tilde{K}$) structure as a pull back from the quotient. Moreover $q^*_{\underline{\underline{Z}}}(\omega_F)$ is surjective and $\tilde{K}$ equivariant. Since both $q^*_{\underline{\underline{Z}}}F$ and $\mathcal{O}_Z \otimes V$ admit representations of $q^*_{\underline{\underline{Z}}}f^*A_{\underline{\underline{X}}/K}(\rho)$, by the result of Grothendieck cited above, there is a line bundle $L$ on $\underline{\underline{\tilde{Z}}}$ so that $q^*_{\underline{\underline{Z}}}F \simeq L \otimes V$. Consider the surjective morphism of coherent sheaves of $\mathcal{O}_{\underline{\underline{Z}}}$-modules

$$\underline{\underline{\mathrm{Hom}}}_{M_n(\mathcal{O}_{\underline{\underline{Z}}})}(\mathcal{O}_{\underline{\underline{\tilde{Z}}}} \otimes V,\ q^*_{\underline{\underline{Z}}} f^*A_{\underline{\underline{X}}/K}(\rho)) \xrightarrow{\ \Phi\ } \underline{\underline{\mathrm{Hom}}}_{M_n(\mathcal{O}_{\underline{\underline{\tilde{Z}}}})}(\mathcal{O}_{\underline{\underline{\tilde{Z}}}} \otimes_k V,\ q^*_{\underline{\underline{Z}}}F), \tag{4}$$

$$\Phi = \mathrm{Hom}(\mathrm{id.},\ q^*_{\underline{\underline{Z}}}\ \omega_F)$$

All the sheaves involved admit $\tilde{K}$ structures and so the Hom's admit natural $\tilde{K}$ structures. Since $q^*_{\underline{\underline{Z}}}(\omega_F)$ is $\tilde{K}$ equivariant $\Phi$ is likewise. Since $q_{\underline{\underline{Z}}}f^*A_{\underline{\underline{X}}/K}(\rho) \simeq M_n(\mathcal{O}_{\underline{\underline{\tilde{Z}}}})$ the first term of (4) is $\tilde{K}$ isomorphic to $\mathcal{O}_{\underline{\underline{\tilde{Z}}}} \otimes V$. Since $q^*_{\underline{\underline{Z}}}F \simeq L \otimes V$, the second term is isomorphic to $L$. Thus $L$ admits a $\tilde{K}$ structure, if not a K structure, and $\Phi$ may be reinterpreted as a surjective $\tilde{K}$ equivariant morphism

$$\mathcal{O}_{\underline{\underline{Z}}} \otimes V \xrightarrow{\ \Phi\ } L \longrightarrow 0\ . \tag{5}$$

Thus $\Phi$ determines a point of $\underline{\underline{X}} \times \mathbb{P}(V)$ in $\underline{\underline{Z}}$. Since $\tilde{K}$ acts on $\mathcal{O}_{\underline{\underline{\tilde{Z}}}} \otimes V$ diagonally and $\Phi$ is equivariant, the point of $\underline{\underline{X}} \times \mathbb{P}(V)$ in $\underline{\underline{Z}}$ is a $\tilde{K}$-equivariant morphism of schemes

$$\varphi:\ \underline{\underline{Z}} \longrightarrow \underline{\underline{X}} \times \mathbb{P}(V) \tag{6}$$

where $\tilde{K}$ acts on $X \times \mathbb{P}(V)$ diagonally. Moreover $\varphi$ is determined by the condition $\varphi^*(\mathcal{O}(1))=L$, and so $\varphi^*(\mathcal{O}(1)\otimes V)\simeq L\otimes V=q^*_Z F$. Now $\varphi$ determines a mapping from $\tilde{Z}/\tilde{K}$ to $X \times \mathbb{P}(V)/\tilde{K}$, i.e. a map $\tilde{f}$: $\underline{\underline{Z}} \to \underline{\underline{X}}\times^K\mathbb{P}(V)$. That $\gamma\circ\tilde{f}=f$ follows from the definition of $\tilde{f}$. Since $\varphi^*(\mathcal{O}(1)\otimes V)$ is $M_n(\mathcal{O}_{\underline{\underline{Z}}})$ isomorphic to $q^*_{\underline{\underline{Z}}}F$ and is equivariantly so, we may apply the functor $(q_{\underline{\underline{Z}}})_{\underline{\underline{\tilde{Z}/K}}}$ to all of the objects involved to obtain the last statement of 6.2. Q.E.D.

6.7 DEFINITION: Let $\rho$ $\underline{\underline{X}}$, $\underline{\underline{Y}}$, q denote what they did in 6.1. Let $\mathbb{B}_{\underline{\underline{X}}/K}(\rho) = \underline{\underline{X}}\times^K\mathbb{P}(V)$. Then $\mathbb{B}_{\underline{\underline{X}}/K}(\rho)$ is called the Brauer-Severi induced by $\rho$ over $\underline{\underline{Y}}$. The natural projection is denoted $\beta\colon \mathbb{B}_{\underline{\underline{X}}/K}(\rho)\to\underline{\underline{Y}}$. The sheaf of modules $F_{\underline{\underline{X}}/K}(\rho)$ is called the generic representation of $A_{\underline{\underline{X}}/K}(\rho)$, while

$$\omega_\rho = \omega_{F_{\underline{\underline{X}}/K}(\rho)}\colon \beta^*(A_{\underline{\underline{X}}/K}(\rho)) \to F_{\underline{\underline{X}}}(\rho)$$

is called the canonical surjection.

Note that while $F_{\underline{\underline{X}}/K}(\rho)$ is called the generic representation of $A_{\underline{\underline{X}}/K}(\rho)$ it is really a representation only of $\beta^*A_{\underline{\underline{X}}/K}(\rho))$.

Taken together theorems 6.1 and 6.2 demonstrate that the quotient field of $\mathbb{B}_{\underline{\underline{X}}/K}(\rho)$ is a "generic splitting field" in a somewhat stronger, and slightly different sense than classically indicated. It is stronger in that it is scheme-theoretically a generic splitting variety; different in that it is only generic for splittings which admit a global sheaf of ideals. Even in the affine case one may have splittings which are not globally split by a map into a Brauer-Severi. Namely, consider an algebra $A$ on $\underline{\underline{X}}$ and a map $q\colon\underline{\underline{Z}} \to \underline{\underline{X}}$ such that $q^*A \cong \underline{\underline{\mathrm{End}}}_{\mathcal{O}_{\underline{\underline{Z}}}} F$ where $F$ is indecomposable. Then for $L$ a line bundle, $F\otimes_{\mathcal{O}_{\underline{\underline{Z}}}} L$ never admits a nowhere vanishing section and so the splitting cannot be induced by a mapping of the type occurring in 6.2.

7. GENERIC ALGEBRAS: In this section we confine our attention to a specific homogeneous space. We shall have reason, occasionally, to think of $M_n(\mathbb{C})$ the n×n matrices over $\mathbb{C}$, as just a vector space. In this case we will write it just as M. In any case it is one and the same vector space as $M_n(\mathbb{C})$. Notice that $Pg\ell(n,\mathbb{C})= \mathrm{Aut}(M_n(\mathbb{C}))$ is a closed subgroup of $G\ell(M)$. Moreover the action of $Pg\ell(n,\mathbb{C})$ on $M_n(\mathbb{C})$ by automorphisms is an Azumaya representation

which we denote $\alpha_n$. We shall call it the standard Azumaya representation of degree n.

7.1 DEFINITION: Let $G=G\ell_{\mathbb{C}}(M)=G\ell(n^2,\mathbb{C})$ and let $K=Pg\ell(n,\mathbb{C})\subset G$. Let $Az(n,\mathbb{C})={}^G/K$. Then $Az(n,\mathbb{C})$ will be called the space of polarized Azumaya algebras of degree n. The induced algebra $A_{G/K}(\alpha_n)$ will be written $A_{n,\mathbb{C}}$ and will be called the generic Azumaya algebra of degree n over $\mathbb{C}$.

It would be opportune at this point to recall certain facts concerning homogeneous bundles. Let G be a linear algebraic group over $\mathbb{C}$ and let $K\subset G$ be a $\mathbb{C}$ closed subgroup. Then, if V is a representation of K the induced representation $I_{G/K}(V)$ admits the following interpretation.

Viewing $\mathcal{O}_G(U)\otimes V$ as the set of algebraic functions from U into V, K operates on it by the equation $(r_x f)(g)=xf(gx)$, for $x\in K(\mathbb{C})$. Then $I_{G/K}(V)$ is just the set of K-invariant functions.

Now assume that the representation of K on V is the restriction of a representation of G on V. Then consider the functions on G/K defined by

$$f_v(g) = g^{-1}v \qquad (v\in V) \tag{2}$$

The functions $f_v$ are K invariant and hence sections in $I_{G/K}(V)$ and if we let $\varepsilon_V(1\otimes v) = f_v$ it defines an isomorphism

$$\varepsilon_V : \mathcal{O}_{G/K} \otimes V \to I_{G/K}(V) . \tag{3}$$

By means of $\varepsilon_V$ we see that when the representation of K in V extends to a representation of G in V, then $I_{G/K}(V)$ is free. When $V=M$, $G=G\ell(n^2,\mathbb{C})$ $K=Pg\ell(n,\mathbb{C})$, the representation $\alpha_n$ is just the restriction of the standard representation of G. Hence $\varepsilon_M$ is an isomorphism from $\mathcal{O}_{G/K}\otimes M$ to $A_{n,\mathbb{C}}$.

7.4 DEFINITION: Let X be a $\mathbb{C}$ scheme and let $\underline{\underline{A}}_X$ be a sheaf of Azumaya algebras over $\mathcal{O}_X$. Then $\underline{\underline{A}}_X$ is said to admit a polarization of degree n if $\underline{\underline{A}}_X$ is free of rank $n^2$ as a sheaf of $\mathcal{O}_X$-modules. An isomorphism $\varphi:\mathcal{O}_X\otimes M \to \underline{\underline{A}}_X$ is called a polarization of $\underline{\underline{A}}_X$ of degree n and the pair $(\underline{\underline{A}}_X,\varphi)$ is called a polarized sheaf of Azumaya algebras of degree n. The isomorphism $\varepsilon_M$ is called the canonical polarization on $A_{n,\mathbb{C}}$.

The notion of a polarization is associated with a natural notion of equivalence.

7.5 DEFINITION: *Let* $(A_{\underline{X}},\varphi)$ *and* $(A'_{\underline{X}},\psi)$ *be two polarized sheaves of Azumaya algebras of degree* n *over* $\mathcal{O}_{\underline{X}}$. *Then* $(A_{\underline{X}},\varphi)$ *and* $(A'_{\underline{X}},\psi)$ *are said to be equivalent if there is an isomorphism of sheaves of Azumaya algebras* $\gamma: A_{\underline{X}} \to A'_{\underline{X}}$ *so that the diagram:*

$$\begin{array}{ccc} & \overset{\varphi}{\nearrow} & A_{\underline{X}} \\ \mathcal{O}_{\underline{X}}\otimes M & & \downarrow \gamma \\ & \underset{\psi}{\searrow} & A'_{\underline{X}} \end{array} \qquad (6)$$

*commutes. The set of equivalence classes of polarized Azumaya algebras of degree* n *over* $\mathcal{O}_{\underline{X}}$ *will be denoted* $PA_{n,\mathbb{C}}(\underline{X})$.

Note that the assignment $\underline{X} \to PA_{n,\mathbb{C}}(\underline{X})$ is a contravariant functor in $\underline{X}$. Namely, if $(A,\varphi)$ is a polarized sheaf of Azumaya algebras of degree n over $\mathcal{O}_{\underline{X}}$, and if $\xi:\underline{Y} \to \underline{X}$ is a $\mathbb{C}$-morphism, then $(\xi^*A,\xi^*(\varphi))$ is a polarized sheaf of Azumaya algebras on $\underline{Y}$. It is equally clear that $\xi^*$ preserves equivalence of polarized Azumaya algebras.

In this way, inverse image gives a natural transformation of functors from points of $Az(n,\mathbb{C})$ to $PA_{n,\mathbb{C}}$. Namely define

by
$$\begin{aligned} \nu_X &: \mathrm{Hom}(X,Az(n,\mathbb{C})) \to PA_{n,\mathbb{C}}(\underline{X}) \\ \nu_X(\xi) &= (\xi^*(A_{n,\mathbb{C}}),\xi^*(\varepsilon_M)) . \end{aligned} \qquad (7)$$

There is a further feature to this correspondence. Namely, since $Az(n,\mathbb{C})$ is a homogeneous space over $G\ell(n^2,\mathbb{C})$, for any $g\in G\ell(n^2,\mathcal{O}\underline{X}) = \mathrm{Hom}_{\mathbb{C}}(\underline{X},G\ell(n^2\mathbb{C}))$, we may define $g\xi$ by $(g\xi)(x)=g(x)\xi(x)$. Further, g is naturally an $\mathcal{O}_{\underline{X}}$-linear invertible transformation from $\mathcal{O}_X\otimes M$ to itself. Hence if $(A,\varphi)$ is a polarized sheaf of Azumaya algebras of degree n, then $\varphi\circ g^{-1}$ is another polarization on $(A,\varphi)$. Thus, defining $g*(A,\varphi)$ by $g*(A,\varphi) = (A,\varphi g^{-1})$, $G\ell(n^2,\mathcal{O}_{\underline{X}})$ acts on $PA_{n,\mathbb{C}}(\underline{X})$ on the left.

7.8 THEOREM: *The natural transformation of functors* $\nu$ *from points of* $Az(n,\mathbb{C})$ *to values of* $PA_{n,\mathbb{C}}$ *is a* $G\ell(n^2)$*-equivariant isomorphism of functors. That is, for* $\underline{X}$ *a* $\mathbb{C}$*-scheme,* $g\in G\ell(n^2,\mathcal{O}_{X)}$ *and* $\xi:\underline{X} \to Az(n,\mathbb{C})$, *then* $\nu_{\underline{X}}(g\xi) = g*\nu_{\underline{X}}(\xi)$, *and* $\nu_{\underline{X}}$ *is bijective for all* $\underline{X}$.

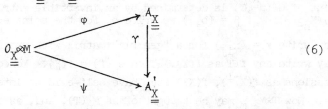

PROOF: Before proceeding we recall some elementary computational devices concerning the general linear group. Let $m_1 \dots m_{n^2}$ be a basis for M over $\mathbb{C}$. Then $G(n^2,\mathbb{C}) = \operatorname{Spec} T$ where $T = \mathbb{C}[x_{11} \dots x_{n^2 n^2}, 1/\delta]$ where $\delta = \det(x_{ij})$. If B is any $\mathbb{C}$ algebra a point of $G\ell(n^2,\mathbb{C})$ is determined by an invertible matrix $(b_{ij}) \in M_{n^2}(B)$. If $\beta = (b_{ij})$ we write $e_\beta$ for the point $e_\beta(x_{ij}) = b_{ij}$. We may write $x = (x_{ij})$ for a "generic" matrix with entries $x_{ij}$ and we may write any $f \in T$ as $f(X)$. Then $e_\beta(f) = f(\beta)$. Moreover such expressions as $f(X^{-1})$, $f(X\beta)$ etc. have self-evident interpretations. Now $T \otimes M_{n,\mathbb{C}}$ may be thought of as $M_n(T)$, and, as in (2) above we may write any $u \in M_n(T)$ as $u(X)$ thinking of it as an $M_n(\mathbb{C})$ valued function on $G\ell(n^2,\mathbb{C})$. Then the diagonal action of $Pg\ell(n,\mathbb{C})$ is given by $(\beta \cdot u)(X) = 1 \otimes \beta(u(X\beta))$. Then $A_{n,\mathbb{C}}$ is the sheaf induced by $A_{n,\mathbb{C}} = \{u : u \in M_n(T), \beta \cdot u = u \forall \beta \in Pg\ell(n,\mathbb{C})\}$. Thus $A_{n,\mathbb{C}} \subset M_n(T)$. Moreover if $S = \{f \in T : f(X\beta) = f(X) \forall \beta \in Pg\ell(n,\mathbb{C})\}$ Spec $S = G\ell(n^2,\mathbb{C})/Pg\ell(n,\mathbb{C})$. Moreover the elements $f_{m_r} = \varepsilon_M(m_r)$ (see (2)) are a basis for $A_{n,\mathbb{C}}$ over S as well as a basis for $M_n(T)$ over T.

We now proceed with the proof. First we show that $\nu_{\underline{\underline{X}}}$ is bijective. Suppose that $(A,\varphi)$ is a polarized sheaf of Azumaya algebras of degree n or $\underline{\underline{X}}$. We shall construct a morphism $\xi : \underline{\underline{X}} \to Az(n,\mathbb{C})$ simultaneously showing that $\xi^*(A_{n,\mathbb{C}}) \cong A$ and that $\xi^*(\varepsilon_M) = \varphi$.

Choose an étale cover $\{h_i : U_i \to \underline{\underline{X}}\}_i$ so that $U_i = \operatorname{Spec}(R_i)$ and $h_i^*(A) \cong M_n(\mathcal{O}_{U_i})$ as an $\mathcal{O}_{U_i}$ algebra. Let $h_i^*(A)$ be the sheaf associated to the algebra $\Omega_i$ and let $\omega_i : \Omega_i \to M_n(R_i)$ be the family of splittings. Write $p^1_{ij}$ and $p^2_{ij}$ for projection of $U_i \times_{\underline{\underline{X}}} U_j$ on the first and second factors respectively. Then $p^{1*}_{ij}\,\Omega_i = p^{2*}_{ij}\,\Omega_j$ and so the composition $\omega_{ij} = p^{1*}_{ij}(\omega_i) \circ p^{2*}_{ij}(\omega_j^{-1})$ is an algebra automorphism of $M_n(\mathcal{O}_{U_i \times_{\underline{\underline{X}}} U_j})$. The obstruction to writing $\omega_{ij}$ as an element of $Pg\ell(n,\mathcal{O}_{U_{ij}})$ is a certain line bundle $L_{ij} \in \operatorname{Pic}(U_{ij})$. Choose a covering $U_{ij} = \bigcup_r U_{ijr}$ so that:

i) $L_{ij}|U_{ijr}$ is trivial

ii) $U_{ijr} = \operatorname{Spec} R_{ijr}$.

Then $\omega_{ijr} = \omega_{ij}|U_{ijr} \in Pg\ell(R_{ijr})$.

Let $\pi : G\ell(n^2,\mathbb{C}) \to Az(n,\mathbb{C})$ be the natural projection. Consider the map $\omega_i \circ h_i^*(\varphi) : M_n(\mathcal{O}_{U_i}) \to M_n(\mathcal{O}_{U_i})$. Let $H_i$ be the matrix of $\omega_i \circ h_i^*(\varphi)$ and put $\tilde{\psi}_i^o = e_{H_i}-1$. Then $\tilde{\psi}_i^o : T \to R_i$ is the ring morphism associated to a map $\tilde{\psi}_i : U_i \to G\ell(n^2,\mathbb{C})$. Put $\psi_i = \pi\circ\tilde{\psi}_i$. Then $\psi_i^o$, the ring morphism is equal to $\tilde{\psi}_i^o|S$.

Write $\beta_{ijr}$ for the matrix of $\omega_{ij}|U_{ijr}$. Since $(\omega_i \circ h_i^*(\varphi)|U_{ijr})^{-1} \circ \omega_{ij}|U_{ijr} = (\omega_j \circ h_j^*(\varphi)|U_{ijr})^{-1}$, we see that $H_i^{-1}\cdot\beta_{ijr} = H_j^{-1}$. Since $\beta_{ijr} \in Pg\ell(R_{ijr})$, this means that $\pi\circ\tilde{\psi}_i\circ p_{ij}^1|U_{ijr} = \pi\tilde{\psi}_j\circ p_{ij}|U_{ijr}$ for all $i,j,r$, whence $\psi_i\circ p_{ij}^1 = \psi_j\circ p_{ij}^2$ for all $i,j$. Consequently the collection $\{\psi_i\}$ descends to a morphism $\psi : \underline{\underline{X}} \to Az(n,\mathbb{C})$.

We must now verify that $(\psi^*(A_{n,\mathbb{C}}),\psi^*(\varepsilon_M))$ is equivalent to $(A,\varphi)$. Consider the diagram:

$$\begin{array}{ccccc}
R_i\otimes M & \xrightarrow{h_i^*(\varphi)} & \Omega_i & \xrightarrow{\omega_i} & M_n(R_i) \\
\uparrow {\scriptstyle e_{H_i}-1\otimes id} & & & & \uparrow {\scriptstyle M_n(e_{H_i}-1)} \\
T\otimes M & & \xrightarrow{\varepsilon_M} & & M_n(T)
\end{array}$$

We wish to establish commutativity. It suffices to do so only for $1\otimes M \subset T\otimes M$. But $e_{H_i}-1\otimes id(1\otimes m) = 1\otimes m$. Moreover, by definition $\varphi\circ\omega_i(1\otimes m) = H_i m$. On the other hand $\varepsilon_M(1\otimes m) = f_m(X) = X^{-1}m$ and $M_n(e_{H_i}-1)(f_m(X)) = h_i^*\varphi(1\otimes m)$. Now put $\overline{U}_i = h_i(U_i) \subset X$ and suppose that $\overline{U}_i = \operatorname{Spec} \overline{R}_i$, $\overline{R}_i \subset R_i$. On the one hand we have commutative diagrams:

$$\begin{array}{ccc}
R_i & \xleftarrow{\tilde{\psi}_i^o} & T \\
\uparrow & & \uparrow \\
R_i & \xleftarrow{\psi_i^o} & S
\end{array}
\qquad (\psi_i^o = (\psi|_{\overline{U}_i})^o \quad )$$

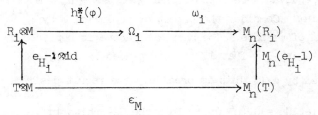

and on the other we have

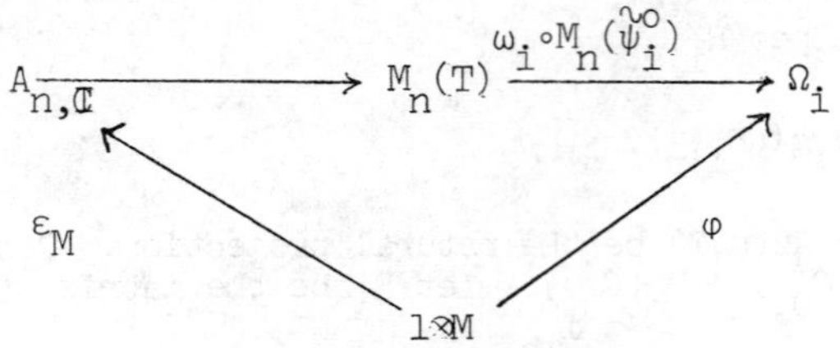


Let $A_i$ be the algebra inducing $A|_{\overline{U}_i}$ over $\overline{R}_i$. Since $A_{n,\mathbb{C}}$ is the S-module generated by $\varepsilon_M(1\otimes M)$, and since $\tilde{\psi}_i^o(S) \subset R_i$, it follows that $\omega_i^{-1} \circ M_n(\tilde{\psi}_i^o)(A_{n,\mathbb{C}})$ is contained in the $R_i$-submodule of $A_i$ generated by $\varphi(1\otimes M)$. That is $\omega_i^{-1} \circ M_n(\tilde{\psi}_i^o)|A_{n,\mathbb{C}}$ is the same as the map from $S\otimes M$ to $\overline{R}_i\otimes M$ given by $\psi_i^o \otimes id_M$. What this shows is the following. Define $\omega : \psi^*(A_{n,\mathbb{C}}) \to A$ by $\omega = \psi^*(\varepsilon_M) \circ \varphi^{-1}$. Then the computations above show that $h_i^*(\omega) = \omega_i^{-1} \circ M_n(\tilde{\psi}^o)|A_{n,\mathbb{C}}$, and hence for each i $h_i^*(\omega)$ is an isomorphism of algebras. It follows immediately that $\omega$ is an isomorphism of algebras inducing an equivalence between $(A,\varphi)$ and $\psi^*(A_{n,\mathbb{C}}), \psi^*(\varepsilon_M))$. This establishes the bijectivity of $\nu_X$.

To prove its equivariance suppose that $g : \mathcal{O}_{\underline{\underline{X}}} \otimes M \to \mathcal{O}_{\underline{\underline{X}}} \otimes M$ is a module isomorphism. Consider the effect of replacing $\overline{\overline{\varphi}}$ by $\varphi g^{-1}$ on the calculation of $\tilde{\psi}^o$ above. The matrix $H_i$ above is replaced by $H_i \cdot g^{-1}$ and $e_{H_i^{-1}}$ by $e_g \cdot e_{H_i^{-1}}$. That is $\tilde{\psi}_i$ is replaced by $g\cdot\tilde{\psi}_i$, and hence $\psi$ is replaced by $g\cdot\psi$. This is precisely the result desired.

Q.E.D.

7.9 COROLLARY. *Let* $G_{\underline{\underline{X}}} = G\ell(n^2, \mathcal{O}_{\underline{\underline{X}}})$. *Then the* $G_{\underline{\underline{X}}}$*-orbits in* $Hom_{\mathbb{C}}(\underline{\underline{X}}, Az(n,\mathbb{C}))$ *correspond bijectively to the Azumaya algebra isomorphism classes of Azumaya algebras which are free sheaves of* $\mathcal{O}_{\underline{\underline{X}}}$ *modules of rank* $n^2$.

PROOF: All that need be observed is that for a given $A$ on $\underline{X}$ any two polarizations differ by an element of $G_{\underline{\underline{X}}}$. Q.E.D.

We now may combine the results of section 6 with 7.8 and 7.9.

7.10 DEFINITION: *Let* $G = G\ell(n^2,\mathbb{C})$, $K = Pg\ell(n,\mathbb{C})$. *We write* $\mathbb{B}_{n,\mathbb{C}}$ *for* $\mathbb{B}_{G/K}(\alpha_n)$. *Then* $\mathbb{B}_{n,\mathbb{C}}$ *is called the generic splitting variety of degree* n *over* $\mathbb{C}$. *We write* $F_{n,\mathbb{C}}$ *for* $F_{G/K}(\alpha_n)$. *It is* *called the generic representation of degree n over* $\mathbb{C}$. *Similarly we write* $\beta_n : \mathbb{B}_{n,\mathbb{C}} \to G/K$ *for the projection and* $\omega_{n,\mathbb{C}}: {}^*_n(A_{n,\mathbb{C}}) \to F_{n,\mathbb{C}}$ *for the canonical surjection.*

There is a sense in which $\mathbb{B}_{n,\mathbb{C}}$ is slightly better behaved than the most general equivariant Brauer-Severi varieties. Let $K = Pg\ell(n,\mathbb{C})$. Then K operates transitively on $\mathbb{P}(V)$ (the space of lines in V, V being the standard representation of $\tilde{K} = G\ell(n,\mathbb{C})$), and so $\mathbb{P}(V) = K/P$ for a certain maximal parabolic subgroup P. Consider the map $\eta: G\times K/P \to G/P$ defined by $\eta(g,xP) = gxP$. Then if $\eta(g,xP) = \eta(h,yP)$ one sees immediately that $h^{-1}g = n\in K$ and $g = hn$, $n^{-1}yP = xP$. That is $G\times^K\mathbb{P}(V) \simeq G/P$. Furthermore the group of characters of P is free on one generator, the generator being n times the (n-1)'th fundamental dominant weight. Hence one$^{n-1}$, sees that $\mathbb{P}(V) = K/P = \mathrm{Proj} \coprod_{r=0}^{\infty} \Gamma(L_{rn\omega_1}))$ where $L_\xi$ denotes the line bundle generated by the character . Notice that $\Gamma(L_{rn\omega_1}) \simeq S^{rn}(V^*)$. Moreover P does not admit a representation on $\mathbb{C}_{-\omega_{n-1}}\otimes V^*$. Let $F = \mathbb{C}_{-\omega_{n-1}}\otimes V^*$. (Here $\mathbb{C}_\xi$ denotes the one dimensional representation of P where P operates via $\xi$.) Then $\Gamma(K/P, I_{K/P}(\mathbb{C}_{-\omega_{n-1}}\otimes V^*)\otimes L_{rn_1}) = V\otimes S^{rn-1}(V^*)$, and so $I_{G/P}(F)$ is quite easily seen to be $F_{n,\mathbb{C}}$.

7.11 PROPOSITION: *Let* $G = G\ell(n^2,\mathbb{C})$, $K = Pg\ell(n,\mathbb{C})$. *Let* P *be the maximal parabolic subgroup of* K *corresponding to the last fundamental dominant weight,* $\omega_{n-1}$ *(the highest weight of the dual of the standard representation.) Then*

$$\mathbb{B}_{n,\mathbb{C}} = G/P$$

*and*

$$F_{n,\mathbb{C}} = I_{G/P}(\mathbb{C}_{-\omega_{n-1}}\otimes V)$$

*while* $\beta_n$ *is just the projection from* G/P *to* G/K *associated to the inclusion* $P\subset K$.

(12) REMARK: Notice that the representation $\mathbb{C}_{-\omega_{n-1}}\otimes V$ makes no particular sense as a representation of P. It is however a perfectly legitimate representation of $\tilde{P}$ the inverse image of P in $G\ell(n,\mathbb{C})$ and as such the scalar matrices act trivially, whence it is after all a representation of P, but P acts on neither factor.

The particular sense in which $\mathbb{B}_{n,\mathbb{C}}$ is the generic splitting variety of degree n over $\mathbb{C}$ is nothing more than a combination of 6.2 and 7.8. Nonetheless it is worth stating it in one place.

7.13 THEOREM: Let $\underline{X}$ be a $\mathbb{C}$ variety and let $A$ be an Azumaya algebra of degree n over $\mathcal{O}_{\underline{X}}$ which is free as an $\mathcal{O}_{\underline{X}}$-module. Let $\varphi:\mathcal{O}_{\underline{X}} \otimes M \to A$ be a polarization and let $\Phi:\underline{X} \to Az(n,\mathbb{C})$ be the corresponding map (i.e. $\nu_{\underline{X}}^{-1}(A,\varphi)$). Let $\mathbb{B}(A) = \mathbb{B}_{n,\mathbb{C}} \times_{Az(n,\mathbb{C})} \underline{X}$ and let $\beta = \beta_n \times id.$ Then

i) $\mathbb{B}(A)$ is the Brauer-Severi variety over $\underline{X}$ associated to the Azumaya algebra $A$.

ii) Let $\eta:\underline{Z} \to \underline{X}$ be a $\mathbb{C}$-morphism of schemes. Then $\eta^*(A) \simeq \underline{\underline{End}}_{\mathcal{O}_Z} G$ for some locally free sheaf $G$ if and only if there is a map $\eta:Z \to \mathbb{B}(A)$ such that $\beta\circ\tilde{\eta} = \eta$. In particular $\beta^*(A) \simeq End(p_1^*(F_{n,\mathbb{C}}))$ where $p_1 : \mathbb{B}(A) \to \mathbb{B}_{n,\mathbb{C}}$ is projection on the first factor.

PROOF: To see i) recall that the Brauer-Severi variety of is the variety of projective fibres over $\underline{X}$ which is pieced together by means of the same non-abelian one cocycle in $Pg\ell(n,\mathbb{C})$ as $A$. A verification of this fact can then be deduced from the computations in the proof of 7.8.

To prove ii) just note that $\eta^*(A) = \Phi\circ\eta^*(A_{n,\mathbb{C}})$. Then apply 6.2 to $\Phi\circ\eta$. If $\eta^*(A)$ splits there is then a mapping $\eta_1:\underline{Z} \to \mathbb{B}_{n,\mathbb{C}}$ such that $\beta_n\circ\eta_1 = (\Phi\circ\eta)$. It follows that there is a map $\tilde{\eta}$ into the Cartesian product, $\mathbb{B}(A)$. The converse is clear. The last statement of the proof is an immediate consequence of 6.2. Q.E.D.

Now we turn to a result which has been implicitly assumed throughout this section. The only reason it must be proven is that $G\ell(n^2,\mathbb{C})$ being reductive but not semi-simple, fails to meet the hypotheses of 5.1.

7.14 LEMMA: The class of $A_{n,\mathbb{C}}$ is of order n in the Brauer group of $Az(n,\mathbb{C})$. Moreover the homogeneous Brauer group of $Az(n,\mathbb{C})$ is a subgroup of the Brauer group of $Az(n,\mathbb{C})$ and is precisely the subgroup generated by the class of $A_{n,\mathbb{C}}$.

PROOF: $G\ell(n^2,\mathbb{C})$ is the semi-direct product $G_{m,\mathbb{C}} \times S\ell(n^2\mathbb{C})$. Moreover $Pg\ell(n,\mathbb{C}) \subset S\ell(n^2,\mathbb{C})$ and so $Az(n,\mathbb{C}) = G_{m,\mathbb{C}} \times S\ell(n^2,\mathbb{C})/Pg\ell(n,\mathbb{C})$. Since $\mu_n$ is the kernel of the projection from $S\ell(n,\mathbb{C})$ to $Pg\ell(n,\mathbb{C})$, 5.1 implies that the map from $\hat{\mu}_n$ (characters on $\mu_n$) into $Br(S\ell(n^2,\mathbb{C})/Pg\ell(n,\mathbb{C}))$ is injective. The result follows at once. Q.E.D.

We close by remarking that the "generic matrices" of the usual theory have not all vanished from the development in this paper. Namely the sections $f_m \in \Gamma(Az(n,\mathbb{C}),A_{n,\mathbb{C}}$ are precisely "generic matrices". In general one may regard the columns of an $n^2 \times n^2$ matrix as generic matrices, whence $Pg\ell(n,\mathbb{C})$ acts by its natural Azumaya representation. If we choose as a basis for $M_{n,\mathbb{C}}$, the usual matrices $e_{ij}$ (one in the i,jth place), then X, the matrix with coefficients which we may now write as $x_{ij,rs}$, is a matrix of indeterminants. Hence $X^{-1}e_{11}$ is the first column of $X^{-1}$. However if the entries of X are indeterminates then those of $X^{-1}$ are as well and it is not at all difficult to check that the product $f_{e_{kj}}(X)\cdot f_{e_{st}}(X)$ in $A_{n,\mathbb{C}}$ correspond precisely to matrix multiplication of the two matrices corresponding to the i,j column and the s,t column in $X^{-1}$. In fact one could obtain traditional generic matrices (X rather than $X^{-1}$) by using the right homogeneous space, $Pg\ell(n,\mathbb{C})\backslash G\ell(n^2,\mathbb{C})$, but it is customary to work on the left.

One may exploit the fact that the $f_m(X)$ are generic matrices to elaborate the relationships between the traditional approach to this subject and the present more group theoretic point of view. This, in part, will be examined in detail in a forthcoming paper.

At this point it seems germaine to indicate two directions of generalization. First one might broaden the notion of polarization. Consider for example a vector space of dimension $r>n^2$, say V. Then one might define an r-polarized Azumaya of degree n as a pair $(A,\varphi)$ where A is Azumaya over $\mathcal{O}_X$ and $\varphi:\mathcal{O}_X\times V\to A$ is a surjective morphism. Then there is a natural notion of equivalence and a corresponding functor.

Then one might consider a surjective mapping $\alpha:V\to M_n\to 0$. Let $W = \ker\alpha$. Then put P equal to the set of $g\in G\ell(V)$ such that $g(W)\subset W$ and the induced morphism $\bar{g}: M_n\to M_n$ is in $Pg\ell(n,\mathbb{C})$. Then P admits a natural Azumaya representation and so $G\ell(V)/P$ has an induced Azumaya algebra $A_{n,r,\mathbb{C}}$ and there is a theorem and corollary analagous to 7.8 and 7.9 showing that $G\ell(V)/P$ equivariantly represents Azumaya algebras generated by r sections. The value of this particular generalization is somewhat vitiated by the following observation. Let $\tilde{P}\subset G\ell(V)$ be the set of g such that $g(W)\subset W$. Then $\tilde{P}/P$ is naturally equivariantly isomorphic to $G\ell(n^2,\mathbb{C})/Pg\ell(n,\mathbb{C})$. There is a natural projection $\pi:G\ell(V)/P\to G\ell(V)/\tilde{P}$ and it is easily seen that $G\ell(V)/P = G\ell(V)\times^{\tilde{P}}\tilde{P}/P$. The functorial properties of $G\ell(V)/P$ then are readily deduced using its structure as a co-fibre product over the Grassman variety $G_{n^2,r} = G\ell(V)/\tilde{P}$.

A second type of generalization may be found by considering an adjoint semi-simple group G. Then let $\mathfrak{g}$ denote its Lie algebra. This G may be identified with the Lie algebra automorphisms of inside of $G\ell(\mathfrak{g})$. Then consider the homogeneous space $G\ell(\mathfrak{g})/G$.

If $\tilde{G}$ is the simply connected group corresponding to G, one may construct the co-fibred product $G\ell(\mathfrak{g})\times^{\tilde{G}} G$ where G acts by conjugation. One may speculate that this space is some sort of "generic" form for G over a k-variety and that $G\ell(\mathfrak{g})/G$ represents such forms. At any rate it is clear that in characteristic zero, it parametrizes forms for $\mathfrak{g}$ which are free as modules and the proof scarcely differs from that of 7.8.

REFERENCES

[1] M. Artin and D. Mumford, "Some elementary examples of unirational varieties which are not rational", Proc. London Math Soc. 3 (1972) pp. 75-95.

[2] M. Auslander and O. Goldman, "The Brauer group of a commutative ring", Trans. Amer. Math. Soc. 97 (1960) pp. 367-409.

[3] W. Browder, "Torsion in H-spaces", Ann. of Math. (2) 74 (1961) pp. 24-51.

[4] E. Cartan, "La topologie des espaces representatifs des groupes de Lie", Ouvres, Part I, v. 2 No. 150 p. 1307.

[5] M. Demazure, A. Grothendieck et.al., Schémas en Groupes I , Lecture Notes in Mathematics v. 151 Springer Verlag, Berlin Heidelberg New York 1970.

[6] O. Gabber, "Some theorems on Azumaya algebras" in Le Groupe de Brauer , Lecture Notes in Mathematics v. 844 Springer Verlag, Berlin Heidelberg New York 1981.

[7] A. Grothendieck, "Le groupe de Brauer I-III" in Dix Exposés sur la Cohomologie des Schemás, Elsevier North Holland, Amsterdam 1968.

[8] R. Hoobler, "When is Br(X)=βr'(X)?" in Brauer Groups in Ring Theory and Algebraic Geometry, Proceedings, Antwerp 1981, F. van Ostaeyen and A. Verschoren ed. Lecture Notes in Mathematics v. 917 Springer Verlag Berlin Heidelberg New York 1982.

[9] B. Iversen, "Brauer group of a linear algebraic group", J. Algebra 42 (1976) pp. 295-301.

# SIMPLE SUBMODULES IN A FINITE DIRECT SUM OF UNIFORM MODULES

HARADA Manabu

Department of Mathematics, Osaka City University

Let R be a ring with identity. We shall study submodules in a direct sum of uniform modules $U_i$ with finite length. Let $S_i$ be the socle of $U_i$. Then we obtain the natural imbedding of $\mathrm{End}_R(U_i)/J(\mathrm{End}_R(U_i))$ into $\mathrm{End}_R(S_i)$, where $J(\mathrm{End}_R(U_i))$ is the Jacobson radical of $\mathrm{End}_R(U_i)$. We shall denote them by $\Delta(U_i)$, $\Delta$, respectively.

In Section 2 we shall give a relationship between the dimension of $\Delta$ over $\Delta(U_i)$ as a left $\Delta(U_i)$-module and direct summands containing a given simple submodule in the direct sum of the $U_i$. In Sections 3 and 4, we shall give some characterization of a ring satisfying d-(**), d-(*), respectively (see §§ 3 and 4). For instance, if R is a commutative artinian ring with d-(**), then R is a serial ring or R/J(R) is a finite ring. Main purpose of Section 4 is to give the complete forms of simple submodules in a direct sum of particular uniform modules as simpler as possible.

Many results in this paper are studied in certain particular cases. However, it is seems to the author that those cases are very natural, because if R is an algebra over an algebraically closed field, then we obtain only the special situation as studied in this paper, provided J(R) is square zero.

## 1 Mini quasiinjective modules.

Let R be a ring with identity and U a uniform module with finite length: $|U| < \infty$. Let S be the unique simple submodule of U and E = E(U) the injective hull of U. We take another submodule U' in E (assume that $|U'| < \infty$). Set $T_E = \mathrm{End}_R(E)$, $T_U = \mathrm{End}_R(U)$,



*F. van Oystaeyen (ed.), Methods in Ring Theory, 145–159.*

$\Delta = \mathrm{End}_R(S)$ and $T_E^{(U,U')} = \{s \in T_E \mid xU \subset U'\}$. We denote $T_E^{(U,U)}$ by $T_E^U$. If we take the restriction of an element in $\mathrm{Hom}_R(U,U')$ to S, then we can obtain the natural mapping $\varphi$ from $\mathrm{Hom}_R(U,U')$ into $\Delta$. Since E is injective, we obtain the following commutative diagram with exact rows:

$$\begin{array}{ccccc} T_E^{(U,U')} & \xrightarrow{\nu} & \mathrm{Hom}_R(U,U') & \longrightarrow & 0 \\ \downarrow i & & \downarrow \varphi & & \\ T_E & \xrightarrow{\varphi_E} & \Delta & \longrightarrow & 0, \end{array}$$

where i is the inclusion mapping and $\nu$ is the natural mapping. It is well known that $\ker \varphi_E$ is the Jacobson radical of $T_E$ and $\Delta(U) = \varphi(T_U)$ is a subdivision ring of $\Delta$ for $|U| < \infty$. We put $\varphi(s) = \bar{x}$ for $x \in T_E$. We shall assume that each element in $T_E$ operates from the left side. Then

$$\Delta(U,U') = \varphi(\mathrm{Hom}_R(U,U')) = \varphi_E\, i(T_E^{(U,U')})$$

is a left $\Delta(U')$-subspace of $\Delta$. We shall denote the dimension of $\Delta$ and $\Delta(U,U')$ over $\Delta(U')$ by $[\Delta : \Delta(U')]$ and $[\Delta(U,U') : \Delta(U')]$, respectively. We note that $\varphi^{-1}(\Delta(U)) \supsetneqq T_E^U$ in general. If we take a representative element $\bar{f}$ in $\Delta(U)$, we always assume that f is an element in $T_E^U$ or $T^U$.

Let $\{U_i\}_{i=1}^{n}$ be a set of uniform modules with finite length. Put $D = \sum_{i=1}^{n} \oplus U_i$. We shall study the property dual to (**) in [7].

d-(**) For every simple submodule S in $D = \sum \oplus U_i$, there exists a non-trivial direct summand U' of D, which contains S (cf. [1]).

Let $D = D_1 \oplus D_2$ and $D_1 \supsetneqq S$. Then we may assume that $D_2$ is also uniform. By Soc(U) (resp. J(U)) we denote the socle (resp. the Jacobson radical) of U. Then $\mathrm{Soc}(U) = \sum \oplus \mathrm{Soc}(U_i)$ and $\mathrm{Soc}(U_i) = S_i$ is simple. Assume that $D = \sum_i \sum_j \oplus U_{ij}$ and $\mathrm{Soc}(U_{ij}) \approx \mathrm{Soc}(U_{ij'})$ and $\mathrm{Soc}(U_{i1}) \not\approx \mathrm{Soc}(U_{i'1})$ if $i \neq j$. Then simple submodule of D is contained in some $\sum_j \oplus \mathrm{Soc}(U_{ij})$. If $i \geqslant 2$, d-(**) is always satisfied. Thus if we study d-(**), we may assume that Soc(D) is homogeneous.

Lemma 1. Let $\{U_i\}_{i=1}^{k'}$ be a set of uniform modules with $|U_i| < \infty$. Then if $\sum_{i \leq k} \oplus U_i$ satisfies d-(**), so does $\sum_{i=1}^{k'} \oplus U_i$ for $k' > k \geqslant 2$.

Proof. Let $D = \sum_{i=1}^{k'} \oplus U_i$ and S a simple submodule of D. Put $D = D_1 \oplus D_2$, where $D_1 = \sum_{i \leq k} \oplus U_i$. Then $S \subset \pi_1(S) \oplus \pi_2(S)$, where the $\pi_i: D \longrightarrow D_i$ are the projections. If either $\pi_1(S) = 0$ or $\pi_2(S) = 0$, we are done. Assume that $\pi_1(S) \neq 0$. Since $\pi_1(S)$ is a simple submodule of $D_1$, there exists a direct summand $D_{11}$ of $D_1$ such that $D_{11} \supseteq \pi_1(S)$. Let $D_1 = D_{12} \oplus D_{11}$ $(D_{12} \neq 0)$. Then $D = D_{11} \oplus D_2 \oplus D_{12}$ and S is contained in $D_{11} \oplus D_2$.

We have defined a maxi quasiprojective module in [4]. From the observation above and [7], we shall define a mini quasiinjective module M dual to a maxi quasiprojective module with $\mathrm{Soc}(M) \neq 0$. If every isomorphism between simple submodules in Soc(M) is extendible to an endomorphism of M, then M is called a mini quasi-injective module. We can discuss dually several results to [8]. However, the argument for mini quasiinjective is much simpler than that for maxi quasiprojcetive.

We quote some result which is dual to [8], Theorem 5. Let $\{M_\alpha\}_I$ be an isomorphism class of indecomposable modules with finite length. By $M_\alpha^{(J_\alpha)}$ we denote the direct sum of $|J_\alpha|$-copies of $M_\alpha$, where $|J_\alpha|$ means the cardinal of $J_\alpha$.

Theorem 1. Put $M = \Sigma \oplus M_\alpha^{(J_\alpha)}$. Then M is mini quasiinjective if and only if each $M_\alpha$ is mini quasiinjective and no simple submodule in $M_\alpha$ is isomorphic to a simple submodule in $M_\beta$ if $\alpha \neq \beta$.

We note that R is right self-miniinjective (see[3]) if and only if R is mini quasiinjective as a right R-module.

We shall come back to the case where indecomposable module U is uniform. In this case U is mini quasiinjective if and only if $\Delta = \Delta(U)$.

Theorem 2. Let U be a uniform module with $|U| < \infty$. Then the following conditions are equivalent:

1) U is mini quasiinjective.
2) $U \oplus U$ has the extending property of simple modules (see [2]).
3) $U \oplus U$ satisfies d-(**).

Proof. This is dual to [8], Theorem 1.

2 Direct sums of uniform modules.

Let U be a uniform module of finite length. Then we are interested in the case $\Delta \neq \Delta(U)$ from Theorem 2. The following theorem is dual to [8], Theorem 2'.

Theorem 3. Let $U_1$ and $U_2$ be uniform modules with finite length and $\mathrm{Soc}(U_1) \approx \mathrm{Soc}(U_2)$. Assume $|U_1| \leq |U_2|$. Then $D = U_2^{(k+1)} \oplus U_1$ satisfies d-(**) if and only if $[\Delta/\Delta(U_1,U_2)] \leq k$.

Proof. Let $E = E(U_2)$. Then we may assume $E \supset U_1$. Put $D = U_2^{(k+1)} \oplus U_1$ and assume that D satisfies d-(**). Let $\{\delta_1, \delta_2, \cdots, \delta_{k+1}\}$ be a set of k+1 elements in $\Delta$, where the $\delta_i$ are in $T_E$. Put $S = \mathrm{Soc}(E)$, then $\delta_i(S) \subset S$ for all i. Take a simple submodule $S^* = \{\delta_1 s + \delta_2 s + \cdots + \delta_{k+1} s + s \mid s \in S\}$ in D. Then $D = D_1 \oplus D_2$, $D_1 \supset S^*$, and $D_2$ is uniform by the assumption. First we assume that $D_2$ is isomorphic to $U_2$ through $\psi$. Let $p_j$ and $i_j$ be the projection and injection on jth component of $D = \sum_{i=1}^{k+2} \oplus U_i$, respectively. Then $1_D = \sum_j i_j p_j$. Let $\pi_2 : D \longrightarrow D_2$ be the projection of $D = D_1 \oplus D_2$. Since $S^* \subset D_1$, $\pi_2(S^*) = 0$. Now $\psi\pi_2 = \sum(\psi\pi_2 i_j)p_j$. Then $y_j = \psi\pi_2 i_j \in T_{U_2}$ for $j \leq k+1$, and $y_{k+2} = \psi\pi_2 i_{k+2} \in \mathrm{Hom}_R(U_1, U_2)$. Since $\psi\pi_2(S^*) = 0$, $\sum y_j \delta_j s + y_{k+2} s = 0$ for all $s \in S$. Therefore, taking an inverse image of $\sum \bar{y}_i \bar{\delta}_i + \bar{y}_{k+2}$ in $T_E$,

$$\sum_{j=1}^{k+1} \bar{y}_j \bar{\delta}_j = -\bar{y}_{k+2} \qquad (1)$$

and $\bar{y}_{k+2}$ is in $\Delta(U_1, U_2)$, and so $[\Delta/\Delta(U_1, U_2) : \Delta(U_2)] \leq k$. Next assume that $D_2 \approx U_1$ through $\psi'$. $y'_j = \psi'\pi_2 i_j \in \mathrm{Hom}_R(U_2, U_1)$ for $j \leq k+1$. If $\bar{y}'_j = 0$ for all $j \leq k+1$, $0 = \sum \bar{y}'_j \bar{\delta}_j = -\bar{y}'_{k+2}$ $(\in \Delta(U_1))$ as in (1). Moreover $\psi'$ is an isomorphism, so $\pi_2 i_j(\mathrm{Soc}(U_2)) = 0$ for $j \leq k+1$ and $\pi_2 i_{k+2}(\mathrm{Soc}(U_1)) = 0$. Hence $\pi_2(\mathrm{Soc}(D)) = 0$, which is a contradiction. Therefore $\bar{y}'_i \neq 0$ for some $i \leq k+1$, so $U_2$ is monomorphic to $U_1$ via $y'_i$. Since $|U_2| \geq |U_1|$, $U_2 \approx U_1$. Hence $[\Delta/\Delta(U_1, U_2) : \Delta(U_2)] \leq k$ from the initial part. Conversely, assume that $[\Delta/\Delta(U_1, U_2)] \leq k$. Let $S^*$ be a simple submodule of D. Then we may assume that $S^* = \{\delta_1 s + \delta_2 s + \cdots + \delta_{k+1} s + s \mid s \in S\}$ as the first part. If $\bar{\delta}_i = 0$ for some i, $S^* \subset \sum_{j \neq i} \oplus U_j$. Assume $\bar{\delta}_i \neq 0$ for all i. There exists a set $\{y_1, y_2, \cdots, y_{k+1}\}$ of elements in $T_{U_2}$ such that $\sum \bar{y}_i \bar{\delta}_i (= \bar{t})$ is in $\varphi(T_E^{(U_1, U_2)})$, and $\bar{y}_i \neq 0$ for some i, say i = 1. Put $D = \sum_{i=1}^{k+2} \oplus V_i$ and $V_i = U_2$ for $i \leq k+1$ and $V_{k+2} = U_1$. We define a homomorphism $\theta$ of $V = \sum_{i=2}^{k+2} \oplus V_i$ to $V_1$ by setting $\theta(\sum_{i \geq 2} x_i) = \sum_{i=2}^{k+1} y_1^{-1} y_i x_i - y_1^{-1} t x_{k+2}$, where $x_i \in V_i$ and $y_1^{-1}$ may be assumed in $T_{U_2}$. Set $V(\theta) = \{-\theta v + v \mid v \in V\} \subset D$. Then $D = V_1 \oplus V(\theta)$. Since $\bar{\delta}_1 s$

$(\bar{y}_1^{-1}\bar{t} - \sum_{i\geq 2} \bar{y}_1^{-1}\bar{y}_i \bar{\delta}_i)s = -\theta(\sum_{i=2}^{k+1}\delta_i s + s)$, $\delta_1 s + \delta_2 s + \cdots + \delta_{k+1}s + s = -\theta(\sum_{i=2}^{k+1}\delta_i s + s) + (\sum_{i=2}^{k+1}\delta_i s + s) \in V(\theta)$. Therefore $V(\theta) \supset S^*$.

Corollary 1. Let U be as above. Then $[\Delta : \Delta(U)]$ is the minimal integer k' such that $U^{(k')}$ satisfies d-(**).

Corollary 2. Let U be as above. Put $D = U^{(m+k)}$. Then $[\Delta : \Delta(U)] = k$ if and only if all simple submodules in D are contained in direct summands of D isomorphic to $U^{(k)}$ but not to $U^{(k-1)}$.

Proof. Assume $[\Delta : \Delta(U)] = k$. Then we can easily show from Lemma 1 and Corollary 1 that every simple submodule is contained in a direct summand of D isomorphic to $U^{(k)}$. We shall give a simple submodule of D not contained in a direct summand of D isomorphic to $U^{(k-1)}$. Let $\{\bar{\delta}_1, \bar{\delta}_2, \cdots, \bar{\delta}_k\}$ be a linearly independent set of $\Delta$ over $\Delta(U)$. Let S be the simple submodule in U. Set $S^* = \{\delta_1 s + \delta_2 s + \cdots + \delta_k s + 0 + \cdots + 0 \mid s \in S\}$ in D. Assume that there exists a decomposition $D = V_1 \oplus V_2 \oplus \cdots \oplus V_t \oplus V_{t+1} \oplus \cdots \oplus V_{m+k}$ such that $S^* \subset V_1 \oplus \cdots \oplus V_t$ and $V_j \approx U$ for all j. Let $\pi_k$ be the projection of D onto $V_k$ in the decomposition above. For the initial decomposition $D = U^{(m+k)}$, we have the projection $p_r$ and the injection $i_r$. Then $\pi_{t+1} = \pi_{t+1}1_D = \sum \pi_{t+1}i_j p_j$. Since $\pi_{t+1}(S^*) = 0$, $\sum \pi_{t+1} i_j \delta_j s = 0$. $\pi_{t+1}i_j$ may be regarded as an element in $T_U$. Hence $\pi_{t+1}i_j$ is not an isomorphism for all $j \leq k$ since $\{\bar{\delta}_i\}$ is independent. On the other hand, $\pi_{t+1} = \sum_{j=1}^{m+k} \pi_{t+1}i_j p_j \pi_{t+1}$. Hence some $\pi_{t+1}i_a p_a \pi_{t+1}$ is an isomorphism of U since $T_U$ is a local ring, and $a > k$ from the fact above. This implies that $D = V_1 \oplus \cdots \oplus V_t \oplus U_a \oplus V_{t+2} \oplus \cdots \oplus V_{m+k}$. Continuing this argument, we can exchange all $V_r$ $(r \geq t+1)$ by $U_{a'}$. Therefore $(m+k-t) \leq m$, so $k \leq t$. The converse is clear from Corollary 1.

If there exists a monomorphism of $U_1$ to $U_2$, we shall denote it by $U_1 \prec U_2$. By making use of the same manner in the proof of Theorem 3, we can show the following results.

Theorem 4. Let $\{U_i\}_{i=1}^{t}$ $(t \geq 2)$ be a set of uniform modules. Assume that $\mathrm{Soc}(U_i) \approx \mathrm{Soc}(U_1)$ and $1 \leq k_i \leq [\Delta : \Delta(U_i)]$ for all i.

If $D = U_1^{(k_1)} \oplus U_2^{(k_2)} \oplus \cdots \oplus U_t^{(k_t)}$ satisfies d-(**), then either $U_i \prec U_j$ or $U_j \prec U_i$ for some pair (i,j).

Proof. Let $\{\bar{\delta}_{i1}, \cdots, \bar{\delta}_{ik_i}\}$ be a set of elements in $\Delta$, which are linearly independent over $\Delta(U_i)$ for each i. Set $S^* = \{\sum \delta_{ij} s \mid s \in S = Soc(U_t)\}$ as in the proof of Theorem 3. Let $D_2$ be the uniform direct summand of D as in the proof of Theorem 3. Assume that $D_2 \approx U_i$ via $\psi$ for some i, say i = 1. Put $y_{pq} = \psi \pi_2 i_{pq} \in Hom_R(U_p, U_1)$. Then $\sum \bar{y}_{ij} \bar{\delta}_{ij} = 0$ from (1). If $\bar{y}_{ij} \neq 0$ for $i \neq 1$, $U_i$ is monomorphic to $U_1$. Assume that all $\bar{y}_{ij} = 0$ for $i \neq 1$. Now $\bar{y}_{1i}$ is an element in $\Delta(U_1)$ and $\sum \bar{y}_{1j} \delta_{1j} = 0$ implies $\bar{y}_{1j} = 0$ for all j, since $\{\bar{\delta}_{11}, \cdots, \bar{\delta}_{1k_1}\}$ is linearly independent, which contradicts the fact: $\pi_2(Soc(D)) \neq 0$ as in the proof of Theorem 3.

Corollary 1. Let $\{U_i\}_{i=1}^{n}$ be a set of uniform modules with $[\Delta : \Delta(U_i)] = k < \infty$ for all i. If all $D = \sum_{i=1}^{k+1} \oplus U'$ satisfy d-(**), then $\{U_i\}$ is linearly ordered with respect to $\prec$. If $\Delta(U_i) = \Delta(U_1)$ for all i, the converse is true, where the $U'_i$ are members in $\{U_i\}$.

Proof. The first assertion is clear from Theorem 4. Conversely assume that $D = U_1 \oplus U_2 \oplus \cdots \oplus U_{k+1}$ and $U_1 \supset U_2 \supset \cdots \supset U_{k+1}$. Let $S^* = \{\delta_1 s + \delta_2 s + \cdots + \delta_{k+1} s \mid s \in S = Soc(U_1)\}$ be a simple submodule of D. Set $\Delta_1 = \Delta(U_1)$. Since $[\Delta : \Delta_1] = k$, there exists a set $\{\bar{y}_i\}$ of elements in $\Delta$ such that $\bar{y}_t \neq 0$ and $\bar{y}_j = 0$ for all $j < t$. $U_t \supset U_{t+p}$ implies $\Delta(U_{t+p}, U_t) \supset \Delta_1$. Set $V = U_1 \oplus U_2 \oplus \cdots \oplus U_{t-1} \oplus U_{t+1} \oplus \cdots \oplus U_{k+1}$. We define $\theta : V \longrightarrow U_t$ by setting $\theta(x_1 + \cdots + x_{t-1} + x_{t+1} + \cdots + x_{k+1}) = \sum_{p=1}^{k-t+1} y_t^{-1} y_{t+p} x_{t+p}$. Then $\delta_1 s + \cdots + \delta_{k+1} s = \delta_1 s + \cdots + \delta_{t-1} s + s + (-y_t^{-1} y_{t+1} + 1) \delta_{t+1} s + \cdots + (-y_t^{-1} y_{k+1} + 1) \delta_{k+1} s = (1-\theta)(\delta_1 s + \cdots + \delta_{t-1} s + s + \delta_{t+1} s + \cdots + \delta_{k+1} s) \in V(\theta)$, so $D = V(\theta) \oplus U_t$ and $V(\theta) \supset S^*$.

From the manner dual to the proof of [8], Theorem 9, we can give examples of such the set $\{U_i\}$ with $\Delta(U_i) = \Delta(U_1)$.

Corollary 2. Let $U_1$ and $U_2$ be uniform modules with finite length and $Soc(U_1) \approx Soc(U_2)$. Assume $[\Delta : \Delta(U_2)] = k$. Then $D = U_2^{(k)} \oplus U_1$ satisfies d-(**) if and only if $U_1 \prec U_2$ or $U_2 \prec U_1$.

Proof. "Only if" part is clear from Theorem 4. Conversely,

assume either $U_2 \supset U_1$ or $U_1 \supset U_2$. Then we shall show that D satisfies d-(**). Assume that $U_2 \supset U_1$. Let S* be any simple submodule of D. Then we may assume as before that $S^* = \{\delta_1 s + \delta_2 s + \cdots + \delta_k s + s \mid s \in Soc(U_1)\}$. Since $[\Delta : \Delta(U_2)] = k$, $\{\bar{\delta}_1, \bar{\delta}_2, \cdots, \bar{\delta}_k, \bar{1}\}$ is linearly dependent over $\Delta(U_2)$, so there exists some $i \leq k$, say $i = 1$ such that $\bar{\delta}_1 = \sum_{i \geq 2}^{k} \bar{y}_i \bar{\delta}_i + \bar{y}_{k+1}$, where the $\bar{y}_j$ are elements in $\Delta(U_2)$. Further, since $U_2 \supset U_1$, $\Delta(U_1, U_2) \supset \Delta(U_2)$. Hence D satisfies d-(**) from the proof of Theorem 3. Finally assume that $U_2 \subset U_1$. If $\bar{y}_{k+1} = 0$ in the above, then we obtain the same situation. If $\bar{y}_{k+1} \neq 0$, $\bar{1} = \sum_{j=1}^{k} \bar{z}_j \bar{\delta}_j$. Since $\Delta(U_2) \subset \Delta(U_2, U_1)$, putting $\theta(\sum_{i=1}^{k} x_i) = \sum z_j x_j \in Hom_R(U_2^{(k)}, U_1)$, we obtain that $D = U_2^{(k)}(\theta) \oplus U_1$ and $U_2^{(k)}(\theta) \supset S^*$.

Corollary 3. Let $\{U_i\}_{i=1}^{t}$ be a set of uniform modules. Assume $|U_i| = |U_1|$, $Soc(U_i) \approx Soc(U_1)$, and $[\Delta : \Delta(U_i)] = k$ for all i. Put $D = \sum_{i=1}^{t} \oplus U_i^{(s_i)}$ $(s_i \geq 1)$, where $k+1 = \sum s_i$. Then D satisfies d-(**) if and only if $U_i \approx U_1$ for all i.

Proof. This is clear from Theorem 4.

3 Commutative rings.

In this section we study d-(**) for direct sums of serial modules with same length.

Lemma 2. Let $\{U_i\}_I$ be a (non-isomorphic) representative set of uniform modules with same length. Assume $Soc(U_i) = S$ for all i. If, for some fixed n, $\sum_{i=1}^{n} \oplus U'_i$ satisfies d-(**) for any $U'_i$ isomorphic to some one in $\{U_i\}_I$, then I is finite.

Proof. This is clear from Thorem 4.

Proposition 1. Let R be a commutative artinian and local ring, and let $\{U_i\}$ be the set of serial modules of length two. Then, in case R/J is an infinite field, all $D = \sum_{i=1}^{n} \oplus U'_i$ satisfy d-(**) for some n if and only if R is a coserial (hence serial) ring. If R/J is a finite field, then $D = \sum_{i=1}^{m} \oplus U'_i$ satisfy d-(**) for some m, where the $U'_i$ are isomorphic to some one in $\{U_i\}$.

Proof. Let E = E(R/J). Then $T_E = R$ by [9]. Hence $\Delta = \Delta(U_i)$ for all i. If R/J is infinite, there exist infinitely many $U_i$ not isomorphic to one another, provided that Soc(E/Soc(E)) is not simple. Assume that every $D = \sum_{i=1}^{n} \oplus U'_i$ satisfies d-(**) for some n. Then Soc(E/Soc(E)) is simple by Lemma 2, so R is coserial by [7], Proposition 1'. Therefore R is serial by [6], Corollary 5. The converse is clear by [4], Theorem 6. If R/J is finite, then there are only finitely many $U_i$, since E is finite. Hence all $D = \sum_{i=1}^{m} \oplus U'_i$ satisfy d-(**) for a large m by Lemma 1 and Corollary 1 to Theorem 3.

4 Property d-(*).

Let R be the right artinian ring as before. We have shown in [6], Corollary 3 that R is right coserial if and only if

d-I Every factor module of a finite direct sum of uniform modules is also a direct sum of uniform modules and

d-II Every indecomposable and injective module is hollow.

It is clear that R is of a right colocal type (see [11] or [10]) if and only if d-I is satisfied for any direct sum of uniform modules and R is a direct sum of uniform modules. In this section, we shall study the right artinian ring satisfying the condition d-(*) below, which is equivalent to d-I (see the proof of Theorem 5).

d-(*) Every factor module,with respect to a simple submodule, of a direct sum of uniform modules is also a direct sum of uniform modules.

The following lemma is dual to [7], Lemma 3.

Lemma 3. Let $\{U_i\}_{i=1}^{t+1}$ be a set of uniform modules such that $|U_i| = t$ for all i. If $D = \sum_{i=1}^{t+1} \oplus U_i$ satisfies d-(*), then D does d-(**) (cf. [1]).

Proof. Let $D = \sum_{i=1}^{t+1} \oplus U_i$ and S a simple submodule of D. We take the exact sequence

$$0 \longrightarrow S \longrightarrow \sum \oplus U_i \xrightarrow{\nu} (\sum \oplus U_i)/S = \sum \oplus U'_i \longrightarrow 0,$$

where $\nu$ is the natural epimorphism and the $U'_i$ are uniform modules from d-(*). Let $\pi_i : D \longrightarrow U_i$ be the projection. If $\pi_i(S) = 0$ for some i, then $S \subset \sum_{j\neq i} \oplus U_j$. Hence we may assume $\pi_j(S) \neq 0$ for all j, so $Soc(U_i) \approx Soc(U_1)$ for all i. $\pi_i(S) \neq 0$ means that $\nu|U_i$ is a monomorphism. Hence we shall denote $\nu(U_i)$ by $U_i$ itself. Then $D' = \sum \oplus U'_i = \sum U_i$. We may assume that $\{U'_i\}$ is arranged as $|U'_i| \leq t$ for $j \leq$ some p and $|U'_{j'}| \geq t+1$ for $j' > p$. If $\sum_{i=1}^{t+1} Soc(U_i) \subset \sum_{j'>p} \oplus U'_{j'}$, $|\sum_{j>p} \oplus U'_{j'}| \geq (t+1)t$ for $k=p \geq |\sum Soc(U_i)| = |\sum \oplus ($

$\mathrm{Soc}(U_i))/S| = t$. However, $|D/S| = (t+1)t-1$. Therefore $\mathrm{Soc}(U_q) \not\subset \sum_{j' \neq p} \oplus U'_{j'}$ for some q. Since $U_q$ and $U'_{j'}$ are uniform, there exists the projection $\pi'_s$ of D' to $U'_s$ ($s \leq p$) such that $\pi'_s|U_q$ is a monomorphism. Therefore $\pi'_s|U_q$ is an isomorphism for $t = |U_q| \geq |U'_s|$, and so $D = U_q \oplus \ker \pi'_s \nu$ and $\ker \pi'_s \nu \supset S$.

Lemma 4 ([7], Proposition 6). Let E be an indecomposable injective module with the socle S. Let $\{U_1, U_2, \cdots, U_t\}$ be a set of submodules of E such that $\delta_k U_k = U_1$ for some $\delta_k$ in $T_E$, and $|U_1| < \infty$. If $|(U_1 \cap U_2 \cap \cdots \cap U_{i-1} \cap U_{i+1} \cap \cdots \cap U_t)/S| = 1$ for all i and $S = U_1 \cap U_2 \cap \cdots \cap U_t$ (is irredundant), then $\{\bar{1}, \bar{\delta}_2, \cdots, \bar{\delta}_t\}$ is linearly independent over $\Delta(U_1)$ and hence $[\Delta : \Delta(U_1)] \geq t$.

Proof. Let $S = U_1 \cap U_2 \cap \cdots \cap U_t$ as in the lemma. Assume that $\{\bar{\delta}_i\}$ is dependent over $\Delta(U_1)$ and take a set $\{\bar{y}_i\}_{i=1}^{t} \neq 0$ in $\Delta(U_1)$ such that $\sum \bar{y}_i \bar{\delta}_i = 0$. Assume $\bar{y}_i \neq 0$. Then $y'_i = (y_i \delta_i)^{-1} y_j \delta_j \in T_E^{(U_j, U_i)}$ and $y'_i = 1_E$. Take an element x in $U_1 \cap \cdots \cap U_{i-1} \cap U_{i+1} \cap \cdots \cap U_t - U_i$. Since $|(U_1 \cap \cdots \cap U_{i-1} \cap U_{i+1} \cap \cdots \cap U_t)/S| = 1$, $S \ni \sum y'_j x = x + \sum_{j \neq i} y'_j x$ and $\sum_{j \neq i} y'_j x \in U_i$. Hence $x = \sum_{j \neq i} y'_j x + s \in U_i$, which is a contradiction. Therefore $[\Delta : \Delta(U_1)] \geq t$.

Lemma 5 ([7], Proposition 10). Let E be an indecomposable injective module with $S = \mathrm{Soc}(E)$. Assume that d-(**) is satisfied for every $\sum_{i=1}^{3} \oplus A_i$, where the $A_i$ are submodules of E with length two. 1) If either $|\mathrm{Soc}(E/S)| \geq 3$ or $\Delta \neq \Delta(A_1)$ for some $A_1$, then all $A_i$ are isomorphic to one another and $[\Delta : \Delta(A_1)] = 2$. 2) $\Delta = \Delta(A_1)$ for some $A_1$ if and only if $|\mathrm{Soc}(E/S)| = 1$ or $\mathrm{Soc}(E/S) = A_1/S \oplus A_2/S$ and $A_1 \approx A_2$.

Proof. 1) If $\Delta = \Delta(A_i)$ for all i, $|\mathrm{Soc}(E/S)| \leq 2$ by Corollary 3 to Theorem 4. Hence $\Delta \neq \Delta(A_1)$ for some $A_1$. Then $[\Delta : \Delta(A_1)] = 2$ by Theorem 3. Hence $A_i \approx A_1$ by Corollary 3 to Theorem 4. The last assertion is also clear from the argument above.

Theorem 5 ([7], Theorem 12). Let R be a right artinian ring, and let $\{E_i\}_{i=1}^{n}$ be a set of indecomposable injective modules with $|E_i| \leq 4$ and $\mathrm{Soc}(E_i/\mathrm{Soc}(E_i)) = E_i/\mathrm{Soc}(E_i)$. The condition d-I is satisfied for all uniform modules $U_i$ whose socles are isomorphic to some of $\{\mathrm{Soc}(E_j)\}$ if and only if one of the following conditions is satisfied for each $E = E_i$:

i) ($|E| = 1$) E is simple.

ii) ($|E| = 2$) E is serial.

iii) ($|E| = 3$) a) The number of isomorphism classes of $\{A_i\}$ is two, i.e., $E/S \approx A_1/S \oplus A_2/S$ and $A_1 \not\approx A_2$, and b) $\Delta = \Delta(A_1)$.

iv) ($|E| = 3$) a) All $A_i$ are isomorphic to one another and b) $[\Delta : \Delta(A_1)] = 2$.

v) ($|E| = 4$) a) All $A_i$ (resp. $B_i$) are isomorphic to one another and b) $[\Delta : \Delta(A_1)] = 2$ (resp. $[\Delta : \Delta(B_1)] = 3$), where the $A_i$ (resp. $B_i$) are submodules in E such that $|A_i| = 2$ (resp. $|B_i| = 3$).

Proof. Assume that Condition d-I is satsfied for uniform modules as in the theorem. Then by Lemma 3 $D = \sum_{i=1}^{3} \oplus A_i$ satisfies d-(**) for any submodules $A_i$ in E with $|A_i| = 2$. Hence i) $\sim$ iv) are obtained from Lemma 5. Finally assume $|E| = 4$. Let $E/S = \bar{A}_1 \oplus \bar{A}_2 \oplus \bar{A}_3$ and $\bar{B} = \bar{A}_1 \oplus \bar{A}_2$, where $S = \mathrm{Soc}(E)$. Then $[\Delta : \Delta(B)] \leqslant 3$ by Lemma 2 and Corollary 1 to Theorem 3. We know by Lemma 5 that $[\Delta : \Delta(A_1)] = 2$ and all $A_i$ are isomorphic to one another. If $\Delta = \Delta(B')$ for every B' with $|B'| = 3$, $\Delta = \Delta(A_1)$ for $A_1 = B' \cap B''$ with $|B''| = 3$. Hence we may assume $\Delta \neq \Delta(B)$. Let $\delta_1$ be in $T_E$ such that $\delta_1 \notin \Delta(B)$. Then, since $B + \delta_1(B) = E$ and $|E/B| = 1$, $|B \cap \delta_1(B) = A| = 2$. Assume $\delta B \supset A$ for all $\delta \in T_E$. Then $\delta A = \delta B \cap \delta\delta_1 B \supset A$, and so $\delta A = A$ for all $\delta$, which is a contradiction. Hence there exists $\delta_2$ in $T_E$ such that $\delta_2(B) \not\supset A$. Then $B \cap \delta_1(B) \cap \delta_2(B) = S$ satisfies the condition in Lemma 4. Therefore $[\Delta : \Delta(B)] \geqslant 3$, so $[\Delta : \Delta(B)] = 3$. Accordingly, all B with $|B| = 3$ are isomorphic to one another by Lemma 3 and Corollary 3 to Theorem 4.

Conversely, let $D = \sum_{i=1}^{k} \oplus C_i$ be a direct sum of uniform modules in the theorem and S* a simple submodule in D. As noted in §1, we may assume $\mathrm{Soc}(C_i) = S$ for all i, where S is the socle of some fixed $E_j = E$. Assume that $C_1 = C_2 = \cdots = C_t = E$ and $C_{t+i} \neq E$ for $i > 0$. Put $D = D_1 \oplus D_2$: $D_1 = \sum_{i \leq t} \oplus C_i$ and $D_2 = \sum_{j>t} \oplus C_j$. S* is expressed as $S^* = \{g(s)+s \mid s \in$ (a simple submodule in $\mathrm{Soc}(D_2))\}$. Since $D_1$ is injective, the mapping is extendible to an element f in $\mathrm{Hom}_R(D_2, D_1)$. Hence $D = D_1 \oplus D_2(f)$ and $D_2(f) \supset S^*$, where $D_2(f) = \{f(x)+x \mid x \in D_2\}$. Therefore $D/S^* = D_1 \oplus D_2(f)/S$. If $D_2$ is simple, then we obtain the same situation by exchanging $D_1$ and $D_2$. Thus, in order to show d-(**), we may assume $C_i = A_1$ or $B_1$. If k =

1, or i) or ii) is occured, we are done by Theorem 2. First we assume cases iii) and iv).

1) $k = 2$.

Let $D = A_1 \oplus A_2$ and $S^* = \{s+f(s) \mid s \in Soc(A_1)\}$, where $f \in T_{Soc(A_1)}$. If f is extendible to an element f in $T_E^{(A_1,A_2)}$, $D = A_1(f) \oplus A_2$ and $S^* \subsetneq A_1(f)$. Hence

$D/S^* \approx A_1 \oplus A_1(f)/S^*$ and $A_1(f)/S^*$ is siomple (2).

Assume f is not extendible and g is an element in $T_E$ such that $g|Soc(A_1) = f$. Then $g(A_1) \neq A_2$, so $E = g(A_1) = A_2$, since $|E| = 3$. We define an epimorphism $\psi$ of $A_1 \oplus A_2$ onto E by setting $\psi(a_1+a_2) = g(a_1)-a_2$. Then $\ker \psi = S^*$ for $|E| = 3$ and $|D| = 4$. Hence

$D/S^* \approx E$ (3).

2) $k = 3$. $D = A_1 \oplus A_2 \oplus A_3$.

We note that there are at most two isomorphism classes of $A_i$. If $\Delta = \Delta(A_1)$, for some simple submodule $S^*$, $D = A'_1 \oplus A'_2 \oplus A'_3$ and $A'_2 \oplus A'_3 \supset S^*$ by Theorem 2. Hence

$D/S^* \approx A_1 \oplus C$, where C is of the form (2) or (3) (4).

Assume $\Delta \neq \Delta(A_1)$ and hence the case iv). Then we have the same situation as above by Corollary 1 to Theorem 3.

Next we assume Case v). Let A and B ($\supset A$) be submodules in E with $|A| = 2$ and $|B| = 3$, respectively.

1) $k = 2$.

(a) $D = A \oplus A \supset S^* = \{s+f_1(s) \mid s \in S, \bar{f}_1 \in \Delta\}$.

If $\bar{f}_1 \in \Delta(A)$, $D = A(f_1) \oplus A$ and $A_1(f) \supset S^*$. Hence

$D/S^* \approx A(f_1)/S^* \oplus A$ ((2)).

Assume $\bar{f}_1 \notin \Delta(A)$. Define a homomorphism $\psi$ of D to E by setting $\psi(a+a') = f_1(a)-a'$. Then $\ker \psi \supset S^*$. Since $|A| = 2$ and $\bar{f}_1 \notin \Delta(A)$, $S = \ker \psi$. Hence

D/S* is isomorphic to the submodule B of E (5).

(b) $D = A \oplus B \supset S^* = \{s+f_1(s) \mid s \in S\}$.

If $\bar{f}_1 \notin \Delta(A,B)$, $f_1(A) + B = E$. Hence

$D/S^* \approx E$ (6).

as above, provided $\bar{f}_1 \notin \Delta(A,B)$.

If $\bar{f}_1 \in \Delta(A,B)$, $D = A(f_1) \oplus B$ and $A(f_1) \supset S^*$. Hence

$D/S^* \approx B \oplus A(f_1)/S^*$ (7).

c) $D = B \oplus B \supset S^* = \{s+f_1(s) \mid s \in S\}$.

If $\bar{f}_1 \in \Delta(B)$, $D = B(f_1) \oplus B$ and $B(f_1) \supset S^*$. Hence

$D/S^* \approx B \oplus B(f_1)/S^*$, where $B(f_1)/S^*$ is semisimple (8).

Assume $\bar{f}_1 \notin \Delta(B)$. Then $E = B + f_1(B)$. Put $A_1 = B \cap f_1(B)$. Take some $A$ such that $E = B + A$ ($B \cap A = S$) and $|A| = 2$. Then there exists $g$ in $T_E^{(A_1,A)}$ by the assumption. Define a homomorphism $\psi$ of $D$ to $E \oplus E/B$ by setting $\psi(b_1+b_2) = (f_1(b_1)-b_2)+\widetilde{g(b_2)}$, where $\widetilde{g(b_2)}$ is the residue class of $g(b_2)$ in $E/B$. Then $\ker \psi = \{b_1+b_2 \mid b_2 = f_1(b_1) \in A_1$ and $g(b_2) \in B\}$. Since $g(A_1) = A$, $\ker \psi = \{b_1+b_2 \mid b_2 = f_1(b_1), b_2 \in S\} = S^*$. Hence $(B \oplus B)/S^*$ is isomorphic to $E \oplus E/B$. Therefore

$D/S^* \approx E \oplus E/B$ (9).

d) $D = B \oplus B \oplus B \supset S^* = \{s+f_1(s)+f_2(s) \mid s \in S\}$.

α) If $\bar{f}_1 \in \Delta(B)$, $D = B(f_1) \oplus B \oplus B$ and $B(f_1) \oplus 0 \oplus B \supset S^*$. Hence

$D/S^* \approx B \oplus C$, where $C$ is of the form (8) or (9) (10).

β) Assume that $\bar{f}_1 \notin \Delta(B)$, $\bar{f}_2 \notin \Delta(B)$ and $\bar{f}_2 = \bar{g}\bar{f}_1$ for some $\bar{g} \in \Delta(B)$. Then $S^* = \{s+f_1(s)+gf_1(s)\}$. Set $B(g) = \{0+b+g(b) \mid b \in B\}$. Then $D = B \oplus B \oplus B(g)$ and $B \oplus 0 \oplus B(g) \supset S^*$. Hence

$D/S^* \approx B \oplus C$ ((10)).

γ) Assume that $\{1, \bar{f}_1, \bar{f}_2\}$ is linearly dependent over $\Delta(B)$. Then we may assume that there exists $\{\bar{y}_1, \bar{y}_2\}$ in $\Delta(B)$ such that $\bar{1} = \bar{y}_1\bar{f}_1+\bar{y}_2\bar{f}_2$ from β) (note that $\bar{f}_i \notin \Delta(B)$). Then $D = B \oplus B(y_1) \oplus B(y_2)$, where $B(y_1) = \{y_1(b)+b+0 \mid b \in B\}$ and $B(y_2) = \{y_2(b)+0+b \mid b \in B\}$ and $s+f_1(s)+f_2(s) = (y_1+1)f_1(s) + (y_2f_2(s)+f_2(s)) \in B(y_1) \oplus By_2)$. Hence

$D/S^* \approx B + C$ ((10)).

δ) Finally assume $\{\bar{1}, \bar{f}_1, \bar{f}_2\}$ is independent over $\Delta(B)$. We shall show in the beginning that there exists $g$ in $T_E$ such that $B \cap g(B) \cap g^2(B) = S$. Let $B = A_1 + A_2$, where $|A_i| = 2$. Assume that $T_E^{(A_1,A_2)} \subset T_E^B$. Take an element $h$ in $T_E^{(A_1,A_2)}$ by the assumption v). Then $T_E^{(A_2)}h \subset T_E^{(A_1,A_2)} \subset T_E^B$. Hence $\Delta(A_2)h \subset \Delta(B)$ so

$\Delta(A_2) \subset \Delta(B)h^{-1} = \Delta(B)$, a contradiction. Hence there exists g in $T_E^{(A_1,A_2)} - T_E^{B}$. Since $|B \cap g(B)| = 2$, $B \cap g(B) = A_2$. If $g^2(B) \supset A_2$, $g^2(B) \supset A_2$, $g(A_2) = g(B) \cap g^2(B) \supset A_2$. Hence $A_2 = g(A_2)$, which contradicts the fact: $g \in T_E^{(A_1,A_2)}$. Therefore $B \cap g(B) \cap g^2(B)$ $(= g^{-2}(B) \cap g^{-1}(B) \cap B) = S$, and so $\{\bar{1}, \bar{g}, \bar{g}^2\}$ is independent over $\Delta(B)$ by Lemma 4. Let $D = B \oplus B \oplus B \supset S'^* = \{s+g(s)+g^2(s) \mid s \in S\}$. We shall define a mapping $\psi$ of D to $E \oplus E$ by setting $\psi(b_1+b_2+b_3) = (b_1-g^{-1}(b_2))+(b_1-g^{-2}(b_3))$. Let $x = b_1+b_2+b_3$ be in $\ker \psi$. Then $b_1 = g^{-1}(b_2) = g^{-2}(b_3) \in B \cap g^{-1}(B) \cap g^{-2}(B) = S$. Hence $\ker \psi = S'^*$. So $D/S'^* \approx E \oplus E$ since $|D/S'^*| = |E \oplus E| = 8$. Now we shall come back to Case v). We have shown that there exists g in $T_E$ such that $B \cap g(B) \cap g^2(B) = S$, and so $\{\bar{1}, \bar{g}, \bar{g}^2\}$ is a basis of $\Delta$ over $\Delta(B)$. Then there exists a unit matrix $(\bar{x}_{ij})$ in $(\Delta(B))_3$ such that

$$\begin{pmatrix} \bar{1} \\ \bar{f}_1 \\ \bar{f}_2 \end{pmatrix} = (\bar{x}_{ij}) \begin{pmatrix} \bar{1} \\ \bar{g} \\ \bar{g}^2 \end{pmatrix}$$

and the $x_{ij}$ are elements in $T_E^B$. Consider the mapping

$$E \xrightarrow{\;G = \begin{pmatrix} 1 \\ g \\ g^2 \end{pmatrix}\;} \begin{pmatrix} E \\ E \\ E \end{pmatrix} \xrightarrow{\;\bar{\Psi} = (x_{ij})\;} \begin{pmatrix} E \\ E \\ E \end{pmatrix}.$$

Since $\bar{\Psi}$ is an isomorphism (note that $J(T_E) = \ker \varphi$), $\bar{\Psi}(D) = D$ and $G(S) = S'^*$. Further $\bar{\Psi}(G(S)) = S^*$ and hence

$$D/S^* \approx E \oplus E \qquad (11).$$

from the above.

e) $D = A \oplus A \oplus B \supset S^* = \{f_2(s)+f_1(s)+s \mid s \in S\}$.

If $\bar{f}_1 = 0$, $S^* \subset A \oplus 0 \oplus B$. Hence

$$D/S^* \approx A \oplus C, \text{ where C is of the form (6) or (7)} \qquad (12).$$

If $\bar{f}_1 \neq 0$ is in $\Delta(A)$, $S^* = \{\bar{f}_1^{-1}\bar{f}_2(s)+s+\bar{f}_1^{-1}(s)\}$ and $\bar{f}_1^{-1} \in \Delta(A,B)$. Hence $D = A \oplus A(f_1^{-1}) \oplus B$ and $S^* \subset A \oplus A(f_1^{-1})$. Thus

$$D/S^* \approx B \oplus C, \text{ where C is of the form (2) or (5)} \qquad (13).$$

Finally assume $\bar{f}_1 \notin \Delta(A)$. Since $[\Delta : \Delta(A)] = 2$, we may assume that there exist $\bar{y}_1, \bar{y}_2$ in $\Delta(A)$ such that $\bar{f}_2 = \bar{y}_1\bar{f}_1+\bar{y}_2$. Then $D = A \oplus A(y_1) \oplus B$, $\bar{f}_2(s)+\bar{f}_1(s)+s = \bar{y}_1(s)+(\bar{y}_1\bar{f}_1(s)+\bar{f}_1(s))+s$, and $\bar{y}_1\bar{f}_1(s)$

$+f_1(s) \in A(y_1)$, where $A(y_1) = \{y_1(x)+x+0 \mid x \in A\}$. Further $\bar{y}_1 \in \Delta(A)$. Therefore

$$D/S^* \approx B \oplus C \qquad ((13)).$$

(f) $D = A \oplus B \oplus B \supset S^* = \{s+f_1(s)+f_2(s) \mid s \in S\}$.

It is **clear** that $\Delta \neq \Delta(A,B) \supset \Delta(A) \cup \Delta(B)$. If $\Delta(A,B) = \Delta(B)$, $\Delta(B) \supset \Delta(A)$, which is a contradiction. Hence $[\Delta / \Delta(A,B) : \Delta(B)] = 1$. Assume $\bar{f}_1 \notin \Delta(A,B)$. Then there exists $\bar{h}$ in $\Delta(B)$ such that $\bar{f}_2 = \bar{h}\bar{f}_1 \pmod{\Delta(A,B)}$. Let $\bar{f}_2 - \bar{h}\bar{f}_1 = \bar{g}$: $g \in T_E^{(A,B)}$. Then $D = A \oplus B(h) \oplus B$, $s+f_1(s)+f_2(s) = s+(\bar{f}_1(s)+\bar{h}\bar{f}_1(s))+\bar{g}(s)$, and $\bar{f}_1(s) + \bar{h}\bar{f}_1(s) \in B(h)$. Further $D = A(g) \oplus B(h) \oplus B$ and $S^* \subset A(g) \oplus B(h)$. Hence

$$D/S^* \approx B \oplus C, \text{ where C is of the form (6) or (7)} \qquad (14).$$

If $\bar{f}_1 \in \Delta(A,B)$, $D = A(f_1) \oplus B \oplus B$ and $A(f_1) \oplus 0 \oplus B \supset S^*$. Hence

$$D/S^* \approx B \oplus C \qquad ((14)).$$

(g) $D = A \oplus A \oplus A \supset S^* = \{s+f_1(s)+f_2(s) \mid s \in S\}$.

Since $[\Delta : \Delta(A)] = 2$ and $|A| = 2$, by Corollary 1 to Theorem 3

$$D/S^* \approx A \oplus C, \text{ where C is of the form (4) or (5)} \qquad (15).$$

3) $k = 4$.

h) $D = A \oplus A \oplus B \oplus B \supset S^* = \{s+f_1(s)+f_2(s)+f_3(s) \mid s \in S\}$, (or $D = A \oplus B \oplus B \oplus B$).

If $\bar{f}_2 \in \Delta(A,B)$, $D = A(f_2) \oplus A \oplus B \oplus B$ and $A(f_2) \oplus A \oplus 0 \oplus B \supset S^*$ ($D = A(f_2) \oplus B \oplus B \oplus B$ and $A(f_2) \oplus B \oplus 0 \oplus B \supset S^*$). Hence

$$D/S^* \approx B \oplus C, \text{ where C is of the form (12) or (13)} \qquad (16).$$
$$(D/S^* \approx B \oplus C, \text{ where C is of the form (14) or (15)} \qquad ((17)).$$

If $\bar{f}_2 \notin \Delta(A,B)$, by the proof of (f) $A(g) \oplus A \oplus B(h) \oplus 0 \supset S^*$ ($D = A(g) \oplus B \oplus B(h) \oplus B$ and $A(g) \oplus B \oplus B(h) \oplus 0 \supset S^*$). Hence

$$D/S^* \approx B \oplus C, \text{ where C is of the form (12) or (13)} \qquad ((16))$$
$$(D/S^* \approx B \oplus C) \qquad ((17)).$$

i) $D = A \oplus B \oplus B \oplus B \supset S^* = \{s+f_1(s)+f_2(s)+f_3(s) \mid s \in S\}$. ($D = A \oplus A \oplus A \oplus B$, $D = A \oplus A \oplus A \oplus A$ or $D = B \oplus B \oplus B \oplus B$).

We can reduce them to the case $k = 3$ by Corollary 1 to Theorem 3 and the proof of Lemma 1.

Thus we can show, by Corollary 1 to Theorem 3 and induction on k, that $E/S^*$ is a direct sum of uniform modules $U_i$ such that

$U_i$ is simple or $Soc(U_i) \approx S$ for all cases. Now we shall show that Condition d-I is satisfied for $U_i$. Let $D = \Sigma \oplus C_i$ and F a submodule of D, where the $C_i$ are uniform and $Soc(C_i)$ is isomorphic to Soc(E), where $E = E_j$ for some j. Let $S_1$ be a simple submodule of F. Then $D/S_1$ is a direct sum of uniform modules $C'_i$ such that $C'_i$ is simple or $Soc(C'_i) \approx Soc(E)$, and $D/S_1 \supset F/S_1$. Take a simple submodule $S_2/S_1$ in $F/S_1$. Then $D/S_2 = (D/S_1)/(S_2/S_1)$ is a direct sum of uniform modules. Repeating this manner, we can show that D/F is a direct sum of uniform modules as above.

References

[1] H.Asashiba: On algebras of second local type, to appear.

[2] M.Harada and K.Oshiro: On extending property on direct sums of uniform modules,1981, Osaka J. Math. 18, pp. 767-785.

[3] M.Harada: On self-miniinjective rings, 1982, ibid, 19, pp. 587-597.

[4] _______ : Uniserial rings and lifting properties,1982, ibid. 19, pp. 217-229.

[5] _______ : On maxi-quasiprojective modules, 1983, J. Austral. Math. Soc (Series A), 35, pp. 357-368.

[6] _______ : Serial rings and decompositions, to appear.

[7] _______ : On maximal submodules of a finite direct sum of hollow modules, I, to appear.

[8] _______ : II, to appear.

[9] E.Matlis: Injective modules over noetherian rings, 1958 Pacific J. Math. 8, pp. 511-528.

[10] T.Sumioka: Tachikawa's theorem on algebras of left colocal type, to appear.

[11] H. Tachikawa: On rings for which every indecomposable right modules has a unique maximal submodules, 1959, Math. Z. 71, pp. 200-222.

# FUNCTORS OF GRADED RINGS

Raymond T. Hoobler

The City College of New York

Suppose F is a functor on rings such that $F(A) \cong F(A[T])$. In the first section it is shown that $F(A) \cong F(A_0)$ for any graded ring A. Several examples of such functors are given of which the most important are the subgroup of the Brauer group consisting of elements of order prime to all of the residue characteristics of A and of commutative, separable A-algebras of rank relatively prime to the residue characteristics of A. In the second section the part of the Brauer group and of the fundamental group of $A[T,T^{-1}]$ of order relatively prime to the residue characteristics is computed.

It is well known that a cone has the homotopy type of its vertex. An algebraic analogue of this fact appears in print in (18, p. 477) and has been discovered independantly by many people during the past several years. No one, however, appears to have fully recognized its power for the study of positively graded rings. Suppose F is a functor on rings such that $F(A) \cong F(A[T])$. We construct a contracting homotopy to show that $F(A) \cong F(A_0)$ for any graded ring A. We then exhibit several functors satisfying this hypothesis. The most interesting one is $Br[p^{-1}]$, the part of the Brauer group of a commutative ring which consists of those elements of order relatively prime to all of the residue characteristics. The proof that this is invariant under polynomial ring extension is a consequence of Gabber's cohomological interpretation of the Brauer



*F. van Oystaeyen (ed.), Methods in Ring Theory, 161–170.*

group and some well known results in etale cohomology. This generalizes results in (4, 10). Finally, in the second part of this note, we investigate the case of $\mathbb{Z}$-graded rings. We obtain a description of the fundamental group and Brauer group of rings of the form $A[T_1,T_1^{-1},\ldots,T_n,T_n^{-1}]$.

Throughout this paper all rings are assumed to be noetherian. Moreover all cohomology groups are taken with respect to the etale topology unless otherwise indicated.

I would like to thank A. Verschoren for several useful references and D. Haile, J. LeBrun, D. Saltman, and A. Verschoren for their comments and suggestions on this material.

## 1. POSITIVELY GRADED RINGS

We begin with the case of positively graded rings where the homotopy result mentioned above may be formulated as follows.

Theorem 1.1(18, p. 477): Let F be a covariant functor from the category of rings to sets. Let A be a graded ring and suppose that the inclusion of A as constants in A[T] gives rise to an isomorphism $F(A) \cong F(A[T])$. Then $F(A_0) \cong F(A)$.

Proof: Since the inclusion of $A_0$ into A admits a section, we need only show that the inclusion induces a surjection from $F(A_0)$ to $F(A)$. Define $H:A \longrightarrow A[T]$ by $H(\Sigma a_i) = \Sigma a_i T^i$ where $a = \Sigma a_i$ is the decomposition of a into a sum of homogeneous elements. Let $e_0$ and $e_1$ be the maps from A[T] to A obtained by evaluating T at 0 and 1 respectively. Then $e_1H$ is the identity map on A and $e_0H$ is the projection from A to $A_0$ followed by the inclusion. But $F(A) \cong F(A[T])$ and so both of these composites define the same map from F(A) to F(A). In particular, the map defined from the inclusion is onto as desired.

There are a wide variety of applications of this theorem:

1) Let $F = K_0$ or $K_1$. Then if A is a left (or right) regular, graded ring, the hypothesis is satisfied(5). Consequently $K_0(A_0) \cong K_0(A)$ and $K_1(A_0) \cong K_1(A)$.

2) If the characteristic of A is $p^n > 0$, then $K_0(A) \longrightarrow K_0(A[T])$ has a p-torsion cokernel(6). The easiest way to see this is to first observe that $K_0(A)$ is

functorially a summand of $K_1(A[T])$. But the cokernel of $K_1(A) \longrightarrow K_1(A[T])$ is generated by unipotent matrices which have p power order in characteristic $p^n > 0$. Consequently $\mathbb{Q} \otimes K_0(A) \cong \mathbb{Q} \otimes K_0(A[T])$.

Bass(6) has introduced a marvelous functor which he denotes M(A). M(A) is defined by introducing an equivalence relation on the set of isomorphism classes of finitely generated, faithful, projective A-modules. The relation is given by setting $P \sim Q$ if there are positive integers m and n and an A-module isomorphism $P \otimes A^m \cong Q \otimes A^n$. The quotient set acquires a monoid structure via $\otimes_A$, and, amazingly enough, it becomes a group with respect to this operation. Bass then shows that M(A) may be identified with $\mathbb{Q} \otimes \tilde{K}_0(A) \oplus (\mathbb{Q} \otimes H(A))^+$ where $\tilde{K}_0(A)$ is the kernel of the rank homomorphism and $(\mathbb{Q} \otimes H(A))^+$ stands for maps from the set of connected components of Spec(A) to $\mathbb{Q}$ which take on everywhere positive values. In particular, M(A) is clearly torsion-free as a consequence of this result.

Let F(A) = M(A). Then, by virtue of the above comments, for any ring of characteristic $p^n > 0$, $M(A) \cong M(A[T])$. Consequently $M(A) \cong M(A_0)$ if A is any graded ring of characteristic $p^n > 0$. In particular, this means that for any faithful, projective A-module P, there are positive integers m and n and a faithful, projective $A_0$-module $P_0$ such that $P \otimes_A A^m \cong P_0 \otimes_{A_0} A^n$!

3) Let F = Pic. Seminormal rings have the property that $\text{Pic}(A) \cong \text{Pic}(A[T])$. This was essentially the original definition of them. A detailed discussion of them can be found in (7). For our purposes the definition of seminormal in (17) is best: A is said to be seminormal if A is reduced and whenever b and c are elements of A with $b^2 = c^3$, there is an element a of A with $a^2 = c$ and $a^3 = b$. We recover a result in (1) since we conclude that $\text{Pic}(A) \cong \text{Pic}(A_0)$ if A is a semi-normal, graded ring. In particular this applies if A is normal.

4) Lindel(13) has proven the Bass-Quillen conjecture that projective A[T]-modules are extended from projective A-modules when A is regular of finite type over a field. This allows us to strengthen our first example. We conclude that if A is a graded, regular ring of finite type over a field, then any projective A-module comes from a projective $A_0$-module by base extension. On the other hand the only examples of graded, regular, commutative rings are $A \cong S(P)$ where S(P) is the symmetric algebra of a

finitely generated, projective $A_0$-algebra and here the result can be established directly by Quillen induction.

5) Since polynomial extensions are acyclic for etale cohomology, we can conclude that the map Spec(A) $\to$ Spec($A_0$) is also acyclic. More specifically, if Y = Spec(A[T]), X = Spec(A), and F is any torsion sheaf whose torsion is prime to the residue characteristics of A, then

$$H^i(X, F) \cong H^i(Y, F|_Y)$$

for any i(14, Ch. VI, 4.20). Thus if A is graded and $X_0$ = Spec($A_0$), we conclude that

$$H^i(X_0, F) \cong H^i(X, F|_X)$$

for any i and any torsion sheaf F whose torsion is prime to the residue characteristics of A.

6) The arguments in (11, 16) show that

$$Br(X) \cong H^2(X, G_m)_{tors}$$

if X = Spec(A) is affine. On the other hand the interpretation of Pic(X) as $H^1(X, G_m)$ is well known. Finally, we note that taking $n^{th}$ roots of units can be done locally in the etale topology and so the Kummer sequence,

$$0 \to \mu_n \to G_m \to G_m \to 0,$$

is exact if n is relatively prime to the residue characteristics of A. The long exact cohomology sequence of this short exact sequence of sheaves then shows that

$$0 \to Pic(A)/nPic(A) \to H^2(A, \mu_n) \to {}_nBr(A) \to 0$$

where ${}_nBr(A)$ denotes the elements of order n in Br(A).

When we investigate the behaviour under polynomial extensions of the three groups in this short exact sequence, we note first that the middle cohomology group is unchanged by 5). But then ${}_nBr(\ )$ is also unchanged since it it a quotient of the cohomology group and the map is always injective. Finally, we conclude that Pic( ) $\otimes$ $\mathbf{Z}/n\mathbf{Z}$ is also unchanged. Note that this implies Pic(A) cannot be finitely generated if A contains $\mathbf{Q}$ and is not a seminormal ring which explains why the definition of seminormality requires that the generic additive singularity not appear. On the other hand, if A is a ring of characteristic $p^n$, then the only change in the torsion in Pic under a polynomial extension has p power order. This also follows from the arguments in 2) since Pic is a summand of $K_0$ by the determinant map.

When we consider the behaviour of Br( ) under polynomial extensions one prime at a time we see that $Br(A)[p^{-1}] \cong Br(A[T])[p^{-1}]$ wwhere $Br( )[p^{-1}]$ denotes the subgroup of the Brauer group consisting of those elements whose order is prime to all of the residue characteristics of A. Accordingly we conclude from the theorem that $Br(A)[p^{-1}] \cong Br(A_0)[p^{-1}]$ if A is any graded ring.

7) As a final example consider the functor F where F(A) is the set of finite, separable, commutative A-algebras of rank relatively prime to the residue characteristics of A. Then the theory of the fundamental group(15) shows that any such A-algebra can be embedded in a Galois A-algebra of rank relatively prime to the residue characteristics of A. But Galois A-algebras are classified by $H^1(A, G)$ where G is the Galois group of the extension. A morphism $f:X \longrightarrow Y$ is said to be 1-asperical if $f^*:H^1(Y, G) \longrightarrow H^1(X, G|_X)$ is an isomorphism for all finite, not necessarily abelian groups G of order relatively prime to the residue characteristics. In (3, Expose XV, Corollaire 2.2) it is shown that the map from $Spec(A[T]) \longrightarrow Spec(A)$ is 1-aspherical. Consequently we conclude that if A is any graded ring and B is a separable, commutative A-algebra of rank relatively prime to the residue characteristics of A, then there is a finite, separable $A_0$-algebra $B_0$ and an isomorphism $B \cong B_0 \otimes_{A_0} A$!

The interested reader should refer to (12) for a more detailed discussion of these examples with elementary proofs.

## 2. $\mathbb{Z}$-GRADED RINGS

In this section we turn our attention to $\mathbb{Z}$-graded rings. If A is generated over a subring $A_0$ by a unit of degree 1, then $A \cong A_0[T,T^{-1}]$. Consequently we direct our attention towwards such rings.

We will approach these rings by using etale cohomology, but most of the analysis does not require any particularly difficult facts beyond the Leray spectral sequence. The best general reference for this in (2) if it can be found or (14). Similar results were obtained in the characteristic 0 case by Tim Ford(8).

Our analysis is based on the following diagram

$$\begin{array}{ccc} \mathrm{Spec}(A[T,T^{-1}]) & \longrightarrow & \mathrm{Spec}(A[T]) \\ & {\searrow}^{j} & \downarrow k \\ & & \mathrm{Spec}(A) \end{array}$$

where A is a reduced, commutative ring and the maps are the obvious ones. Since A is reduced, the units of A[T] coincide with the units of A(6, p. 673). Moreover since an etale extension of a reduced ring is reduced, we conclude that this statement holds on the level of functors. Thus if we let $G_m$ stand for the sheaf of units, we get a short, split-exact sequence of sheaves on $A_{et}$

$$0 \longrightarrow k_* G_m \longrightarrow j_* G_m \xrightarrow{d} \mathbb{Z} \longrightarrow 0 \quad (2.1)$$

where d is the "degree" map from the units of $A[T,T^{-1}]$ to the integers and $k_* G_{m,A[T]} = G_{m,A}$.

Let us begin our study with the Picard group of such rings.

Proposition 2.1: If A is a normal ring, then Pic(A) $\cong$ Pic($A[T,T^{-1}]$).

Proof: Since the inclusion of A in $A[T,T^{-1}]$ has a section, we need only show that the map Pic(A) $\longrightarrow$ Pic($A[T,T^{-1}]$) is onto. If L is in Pic($B[T,T^{-1}]$), then there is a reflexive, rank one B[T]-module M such that $M \otimes_{B[T]} B[T,T^{-1}] \cong L$ since the divisor class group of B[T] maps onto the divisor class group of $B[T,T^{-1}]$ and the divisor class group may be interpreted as the group of isomorphism classes of reflexive, rank one modules. Now Samuel has shown that any reflexive, rank one B[T]-module is extended from B(9). Consequently there is a B-module N such that $N \otimes_B B[T,T^{-1}] \cong L$. Since $B[T,T^{-1}]$ is a faithfully flat extension of B, N is invertible and so belongs to Pic(B) as desired.

Corollary 2.2: If A is a normal ring then $R^1 k_* G_m = R^1 j_* G_m = 0$.

Proof: We compute the stalks of these two sheaves by computing $H^1(B[T], G_m)$ and $H^1(B[T,T^{-1}], G_m)$ respectvely where B is the strict henselization of A at a prime ideal of A. Since A is normal so is B. Moreover B is local and thus Pic(B) and so Pic(B[T]) and Pic($B[T,T^{-1}]$) vanish.

We are now in a position to calculate Br($A[T,T^{-1}]$). We

will need the following definition.

Definition 2.3: If B is an A-algebra, then

$$Br_{sp}(B) = Ker[Br(B) \to \prod Br(B \otimes_A A_m^{hs})]$$

where the product is taken over all maximal ideals m in A and $A_m^{hs}$ is the strict henselization of A with respect to m.

Our theorem alloww s us to calculate $Br_{sp}(A[T,T^{-1}])$.

Theorem 2.4: Let A be a normal ring. Then there is a split exact sequence

$$0 \to Br(A) \to Br_{sp}(A[T,T^{-1}]) \to Hom_{cont}(\pi^1(A), \mathbb{Q}/\mathbb{Z}) \quad 0$$

Moreover, in the notation of the first section,

$$Br_{sp}(A[T,T^{-1}])[p^{-1}] = Br(A[T,T^{-1}])[p^{-1}].$$

Proof: The cohomology of the split exact sequence (2.1) is a long exact sequence which splits into a collection of short exact sequences. The one of interest to us is

$$0 \to H^2(A, G_m) \to H^2(A, j_*G_m) \to H^2(A, \mathbb{Z}) \to 0$$

Since the cohomology sequence above is split exact, it remains so if we take torsion subgroups. $_2Br(A)$ is identified with the torsion subgroup of $H^2(A, G_m)$. $H^1(A, \mathbb{Q})$, being a $\mathbb{Q}$-vector space, is divisible and torsion free. Thus $H^1(A, \mathbb{Q}/\mathbb{Z})$, being entirely torsion, is equal to the torsion subgroup of $H^2(A, \mathbb{Z})$. But $H^1(A, \mathbb{Q}/\mathbb{Z})$ classifies etale covering spaces with cyclic group actions and so is given by $Hom_{cont}(\pi^1(A), \mathbb{Q}/\mathbb{Z})$.

Finally we must interpret the middle group. $R^1j_*G_m = 0$ since A is normal. Thus the short exact sequence of low degree terms for the Leray spectral sequence of j becomes

$$0 \to H^2(A, j_*G_m) \to H^2(A[T,T^{-1}, G_m) \to H^0(A, R^2j_*G_m)$$

and this, on the torsion subgroups, is precisely the definition of $Br_{sp}(A[T,T^{-1}])$.

The last assertion of the theorem requires the evaluation of the stalks of $R^2 j_* G_m$; that is, we must identify the torsion prime to the residue characteristics of A in $H^2(A[T,T^{-1}], G_m)$ when A is a strictly local ring. But, just as before, the Kummer sequence comes to our rescue. It describes the n-torsion as $H^2(A[T,T^{-1}], \mu_n)$ since $Pic(A[T,T^{-1}]) = 0$. The general yoga of smooth base change (14) shows that

$$H^2(A[T,T^{-1}], \mu_n) \cong H^2(k(A)[T,T^{-1}], \mu_n)$$

where k(A) is the residue field of the strictly local ring A. Since k(A)(T) is a $C_1$ field, $Br(k(A)[T,T^{-1}])$, being contained in Br(k(A)(T)), vanishes.

It seems reasonable to conjecture that, for A a strictly local ring,

$$Br(A[T]) \longrightarrow Br(k(A)[T]) \text{ and}$$

$$Br(A[T,T^{-1}]) \longrightarrow Br(k(A)[T,T^{-1}])$$

are monomorphisms, even isomorphisms, without the assumption that we are dealing with torsion of order relatively prime to the residue characteristics. If this were correct, we would conclude that $Br(A) \cong Br(A[T])$ and $Br_{sp}(\mathbb{A},T^{-1}]) \cong Br(A[T,T^{-1}])$ if A is of finite type over the prime ring since the Nullstellensatz would then tell us that the residue fields of all the strict henselizations of A at maximal ideals are algebraically closed.

We need to determine what happens to the fundamental group of $A[T,T^{-1}]$ in order to compute the Brauer group of a Laurent polynomial ring in several variables.

Proposition 2.5: Let A be a connected, normal ring. If n is relatively prime to the residue characteristics of A, then there is a split-exact sequence

$$0 \longrightarrow H^1(A, \mathbb{Z}/n) \longrightarrow H^1(A[T,T^{-1}], \mathbb{Z}/n) \longrightarrow \mu_n^{-1} \longrightarrow 0$$

where $\mu_n^{-1} = Hom(\mu_n, \mathbb{Z}/n)$.

Proof: We get the diagram below by looking at multiplication by n on our basic exact sequence of sheaves. The middle column is $j_*$ applied to the Kummer sequence of $A[T,T^{-1}]$.

$$\begin{array}{ccccccccc}
 & & 0 & & 0 & & & & \\
 & & \downarrow & & \downarrow & & & & \\
 & & \mu_n & \cong & j_*\mu_n & & & & \\
 & & \downarrow & & \downarrow & & & & \\
0 & \to & G_m & \to & j_*G_m & \to & \mathbb{Z} & \to & 0 \\
 & & \downarrow & & \downarrow & & \downarrow & & \\
0 & \to & G_m & \to & j_*G_m & \to & \mathbb{Z} & \to & 0 \\
 & & \downarrow & & \downarrow & & \downarrow & & \\
 & & 0 & & R^1 j_*\mu_n & \cong & \mathbb{Z}/n & & \\
 & & & & \downarrow & & \downarrow & & \\
 & & & & 0 & & 0 & &
\end{array}$$

Consequently $H^0(A, R^1 j_*\mu_n) = \mathbb{Z}/n$ since A is connected. Hence we arrive at the exact sequence below from the Leray spectral sequence for j since the inclusion of A into $A[T,T^{-1}]$ has a section.

$$0 \to H^1(A, \mu_n) \to H^1(A[T,T^{-1}], \mu_n) \to \mathbb{Z}/n \to 0$$

If we tensor this sequence with $\mu_n^{-1}$, we get the desired result.

Corollary 2.6: Let $B = A[T_1, T_1^{-1}, \ldots, T_r, T_r^{-1}]$ where A is a connected, normal ring. Suppose n is an integer relatively prime to the residue characteristics of A. Then

$$H^1(B, \mathbb{Z}/n) = H^1(A, \mathbb{Z}/n) \oplus (\oplus^r \mu_n^{-1}).$$

and

$${}_n\mathrm{Br}(B) = {}_n\mathrm{Br}(A) \oplus (\oplus^r H^1(A, \mathbb{Z}/n)) \oplus (\oplus^{r(r-1)/2} \mu_n^{-1})$$

In particular, ${}_n\mathrm{Br}(B) = {}_n\mathrm{Br}(A)$ if A is simply connected and contains no $n^{th}$ roots of unity.

Proof: This is a straightforward induction argument once we note that $A[T,T^{-1}]$ is normal.

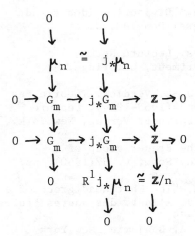

BIBLIOGRAPHY

1. D.D. Anderson and D.F. Anderson, "Divisorial Ideals and Invertible Ideals in a Graded Integral Domain", J. of Algebra, v. 76(1982), pp.549-569.

2. M. Artin, Grothendieck Topologies, Lecture Notes, Harvard University Mathematics Department, Cambridge, Mass., 1962.

3. M. Artin, A. Grothendieck, and J.-L. Verdier, Theorie des topos et cohomologie etale des schemas(1963-1964), Lecture Notes in Mathematics 269, 270, 305,Springer Verlag, New York, 1972-1973.

4. M. Auslander and O. Goldman, "The Brauer group of a commutative ring", Trans. Amer. Math. Soc., v. 97(1960), pp. 367-409.

5. H. Bass, A. Heller, and R. Swan, "The Whitehead group of a polynomial extension", Publ. Math. de l'Inst. des Hautes Etudes Scient., Paris, v. 22(1964), pp. 61-79.

6. H. Bass, Algebraic K-Theory, W. A. Benjamin, New York, 1968.

7. D. Costa, "Semi-normalty and projective modules", Seminaire d'Algebre Paul Dubreil et Marie Paule Mailliavin, Lecture Notes in Mathematics 924, Springer Verlag, New York, 1982, pp. 400-412.

8. T. Ford, thesis, "Every finite group is the Brauer group of a ring", 1981.

9. R. Fossum, The Divisor Class Group of a Krull Domain, Springer Verlag, New York, 1973.

10. P. Griffiths, "The Brauer group of A[T]", Math. Zeit., v. 147(1976), pp. 79-86.

11. R. Hoobler, "When is Br(X) = Br'(X)?", Brauer Groups in Ring Theory and Algebraic Geometry, Lecture Notes in Mathematics 917, Springer Verlag, New York, 1982, pp. 231-245.

12. R. Hoobler, in preparation.

13. H. Lindel, "On the Bass-Quillen conjecture concerning projective modules over polynomial rings", Invent. Math., v. 65(1981), pp. 319-323.

14. J. Milne, Etale Cohomology, Princeton Mathematical Series, no. 33, Princeton University Press, Princeton, New Jersey.

15. J. Murre, Lectures on an introduction to Grothendieck's theory of the fundamental group, Lecture notes, Tata Institute of Fundamental Research, Bombay, 1967.

16. O. Gabber, "Some theorems on Azumaya algebras", Le Groupe de Brauer, Lecture Notes in Mathematics 844, Springer Verlag, New York, 1981.

17. R. Swan, "On seminormality", J. of Algebra, v.67(1980), pp. 210-229.

18. C. Weibel, "Mayer-Vietoris sequences and module structures on $NK_*$", Algebraic K-Theory: Evanston 1980, Lecture Notes in Mathematics 854, Springer Verlag, New York, 1981, pp. 466-493.

SUR UNE CLASSE D'ALGEBRES FILTREES

A. Hudry

Université Claude Bernard - Lyon I
69622 Villeurbanne Cedex, France

Abstract : Gelfand-Kirillov dimension is used in the study of some filtered algebras to find the two-sided ideals that can be decomposed as a finite intersection of primary two-sided ideals.

1. INTRODUCTION.

Soit k un corps commutatif. Dans tout ce qui suit A désigne, suivant la terminologie de [17], une k-algèbre presque commutative c'est-à-dire une k-algèbre de type fini dont l'anneau gradué associé à une filtration standard définie par un ensemble fini de générateurs est commutatif ; ces algèbres ont été considérées par D. Quillen [16], J. Dixmier et G. Cauchon [3] ; les algèbres enveloppantes des k-algèbres de Lie de dimension finie, les k-algèbres de Poincaré [12] de type fini, les algèbres de Mc Connell [4], les algèbres de Weyl, sont de telles algèbres ; le quotient, le produit tensoriel de k-algèbres presque commutatives, sont encore des k-algèbres presque commutatives. Ici la question suivante est considérée ;

Question : Quels sont les idéaux bilatères de A qui peuvent être décomposés en une intersection finie d'idéaux bilatères primaires d'un côté ?

Ici primaire signifie primaire au sens de O. Goldman [6] ou encore tertiaire, car dans un anneau noethérien des deux côtés, il y a équivalence pour un idéal bilatère entre être primaire au sens de O. Goldman d'un côté et être tertiaire du même côté (bien sûr ceci est faux pour un idéal d'un côté). Dans le cas d'un anneau noethérien totalement borné la question précédente a été résolue par



F. van Oystaeyen (ed.), Methods in Ring Theory, 171–179.

R. Gordon [7]. Dans le cadre d'une k-algèbre presque commutative A, la réponse à cette question n'est pas aussi simple car un idéal bilatère de A n'est pas nécessairement intersection d'un nombre fini d'idéaux bilatères de A primaires d'un côté : en effet d'après [2], tout idéal bilatère de l'algèbre enveloppante $U\mathfrak{g}$ d'une $\mathbb{C}$-algèbre de Lie $\mathfrak{g}$ de dimension finie, admet une décomposition primaire bilatère si et seulement si $\mathfrak{g}$ est résoluble. Dans un article à paraître [9] nous donnons une réponse partielle à la question précédente ; ici nous nous proposons d'en donner une réponse plus complète. Bien sûr les résultats de [9] utilisés ici sont cités sans démonstration.

## 2. DIMENSION DE GELFAND-KIRILLOV ET LOCALISATION.

Tout d'abord il peut être observé que l'existence d'une décomposition d'un idéal bilatère de A en une intersection finie d'idéaux bilatères primaires dépend de l'existence de "bons" idéaux bilatères inter-irréductibles ; un idéal bilatère B de A inter-irréductible dans le treillis des idéaux bilatères de A, est dit bon s'il satisfait aux deux conditions suivantes :

i) B est primaire à droite et à gauche,
et ii) $A/_B$ a un anneau classique de fractions artinien.

Exemples.
1) Soient, U l'algèbre enveloppante de $sl(2,\mathbb{C})$, $U_+$ l'idéal d'augmentation de U et q l'élément de Casimir de $sl(2,\mathbb{C})$ ; l'idéal bilatère $B = q\,U_+ = U_+ q$ est inter-irréductible et ne satisfait ni à la condition i), ni à la condition ii), donc n'est pas bon.
2) Soit $\mathfrak{g}$ une k-algèbre de Lie de dimension finie résoluble ; alors tous les idéaux bilatères inter-irréductibles de $U\mathfrak{g}$ sont bons.

Pour répondre à la question posée, il nous faut donc caractériser les bons idéaux bilatères inter-irréductibles ; au vu de la condition i), quelques résultats concernant les localisations artiniennes sont nécessaires. Pour cela les notations suivantes sont adoptées : pour tout A-module à droite M de type fini, on note GK(M) la dimension de Gelfand-Kirillov de M (d'après la proposition 4.6 de [19], GK(M) est l'entier naturel qui est le degré du polynôme de Hilbert-Samuel associé à une filtration standard de M) ; si M est un (A ; A)-bimodule de type fini à droite et à gauche, d'après [14] sa dimension de Gelfand-Kirillov à droite est la même que celle à gauche et la notation GK(M) n'est pas ambigüe ; pour tout entier $m < GK(A)$, l'ensemble topologisant et idempotent des idéaux à droite I de A tels que $GK(A/_I) < m$ est noté $\tau_m$ et pour tout A-module à droite M on pose, $\tau_m(M) = \{x \in M \mid r_A(x) \in \tau_m\}$ ; enfin un idéal bilatère B de A est dit localisable si la partie multiplicative $\mathscr{C}(B)$ des éléments de A réguliers modulo B vérifie la condition de Ore, et dans ce cas le localisé correspondant de A est noté $A_{\mathscr{C}(B)}$.

Proposition 1. - *Soit* n = GK(A). *Alors on a :*

a) $\{P \in \mathrm{Spec}(A) \mid GK(A/P) = n\}$ *est un ensemble fini d'idéaux premiers minimaux et l'ensemble topologisant et idempotent des idéaux à droite associé à l'idéal semi-premier*

$$K = \bigcap_{P \in \mathrm{Spec}(A), GK(A/P) = n} P$$

*est précisément* $\tau_{n-1}$.

b) $\mathscr{C}(K)$ *vérifie la condition de Ore à droite et à gauche.*

c) $A_{\mathscr{C}(K)}$ *est artinien et c'est l'anneau classique des fractions de* $A/\tau_{n-1}(A)$.

Démonstration. - La démonstration résulte du lemme 2 que nous avons donné dans [9] et du fait que les sous-catégories localisantes $C_m$, associées aux ensembles topologisants et idempotents $\tau_m$, sont invariantes par idéaux bilatères (cette dernière remarque a été faite dans le cadre des algèbres enveloppantes des $\mathbb{C}$-algèbres de Lie de dimension finie dans le théorème 1.8 de [13]).

Rappelons que suivant [11] un A-module à droite M est dit GK-homogène si pour tout sous-module non nul N de M on a GK(M) = GK(N). Si on applique la proposition 1 dans le cas particulier où A est un quotient GK-homogène de l'algèbre enveloppante d'une $\mathbb{C}$-algèbre de Lie de dimension finie, on obtient alors comme conséquence le théorème 2.4 de [13]. Le lemme suivant, bien que facile à établir, est utile.

Lemme 2. - *Pour tout* (A ; A)-*bimodule* M *de type fini à droite et à gauche, les propriétés suivantes sont équivalentes :*

a) $M_A$ *est* GK-*homogène ;*

b) *Pour tout sous-*(A ; A)-*bimodule non nul* N *de* M *on a,* GK(N) = GK(M) ;

c) ${}_AM$ *est* GK-*homogène ;*

d) *Pour tout* $P \in \mathrm{Ass}_r(M)$ *on a,* GK(M) = GK(A/P) ;

e) *Pour tout* $Q \in \mathrm{Ass}_\ell(M)$ *on a,* GK(M) = GK(A/Q) ;

*où* $\mathrm{Ass}_r(M)$ *(resp.* $\mathrm{Ass}_\ell(M)$*) désigne l'ensemble des idéaux premiers associés à* $M_A$ *(resp.* ${}_AM$*).*

Définition. - Un (A ; A)-bimodule de type fini à droite et à gauche est dit GK-homogène s'il vérifie l'une des conditions équivalentes du lemme 2 ; un idéal bilatère B de A est dit GK-homogène dans A si le (A ; A)-bimodule A/B est GK-homogène.

Exemples.

1) Tout idéal bilatère de A qui est premier ou plus généralement semi-premier de la forme $K = P_1 \cap \ldots \cap P_s$ avec $P_1, \ldots, P_s \in \mathrm{Spec}(A)$ tels que $GK(A/P_1) = \ldots = GK(A/P_s)$, est GK-homogène dans A.

2) Il va résulter de la proposition 4 que tout idéal bilatère de A primaire d'un côté est GK-homogène dans A. Dans le cas particulier où A est le quotient de l'algèbre enveloppante d'une $\mathbb{C}$-algèbre de Lie de dimension finie, cet exemple implique le théorème 4.1 de [13].

Lemme 3. - *Si* M *est un* A*-module à droite finiment annulé (c'est le cas si* M *est un sous-*A*-module à droite d'un* (A ; A)*-bimodule de type fini à gauche) et extension essentielle d'un objet de* $C_m$, *alors* M *est un objet de* $C_m$.

Démonstration. - Par hypothèse $\tau_m(M)$ est essentiel dans M. Supposons que l'on ait $r_A(M) \notin \tau_m$. On peut alors démontrer que si P est maximal dans l'ensemble des annulateurs $r_A(N)$ des sous-modules non nuls N de $A/_{r_A(M)}$ vérifiant $r_A(N) \notin \tau_m$, alors P est premier ; comme P est de la forme $r_A(B/_{r_A(M)})$ où B est un idéal bilatère de A, il existe un monomorphisme de $A/_P$ dans une somme directe d'un nombre fini de copies de $B/_{r_A(M)}$. Or M étant finiment annulé, $\tau_m(A/_{r_A(M)})$ est essentiel dans $A/_{r_A(M)}$ ce qui implique d'après ce qui précède que $\tau_m(A/P)$ est essentiel dans $A/_P$. Ceci contredit alors le fait que $P \notin \tau_m$. Le lemme 3 en résulte.

Remarque. - En particulier, si M est un sous-A-module d'un A-module à droite projectif de type fini et si M est extension essentielle d'un module de Bernstein (c'est-à-dire d'un module dont la GK-dimension est minimum dans l'ensemble des GK-dimensions des A-modules à droite non nuls) alors M est de Bernstein.

Proposition 4. - *Soit* M *un* (A ; A)*-bimodule de type fini à droite et à gauche tel que* $\mathrm{Ass}_r(M) = \{P\}$. *Alors* M *est* GK*-homogène.*

Démonstration. - Posons $m = GK(A/P)$. Comme $P = r_A(\ell_M(P))$, il existe un monomorphisme de A-modules à droite de $A/_P$ dans un nombre fini de copies de $\ell_M(P)$ ce qui implique $m = GK(\ell_M(P))$. Compte tenu de ce que $\ell_M(P)$ est essentiel dans le A-module à droite finiment annulé M, le lemme 3 implique $\tau_m(M) = M$ et finalement $m = GK(M)$. Le lemme 2 permet alors de conclure.

Dans un anneau noethérien à droite et à gauche, on ne sait pas en général si un idéal premier vérifiant la condition A.R. est localisable. Le résultat suivant montre que la réponse est affirmative dans une k-algèbre presque commutative A et ceci non seulement pour un idéal premier, mais pour tout idéal GK-homogène dans A.

Proposition 5. - *Pour tout idéal bilatère* B GK-*homogène dans* A, *et tout entier* $p \geq 1$, $B/_{B^p}$ *est localisable dans* $A/_{B^p}$. *Si de plus* B *vérifie la condition* A.R., *alors* B *est localisable dans* A.

Remarque. - Dans le cas particulier où A est le quotient de l'algèbre enveloppante d'une $\mathbb{C}$-algèbre de Lie de dimension finie et où B est un idéal premier, la proposition 5 redonne le théorème 3.1 de [13].

## 3. CARACTERISATION DES BONS IDEAUX INTER-IRREDUCTIBLES.

Un (A ; A)-bimodule non nul M est dit GK-critique si on a $GK(M/N) < GK(M)$ pour tout sous-bimodule non nul N de M. (Pour une définition de cette notion dans le cadre des A-modules d'un côté voir [11] ou [13]).

Lemme 6. - *Soit* M *un* (A ; A)-*bimodule* GK-*critique de type fini à droite et à gauche. Alors on a :* $\mathrm{Ass}_r(M) = \{r_A(M)\}$ *et* $\mathrm{Ass}_\ell(M) = \{\ell_A(M)\}$, *avec* $GK(A/_{r_A(M)}) = GK(A/_{\ell_A(M)}) = GK(M)$.

Démonstration. - Voir [9].

Un bimodule du type précédent est alors primaire au sens de O. Goldman, à droite et à gauche et le lemme suivant montre qu'il en existe assez.

Lemme 7. - *Tout* (A ; A)-*bimodule non nul de type fini à droite et à gauche contient un sous-bimodule* GK-*critique.*

Démonstration. - Elle est basée sur un argument standard utilisant la multiplicité de M, (cet argument est décrit par exemple dans [13] et [18]).

Proposition 8. - *Soit* B *un idéal bilatère de* A, *inter-irréductible dans le treillis des idéaux bilatères de* A ; *alors les conditions suivantes sont équivalentes :*

a) B *est bon ;*
b) B *est primaire à droite dans* A ;
c) B *est* GK-*homogène dans* A ;
d) $A/_B$ *a un anneau classique de fractions artinien ;*
e) B *est primaire à gauche dans* A.

Démonstration.

(b) ⇒ (c) : Il existe d'après le lemme 7 un sous-bimodule GK-critique X de $A/_B$. Posons $m = GK(X)$. Compte tenu des lemmes 6 et

3, il peut être démontré que l'on a $\tau_m(A/_B) = A/_B$ et $\tau_{m-1}(A/_B) = 0$. Le c) en résulte.

(c) ⇒ (d) : Résulte de la proposition 1.

(d) ⇒ (b) : D'après le lemme 7, $A/_B$ contient un sous-bimodule X GK-critique. Soit $P = r_A(X)$. D'après le lemme 6, il vient $P \in Ass_r(A/_B)$. Soit $Q \in Ass_r(A/_B)$. Il existe un sous-bimodule non nul Y de $A/_B$ tel que $Q = r_A(Y)$. $Y \cap X$ est un sous-bimodule de $A/_B$ GK-critique tel que $GK(X) = GK(Y \cap X)$. Compte tenu du lemme 6 on a, $GK(A/_{r_A(Y \cap X)}) = GK(A/_P)$. Par suite l'inclusion $P \subset r_A(Y \cap X)$ implique $P = r_A(Y \cap X)$. On a donc les inclusions $B \subset Q \subset P$ avec $P/_B \in Ass_r(A/_B)$ ; or un résultat attribué à Warefield dans [5], implique que P est un idéal premier minimal au-dessus de B, donc que l'on a P = Q.

Les autres implications sont claires.

Remarque. - Si B est un idéal bilatère inter-irréductible dans le treillis des idéaux bilatères et de plus Krull-homogène à droite dans A, la proposition 8 montre que B est bon. Si A était Krull-symétrique, la réciproque serait vraie ; mais même dans le cas où A est l'algèbre enveloppante d'une k-algèbre de Lie $\mathfrak{g}$ de dimension finie, on ne sait pas si A est Krull-symétrique : le seul résultat dans cette direction est dû à A.G. Heinicke [8] et ne concerne que la propriété de Krull-symétrie faible lorsque $\mathfrak{g}$ est résoluble et caract(k) = 0.

Dans [9], nous avons montré comment les lemmes 6 et 7 permettent d'obtenir une décomposition primaire pour les (A ; A)-bimodules de type fini GK-homogènes ; compte tenu de la proposition 8, le résultat correspondant concernant les idéaux bilatères devient :

Proposition 9. - *Soit* B *un idéal bilatère* GK*-homogène dans* A *avec* $GK(A/_B) = m$. *Alors il existe un nombre fini de bons idéaux bilatères inter-irréductibles (dans le treillis des idéaux bilatères de* A) $B_1, B_2, \ldots, B_s$ *tels que les conditions suivantes soient réalisées :*

a) $GK(A/_{B_i}) = m$ *pour* $i = 1, \ldots, s$.

b) $Ass_r(A/_B) = \bigcup_{i=1}^{s} Ass_r(A/_{B_i})$ *et* $Ass_\ell(A/_B) = \bigcup_{i=1}^{s} Ass_\ell(A/_{B_i})$.

c) $B = \bigcap_{i=1}^{s} B_i$ *est une décomposition irredondante (i.e. pour tout* $i \in \{1, \ldots, s\}$ *on a* $\bigcap_{j \neq i} B_j \not\subset B_i$*) de* B *en idéaux bilatères primaires à droite et à gauche.*

d) *Si* $\{Q_{j_i}\} = Ass_\ell(A/_{B_i})$ *et* $\{P_{k_i}\} = Ass_r(A/_{B_i})$, *il existe suivant la terminologie de* [10] *un lien* $Q_{j_i} \hookrightarrow P_{k_i}$.

e) *La décomposition* c) *peut être réduite en une intersection finie d'idéaux bilatères primaires à droite (resp. à gauche).*

Avec les notations précédentes $A/_B$ a un anneau classique de fractions artinien d'après la proposition 1. Le corollaire suivant généralise donc la proposition 9.

Corollaire 10. - *Tout idéal bilatère* B *de* A *pour lequel* $A/_B$ *admet un anneau classique de fractions artinien, est intersection d'un nombre fini de bons idéaux bilatères inter-irréductibles (dans le treillis des idéaux bilatères de* A*).*

Remarque. - Dans le cas où A serait Krull-symétrique, ce corollaire pourrait être obtenu à partir du théorème 16 de [15], mais ici on ne fait pas d'hypothèse de Krull-symétrie.

Exemple. - Supposons que A soit un ordre maximal dans un anneau de fractions (ou un ordre maximal généralisé dans une extension rationnelle bilatère) : c'est le cas des algèbres enveloppantes des k-algèbres de Lie de dimension finie ou plus généralement des k-algèbres de Poincaré de type fini. Soit alors B un idéal bilatère réflexif de A. On peut démontrer que $A/_B$ admet un anneau classique de fractions artinien et qu'il existe un nombre fini de bons idéaux bilatères inter-irréductibles $B_i$ $(i = 1,\ldots,s)$, qui sont réflexifs, localisables, classiquement $P_i$-primaires à droite et à gauche tels que $B = \bigcap_{i=1}^{s} B_i$ ; B lui-même est alors localisable.

4. CONCLUSION.

La première partie du résultat suivant montre que le problème de la décomposition primaire bilatère pour une k-algèbre presque commutative se réduit au problème de l'existence d'un anneau classique de fractions artinien pour une telle algèbre, et la deuxième partie montre comment ce dernier problème peut être résolu.

Proposition 11.

a) *Un idéal bilatère* B *de* A *admet une décomposition comme intersection d'un nombre fini d'idéaux bilatères primaires d'un côté si et seulement si* $A/_B$ *est une somme sous-directe d'un nombre fini de* k*-algèbres presque commutatives admettant un anneau classique de fractions artinien.*

b) *Une* k*-algèbre presque commutative* A' *admet un anneau classique de fractions artinien si et seulement si pour tout entier* m *tel que* $\mathrm{Min}\{i \mid \tau_i(A') \neq 0\} < m < GK(A')$ *on a,*

$$GK(A'/\tau_m(A')) + r_A(\tau_m(A')) < m .$$

Démonstration. - La partie a) résulte de la proposition 8 et du corollaire 10. Pour démontrer la partie b), il suffit d'utiliser les idées développées dans [5], et [15] et d'employer la dimension de Gelfand-Kirillov au lieu de la dimension de Krull.

Corollaire 12. - *Si* A *est une somme sous-directe d'un nombre fini de* k*-algèbres presque commutatives admettant un anneau classique de fractions artinien, alors pour tout idéal bilatère* B *de* A *essentiel dans le treillis des idéaux bilatères de* A, *on a*,

$$GK(A) = GK(B) .$$

Exemple. - On dit qu'une k-algèbre presque commutative A est GK-lisse si pour tout A-module à droite de type fini X extension essentielle d'un A-module à droite M, on a,

$$GK(A/_{r_A(M)}) = GK(A/_{r_A(X)})$$

(voir une définition analogue utilisant la dimension de Krull dans [1]) ; on peut démontrer que si A est poly-AR (c'est le cas pour $U\mathfrak{g}$ où $\mathfrak{g}$ est une k-algèbre de Lie de dimension finie résoluble) alors A est GK-lisse ; on peut encore remarquer que si $\mathfrak{g}$ est une $\mathbb{C}$-algèbre de Lie de dimension finie alors $U\mathfrak{g}$ est GK-lisse si et seulement si $\mathfrak{g}$ est résoluble. Compte tenu de la proposition 11, on peut démontrer que si A est GK-lisse alors tout quotient de A est une somme sous-directe d'un nombre fini de k-algèbres presque commutative admettant un anneau classique de fractions artinien, et par suite vérifie les conditions du corollaire 12. Cet exemple implique en particulier le corollaire 4.2 et le lemme 4.3 de [13].

## 5. REFERENCES.

[1] K.A. Brown, "Module extensions over noetherian rings", J. Algebra 69, 1981, pp. 247-260.

[2] K.A. Brown and T.H. Lenagan, "Primary decomposition in enveloping algebras", Proc. Edin. Math. Soc., 1981, pp. 1-4.

[3] G. Cauchon, "Commutant des modules de longueur finie sur certaines algèbres filtrées", Lecture Notes in Math. 825, Ring theory Antwerp 1980, pp. 10-18.

[4] J.C. Mc Connell, "Representations of solvable Lie Algebras II : twisted group-rings", Ann. Sc. Ec. Norm. Sup., série 4, t. 8, pp. 157-178.

[5] A.W. Goldie and G. Krause, "Artinian quotient rings of ideal invariant noetherian rings", J. Algebra 63, 1980, pp. 374-388.

[6] O. Goldman, "Rings and modules of quotient", J. Algebra 13, 1969, pp. 10-47.

[7] R. Gordon, "Primary decomposition in right noetherian rings", Comm. in Algebra, 2 (6), 1974, pp. 491-524.

[8] A.G. Heinicke, "On the Krull symmetry of enveloping algebra", J. London Math., Soc. (2) 24, 1981, pp. 109-112.

[9] A. Hudry, Dimension de Gelfand-Kirillov et décomposition primaire bilatère dans certaines algèbres filtrées", à paraître Comm. in Algebra.
[10] A.V. Jategaonkar,"Injective modules and classical localization in noetherian rings", Bull. of the Amer. Soc. 79 (1), 1981, pp. 152-157.
[11] A. Joseph, "Dimension en algèbre non commutative", Cours de 3-ème cycle, Université Paris VI, 1980.
[12] G. Kempf, "Poincaré and Clifford Algebra", J. Algebra 52, 1978, pp. 547-561.
[13] T.H. Lenagan, "Gelfand-Kirillov dimension in enveloping algebras", Quart. J. Math. Oxford (2) 32, 1981, pp. 63-80.
[14] T.H. Lenagan, "Gelfand-Kirillov dimension and affine PI-rings", Comm. in Algebra 10 (1), 1982, pp. 87-92.
[15] T.H. Lenagan and P.B. Moss, "K-symmetric rings", J. London Math. Soc. (2) 21, 1980, pp. 45-52.
[16] D. Quillen, "On the endomorphism ring of a simple module over an enveloping algebra", Proc. Amer. Math. Soc. 21, 1969, pp. 171-172.
[17] G. Renault, "Algèbre non commutative", Gauthier-Villars éditeur, 1975.
[18] S.P. Smith, "Krull dimension of the enveloping algebra of $sl(2,\mathbb{C})$", J. Algebra 71, 1981, pp. 189-194.
[19] P. Tauvel, "Sur la dimension de Gelfand-Kirillov", Comm. in Algebra 10 (9), 1982, pp. 939-963.

# SOME SPECIAL CLASS OF ARTIN RINGS OF FINITE TYPE

Yasuo Iwanaga

Fac. of Education, Shinshu Univ., Nagano 380

In this note, we will investigate on an artinian QF-3 ring A which has an indecomposable module U such that every indecomposable A-module is a subquotient of U.

In representations of artin rings, the classification of indecomposable modules is one of the most important problems and especially, we'd like to know the 'types' of all indecomposables. Although, in case of finite (representation) type, we have had some general results on this problem, it seems to me that there still are fascinating topics in some special cases. For instance, Dlab-Ringel in the Proceedings of Antwerp Conference in 1978 showed that if a hereditary, connected, finite dimensional algebra A over an infinite field is of finite type, then A has a finitely generated indecomposable module U such that every indecomposable A-module is a subquotient of U, i.e. homomorphic image of some submodule of U. This is the special case of finite type but is interesting in the sense that we can have one important, big, finitely generated indecomposable module concretely and other indecomposables are obtained by taking subquotient of it. Thus, we call such indecomposable module <u>dominant</u>, and we will consider the following problem.

<u>PROBLEM 1.</u> When does an artin ring have a dominant indecomposable module?

Now, we have the following remarks:

Let A be an artin ring with a dominant indecomposable left module U, then

(1) For any indecomposable left A-module X, $|X| \leq |U|$ and so A is of finite type by Roiter-Auslander's theorem, where $|X|$ stands for the composition length of X;



*F. van Oystaeyen (ed.), Methods in Ring Theory, 181–183.*

(2) For an indecomposable left A-module M, $|M|=|U|$ implies $M\cong U$;

(3) A dominant indecomposable module is unique up to isomorphic;

(4) For any projective indecomposable left A-module P and any injective indecomposable left A-module E, P is monomorphic to U, U is epimorphic to E and hence U is faithful, sincere.

If A is a finite dimensional algebra over a field K, then the equalities on composition length above may be replaced by the K-dimension.

Next, we give some examples of artin rings with dominant indecomposables.

EXAMPLES (1) A balanced ring (e.g. uniserial ring) has a dominant incecomposable module. (Dlab-Ringel)

(2) Let $\mathbb{Z}$ be a ring of rational integers, p a prime in $\mathbb{Z}$ and

$$A = \begin{pmatrix} \mathbb{Z}/(p^2) & 0 \\ \mathbb{Z}/(p^2) & \mathbb{Z}/(p^2) \end{pmatrix}$$

then A has a dominant indecomposable. Similarly, if we take A as an Auslander algebra of $K[X]/(X^2)$ with K a field, A has a dominant indecomposable.

The second example above is so called a QF-3 ring, thus in the following, we consider in case of artinian QF-3 rings. An artin ring A is QF-3 if A has a faithful, injective left(or right) ideal or equivalently, injective hull $E({}_AA)$ (or $E(A_A)$) of ${}_AA$ (or $A_A$) is projective. Furthermore, if every projective indecomposable left module over an artin ring A has simple socle, then A is called left QF-2 and right QF-2 ring is defined similarly. Though QF-2 rings are not left-right symmetric, left and right QF-2 rings are QF-3. (Fuller)

Now we get

PROPOSITION 1. Let A be an artin ring with a dominant indecomposable left module U and assume there exists a (nonzero) projective-injective left A-module (for example, let A be a QF-3 ring). Then U is projective-injective and all projective-injective indecomposables are isomorphic to U. In particular, A is left and right QF-2.

Proof. For any projective-injective indecomposable module E, E is a homomorphic image of U by remark (4) and hence E is isomorphic to U since E is projective, too. Therefore U is a (unique) minimal faithful left module over A, i.e. A is QF-3, and U is injective indecomposable, especially, U has a simple socle. Moreover, since every projective indecomposable left module is monomorphic to U with Soc(U) simple by remark (4), A is left QF-2. All projective indecomposables and injective indecomposables are local-colocal, so A is right QF-2.

Next, we consider an Auslander ring.

PROPOSITION 2. (Auslander ring) Let A be an artin ring with dom.dim ${}_AA \geqq 2$ and gl.dim $A \leqq 2$. If A has a dominant indecomposable left module, then there exists a primitive idempotent e in A such that eAe is a uniserial ring with cube-zero radical.

Proof. Recall theorem by Auslander-Ringel-Tachikawa on 1-1 correspondence between the Morita equivalence classes of artin rings of finite type and of Auslander rings.

Finally,

PROPOSITION 3. (1) Let A be a serial ring with dominant indecomposable left module, then A has a Kupisch series $Ae_1, \ldots, Ae_n$ (each $e_i$ is a primitive idempotent) satisfying $|Ae_{i+1}| = |Ae_i| + 1$ for $i=1,\ldots,n-1$ and $Ae_n$ is a dominant indecomposable.

(2) If a quasi-Frobenius ring A has a dominant indecomposable, then A has to be uniserial and ${}_AA$ is a dominant indecomposable.

Proof. (1) Let U be a dominant indecomposable left module over A, then there is a projective indecomposable A-module $P_0$ such that $P_0$ is epimorphic to U and $|P_0| \geqq |U|$. On the one hand, $|P_0| \leqq |U|$ by remark (1). Hence $|P_0| = |U|$ and $P_0 \cong U$. Furthermore, since $P_0$ is projective indecomposable of the largest composition length, $P_0$ is injective.

Next, for any projective indecomposable P, there is a submodule M of $P_0$ such that M is epimorphic to P and it splits, so P is monomorphic to $P_0$. Therefore $P_0$ is the only projective-injective indecomposable over A. Now, if n=the number of non-isomorphic simple A-modules is at least two, let $Ae_1, \ldots, Ae_n$ with $Ae_n = P_0$ be a Kupisch series of A, then $|Ae_i| < |Ae_{i+1}|$ and so $|Ae_{i+1}| = |Ae_i| + 1$ for $i=1,\ldots,n-1$. Hence $Ae_i \cong Je_{i+1}$ as required, where J is the radical of A.

The proof of (2) is easy.

We complete this note by giving the following problem related to a hereditary case.

PROBLEM 2. When does a tilted algebra have a dominant indecomposable module?

# GROUP RINGS AND MAXIMAL ORDERS

E. Jespers and P.F. Smith

Katholieke Universiteit Leuven, Department Wiskunde, Celestijnenlaan 200B, B-3030 Leuven (België).
Department of Mathematics, University of Glasgow, Glasgow G12 8QW, Scotland, U.K.

A ring Q with identity is a quotient ring provided every regular element (i.e. non-zero-divisor) is a unit. For example Artinian rings are quotient rings. A ring R is an order in a quotient ring Q provided

(i) R is a subring of Q,

(ii) every regular element of R is a unit of Q, and

(iii) for every element q of Q there exist elements $r_i$, $c_i \in R$ with $c_i$ regular (i = 1, 2) such that $q = c_1^{-1}r_1 = r_2c_2^{-1}$.

Orders R, S in Q are called equivalent, written $R \sim S$, provided there exist units $q_i$ of Q ($1 \leq i \leq 4$) such that

$$q_1Rq_2 \leq S \text{ and } q_3Sq_4 \leq R.$$

An order R in Q is maximal if the only order S in Q such that $R \sim S$ and $R \leq S$ is $S = R$.

Let R be an order in a quotient ring Q. A left R-submodule A of Q is called a (fractional) left R-ideal provided

(i) A contains a unit of Q, and

(ii) $Aq \leq R$ for some unit q of Q.

Right R-ideals are defined analogously. By an R-ideal of Q we mean a left R-ideal which is also a right R-ideal and by integral



F. van Oystaeyen (ed.), Methods in Ring Theory, 185–195.

R-ideal an R-ideal contained in R.

Let R be an order in a quotient ring Q and A an R-ideal of Q. Define

$$O_\ell(A) = \{q \in Q : qA \le A\} \text{ and } O_r(A) = \{q \in Q : Aq \le A\}.$$

It is routine to prove that $O_\ell(A)$ and $O_r(A)$ are both orders in Q equivalent to and containing R (see [14, p.5 Proposition 2.6]).

<u>Lemma 1.</u> Let R be an order in a quotient ring Q. Then R is a maximal order in Q if and only if $O_\ell(A) = O_r(A) = R$ for every integral R-ideal A.

Proof. See [14, p.7 Proposition 3.1].

Maximal orders in simple Artinian rings have attracted considerable attention recently (see for example [2] - [5], [7] - [14], [16]). Our purpose is to consider the following question to be found in [14, p.181 Problem 7]:

> Let J be a ring and G a group such that the group ring J[G] is an order in a quotient ring Q. When is J[G] a maximal order in Q?

In this note we shall restrict ourselves to considering Abelian-by-finite groups, i.e. groups G containing an Abelian normal subgroup of finite index.

Let J be a ring. If G, H, ... are groups then the augmentation ideals of the group rings J[G], J[H], ... will be denoted $\underline{\underline{g}}$, $\underline{\underline{h}}$, ..., respectively.

<u>Lemma 2.</u> Let J be a ring and G a group such that the group ring R = J[G] is a maximal order in a quotient ring Q. Then G does not contain an infinite dihedral normal subgroup.

Proof. Suppose (on the contrary) that G contains an infinite dihedral normal subgroup N. There exist x, y ∈ N such that

$$N = \langle x, y \, ; \, x^y = x^{-1}, y^2 = 1 \rangle.$$

Let I be the ideal $\underline{\underline{n}}$ R of R. Note that I is an integral R-ideal of Q because I contains the regular element x - 1 of R. Let

$$q = (x - 1)^{-1} (y + 1) \in Q.$$

Since I = (x - 1)R + (y - 1)R it follows that

$$\begin{aligned} qI &\leq (x-1)^{-1}(y+1)(x-1)R = (x-1)^{-1}\{(x^{-1}-1)y + (x-1)\}R \\ &= (-x^{-1}y + 1)R = \{(1-x^{-1})y + (1-y)\}R \leq I. \end{aligned}$$

Thus $q \in O_\ell(I)$. Suppose $q \in R$. Then y + 1 = (x - 1)r for some r in R. Since R is a free left J[N]-module it follows that $r \in J[N]$ without loss of generality. Let $H = \langle x \rangle$. Then r = a + by for some a, $b \in J[H]$. Then 1 = (x - 1)a, a contradiction. Thus $q \notin R$ and hence $R < O_\ell(I)$, contradicting Lemma 1.

Let J be a maximal order in a simple Artinian ring and G a polycyclic-by-finite group. By [15, Lemma 10.2.5] G contains a normal subgroup N of finite index such that N is a poly-(infinite cyclic) group. A result of Chamarie [2, Proposition 2.2.1] then gives that J[N] is a maximal order in a simple Artinian ring. Thus the problem of deciding whether J[G] is a maximal order reduces to passing from J[N] to J[G]. A first case is when N is free Abelian of finite rank. In this way one is led to consider Abelian-by-finite groups. The next lemma is more general than we really need.

<u>Lemma 3</u>. Let J be a ring and H a normal subgroup of finite index in a group G such that the group ring S = J[H] is a maximal order in an Artinian ring D. Then the group ring R = J[G] is an order in an Artinian ring Q. Moreover, if I is an integral R-ideal of Q such that $rI \leq (I \cap S)R$ for some r in R then $rO_\ell(I) \leq R$.

Proof. By [15, Lemma 3.5 (iii) and its proof] R is an order in an Artinian ring Q and every element of Q can be written in the form $c^{-1}a$ where $a \in R$, $c \in S$ and c is regular. Let $q \in O_\ell(I)$. Then $rq = v^{-1}u$ for some $u \in R$, $v \in S$ with v regular. Let T be a transversal to the cosets of H in G. Then

$$u = \sum_{t \in T} u_t t$$

for some $u_t \in S(t \in T)$. Now $rI \leq (I \cap S)R$ implies

$$rqI \leq rI \leq (I \cap S)R$$

and hence

$$u(I \cap S)R \leq uI = vrqI \leq v(I \cap S)R.$$

Since $(I \cap S)^t = I \cap S$ for all t in T it follows that

$$u_t(I \cap S) \leq v(I \cap S) \quad (t \in T).$$

The ideal I contains a regular element d of R and as we remarked above $d^{-1} = c^{-1}a$ for some $a \in R$, $c \in S$ with c regular. Thus $c = ad \in I \cap S$. It follows that $I \cap S$ is an integral S-ideal of D and

$$v^{-1}u_t (I \cap S) \leq I \cap S \quad (t \in T).$$

The hypothesis that S is a maximal order in D implies

$$v^{-1}u_t \in S \quad (t \in T)$$

and hence

$$rq = v^{-1}u = \sum_{t \in T} v^{-1}u_t t \in R.$$

It follows that $rO_\ell(I) \leq R$.

Let A be an Abelian normal subgroup of a group G. Let $x \in G$. If $a \in A$ then $[a, x]$ denotes the commutator $a^{-1}x^{-1}ax$ which is an element of A. Define

$$[A, x] = \{[a, x] : a \in A\}.$$

If $a, b \in A$ then

$$[ab, x] = [a, x]^b [b, x] = [a, x][b, x].$$

Hence $[A, x]$ is a subgroup of A. The proof of the next result uses an idea taken from the proof of [15, Lemma 7.4.9].

<u>Lemma 4</u>. Let J be a ring A a non-central Abelian normal subgroup of a group G such that the centralizer C of A in G has finite

index. Let $S = J[C]$, $T = \{x_0 = 1, x_1, \ldots, x_n\}$ be a transversal to the cosets of C in G and $A_i = [A, x_i]$ $(1 \le i \le n)$. Let I be an ideal of $R = J[G]$. Then $(\underline{\underline{a}}_1 \ldots \underline{\underline{a}}_n)^{n+1} I \le (I \cap S)R$.

Proof. Let $r \in I$. Then

$$r = s_0 x_0 + \ldots + s_n x_n \tag{1}$$

for some $s_i \in S(0 \le i \le n)$. Let $d(r)$ denote the number of non-zero elements $s_i (0 \le i \le n)$ in (1). We claim

$$(\underline{\underline{a}}_1 \ldots \underline{\underline{a}}_n)^{d(r)+1} r \le (I \cap S)R.$$

The proof is by induction on $d(r)$. If $d(r) = 0$ then there is nothing to prove because $r = 0$. Suppose $d(r) > 0$. There exists $0 \le i \le n$ such that $s_i \ne 0$. Then

$$r x_i^{-1} = s_i x_0 + \bar{s}_1 x_1 + \ldots + \bar{s}_n x_n$$

for some $\bar{s}_i \in S(1 \le i \le n)$ and $d(r x_i^{-1}) = d(r)$. Thus without loss of generality we can suppose $s_0 \ne 0$. Let $a \in A$. Then

$$\begin{aligned} r^a &= s_0 x_0 + s_1 x_1^a + \ldots + s_n x_n^a \\ &= s_0 x_0 + s_1 [a, x_1^{-1}] x_1 + \ldots + s_n [a, x_n^{-1}] x_n \end{aligned}$$

where $[a, x_i^{-1}] \in A \le S$ $(1 \le i \le n)$. It follows that

$$\bar{r} = [a, x_n^{-1}] r - r^a = ([a, x_n^{-1}] - 1) s_0 x_0 + s_1' x_1 + \ldots + s_{n-1}' x_{n-1}$$

for some $s_i' \in S(1 \le i \le n - 1)$. Note that $\bar{r} \in I$. Repeating this argument we have

$$([a_1, x_1^{-1}] - 1) \ldots ([a_n, x_n^{-1}] - 1) s_0 x_0 \in I$$

for all $a_i \in A(1 \le i \le n)$. Thus

$$\underline{\underline{a}}_1 \ldots \underline{\underline{a}}_n s_0 \le I \cap S,$$

because $\underline{\underline{a}}_i \le S(1 \le i \le n)$. Let $v \in \underline{\underline{a}}_1 \ldots \underline{\underline{a}}_n$. Then

$$vr = vs_0 x_0 + vs_1 x_1 + \ldots + vs_n x_n$$

and hence $r^* = vr - vs_0x_0 = s_1^*x_1 + \dots + s_n^*x_n$, where $s_i^* \in S(1 \leq i \leq n)$. Note that $d(r^*) < d(r)$ and hence by induction

$$(\underline{\underline{a}}_1 \dots \underline{\underline{a}}_n)^{d(r^*)} r^* \leq (I \cap S)R.$$

Thus, if $d = d(r)$ then

$$(\underline{\underline{a}}_1 \dots \underline{\underline{a}}_n)^d vr = (\underline{\underline{a}}_1 \dots \underline{\underline{a}}_n)^d (r^* + vs_0) \leq (I \cap S)R.$$

It follows that

$$(\underline{\underline{a}}_1 \dots \underline{\underline{a}}_n)^{d+1} r \leq (I \cap S)R,$$

as required. The result follows.

Combining Lemmas 3 and 4 we have:

Corollary 5. Let J be a ring and A a non-central Abelian normal subgroup of a group G such that the centralizer C of A in G has finite index and J[C] is a maximal order in an Artinian ring. Then R = J[G] is an order in an Artinian ring Q and for every integral R-ideal I of Q there exist a positive integer n and elements $x_i (1 \leq i \leq n)$ in G but not C such that $\underline{\underline{a}}_1 \dots \underline{\underline{a}}_n O_\ell(I) \leq R$ where $A_i = [A, x_i] \; (1 \leq i \leq n)$.

Lemma 6. Let J be a ring and G a group such that R = J[G] is an order in a quotient ring Q. Let A be an Abelian subgroup of G of rank > 1. Then $R = \{q \in Q : \underline{\underline{a}}q \leq R\}$.

Proof. Suppose $q \in Q$ and $\underline{\underline{a}}q \leq R$. Let $a, b \in A$ have infinite order and satisfy $\langle a \rangle \cap \langle b \rangle = 1$. Then $(a - 1)q \in R$ implies $q = (a - 1)^{-1}r$ for some r in R. Similarly, $q = (b - 1)^{-1}s$ for some s in R. Since ab = ba it follows that

$$(b - 1)r = (a - 1)(b - 1)q = (a - 1)s. \qquad (2)$$

Let T be a transversal to the right cosets of A in G. Then there exist a finite subset T' of T and elements $r_t, s_t \in J[A] (t \in T')$ such that

$$r = \sum_{t \in T'} r_t t \quad \text{and} \quad s = \sum_{t \in T'} s_t t.$$

Then (2) gives

$$(b - 1)r_t = (a - 1)s_t \quad (t \in T').$$

Since $b - 1$ is a regular element of the ring $J[A]$ modulo the ideal $(a - 1)J[A]$ it follows that $r_t = (a - 1)\overline{r}_t$ for some $\overline{r}_t \in J[A]$ for each $t$ in $T'$. Then

$$r = (a - 1) \sum_{t \in T'} \overline{r}_t t$$

and $q = (a - 1)^{-1}r = \sum_{t \in T'} \overline{r}_t t \in R$, as required.

Proposition 7. Let J be a maximal order in a simple Artinian ring and G a group containing an Abelian normal subgroup A of finite index such that rank $[A, x] \neq 1$ for all x in G. Then $J[G]$ is a maximal order in a simple Artinian ring if and only if G contains no non-trivial finite normal subgroup.

Proof. If $J[G]$ is an order in a simple Artinian ring then $J[G]$ is a prime ring (see for example [6, Theorem 1.28]) and hence G contains no non-trivial finite normal subgroup by [15, Theorem 4.2.10]. Conversely, suppose G is non-trivial and contains no non-trivial finite normal subgroup. Then A is torsion-free Abelian. Let C denote the centralizer of A in G. Then C is centre-by-finite and by a theorem of Schur (see [15, Lemma 4.1.4]) it follows that C' is finite and hence trivial. Thus C is Abelian and by hypothesis torsion-free. By [1, Theorem A] and [15, Theorem 4.2.10] $R = J[G]$ and $S = J[C]$ are orders in simple Artinian rings Q, Q', respectively. Moreover, by [2, Proposition 2.2.1], S is a maximal order in Q'.

Suppose $C \neq G$ (otherwise $R = S$ and the proof is complete). Let I be an integral R-ideal of Q. By Corollary 5 there exist a positive integer n and elements $x_i$ $(1 \leq i \leq n)$ in G but not C such that

$$\mathfrak{a}_1 \ldots \mathfrak{a}_n O_\ell(I) \leq R$$

where $A_i = [A, x_i]$ $(1 \leq i \leq n)$. By hypothesis rank $A_i > 1$ $(1 \leq i \leq n)$ and hence $O_\ell(I) \leq R$. Thus $O_\ell(I) = R$ and similarly $O_r(I) = R$. By Lemma 1 R is a maximal order in Q.

In view of Proposition 7 we are interested in a torsion-free Abelian normal subgroup A of a group G and an element x of G such that rank $[A, x] = 1$. Note that if B is a torsion-free Abelian group of rank 1 and $\phi$ is an automorphism of B of finite order then $\phi$ induces an automorphism of finite order of the one-dimensional Q-vector space $Q \otimes B$ and hence $\phi^2 = 1$. Applying this fact to A and x it follows that if $x^k \in \mathbb{C}_G(A)$ for some positive integer k then k is even. Suppose on the contrary that $x^k \in \mathbb{C}_G(A)$ for some odd positive integer k. Conjugation by x induces an automorphism of finite order of $[A, x]$ and hence $x^2$ centralizes $[A, x]$. But $x^k$ centralizes $[A, x]$ so that x centralizes $[A, x]$. For any element a in A we have

$$1 = [a, x^k] = [a, x^{k-1}][a, x]^{x^{k-1}} = [a, x^{k-1}][a, x],$$

and by induction

$$1 = [a, x]^k.$$

Since A is torsion-free it follows that $[a, x] = 1$. Therefore $[A, x] = 1$. Thus if G/A is a finite group of odd order then rank $[A, x] > 1$ for all x in G but not $\mathbb{C}_G(A)$. Combining this fact with Proposition 7 we have at once:

<u>Theorem 8</u>. Let J be a maximal order in a simple Artinian ring and G an Abelian-by-(finite of odd order) group. Then the group ring $J[G]$ is a maximal order in a simple Artinian ring if and only if G contains no non-trivial finite normal subgroup.

A second application of Proposition 7 is given in the next result.

Theorem 9. Let J be a maximal order in a simple Artinian ring and G the wreath product $A \wr F$ of a torsion-free Abelian group A of rank > 1 and a finite group F. Then the group ring J[G] is a maximal order in a simple Artinian ring.

Proof. It is not difficult to prove that G contains no non-trivial finite normal subgroup. The group G contains an Abelian normal subgroup N such that G = NF and N is a direct product

$$N = \prod_{f \in F} A_f \tag{3}$$

of subgroups $A_f$ where for each f in F there exists an isomorphism $a \longmapsto a_f$ $(a \in A)$ of A onto $A_f$ and

$$a_f^h = a_{(fh)} \quad (a \in A,\ f,\ h \in F).$$

Suppose there exists $g \in G$ such that rank $[N, g] = 1$. Then there exists $f \in F$ such that $[N, g] = [N, f]$ and clearly $f \neq 1$. Define $\phi : A \longrightarrow [N, f]$ by

$$\phi(a) = [a, f] \quad (a \in A).$$

Then $\phi$ is a homomorphism with kernel $K = \{a \in A : [a, f] = 1\}$. It is easy to see that K = 1 and hence rank $A \leq 1$, a contradiction. Thus rank $[N, g] \neq 1$ for all g in G. The result follows by Proposition 7.

A special case is given in the final result.

Corollary 10. Let J be a maximal order in a simple Artinian ring and G the wreath product $A \wr F$ of a non-trivial free Abelian group A and a finite group F of order n. Then the group ring J[G] is a maximal order in a simple Artinian ring if and only if A is not infinite cyclic or $n \neq 2$.

Proof. Suppose A is infinite cyclic and n = 2. Then there exist elements a, b, y in G such that

$$G = \langle a, b, y \; ; \; [a, b] = 1, a^y = b, b^y = a, y^2 = 1 \rangle .$$

Let $x = ab^{-1}$. Then $x^y = x^{-1}$. If $N = \langle x, y \rangle$ then N is an infinite dihedral normal subgroup of G. By Lemma 2 $J[G]$ is not a maximal order in a simple Artinian ring.

Conversely, suppose G is not infinite cyclic or $n \neq 2$. We adopt the notation of the proof of Theorem 9. If $n = 1$ then G is free Abelian and $J[G]$ is a maximal order in a simple Artinian ring by [2, Proposition 2.2.1]. Suppose $n \geq 3$. Let $g \in G$ satisfy rank $[N, g] = 1$. Then there exists $f \in F$ such that $[N, f] = [N, g]$. Let $h \in F$, $h \neq 1, f$. Let $a_1 \in A_1$, $a_1 \neq 1$. Then $[N, f]$ contains the non-trivial elements $[a_1, f] = a_1^{-1}a_f$ and $[a_h, f] = a_h^{-1}a_{(hf)}$. Because rank $[N, f] = 1$, there exist non-zero integers s, t such that

$$\left(a_1^{-1}a_f\right)^s = \left(a_h^{-1}a_{(hf)}\right)^t$$

and this contradicts (3). If $n \neq 2$ then rank $[N, g] \neq 1$ for all g in G and $J[G]$ is a maximal order in a simple Artinian ring by Proposition 7. If $n = 2$ then G is not infinite cyclic and $J[G]$ is a maximal order in a simple Artinian ring by Theorem 9.

Acknowledgement

The first author would like to thank the British Council for financial support and the University of Glasgow Mathematics Department for their hospitality during the preparation of this paper.

## REFERENCES

1. K.A. Brown, Artinian quotient rings of group rings, J. Algebra 49 (1977), 63-80.

2. M. Chamarie, Sur les ordres maximaux au sens d'Asano, Vorlesungen aus dem Fachbereich Mathematik der Universität Essen Heft 3 (1979).

3. M. Chamarie, Anneaux de Krull non commutatif, J. Algebra 72 (1981), 210-222.

4. M. Chamarie, Modules sur les anneaux de Krull non commutatif, preprint.

5. A.W. Chatters, Non-commutative unique factorization domains, preprint.

6. A.W. Chatters and C.R. Hajarnavis, Rings with chain conditions (Pitman, 1980).

7. C.R. Hajarnavis and J.C. Robson, Decomposition of maximal orders, Bull. London Math. Soc. 15 (1983), 123-125.

8. C.R. Hajarnavis and S. Williams, Maximal orders in Artinian rings, preprint.

9. E. Jespers, L. Le Bruyn and P. Wauters, $\Omega$ -Krull rings I, Comm. in Algebra 10 (1982), 1801-1818.

10. H. Marubayashi, Noncommutative Krull rings, Osaka J. Math. 12(1975), 703-714.

11. H. Marubayashi, On bounded Krull prime rings, ibid 13 (1976), 491-501.

12. H. Marubayashi, Remarks on ideals of bounded Krull prime rings, Proc. Japan Acad. 53 (1977), 27-29.

13. H. Marubayashi, A characterization of bounded Krull prime rings, Osaka J. Math. 15 (1978), 13-20.

14. G. Maury and J. Raynaud, Ordres maximaux au sens de K. Asano, Springer Lecture Notes in Mathematics 808 (1980).

15. D.S. Passman, The algebraic structure of group rings (Wiley-Interscience, 1977).

16. J.T. Stafford, Modules over prime Krull rings, preprint.

A note on infinite torsion primes of a commutative ring

Teruo Kanzaki
Osaka Women's University
Daisen-cho Sakai, Osaka 590 Japan

Let $R$ be a commutative ring with identity 1. In this note, we show that a notion of infinite torsion prime with a bounded level is included in a notion of local signature.

By [1], a subset $P$ of $R$ is called an infinite preprime, if it satisfies 1) $P + P \subset P$, 2) $P \cdot P \subset P$, 3) $1 \in P$ and $-1 \notin P$. And, a maximal preprime is called an infinite prime. If an infinite prime holds that for each $a \in R$, there is a positive integer $n$ satisfying $a^{2^n} \in P$, then $P$ is called an infinite torsion prime. We set $R^n = \{a^n \mid a \in R\}$. We say an infinite torsion prime $P$ to be bounded level, if there is a positive integer $n$ with $R^{2^n} \subset P$. Let $GF(q)$ be a finite field with characteristic not 2. In [3], a notion of a local signature is defined as follows: A map $\sigma\colon R \longrightarrow GF(q)$ is called a local signature of $R$ over $GF(q)$, if the following conditions hold; 1) $\sigma(-1) = -1$, 2) $\sigma(ab) = \sigma(a)\sigma(b)$ and 3) either $\sigma(a) = 0$ or $\sigma(a) = \sigma(b)$ implies $\sigma(a + b) = \sigma(b)$, for $a, b \in R$. For a local signature



F. van Oystaeyen (ed.), Methods in Ring Theory, 197–200.

$\sigma$ of $R$ over GF($q$), we set $p_\alpha(\sigma) = \{a \in R \mid \sigma(a) = \alpha\}$ for $\alpha \in$ GF($q$), and $P(\sigma) = p_0(\sigma) \cup p_1(\sigma)$. In [3], (1.4), it was shown that $P(\sigma)$ is an infinite preprime with $R^{q-1} \subset P(\sigma)$.

Theorem. *For any infinite torsion prime $P$ with a bounded level, there is a local signature $\sigma$ of $R$ over* GF($q$) *with $P(\sigma) = P$ for a suitable prime number $q \neq 2$.*

To prove this theorem we prepare the following lemmata in which some of them are already known.

Lemma 1. ([3], (1.3)) *Let* GF($q$) *be a finite field with characteristic not 2. For a given family $\{p_\alpha \mid \alpha \in \text{GF}(q)\}$ of subsets $p_\alpha$ of $R$, there is a local signature $\sigma$ of $R$ over* GF($q$) *satisfying $p_\alpha(\sigma) = p_\alpha$ for all $\alpha \in$* GF($q$), *if and only if the following conditions hold;*

1) $R = \bigcup_{\alpha \in \text{GF}(q)} p_\alpha$ , *and* $p_\alpha \cap p_\beta = \phi$ *for* $\alpha \neq \beta$ *in* GF($q$).

2) $-1 \in p_{-1}$ .

3) $p_\alpha p_\beta \subset p_{\alpha\beta}$ *for* $p_\alpha \neq \phi$ *and* $p_\beta \neq \phi$.

4) $p_\alpha + p_\alpha \subset p_\alpha$ *and* $p_0 + p_\alpha \subset p_\alpha$ *for* $p_\alpha \neq \phi$.

Lemma 2. ([2], Theorem 1) *If $P$ is an infinite torsion prime, then $P \cap -P$ is a prime ideal of $R$.*

Lemma 3. ( cf. [2], Lemma 1) *If $P$ is an infinite prime of $R$, then for any $a \in R$, $a^2 \in P$ implies $a \in P \cup -P$.*

Lemma 4. (K. Kitamura) *Let $P$ be an infinite torsion prime satisfying $R^{2^n} \subset P$ and $R^{2^{n-1}} \not\subset P$, and $\omega$ an element of $R$ with $\omega^{2^{n-1}} \notin P$. Then, for every $a \in R$, there exists an integer $m > 0$ with $\omega^m a \in P$.*

*Proof*. Let $a$ be an element of $R$ with $a \notin P$. There is a positive integer $k$ such that $k \leq n$, $a^{2^k} \in P$ and $a^{2^{k-1}} \notin P$. By Lemma 3, $a^{2^{k-1}}$

and $\omega^{2^{n-1}}$ belong to $-P$. Hence we get $\omega^{2^{n-1}} \cdot a^{2^{k-1}} = (\omega^{2^{n-k}} \cdot a)^{2^{k-1}} \in P$. By the induction on $k$, we get $\omega^{2^l} \cdot a \in P$ for some integer $l \geqq 0$.

Lemma 5. ([1], cf. [2], (0.2)) *Let $P$ be an infinite prime, and set $p_0 = P \cap -P$ and $p_1 = \{x \in P \mid x \notin p_0\}$. If $xa$ belongs to $P$ (resp. $p_1$) for some $a \in p_1$, then so is $x$.*

Lemma 6. *Let $P$ be an infinite prime of $R$, and $p_0$ and $p_1$ as in Lemma 5. Suppose that there is an element $\omega$ of $R$ satisfying $\omega^{2^n} \in P$ and $\omega \notin p_0$. Then, we have the following;*

1) *The minimal positive integer $k$ with $\omega^k \in P$ is a factor of $2^n$.*

2) *If $\omega^{2^{n-1}} \notin P$, then $k = 2^n$, that is, $\omega^i \notin P$ for all $i$; $0<i<2^n$.*

*Proof*. 1) Let $k$ be the minimal positive integer with $\omega^k \in P$. For integers $q$ and $r$ satisfying $2^n = qk + r$ and $0 \leqq r < k$, we have $(\omega^k)^q \omega^r \in P$ and $(\omega^k)^q \in p_1$. By Lemma 5, it follows that $\omega^r \in P$ and $r = 0$. 2) If $\omega^{2^{n-1}} \notin P$, then by 1) $k$ must be $2^n$.

Lemma 7. *For any infinite prime $P$ of $R$, the subsets $p_0 = P \cap -P$ and $p_1 = \{x \in P \mid x \notin p_0\}$ verify $p_1 + p_1 \subset p_1$ and $p_0 + p_1 \subset p_1$.*

*Proof*. Suppose $x \in P$ and $y \in p_1$. If $x + y$ is in $p_0$, then $y = (x + y) - x$ belongs to $p_0$ which is a contradiction. Hence $x + y \in p_1$.

*Proof of Theorem*. Suppose that $P$ is an infinite torsion prime with $R^{2^n} \subset P$ and $R^{2^{n-1}} \not\subset P$. Then, by theorem of Dirichlet, there is a prime number $q = 2^n m + 1$ for a suitable large integer $m > 0$. We consider a prime field GF($q$) with characteristic $q \neq 2$. There exists a cyclic subgroup $G = \langle\eta\rangle$ with order $|G| = 2^n$ of the multiplicative group GF($q$)* of the field GF($q$). By Lemma 2, the set $p_0 = P \cap -P$ is a prime ideal of $R$. Since $R^{2^{n-1}} \not\subset P$, there is an element $\omega$ in $R$ with $\omega^{2^{n-1}} \notin P$, and so $\omega \notin p_0$. For the set $p_1 = \{x \in P \mid x \notin p_0\}$, we consider sets $p_{\eta^i} = \{x \in R \mid \omega^{2^n - i} x \in p_1\}$ for each $\eta^i \in G$, especially we have $p_{\eta^0} = p_1$. For the orther $\alpha$ in GF($q$) with $\alpha \notin G \cup \{0\}$, we set $p_\alpha = \phi$. Then, the family $\{p_\alpha \mid \alpha \in \mathrm{GF}(q)\}$ verifies the conditions in Lemma 1: 1) For any $x \in R$, by Lemma 4 there is an integer $l > 0$ with $\omega^l x \in P$. If $x \notin p_0$, then $\omega^l x$ belongs to $p_1$. Let $l = 2^n k + r$ for integers $k$ and $r$ with $0 \leqq r < 2^n$. By Lemma 5, it follows that $\omega^l x = (\omega^{2^n})^k (\omega^r x) \in p_1$ yields $\omega^r x \in p_1$, hence $x \in p_{\eta^{2^n - r}}$. Thus, we have $R = \bigcup_{\alpha \in \mathrm{GF}(q)} p_\alpha$. By Lemma 2, $p_0 \cap p_{\eta^i} = \phi$ is easily seen. For integers $i$ and $j$ with $0 \leqq i < j < 2^n$, if $x \in p_{\eta^i} \cap p_{\eta^j}$ then by Lemma 5, $\omega^{2^n - i} x \in p_1$ and $\omega^{2^n - j} x \in p_1$ yield $\omega^{j-i} \in p_1$ which is contrary to Lemma 6. Hence, $p_{\eta^i} \cap p_{\eta^j} = \phi$. The condition 2) directly follows from the definition of $p_1$. The condition 3) is also immediate. The condition 4) follows from Lemma 7. Thus, the family $\{p_\alpha \mid \alpha \in \mathrm{GF}(q)\}$ defines a local signature $\sigma : R \longrightarrow \mathrm{GF}(q)$ with $p_\alpha = p_\alpha(\sigma)$ for all $\alpha \in \mathrm{GF}(\sigma)$, and $\mathrm{P}(\sigma) = P$ holds. The proof of Theorem is completed.

I would like to express my thanks to Professor Akira Hattori for his useful advice.

## References

[1] D.K.Harrison: Finite and infinite primes for rings and fields, Memoirs Amer. Math. Soc. 68 (1966).

[2] T.Kanzaki and K.kitamura: Notes on infinite primes of a commutative ring, to appear in Mathematica Japonica Vol. 28, No.5.

[3] T.Kanzaki: On signatures on rings, preprint.

# APPLICATIONS OF KUMMER THEORY WITHOUT ROOTS OF UNITY

I. Kersten and J. Michaliček

Univ. Regensburg and Univ. Hamburg - FRG

In this note we give some examples and applications of the Kummer theory for Galois extensions of prime order $p > 2$ developed in [2] and [3].

First let us briefly recall the results of [2] and [3]. Let R be a commutative ring and

$$G = (\sigma^i)_{i \in \mathbb{Z}/p\mathbb{Z}}$$

a group of order p. Our main purpose is to give an explicit description of the group

H(R,G) consisting of the isomorphism classes of Galois extensions with Galois group G and with normal basis over R,

(see [1], sec. 1 for the definition of a Galois extension and the product in H(R,G)).

Let A be a Galois extension of R with group G. Then the group ring A[G] is a Galois extension of R[G] with group G where G acts on the coefficients. The following lemma is easily proved.

<u>Lemma 1.</u> Set $X = \sum_{i=0}^{p-1} X_i \sigma^{-i}$ with $X_i \in A$. Then the following conditions are equivalent.

(i) $X_i = \sigma^i(X_o)$ for $0 \leq i \leq p-1$, and $X_o, \ldots, X_{p-1}$ is an R-basis of A.

(ii) X is a unit in A[G] and $\sigma(X) = \sigma \cdot X$.

In accordance with this lemma we call an element $X \in A[G]$ a normal basis of A over R, if X is a unit and $\sigma(X) = \sigma \cdot X$.



*F. van Oystaeyen (ed.), Methods in Ring Theory, 201–205.*

Let us define A-algebra homomorphisms

$$\nu_i : A[G] \longrightarrow A[G] \text{ by setting } \nu_i(\sigma) = \sigma^i$$

for $i \in \mathbb{Z}/p\mathbb{Z}$.
An element $\alpha \in A[G]$ is called normalized if

$$\alpha\, \nu_{-1}(\alpha) = 1 \text{ and } \nu_o(\alpha) = 1.$$

If $Z \in A[G]$ is a normal basis of A over R then $X = (Z\, \nu_{-1}(Z^{-1}))^{\frac{p+1}{2}}$ is normalized and is a normal basis of A over R [2], Satz 1. In [3] we defined a commutative ring $\Lambda(R)$ such that the normalized units in $\Lambda(R)$ correspond to the Galois extensions with Galois group G and with normal basis over R. We have $\Lambda(R) = R^p$ as an R-module, and the map

$$\varphi: \Lambda(R) \longrightarrow R[G],\ (x_T, T_1, \ldots, T_{p-1}) \longmapsto x_T + p \sum_{i=0}^{p-1} T_i \sigma^{-i},$$
$$\text{where } T_o = -\sum_{i=1}^{p-1} T_i,$$

is a ring homomorphism [3], p. 697/98. The unit element in $\Lambda(R)$ is $(1,0,\ldots,0)$. We denote by $C(R)$ the group of normalized units in $\Lambda(R)$, that is

$$C(R) = \{T = (x_T, T_1, \ldots, T_{p-1}) \mid x_T = 1 \text{ and } T\, \nu_{-1}(T) = (1,0,\ldots,0)\},$$

where $\nu_{-1}(x_T, T_1, \ldots, T_{p-1}) = (x_T, T_{p-1}, T_{p-2}, \ldots, T_1)$.

If S is a unit in $\Lambda(R)$ then $T = S\, \nu_{-1}(S^{-1})$ is in $C(R)$. Finally, we have a multiplicative map

$$g : A[G] \rightarrow \Lambda(A) \text{ such that } (\varphi \circ g)(x) = x^p$$

for all $x \in A[G]$, [3] p. 698. Furthermore, g commutes with $\nu_{\frac{1}{i}}$, where $\nu_{\frac{1}{i}}(x_T, T_1, \ldots, T_{p-1}) = (x_T, T_i, T_{2i}, T_{(p-1)i})$ for $i \in \{1, \ldots, p-1\}$ and all indices are in the field $\mathbb{Z}/p\mathbb{Z}$, which we identify with $\{0,1,\ldots,p-1\}$.

<u>Theorem 1.</u> ([3], Satz 3.3 and Satz 4.1).

i) Given any $T \in C(R)$ there exists a Galois extension $R_T$ over R with group G such that $R_T$ has a normal basis $X \in R_T[G]$ which is normalized and satisfies

$$g(X) = \prod_{i=1}^{p-1} \nu_{-\frac{1}{i}}(T^i) =: \gamma(T).$$

ii) The correspondence $T \mapsto R_T$ induces a group isomorphism $C(R)/C'(R) \longrightarrow H(R,G)$, where $C'(R) = \{T \in C(R) \mid \exists \alpha \in R[G]$ such that $\alpha$ is normalized and $g(\alpha) = \gamma(T)\}$.

Examples.

1) If the characteristic of R is equal to p, then $R_T$ for $T = (1, T_1, \ldots, T_{p-1}) \in C(R)$ is isomorphic to the Artin-Schreier extension $R[z]/(z^p - z + t)$, where

$$t = -\sum_{i=1}^{p-1} i\, T_i$$ [6], Satz 2, p. 27.

2) If p is a unit in R, then the ring homomorphism $\varphi : \Lambda(R) \to R[G]$ is an isomorphism and we replace the ring $\Lambda(R)$ by the group ring $R[G]$ and the map g by the map $x \mapsto x^p$.

Let R = K be a field of characteristic different from p and $\zeta$ a primitive p-th root of unity in K. If we define a K-algebra homomorphism $\chi : K[G] \to K$ by setting $\chi(\sigma) = \zeta$, then the Galois extension $K_T$ for a normalized $T \in K[G]$ is isomorphic to the Kummer extension $K_t := K[z]/(z^p - t)$, where $t = \chi(\gamma(T))$ and $\sigma(\bar z) = \zeta \cdot \bar z$ for $\bar z = z \bmod (z^p - t)$. If $X \in K_T[G]$ is a normalized normal basis of $K_T$ over K satisfying $X^p = \gamma(T)$, then the isomorphism $K_t \to K_T$ is given by $\bar z \mapsto \chi(X)$.

3) Set $R = \mathbb{Q}$, where $\mathbb{Q}$ is the field of rational numbers, and set $T = \sigma$. Then $\mathbb{Q}_1 := \mathbb{Q}_T$ has a normalized normal basis $X^{(1)}$ over $\mathbb{Q} = \mathbb{Q}_o$ such that $X^{(1)p} = \gamma(\sigma) = \sigma =: X^{(o)}$. For $n \geq 1$ we define $\mathbb{Q}_{n+1} = (\mathbb{Q}_n)_{X(n)}$, where $X^{(n)}$ is a normalized normal basis of $\mathbb{Q}_n$ over $\mathbb{Q}_{n-1}$ such that $X^{(n)p} = \gamma(X^{(n-1)})$. We then get a tower of fields

$$\mathbb{Q}_o \subset \mathbb{Q}_1 \subset \ldots \subset \mathbb{Q}_n \subset \ldots \subset \mathbb{Q}_\infty ,$$

where $\mathbb{Q}_n$ is a cyclic extension of degree $p^n$ over $\mathbb{Q}$ and $\mathbb{Q}_\infty$ is the cyclotomic $\mathbb{Z}_p$-extension over $\mathbb{Q}$. This can be shown by adjoining the p-th roots of unity to $\mathbb{Q}$ or directly using the proof of Satz 2 in [5].

Applications of Theorem 1.

1) Let I be an ideal in R. If one can lift units in $\Lambda(R/I) = \Lambda(R)/I\,\Lambda(R)$ to units in $\Lambda(R)$, then one can lift Galois extensions with group G over R/I to Galois extensions over R.

Proposition 1. ([6] Satz 1, p. 89). If R is semilocal or if I is contained in the Jacobson radical of R then the canonical homomorphism

$$H(R,G) \longrightarrow H(R/I,G),\ A \longmapsto A \otimes_R R/I,$$

is surjective.

2) Proposition 2. ([4], Theorem 1) Let R be semilocal and connected, and let A be a separable closure of R. If R' is a covering of R with $R \subset R' \subset A$ and $1 < [A : R'] < \infty$ then $[A : R'] = 2$.

Proposition 2 was conjectured by Knebusch [7], Conjecture 10.11. Knebusch's theory [7] and Proposition 2 yield a complete characterization of all real closures over R. This generalizes the well-known theorem of Artin-Schreier about real closures of fields.

3) We give a nice description of all cubic Galois field extensions.

Let K be a field of characteristic different from 3. For $k \in K$ we define a polynomial

$$f_k(z) = z^3 + 3kz^2 - 9z - 3k.$$

<u>Proposition 3.</u> a) If $f_k(z)$ is irreducible over K, then $L = K[z]/(f_k(z))$ is a Galois field extension of K with group $G = \{1, \sigma, \sigma^2\}$, where G acts on L via

$$\sigma(\bar{z}) = \frac{\bar{z}-3}{\bar{z}+1}, \quad \bar{z} := z \bmod (f_k(z)).$$

b) Conversely, if L is a Galois field extension of K with group $G = \{1, \sigma, \sigma^2\}$, then there is an element $k \in K$ such that L is isomorphic to $K[z]/(f_k(z))$.

<u>Proof.</u> a) One easily checks $f_k(z) = \prod_{i=0}^{2} (z - \sigma^i(\bar{z}))$ where $\sigma^2(\bar{z}) = \frac{-\bar{z}-3}{\bar{z}-1}$.

b) We identify L with $K_T$ where $T = (1, T_1, T_2) \in C(K)$. The condition $T\, \nu_{-1}(T) = (1,0,0)$ is equivalent to

$$(1) \quad T_1 + T_2 = 3(T_1^2 + T_2^2 + T_1 T_2).$$

If $X = \sum_{i=0}^{2} X_i \sigma^{-i}$ is a normalized normal basis of $K_T$ over K such that $X^3 = \varphi(T) = 1 - 3T_1 - 3T_2 + 3T_1\sigma^2 + 3T_2\sigma$, then we have $X_o + X_1 + X_2 = 1$, $X_o X_1 + X_1 X_2 + X_2 X_o = 0$ and

$$T_1 = \text{trace}\,(X_o^2 X_1), \quad T_2 = \text{trace}\,(X_1^2 X_o).$$

These equations imply $f_k(z) = \prod_{i=0}^{2} (z - \sigma^i(\frac{X_2 - X_1}{X_1 + X_2}))$, where

$$k = \frac{T_1 - T_2}{T_1 + T_2}.$$

Since L is a field it follows from (1) that $T_1 + T_2 \neq 0$. We now have an isomorphism

$$\varphi : K[z]/(f_k(z)) \longrightarrow L, \quad \bar{z} \mapsto \frac{X_2 - X_1}{X_1 + X_2},$$

and $\varphi(\frac{\bar{z}-3}{\bar{z}+1}) = \sigma(\frac{X_2 - X_1}{X_1 + X_2})$.

REFERENCES.

[1] D.K.Harrison, Abelian extensions of commutative rings, Mem. Amer. Math. Soc. 52 (1965), 1-14.

[2] I. Kersten and J. Michaliček, Kubische Galoiserweiterungen mit Normalbasis, Commun. Algebra 9 (1981), 1863-1871.

[3] I. Kersten and J. Michaliček, Galoiserweiterungen der Ordnung p mit Normalbasis, Commun. Algebra 10 (1982), 695-718.

[4] I. Kersten and J. Michaliček, Eine Charakterisierung der reellen Abschlüsse von semilokalen Ringen, Commun. Algebra 11 (1983), 581-593.

[5] I. Kersten and J. Michaliček, Das Einbettungsproblem für zyklische $p^n$-Erweiterungen über einem semilokalen Ring, J. Algebra 82 (1983), 275-281.

[6] I. Kersten, Eine neue Kummertheorie für zyklische Galoiserweiterungen vom Grad $p^2$, Algebra Berichte 45, Math. Inst. Univ. München, 1983.

[7] M. Knebusch, Real closures of commutative rings II, J. reine angew. Math. 286/87 (1976), 278-313.

# THE INDEX OF NILITY OF A MATRIX RING OVER A RING WITH BOUNDED INDEX

Abraham A. Klein

School of Mathematical Sciences, Tel-Aviv, Israel

Let R be a nil ring with bounded index. It is known that the ring of polynomials R[t] is nil with bounded index, and a bound for its index can be given. This is used to get a bound for the index of the matrix ring $M_k(R)$. If R is not necessarily nil but has bounded index and satisfies a multilinear monic identity, then $M_k(R)$ has bounded index, and a bound for its index is given.

If R is a ring with bounded index, the matrix ring $M_k(R)$, $k \geq 2$, does not necessarily have bounded index (2). Necessary and sufficient conditions under which $M_k(R)$ will have bounded index are not yet known. The problem that we consider in this note is to find bounds for the indices of nility of $M_k(R)$ when these can be shown to be bounded. For instance, if R is nil with bounded index n, then $M_k(R)$ is nil with bounded index (5), however no explicit bound (in terms of n and k) seems to appear somewhere. It has been pointed out by A. Braun that applying a theorem of Shirshov (1) one may obtain a bound for the index of nility of $M_k(R)$. Such a bound will be huge even for small values of n and k. In (3) we gave bounds for the indices of nility of $M_k(R)$ when $n \leq 5$. In this note we present a completely different method by which we get bounds for the indices of $M_k(R)$ for all n. Let us describe the idea that helped us to find this method.

If R is nil with bounded index, then R[t] is nil with bounded index (2). If R is merely assumed to be nil, then it is not known whether R[t] or $M_k(R)$ are nil. The implication "If R is nil then $M_k(R)$ is nil" is equivalent to the Köthe Conjecture (4). If R[t] is nil then $M_k(R)$ is nil. Indeed, $M_k(R[t])$ is quasi-regular and since $M_k(R[t]) = (M_k(R))[t]$, it follows that $M_k(R)$ is nil (see the



*F. van Oystaeyen (ed.), Methods in Ring Theory, 207–210.*

lemma). Considering this fact, we looked for a bound for the index of $M_k(R)$, and to get it we used the fact that $R[t]$ has bounded index when R has bounded index.

We first prove the following easy result:

Lemma: Let $S[t]$ be the ring of polynomials in one indeterminate t over a ring S. If $c \in S$ is such that ct has a quasi-inverse of degree m, then $c^{m+1}=0$.

Proof: Let $q(t)$ be the quasi-inverse of ct, then $ct + q(t) + ct\,q(t) = 0$, so $q(0) = 0$ and $q(t) = -ct - ct\,q(t)$. It follows that

$$q(t) = \sum_{i=1}^{m} (-ct)^i + (-ct)^m q(t).$$

Since $\deg q(t) = m$ we get $(-ct)^m q(t) = 0$ and $q(t) = \sum_{i=1}^{m} (-ct)^i$, so $c^{m+1} = 0$.

Our main result is:

Theorem 1. Let R be a nil ring with index n and let r be the index of $R[t]$. Then $M_k(R)$ has index which is bounded by $(n+1)(r+1)^{k-1}$.

Proof. Let $S = M_k(R)$. The quasi-inverse of an element $A \in S[t]$ will be denoted by $A'$. By the result of the lemma, a bound for the degrees of $(Ct)'$, $C \in S$, will yield a bound for the index of S.

Let $C = (c_{ij})$, $c_{ij} \in R$, $1 \leq i,j \leq k$ and we proceed to find $(Ct)'$.

First write $Ct = \sum_{i=1}^{k} A_{i1}$ with $A_{i1} = \sum_j c_{ij} t e_{ij}$, $i=1,\ldots,k$. Since $c_{11}^n = 0$, it follows that $A_{11}^{n+1} = 0$, so $\deg A'_{11} \leq n$.

Next we use the "circle operation" and get $(Ct)\circ A'_{11} = \sum_{i=2}^{k} A_{i2}$ with $A_{i2} = A_{i1}A'_{11} + A_{i1} = \sum_j f_{ij}(t)\, e_{ij}$, $i = 2,\ldots, k$. So $\deg A_{i2} \leq \deg A_{i1} + \deg A'_{11} \leq n + 1$. Since r is the index of $R[t]$ we have $f_{22}(t)^r = 0$ and this implies that $A_{22}^{r+1} = 0$, so $\deg A'_{22} \leq r \deg A_{22} \leq r(n+1)$.

Take m such that $2 \leq m < k$ and assume by induction that $A_{pq}$ have been defined for $q \leq m$, $p \geq q$ and

$$\deg A_{im} \leq (r + 1)^{m-2}(n + 1), \quad \deg A'_{mm} \leq r(r + 1)^{m-2}(n + 1).$$

Assume also that $(Ct)\circ A'_{11}\circ\ldots\circ A'_{mm} = \sum_{i=m+1}^{k} A_{i,m+1}$ with $A_{i,m+1} = A_{im}A'_{mm} + A_{im} = \sum_j g_{ij}(t)e_{ij}$, $i = m+1,\ldots,k$. So we have

$$\deg A_{i,m+1} \leqslant \deg A_{im} + \deg A'_{mm} \leqslant (r+1)^{m-1}(n+1).$$

As above, since the index of $R[t]$ is $r$ we get $A^{r+1}_{m+1,m+1} = 0$, so

$$\deg A'_{m+1,m+1} \leqslant r \deg A_{m+1,m+1} \leqslant r(r+1)^{m-1}(n+1).$$

and it is easy to see that $(Ct)\circ A'_{11}\circ\ldots\circ A'_{m+1,m+1} = \sum_{i=m+2}^{k} A_{i,m+2}$ with $A_{i,m+2} = A_{i,m+1}A'_{m+1,m+1}+A_{i,m+1} = \sum_j h_{ij}(t)e_{ij}$, $i = m+2,\ldots,k$.

Finally we get $(Ct)\circ A'_{11}\circ\ldots\circ A'_{kk} = 0$, so $(Ct)' = A'_{11}\circ\ldots\circ A'_{kk}$ and

$$\deg (Ct)' \leqslant \sum_{m=1}^{k} \deg A'_{mm} \leqslant n + \sum_{m=2}^{k} r(r+1)^{m-2}(n+1) =$$
$$= n + (n+1)((r+1)^{k-1}-1) = (n+1)(r+1)^{k-1} - 1.$$

It follows that $M_k(R)$ is nil of index $\leqslant (n+1)(r+1)^{k-1}$.

By (2,Th 1) we have $r \leqslant \frac{2}{3}n!$ when $n \geqslant 3$. If $n = 2$ then $r = 2$ so the above proof shows that $M_k(R)$ is nil with index $\leqslant 3^k$. But by (3, Th.12,Cor.) $M_k(R)$ is nil with index $\leqslant k^2 + 1$. Since $k^2+1 < 3(\frac{4}{3} + 1)^{k-1}$ we conclude:

<u>Theorem 2</u>. Let R be a nil ring with index n and let k be a positive integer. Then $M_k(R)$ is nil with index $\leqslant (n+1)(\frac{2}{3}n!+1)^{k-1}$.

If we compare the last result with the results in (3) for $n \leqslant 5$, we observe that the bounds given there are smaller with one exception when $n = 4$ and $k = 2$. There we obtained 155 and here we have 85. For the convenience of the reader we write here the results of (3). The case $n = 2$ has been already quoted. If $n = 3$ then $M_k(R)$ has index $\leqslant 3k^2+1$ (3, Th.13,Cor.). If $n = 4$ then $M_k(R)$, $k \geqslant 3$, has index $\leqslant 12k^4 - 5k^2+1$ (3,Th.16. Cor.). Finally, if $n = 5$ then $M_2(R)$ has index $\leqslant 355$ and $M_k(R)$, $k \geqslant 3$, has index $\leqslant 28k^4-13k^2+1$ (3, Th.19, Cor.).

Our previous results may be used as in (2, Th.2) to find a bound for the index of $M_k(R)$ in case R is assumed to be a PI ring with bounded index. If this is done we get:

<u>Theorem 3</u>. Let R be a PI ring satisfying a multilinear monic identity of degree d. If R has index n then $M_k(R)$ has index

$\leqslant k[d/2](n+1)(\frac{2}{3}\, n!+1)^{k-1}$.

Note that a similar result can be obtained under suitable conditions on R that yield R/B(R) to be embeddable in a direct sum of matrix rings of order ⩽ m over division rings. What we get is that in the previous result [d/2] is replaced by m. For instance, if R/B(R) is Goldie then such an embedding exists, but we do not know neither under what conditions R/B(R) is Goldie nor under what conditions on a semiprime Goldie ring a bound for m can be found.

References

(1) Jacobson, N., "Shirshov's theorem on local finiteness", 1978, Lecture Notes in Math. Springer Verlag 697, pp. 25-46.
(2) Klein, A.A., "Rings with bounded index of nilpotence", 1982, Contemporary Math. 13, pp. 151-154.
(3) Klein, A.A., "Upper bounds for the indices of nilpotence of certain rings and ideals", Communications in Algebra (to appear).
(4) Krempa, J., "Logical connections between some open problems concerning nil rings", 1972, Fund, Math. 76, pp. 121-130.
(5) Tominaga, H., "Some remarks on π-regular rings of bounded index", 1955, Math. J. Okayama Univ. 4, pp. 135-141.

ON ALGEBRAIC DERIVATIONS OF PRIME RINGS

J.Krempa and J.Matczuk
Institute of Mathematics
00-901 University of Warsaw

1. Preliminaries. It is well known that if $R$ is a ring of characteristic $p>0$ and $d$ is a derivation of $R$, then $d^p$ is also a derivation. On the other hand, for a prime ring $R$, powers, less than char $R$, of inner derivations which are inner derivations were investigated in [3]. It appeared in particular that elements which determined such derivations have to be algebraic and the power of a derivation is not often a derivation.

Our main objective is to extend the results of Martindale ([3]) to arbitrary algebraic elements. Moreover, besides powers of derivations we will consider polynomials in derivations, providing natural assumptions on degree of polynomials and characteristic of a ring. Analogously as in [3] we will concentrate on prime rings only.



*F. van Oystaeyen (ed.), Methods in Ring Theory, 211–229.*

Notice that if $d$ is a derivation of a prime ring $R$ then, by passing to the skew polynomial ring $R[y,d]$, we may assume that $d$ is an inner derivation. Further, if we pass to the central closure of R, we may assume that $R$ is centrally closed. Henceforth $R$ will denote prime, centrally closed ring with the center $C$ and $d$ will stand for the inner derivation of $R$ adjoint to an element $a$ of R. This assumption is very useful because of the following lemma of Martindale (for the proof see [2]).

Lemma 1.1. Suppose that $a_i, b_i$'s are non-zero elements in a prime ring R such that $\sum a_i x b_i = 0$ for all $x \in R$. Then $a_i$'s are linearly dependent over C and $b_i$'s are linearly dependent over C.

In other words, the above lemma says that R is a faithful $R \otimes_C R^o$ module with natural action of $R \otimes_C R^o$ on R.

For convenience, we will take char $R = \infty$ if the additive group of R is torsion-free.
For any $x,y \in R$ we will take $[x,y] = xy-yx$.

One can easily compute by induction that for any $k \geqslant 1$ and $x \in R$ we have

$$\text{(I)} \qquad d^k(x) = \sum_{i=0}^{k} (-1)^i \binom{k}{i} a^{k-i} x a^i ,$$

and consequently if $f(t) = \sum_{k=0}^{n} c_k t^k \in C[t]$, then

$$\text{(II)} \qquad f(d)(x) = \sum_{k=0}^{n} c_k \sum_{i=0}^{k} (-1)^i \binom{k}{i} a^{k-i} x a^i .$$

Lemma 1.2. Let $f(t) = \sum_{k=0}^{n} c_k t^k \in C[t]$. Then $f(d)$ is a derivation of R if and only if for any $x,y \in R$ the following identity holds:

$$0 = \sum_{i=1}^{n-1} (-1)^i \Big[\sum_{k=i+1}^{n} c_k \binom{k}{i} a^{k-i}, x\Big] y a^i + \big([f(a), x] - f(d)(x)\big)\, y.$$

Moreover $c_o = 0$ in this case.

Proof. By changing the order of the sum in (II) we get $f(d)(x) = \sum_{i=0}^{n} (-1)^i \sum_{k=i}^{n} c_k \binom{k}{i} a^{k-i} x a^i$. Since $f(d)$ is a derivation if and only if $f(d)(xy) - xf(d)(y) - f(d)(x)\,y = 0$ for any $x,y \in R$, the last identity gives us: $f(d)$ is a derivation if and only if

$$0 = \sum_{i=0}^{n} (-1)^i \Big[\sum_{k=i}^{n} c_k \binom{k}{i} a^{k-i}, x\Big] y a^i - f(d)(x)\, y =$$

$$= \sum_{i=1}^{n-1} (-1)^i \Big[\sum_{k=i+1}^{n} c_k \binom{k}{i} a^{k-i}, x\Big] y a^i + \big([f(a), x] - f(d)(x)\big)\, y.$$

This gives the first statement of the lemma. Now, if $f(d)$ is a derivation, then by substituting $x = 1$ in (II) we obtain $c_o = 0$.

Theorem 1.3. Suppose that char $R > n > 1$ and $f \in C[t]$ is a polynomial of degree n. If $f(d)$ is a derivation of R, then a is algebraic over C of degree less than or equal to $(n+1)/2$ .

Proof. If $d = 0$, then obviously $a \in C$, so a is

algebraic of degree $1 \leq {}^{(n+1)}/_2$ . Let us suppose that $d \neq 0$. Since $f(d)$ is a derivation, thus Lemma 1.2 implies that for some suitable $f_i \in C[t]$, with $\deg f_i = n-i$, $i = 1,\dots,n-1$

$$0 = \sum_{i=1}^{n-1} [f_i(a),x]ya^i + \big([f(a),x] - f(d)(x)\big)y .$$

Because $d \neq 0$ and $n < \operatorname{char} R$, we also have $[f_{n-1}(a),x] = (-1)^{n-1} c_n nd(x) \neq 0$ for some $x \in R$. Therefore in the presence of arbitrariness of $y \in R$, Lemma 1.1 implies that a is algebraic over C of degree less than or equal to n-1. Let $k = \left[{}^{n+1}/_2\right]$ and let m be the smallest natural number such that $m \geq k$ and for $i = 0,\dots,m$ there exists $g_i \in C[t]$ with $\deg g_i = n-i$ such that

$$\text{(III)} \quad 0 = \sum_{i=1}^{m} [g_i(a),x]ya^i + \big([g_o(a),x] - f(d)(x)\big)y$$

for any $x,y \in R$.

If $[g_m(a),x] = 0$ for all $x \in R$, then a is algebraic over C of degree less than or equal to $n-m \leq k$. This gives the thesis. Now let us suppose that for some $x \in R$ $[g_m(a),x] \neq 0$. Then it follows from Lemma 1.1 and arbitrariness of $y \in R$ that $1,a,\dots,a^m$ are linearly dependent over C. If $m = k$ we have the thesis, so let us suppose that $m > k$, and let $a^m = \sum_{i=0}^{m-1} \lambda_i a^i$ where $\lambda_i \in C$ for all i's. By substituting the formula on $a^m$ in (III) we obtain

$$0 = \sum_{i=1}^{m-1} [g_i(a),x]ya^i + [g_m(a),x]y \sum_{i=0}^{m-1} \lambda_i a^i +$$
$$+ \big([g_o(a),x] - f(d)(x)\big)y .$$

Hence

$$0 = \sum_{i=1}^{m-1} [g_i(a)+ \lambda_i g_m(a),x]ya^i + \big([g_o(a) + \lambda_o g_m(a),x] -$$
$$- f(d)(x)\big)y,$$ and obviously $\deg (g_i + \lambda_i g_m) = n-i$, since $\deg g_m < \deg g_i$ for $i \leq m-1$. This contradicts the choice of m.

The above result is a stronger version of Lemma 1 from [3].

Note that the bound of a degree of a over C in the above theorem cannot be reformed, since if a is a nilpotent element of index k then $d^{2k-1} = 0$.

Throughout the paper for $f = \sum_{k=0}^{n} c_k t^k \in C[t]$ we put $f_+ = \sum_{k=0}^{[n+1/2]} c_{2k} t^{2k}$.

In the case a is a zero divisor Lemma 1.2 takes the following simpler form:

<u>Lemma 1.4</u>. Suppose that a is two-sided zero divisor and $f = \sum_{k=1}^{n} c_k t^k \in C[t]$. Then:

1. If f(d) is a derivation then f(d) is the inner derivation adjoint to f(a).

2. If char R $\neq$ 2 then

f(d) is a derivation if and only if the following

conditions are satisfied

a/ $\sum_{k=1}^{n} c_k \sum_{i=1}^{k-1}(-1)^i\binom{k}{i} a^{k-i}xa^i = 0$ for all $x \in R$.

b/ $f_+(a) = 0$.

<u>Proof</u>. 1. Let us suppose that f(d) is a derivation. Thus from Lemma 1.2 we obtain

$$0 = \sum_{i=1}^{n-1}(-1)^i[\sum_{k=i+1}^{n} c_k\binom{k}{i}a^{k-i},x]ya^i + ([f(a),x]-f(d)(x))y$$

for any $x,y \in R$. Let $0 \neq b \in R$ be such that $ab = 0$. By multiplying the last identity by b on the right we get $([f(a),x] - f(d)(x))yb = 0$ for any $x,y \in R$. Now, primeness of R gives 1.

2. Suppose that f(d) is a derivation. Therefore, by part 1, $f(d)(x) = [f(a),x]$ for all $x \in R$. Thus, by using (II) we obtain

$$f(d)(x) = \sum_{k=1}^{n} c_k \sum_{i=0}^{k}(-1)^i\binom{k}{i}a^{k-i}xa^i = [f(a),x].$$

Hence

$$\sum_{k=1}^{n} c_k \sum_{i=1}^{k}(-1)^i\binom{k}{i}a^{k-i}xa^i = -xf(a)$$

and consequently

$$\text{(IV)} \quad \sum_{k=1}^{n} c_k \sum_{i=1}^{k-1}(-1)^i\binom{k}{i}a^{k-i}xa^i = -2xf_+(a)$$

There exists an element $0 \neq b' \in R$ such that $b'a = 0$. If we multiply the identity (IV) by $b'$ on the left, we obtain $2b'xf_+(a) = 0$ for all $x \in R$. Since R is prime and char $R \neq 2$, we have $f_+(a) = 0$. This and the identity (IV) give a/ and b/ of 2.

If a/ and b/ of 2 are fulfilled then it is straightforward to verify that $f(d)(x) = [f(a),x]$ for all $x \in R$.

The last lemma and Theorem 1.3 become very useful in case the base field C is algebraically closed. Let $\bar{C}$ denote the algebraic closure of C and let $\bar{R} = R \otimes_C \bar{C}$. It is known ([1]) that $\bar{R}$ is a prime, centrally closed ring and $\bar{C}$ is the center of $\bar{R}$. Moreover $d \otimes 1$ is a derivation of $\bar{R}$ adjoint to $a \otimes 1$ and for $f \in C[t]$ $f(d)$ is a derivation of R if and only if $f(d \otimes 1) = f(d) \otimes 1$ is a derivation of $\bar{R}$. Also, the degrees of a over C and $\bar{C}$, respectively, are equal. Therefore till the end of the paper we will assume that C is an algebraically closed field.

<u>Theorem 1.5</u>. If $f(d)$ is a derivation for some $f \in C[t]$, then $f(d)$ is the inner derivation adjoint to $f(a-\lambda)$ for some $\lambda \in C$.

<u>Proof</u>. Because C is algebraically closed, the proof is a direct consequence of Theorem 1.3 and Lemma 1.4.

<u>2. Separable elements</u>. Throughout this part we will assume that a is algebraic over C and the minimal polynomial for a has simple roots. i.e. a is a separable element. Since C is algebraically closed

thus, by Chinse Remainder Theorem, there exist ortogonal idempotents $e_1,\dots,e_q \in R$ and elements $\lambda_1,\dots,\lambda_q \in C$ such that

(V) $a = \sum_{i=1}^{q} \lambda_i e_i, \ \sum_{i=1}^{q} e_i = 1$ and $\lambda_i \neq \lambda_j$ for $i \neq j$ .

For $1 \leqslant i,j \leqslant q$ we put $\lambda_{i,j} = \lambda_i - \lambda_j$ .

In our notation, by using (V), (II) takes the form of

(VI) $$f(d)(x) = \sum_{k=0}^{n} c_k \sum_{l,m=1}^{q} \sum_{i=0}^{k} (-1)^i \binom{k}{i} \lambda_l^{\,k-i} \lambda_m^{\,i} e_l x e_m =$$

$$= \sum_{l,m=1}^{q} \sum_{k=0}^{n} c_k (\lambda_{l,m})^k e_l x e_m = \sum_{l,m=1}^{q} f(\lambda_{l,m}) e_l x e_m$$

<u>Theorem 2.1</u>. Let $f \in C[t]$. Then $f(d) = 0$ if and only if $f(\lambda_{l,m}) = 0$ for any $1 \leqslant l,m \leqslant q$. In particular d is algebraic over C and if g is the minimal polynomial for d over C, then g is separable and all different $\lambda_{l,m}$'s are precisely its roots.

<u>Proof</u>. Let $f \in C[t]$. If $f(\lambda_{l,m}) = 0$ for all $1 \leqslant l,m \leqslant q$ then, by (VI), $f(d) = 0$.

Now, if $f(d) = 0$ then, by multiplying (VI) by $e_l$ and $e_m$ on left and right respectively, we get $f(\lambda_{l,m}) e_l x e_m = 0$ for all $x \in R$. Since R is prime, $f(\lambda_{l,m}) = 0$. This gives the proof of the first part of the theorem. The second part is a direct consequence of the first one.

<u>Corollary 2.2</u>. Let g be the minimal polynomial for d over C. Then:

1. If char $R \neq 2$, then $2q-1 \leqslant \deg g \leqslant q(q-1)+1$ .

2. If char $R = 2$, then $q \leqslant \deg g \leqslant {}^{q(q-1)}/_{2} +1$ .

The proof of the above corollary is a consequence of Theorem 2.1 by making use of the following equations $\lambda_{i,i} = 0$ for $1 \leqslant i \leqslant q$ and $\lambda_{i,j} = -\lambda_{j,i}$ for $i \neq j$.

The following example shows that the lower and upper bounds, which appeared in Corollary 2.2, can be obtained.

Example 2.3. Let $R = M_q(\mathbf{C})$ be the ring of all $q \times q$ matrices with entries from $\mathbf{C}$ - the field of complex numbers.

1. Let $a = \mathrm{diag}(1,2,\dots,q)$. Then $\lambda_m = m$, so $\lambda_{l,m} = l-m$ for any $1 \leqslant l,m \leqslant q$ . Therefore we have $2q-1$ different elements $\lambda_{l,m}$, and thus Theorem 2.1 implies that the degree of the minimal polynomial for d is equal to $2q-1$.

2. Because $\mathbf{C}$ is infinite one can easily pick such elements $\lambda_1,\dots,\lambda_q \in \mathbf{C}$ that $\lambda_{i,j} = \lambda_{l,m}$, with $i \neq j$, yields $i = l$ and $j = m$. Let $a = \mathrm{diag}(\lambda_1,\dots,\lambda_q)$. Then we have exactly $q(q-1)+1$ different values of $\lambda_{i,j}$'s for $1 \leqslant i,j \leqslant q$. Therefore, by Theorem 2.1, $q(q-1)+1$ is the degree of the minimal polynomial for d.

Now, if we pass to polynomials in a derivation which are derivations, we have the following generalization of Corollary 3 from [3].

Lemma 2.4. Suppose that char R $\neq$ 2 and $f = \sum_{k=1}^{n} c_k t^k \in C[t]$. Then f(d) is a derivation if and only if

a/ $f(\lambda_{l,m}) - f(\lambda_{l,s}) - f(\lambda_{s,m}) = 0$ for all $1 \leq l,m,s \leq q$

and

b/ $f_+(d) = 0$.

Proof. Let $1 \leq m \leq q$. Then, by (V), $a-\lambda_m = \sum_{i=1}^{q} \lambda_{i,m} e_i$.

Obviously $a-\lambda_m$ is a two sided zero divisor and d is adjoint to $a-\lambda_m$. Let us suppose that f(d) is a derivation. By Lemma 1.4, we have

$$0 = \sum_{k=1}^{n} c_k \sum_{i=1}^{k-1} (-1)^i \binom{k}{i} (a-\lambda_m)^{k-i} x (a-\lambda_m)^i \quad \text{for any}$$

$x \in R$. Therefore

$$0 = \sum_{k=1}^{n} c_k \sum_{i=1}^{k-1} (-1)^i \binom{k}{i} \sum_{l,s=1}^{q} \lambda_{l,m}{}^{k-i} \lambda_{s,m}{}^{i} e_l x e_s =$$

$$= \sum_{l,s=1}^{q} \Big\{ \sum_{k=1}^{n} c_k \Big( \sum_{i=0}^{k} (-1)^i \binom{k}{i} \lambda_{l,m}{}^{k-i} \lambda_{s,m}{}^{i} - \lambda_{l,m}{}^{k} - \lambda_{m,s}{}^{k} \Big) \Big\} e_l x e_s$$

This yields

$$0 = \sum_{l,s=1}^{q} \big( f(\lambda_{l,s}) - f(\lambda_{l,m}) - f(\lambda_{m,s}) \big) e_l x e_s$$

for any $1 \leq l,m,s \leq q$ and $x \in R$.

Now, primenes of R implies that for all $1 \leq l,m,s \leq q$ $f(\lambda_{l,s}) - f(\lambda_{l,m}) - f(\lambda_{m,s}) = 0$. This gives the condition a/ of the lemma. Moreover, since f(d) is a derivation, Lemma 1.4 also gives $f_+(a-\lambda_m) = 0$ for any $1 \leq m \leq q$. It means that

$$0 = \sum_{k=1}^{[n+1/2]} c_{2k}\Big(\sum_{l=1}^{q} \lambda_{l,m}e_l\Big)^{2k} = \sum_{l=1}^{q} f_+(\lambda_{l,m})e_l$$

and consequently, by using (V), $f_+(\lambda_{l,m}) = 0$ for all $1 \leq l,m \leq q$. Now, Theorem 2.1 implies $f_+(d) = 0$.

By making use of Lemma 1.4 and Theorem 2.1 it is straightforward to verify that conditions a/ and b/ of the lemma imply that f(d) is a derivation.

Let g be the minimal polynomial for d and let $f \in C[t]$. Then $f = hg + r$ where $r,h \in C[t]$ and $\deg r < \deg g$. Hence f(d) is a derivation if and only if r(d) is a derivation. Thus, in our considerations, we may assume that $\deg f < \deg g$.

Corollary 2.5. Suppose that char $R \neq 2$ and $f \in C[t]$ is such that $\deg f < \deg g$. Then f(d) is a derivation if and only if

a/ $f(\lambda_{l,m}) - f(\lambda_{l,s}) - f(\lambda_{s,m}) = 0$
for all $1 \leq l,m,s \leq q$,

b/ $f_+ = 0$.

We will show later that the polynomial f from the above corollary does not have to be of the form $\lambda t$ where $\lambda \in C$.

Now we are in position to discuss powers of derivations which are derivations.

It was shown in [3], provided that $a \in R$ is an algebaic, not separable element, that if $d^n$ is a derivation for some $1 < n < \mathrm{char}\, R$ then $a = \lambda + b$ for some $\lambda \in C$, $b \in R$ and $b^{[n+1/2]} = 0$, so $d^n = 0$. On the other hand, if $a^2 = \lambda a + \mu$ for some $\lambda, \mu \in C$ then it is easy to compute that $d^3 = (\lambda^2 + 4\mu)d$, so $d^3, d^5, \ldots$ are derivations without any assumptions on separability of a.

Proposition 2.6. Suppose that $R = M_3(\mathbf{C})$ and $a = \mathrm{diag}\ 0,1,\lambda$ where $\lambda \neq 0,1$. Then for $n > 1$ we have: $d^n$ is a derivation if and only if n is odd and $(1-\lambda)^n = 1 - \lambda^n$.
In particular $d^5$ is a derivation if and only if $d^7$ is a derivation.

Proof. The proof of the first part of the proposition is a consequence of Lemma 2.4 and Theorem 2.1. One can see that the sets of all different roots of equations $(t-1)^5 = t^5 - 1$ and $(t-1)^7 = t^7 - 1$ are equal to $\{0, 1, \varepsilon_6, \varepsilon_6^5\}$ where $\varepsilon_6$ is the 6-th primitive root of unit. This finishes the proof because of the first statement from the proposition.

Example 2.7. Let $R = M_3(\mathbf{C})$.

1. If we take $a = \mathrm{diag}\ (0,1,2)$, then, by Proposition 2.6, $d^n$ is not a derivation for any $n > 1$.

2. If we take $a = \operatorname{diag}(0,1,\varepsilon_6)$, then, by Proposition 2.6, $d^5$ is a derivation. Moreover $d^5 \neq \lambda d$ for any $\lambda \in C$, since by Theorem 2.1 and Corollary 2.2 the degree of minimal polynomial for d is equal to 7.

3. Let $\lambda \in \mathbf{C}$ be a root of the eqation $(t-1)^9 = t^9-1$ different from $0,1,\varepsilon_6,\varepsilon_6{}^5$ and let $a = \operatorname{diag}(0,1,\lambda)$. Then, by Proposition 2.6, $d^9$ is a derivation and $d^n$ is not a derivation for $1<n<9$.

4. Prof. Schinzel kindly informed us that there exists a natural number N such that for any odd $n>N$ there exists such $\lambda_n \in C$ that $(1-\lambda_n)^n = 1-\lambda_n{}^n$ and $(1-\lambda_n)^m \neq 1-\lambda_n{}^m$ for $1<m<n$. If we put $a = \operatorname{diag}(0,1,\lambda_n)$, then we get, by Proposition 2.6, that $d^n$ is a derivation but $d^m$ is not a derivation for all $1<m<n$.

Now we return to the general situation. The last example and the remark before Proposition 2.6 suggest the following conjecture:

Let a be a separable , algebraic element of degree greater than or equal to 4. Then for any $1<n<\operatorname{char} R$ $d^n$ is not a derivation.

In the terminology of a system of equations over C this conjecture may be formulated as follows:

Let $q \geq 4$ and let n be odd number such that $1<n<\operatorname{char} R$. If $(\lambda_1,\ldots,\lambda_q) \in C^q$ is a salution of the system of equations

$$(t_i - t_j)^n + (t_j - t_k)^n + (t_k - t_i)^n = 0$$

where $1 \leq i,j,k \leq q$, then $\lambda_i = \lambda_j$ for some $i \neq j$.

3. General case. In this part $a \in R$ will be an arbitrary algebraic element. Analogously as in part 2, because C is algebraically closed, if we use Chinese Remainder Theorem we obtain:
there exist ortogonal idempotents $e_1, \ldots, e_q \in R$, nilpotent elements $b_1, \ldots, b_q \in R$ of index of nilpotency $r_1, \ldots, r_q$ respectively, and $\lambda_1, \ldots, \lambda_q \in C$ such that

(VII) $a = \sum_{i=1}^{q} \lambda_i e_i + b_i, \; 1 = \sum_{i=1}^{q} e_i, \; b_j e_i = e_i b_j = \delta_{ij} b_j$ .

We put $a_o = \sum_{i=1}^{q} \lambda_i e_i$, $b = \sum_{i=1}^{q} b_i$ and $\lambda_{i,j} = \lambda_i - \lambda_j$ .

Theorem 3.1. Let $f \in C[t]$. Then:

1. If for all $1 \leq l,m \leq q$ $\lambda_{l,m}$ is a root of f of multiple greater than or equal to $r_l + r_m - 1$, then $f(d) = 0$.

2. If $f(d) = 0$, then all $\lambda_{l,m}$'s are roots of f.

3. If char $R > 2 \deg(a) - 1$, then f is the minimal polynomial for d if and only if all different $\lambda_{l,m}$'s are precisely the roots of f and their multiple is equal to $\max_{i,j:\lambda_{ij}=\lambda_{l,m}} (r_i + r_j - 1)$, respectively.

In particular, if a is not a separable element, then f is not separable, either.

Proof. From the formula (VII) it follows that $e_l R e_m$ is d-invariant subspace of R for any $1 \leq l,m \leq q$. Therefore if we set $\varphi_{l,m} : e_l R e_m \to e_l R e_m$ according to the rule $\varphi_{l,m}(x) = d(x) - \lambda_{l,m} x$, then $\varphi_{l,m}$ is well defined. Moreover, for any $x \in R$ we have

$$\varphi_{l,m}(e_l x e_m) = b_l e_l x e_m - e_l x e_m b_m = e_l(b_l x - x b_m)e_m ,$$

and hence for any $k \geq 1$

$$\text{(VIII)} \quad \varphi_{l,m}{}^k(e_l x e_m) = e_l \sum_{i=1}^{k} (-1)^i \binom{k}{i} b_l{}^{k-i} x b_m{}^i e_m .$$

Our choice of $r_l$ and $r_m$ and the above formula yield

$$\text{(IX)} \qquad \varphi_{l,m}{}^{r_l+r_m-1} = 0$$

Now we are in position to finish the proof.

1. Suppose that $f \in C[t]$ and for all $1 \leq l,m \leq q$ f has $\lambda_{l,m}$ as a root of multiple not less than $r_l+r_m-1$. Thus there exist polynomials $h_{l,m}$'s $\in C[t]$ such that $f(t) = h_{l,m}(t)(t - \lambda_{l,m})^{r_l+r_m-1}$. Therefore

$$f(d)(e_l R e_m) = h_{l,m}(d)(d - \lambda_{l,m})^{r_l+r_m-1}(e_l R e_m) =$$

$$= h_{l,m}(d)\varphi_{l,m}{}^{r_l+r_m-1}(e_l R e_m) = 0$$

because of (IX). Now, arbitrariness of l,m gives us $f(d) = 0$.

2. Suppose that $f(d) = 0$. Let $d_o$, $d_1$ be inner derivations adjoint to $a_o$ and b, respectively. Obviously, by (VII), $d_o$ and $d_1$ commute and $d = d_o+d_1$. Hence $0 = f(d) = f(d_o+d_1) = f(d_o) + d_1 h(d_o,d_1)$ for some $h \in C[t_o,t_1]$. It means that $f(d_o) = -d_1 h(d_o,d_1)$.

Since $d_1$ is nilpotent, the last equality implies that $f(d_o)^r = 0$ for some natural number r. So if g is the minimal polynomial for $d_o$ then g divides $f^r$. In other words, every root of g is a root of f. Now, since $a_o$ is a separable element, Theorem 2.1 ends the proof of 2.

3. The formula (VIII) and our assumption on characteristic of R gives us :

$$\varphi_{l,m}^{r_l+r_m-2}(x) = (-1)^{r_m-1}\binom{r_m+r_l-2}{r_m-1} b_l^{r_l-1} x b_m^{r_m-1} \neq 0$$

for some $x \in R$. It follows from this and (IX) that $(t - \lambda_{l,m})^{r_l+r_m-1}$ is the minimal polynomial for $d|_{e_lRe_m}$. Let us suppose that f is the minimal polynomial for d and let $1 \leqslant l,m \leqslant q$. Because $f(d)(e_lRe_m) = 0$ and $(t - \lambda_{l,m})^{r_l+r_m-1}$ is the minimal polynomial for $d|_{e_lRe_m}$, $\lambda_{l,m}$ is a root of f of multiple not less than $r_l+r_m-1$. Now the part 1 of the theorem gives the proof of 3.

Note that if char $R = p < \infty$ and $p \leqslant r_l$, $p \leqslant r_m$, then $\varphi_{l,m}^p = 0$. This ilustrates the necessity of our assumption on characteristic.

<u>Corollary 3.2</u>. Let a be an algebraic element of degree m over C and let g be the minimal polynomial for d. Then :

1. $\deg g \leqslant m(m-1)+1$ .

2. If char R $\geqslant$ 2m-1, then deg g $\geqslant$ 2m-1 .

Proof. 1. Since a is algebraic of degree m $\sum_{i=1}^{q} r_i = m$ (see (VII)). In virtue of Theorem 3.1

$$\deg g \leqslant \sum_{i,j=1,\ i\neq j}^{q} (r_i+r_j-1) + 2\max_i(r_i) - 1 =$$

$$= \sum_{i=1}^{q}((q-2)r_i+m) - q(q-1) + 2\max_i(r_i) - 1 \leqslant$$

$$\leqslant (2m-q)(q-1) + 2(m-q) + 1 = -q^2 + (2m-1)q + 1 .$$

Thus, to prove 1, it is enough to show that

$$-q^2 + (2m-1)q + 1 \leqslant m(m-1) + 1$$

for any $1 \leqslant q \leqslant m$. But this is obvious because the equality holds for q = m and q = m-1.

2. This follows immediately from Theorem 1.3.

At the end we give some remarks connected with Theorem 3.1. It is well known that if $a = \lambda + b$, where $\lambda \in C$ and b is a nilpotent element of index n, then $d^{2n-1} = 0$ is a derivation. We also know that if a is an algebraic element of degree 2, then $d^3$ is a derivation.

Proposition 3.3. Suppose that a is an algebraic element of degree 3. Then there exists polynomial f of degree 5 such that f(d) is a derivation.

Proof. Because of the above observation we may assume that the minimal polynomial for a has no less than two different roots. Moreover, since C is algebraically closed, by substituting a by a $-\mu$ for some suitable $\mu \in C$, and by multiplying a by an element from C, we may also assume that the minimal polynomial for a has roots 0, 1, $\lambda$. Thus $a^3 = (\lambda+1)a^2 - \lambda a$. Let $f(t) = 3t^5 - 5(\lambda^2 - \lambda + 1)t^3$. Then, by making use of Lemma 1.4, it is easy to check that f(d) is a derivation for any $\lambda \in C$.

For the polynomial f from the above proposition, by Theorem 3.1 we have $f(d) \neq \mu d$ for any $\mu \in C$ if and only if $\lambda \neq 2$.

Because of Proposition 3.3 and the proceding remarks, one can think that if a is an algebraic element of degree n, then there always exists polynomial f of degree 2n-1 such that f d is a derivation. As the following example shows, it is not true even for n = 4.

<u>Example 3.4</u>. Let a be an algebraic of degree 4, which satisfies the identity $a^4 = a^3$. Let us suppose that there exists polynomial f of degree 7 such that f(d) is a derivation. Without loosing generality we may assume that $f = \sum_{i=2}^{7} c_i t^i$ and $c_7 = 1$. It follows from Chinse Remainder Theorem that $a = e + b$ where $e^2 = e \neq 0$, $b^2 \neq 0$, $b^3 = 0$ and $eb = be = 0$. Hence it is easy to see that

$R'=(1-e)R(1-e)$ is d-invariant subalgebra of R. So $f(d)|_{R'}$ is a derivation of $R'$. Moreover $d|_{R'}$ is the inner derivation adjoint to b. Now, by Theorem 1.3, since b is nilpotent of index 3, $t^5$ divides f and, consequently, $f_+$ is a monomial. By Lemma 1.4 $0 = f_+(a) = f_+(e)$, thus $f_+ = 0$. Therefore f is of the form $t^7 + ct^5$. Because f(d) is a derivation, so

$$\sum_{i=1}^{6} (-1)^i \binom{7}{i} a^{7-i} x a^i + c \sum_{i=1}^{4} (-1)^i \binom{5}{i} a^{5-i} x a^i = 0$$

for any $x \in R$. Thus

$$(-7a^6 - 5ca^4)xa + (21a^5 + 10ca^3)xa^2 + \sum_{i=3}^{6}(-1)^i \binom{7}{i} a^{7-i} x a^i + c \sum_{i=3}^{4} (-1)^i \binom{5}{i} a^{5-i} x a^i = 0.$$

Now, by using the equality $a^4 = a^3$, we have $(-7-5c)a^3xa + (21+10c)a^3xa^2 + g(a)xa^3 = 0$ for some $g \in C[t]$. Since $a, a^2, a^3$ are linearly independent over C, Lemma 1.1 implies that $(7+5c)a^3 = 0$ and $(21+10c)a^3=0$. These equalities yield $a^3 = 0$, which is imposible.

## REFERENCES

[1] T.S. Erickson, W.S. Martindale, 3rd, and J.M. Osborn, Prime Nonassociative Rings, Pac. J. Math. 60 (1975), 49-63.

[2] I.N Herstein, Rings with Involution, Univ. of Chicago Press, Chicago, 1976.

[3] W.S. Martindale, 3rd, C.R. Miers, On The Iterates of Derivations of Prime Rings, Pac. J. Math. 104 (1983), 179-190.

# SMOOTH MAXIMAL ORDERS IN QUATERNION ALGEBRAS, I.

Lieven Le Bruyn (*)
University of Antwerp, UIA

## 0. Introduction.

In their fundamental paper on maximal orders[ 4 ] , M. Auslander and O. Goldman studied the structure theory of maximal orders in two specific situations : over Dedekind domains in arbitrary central simple algebras and over regular domains in full matrix rings.
Whereas L. Silver [ 25 ] and r. Fossum [ 7 ] generalized some results to maximal orders over normal (resp. Krull) domains, M. Ramras [ 21,22,23 ] continued on the path taken by Auslander and Goldman, i.e. the study of maximal orders (prefarably with finite global dimension) over regular local domains in arbitrary central simple algebras.
Renewed interest in this rather restricted but important class of maximal orders came with the publication of two recent papers by M. Artin [ 2 ] , [ 3 ] . The first deals with the Zariski local structure of maximal orders over a smooth surface (i.e. the study of the number of conjugacy classes) whereas the second describes the étale local structure of the Brauer-Severi scheme associated to a maximal order over a Dedekind domain. Smooth (or strongly regular) maximal orders were introduced and studied jointly with M. Van den Bergh [ 15 ] in an attempt to grasp the contents of [ 2 ] . Later, it turned out that for this class of maximal orders one can easily describe the étale local structure of the (weak) Brauer-Severi scheme.
In this paper I have chosen to restrict attention to smooth maximal orders in quaternion algebras in order to make matters as concrete as possible. Further, it sometimes simplifies notation considerably while

(*) work supported by an NFWO/FNRS -/grant



*F. van Oystaeyen (ed.), Methods in Ring Theory, 231–264.*

preserving the heart of the more general arguments which will appear elsewhere [ 14 ] [ 15 ] .

Much of what follows is joint work with M. Van den Bergh.

1. Smooth maximal orders in quaternion algebras

In this section we will introduce smooth maximal orders and study their Zariski local structure in quaternion algebras. Since the definition differs slightly from that of strongly regular orders in [ 15 ] we will briefly recall both definitions.

Let $\Lambda$ be a tame order over a normal domain R, i.e. a reflexive R-order such that $\Lambda_p$ is hereditary for every height one prime p of R, in some central simple K-algebra $\Sigma$, K being the field of fractions.

Let $D = \{D_1,\ldots,D_n\}$ be a set of Weil divisors of $\Lambda$, i.e. a set of divisorial $\Lambda$-ideals, then we define the Rees ring of $\Lambda$ associated to the set of Weil divisors $D$, $\Lambda[D]$ to be the $\mathbb{Z}^{(n)}$-graded subring of $\Sigma[X_1,X_1^{-1},\ldots,X_n,X_n^{-1}]$, where $\deg(X_i) = (0,\ldots,1,\ldots,0)$, whose part of degree $(m_1,\ldots,m_n)$ is given by

$$\Lambda[D](m_1,\ldots,m_n) = (D_1^{m_1}\star\ldots\star D_n^{m_n})X_1^{m_1}\ldots X_n^{m_n}$$

These Rees rings were first introduced by F. Van Oystaeyen in the $\mathbb{Z}$-graded commutative case [ 27 ] and were subsequently generalized in e.g. [ 11,12,16 ] .

Definition 1. [ 15 ] with notation as above. $\Lambda$ is said to be a strongly regular order iff there exists a set of Weil divisors $D$ such that $\Lambda[D]$ is an Azumaya algebra over a regular center and every $D_i \in D$ is an invertible $\Lambda$-ideal.

For a more intrinsic characterization of strongly regular orders and the relation to the other regularity conditions, the reader is referred to [ 15 ] . Let us now turn to the archetype-example : let $P_c = \{p_1,\ldots,p_n\}$ be the finite set of height one prime ideals of R which ramify in the maximal (!) order $\Lambda$ with ramification indices say $\{e_1,\ldots,e_n\}$ and let

$P = \{P_1,\dots,P_n\}$ be the uniquely determined height one prime ideals of $\Lambda$ which lie over the ramified central primes. It is now fairly easy to compute the center of $\Lambda[P]$. It turns out to be the $\mathbb{Z}^{(n)}$-graded subring $R[P]$ of $K[X_1,X_1^{-1},\dots,X_n,X_n^{-1}]$ whose part of degree $(m_1,\dots,m_n)$ is given by

$$R[P](m_1,\dots,m_n) = (p_1^{[\frac{m_1}{e_1}]} \star\dots\star p_u^{[\frac{m_n}{e_n}]})\, X_1^{m_1}\dots X_n^{m_n}$$

where $[\frac{a}{b}]$ denotes the least integer $\geqslant \frac{a}{b}$.

Throughout we will assume that the maximal order $\Lambda$ satisfies :

$(Et_1)$ : For every height one prime ideal p of R there exists an étale extensions $R_p \rightarrow S(p)$ which splits $\Sigma$.

This condition is always satisfied in case the residue fields $\mathbb{K}(p)$ are perfect, cfr. [ 24 ] .

The main application of the graded construction given above to the theory of maximal orders satisfying $(Et_1)$ is that it basicly reduces questions about these rather arbitrary maximal orders to graded questions about graded reflexive Azumaya algebras [ 19 ] i.e. graded algebras A over a graded normal domain S such that

$$(A \otimes_S A^{opp})^{**} \cong END_S(A)$$

where the tensorproduct and $END_S(A)$ is graded in the obvious way (cfr. [ 18 ] ) and the isomorphism is gradation preserving. This fact follows from :

Proposition 1 : [ 12 ] , [ 13 ]

With notations and assumptions as above, we have :

(1) : $\Lambda[P]$ is a graded reflexive Azumaya algebra with center $R[P]$ which is a normal domain.

(2) : there is an equivalence of categories between left reflexive $\Lambda$-modules and graded reflexive $\Lambda[P]$ - modules.

(3) : if every $P_i \in P$ is an invertible Λ-ideal, then there is an equivalence of categories between Λ-mod and $\Lambda[P]$-gr, the category of all $\mathbb{Z}^{(n)}$-graded left $\Lambda[P]$-modules.

Definition 2 : A maximal order Λ over a regular domain R is said to be smooth iff Λ is strongly regular with respect to the set of Weil divisors $P$.

Using Prop. 1.3. it is clear that both smooth and strongly regular orders have finite global dimension. If Λ is a maximal order over a regular local domain of gldim(R) ≤ 2 every $P_i \in P$ is invertible because $P_i$ is a reflexive R- module, hence free whence projective as a Λ-module by [21, Prop.3.5]. Let us give an example due to M. Van den Bergh showing that a strongly regular maximal order need not be smooth.

Example 1 : let R be the affine cone $\mathbb{C}[X,Y,Z]/(XY-Z^2)$, then the classgroup of R is $\mathbb{Z}/2\mathbb{Z}$ and is generated by the ruling p = (Y,Z). Let Λ be the reflexive Azumaya algebra over R

$$\Lambda = \mathrm{End}_R(R \oplus p) \cong \begin{pmatrix} R & p \\ p^{-1} & R \end{pmatrix}$$

and let $D = \left\{ \begin{pmatrix} p & p{\star}p \\ p^{-1}{\star}p & p \end{pmatrix} = \Lambda.\begin{pmatrix} 0 & y \\ 1 & 0 \end{pmatrix} \right\}$ then $\Lambda[D]$ is the $\mathbb{Z}$ graded ring :

$$\Lambda[D] = \Lambda\left[\begin{pmatrix} 0 & Y \\ 1 & 0 \end{pmatrix} X_1, \begin{pmatrix} 0 & 1 \\ Y^{-1} & 0 \end{pmatrix} X_1^{-1}\right]$$

which is readily checked to be an Azumaya algebra with a regular center

$$R[D] : \quad \ldots \oplus (Y^{-1})X_1^{-2} \oplus p^{-1}X_1^{-1} \oplus R \oplus pX_1 \oplus (Y)X_1^2 \oplus p(Y)\,X_1^3 \oplus \ldots$$

Therefore, Λ is strongly regular but not smooth since R is singular.

From now on, we will restrict attention to smooth maximal orders over a regular domain R in a quaternion algebra Σ. First, we aim to study the Zariski local structure. For simplicity's sake we will assume that height one prime ideals of $\Lambda$ lying over ramified prime ideals are generated by a normalizing element. However, we do not know whether this condition is always satisfied.

Lemma 1 : If $\Lambda$ is a smooth maximal order over a regular local domain R in $\Sigma$, then $\# P \leq 2$.

Proof.

Let $n = \# P$, then, because $\Lambda [P]$ is a $\mathbb{Z}^{(n)}$-graded Azumaya algebra over $R[P]$, which is a graded local domain with unique graded maximal ideal $m[P] = \sum_{\sigma \in H} m.R[P]_\sigma + \sum_{\tau \in G \setminus H} R[P]_\tau$ where $G = \mathbb{Z}^{(n)}$ and $H = 2\mathbb{Z} \oplus \ldots \oplus 2\mathbb{Z}$, we must have that $\Lambda[P]/\Lambda[P].m[P]$ is a $\mathbb{Z}^{(n)}$-graded central simple algebra of dimension 4 over the $\mathbb{Z}^{(n)}$-graded field

$$R[P]/m[P] = R/m[X_1^2, X_1^{-2}, \ldots, X_n^2, X_n^{-2}]$$

Because every prime ideal $P_i$, $1 \leq i \leq n$, is supposed to be generated by a normalizing element, an easy calculation shows that

$$\Lambda[P]/\Lambda[P].m[P] \cong \bigoplus_{0 \leq i_j < 2} \Lambda/(\Lambda m + P_1 + \ldots + P_n)\, X_1^{i_1} \ldots X_n^{i_n}$$

the isomorphism being one of graded $R/m[X_1^2, X_1^{-2}, \ldots, X_n^2, X_n^{-2}]$-modules.

Calculating dimensions on both sides yields :

$$(F_1) \quad 4 = 2^n . \dim_{R/m} (\Lambda/(\Lambda m + P_1 + \ldots + P_n))$$

This equality immediately implies that $n \leq 2$;

This lemma also holds for smooth maximal orders in division algebras of dimension $p^2$, p a prime number. Therefore, it seems to me that for high ($\geq 3$) dimensions, smooth maximal orders are only a first approximation for "nice" regular maximal orders.

If $\dim(\Sigma) \neq p^2$ one can have a worse ramification divisor. E.g. it is perfectly possible to have smooth maximal orders over regular local domains of dimension 4 in central simple algebras of dimension 16 with 4 central ramified height one primes, each having ramification index 2. Let us recall the definition of a set of regular divisors with normal crossings [10]. We say that a set of Weil divisors $\mathcal{D} = \{D_i ; i \in I\}$ has strictly normal crossings if for every prime ideal $q \in R$ lying in $\cup \operatorname{supp}(D_i)$ we have : if $I_q = \{i ; s \in \operatorname{supp}(D_i)\}$, then for $i \in I_q$ we have that $D_i = \sum_\lambda \operatorname{div}(x_{i,\lambda})$ with $x_{i,\lambda} \in R_q$ and $\{(x_{i,\lambda})_{i,\lambda}\}$form part of a regular system of parameters in $R_q$. We say that the set $\mathcal{D} = \{D_i, i \in I\}$ has normal crossings if for every $q \in \cup \operatorname{supp}(D_i)$ there exists an étale neighbourhood Spec(S) of q in Spec(R) such that the family of inverse images of the $\{D_i ; i \in I\}$ on S have strictly normal crossings. Finally, a divisor D of R is called regular at $q \in \operatorname{supp}(D)$ if the subscheme D of Spec(R) is regular at q. The divisor D is called regular if it is regular everywhere.

Lemma 2 : If $\Lambda$ is a smooth maximal order over a regular domain R, then $\mathcal{P}_c$ is a set of regular divisors with normal crossings.

Proof.

Let $\mathcal{P}_c = \{p_1, \dots, p_n\}$, then by lemma 1 we know that for every prime ideal $m \in \operatorname{Spec}(R)$, $N_m = \{i \leq i \leq n \mid m \in \operatorname{supp}(p_i)\} \leq 2$.

Let us first consider the case : $N_m = 1$. Then $R[\mathcal{P}]_m$ must be a regular and graded domain. After a possible renumeration we have :

$$R[\mathcal{P}]_m \cong R[\{p_1\}][X_2, X_2^{-1}, \dots, X_n, X_n^{-1}]_m$$

so we may assume that :

$$\hat{R} : (p^{-1})X^{-2} \oplus R_m X^{-1} \oplus R_m \oplus (p) X \oplus (p) X^2 \oplus X^3 \oplus$$

is a regular graded domain which is graded local with unique maximal graded ideal

$$\hat{m} : (p^{-1} m)_m X^{-2} \oplus R_m X^{-1} \oplus m_m \oplus (p)_m X \oplus (pm)_m X^2 \oplus \dots$$

By a result of [ 15 ], this is equivalent with

$$\mathrm{gr.dim}_{\hat{R}/\hat{m}}(\hat{m}/\hat{m}^2) = \mathrm{gldim}(R_m)$$

Calculating $(\hat{m})^2$ gives us :

$$(\hat{m})^2 : (m^2+p)^{-1}p_m X^{-2} \oplus m_m X^{-1} \oplus (m^2+p)_m \oplus (mp)_m X \oplus (m^2+p)p_m X^2 \oplus \dots$$

and therefore

$$\hat{m}/(\hat{m})^2 \cong m/m^2+p\,[X^2,X^{-2}] \oplus R/m\,[X^2,X^{-2}]X^{-1}$$

as graded $\hat{R}/\hat{m} = R/m\,[X^2,X^{-2}]$ -modules. Hence, we must have :

$$\dim_{R/m}(m/m^2+p) = \mathrm{gldim}(R_m)-1$$

and this is equivalent to saying that $(p) \not\subset m^2$, i.e. m is a regular point on the subscheme determined by (p).

Let us now turn to the case : $N_m = 2$. Again, after a possible renumeration, we have :

$$R\,[P]_m \cong R\,[\{p,q\}]\,[X_3,X_3^{-1},\dots,X_n,X_n^{-1}]$$

so we may assume that $\hat{R} = R\,[\{p,q\}]_m$ is a regular domain which is graded local with unique maximal ideal $\hat{m}$ :

$$\begin{array}{ccccccccc}
mp^{-1}q & \oplus & q & \oplus & mq & \oplus & pq & \oplus & mpq \\
\oplus & & \oplus & & \oplus & & \oplus & & \oplus \\
p^{-1}q & \oplus & q & \oplus & q & \oplus & pq & \oplus & pq \\
\oplus & & \oplus & & \oplus & & \oplus & & \oplus \\
mp^{-1} & \oplus & R & \oplus & m & \oplus & p & \oplus & mp \\
\oplus & & \oplus & & \oplus & & \oplus & & \oplus \\
p^{-1} & \oplus & R & \oplus & R & \oplus & p & \oplus & p \\
\oplus & & \oplus & & \oplus & & \oplus & & \oplus \\
mp^{-1}q^{-1} & \oplus & q^{-1} & \oplus & mq^{-1} & \oplus & pq^{-1} & \oplus & mpq^{-1}
\end{array}$$

Again, calculating $(\hat{m})^2$ gives us :

$$\begin{array}{ccccccccc}
Ap^{-1} & \oplus & mq & \oplus & Aq & \oplus & mpq & \oplus & Apq \\
\oplus & & \oplus & & \oplus & & \oplus & & \oplus \\
mp^{-1}q & \oplus & q & \oplus & mq & \oplus & pq & \oplus & mpq \\
\oplus & & \oplus & & \oplus & & \oplus & & \oplus \\
Ap^{-1} & \oplus & m & \oplus & A & \oplus & mp & \oplus & Ap \\
\oplus & & \oplus & & \oplus & & \oplus & & \oplus \\
mp^{-1} & \oplus & R & \oplus & m & \oplus & p & \oplus & mp \\
\oplus & & \oplus & & \oplus & & \oplus & & \oplus \\
Ap^{-1}q^{-1} & \oplus & mq^{-1} & \oplus & Aq^{-1} & \oplus & mpq^{-1} & \oplus & Apq^{-1}
\end{array}$$

where $A = m^2+p+q$, $R = R_m$ and we have omitted all powers of $X_1$ and $X_2$. From this description it follows that :

$$\begin{aligned}
\hat{m}/\hat{m}^2 \cong m/m^2+p+q\ & [X_1^2, X_1^{-2}, X_2^2, X_2^{-2}] \\
\oplus\ R/m\ & [X_1^2, X_1^{-2}, X_2^2, X_2^{-2}]\ X_1 \\
\oplus\ R\ m\ & [X_1^2, X_1^{-2}, X_2^2, X_2^{-2}]\ X_2
\end{aligned}$$

the isomorphism being one as graded $\hat{R}/\hat{m} = R/m\,[X_1^2, X_1^{-2}, X_2^2, X_2^{-2}]$-modules. hence, we must have :

$$\dim_{R/m}(m/m^2+p+q) = \text{gldim}\,(R/Rm)-2$$

or equivalently that $\{p,q\}$ is a part of a regular m-sequence in $R_m$. In particular it follows that m is a regular point on the subscheme determined by (p) (resp. by (q)). Finally, [ 10, lemma 1.8.4.] finishes the proof.

Definition 3 : If R is a regular domain with field of fractions K, if $\Sigma$ is a quaternion algebra over K and if $P_c$ is the set of ramified height one primes of R in $\Sigma$, we say that R has a regular ramification divisor with normal crossings in $\Sigma$ iff $P_c$ is a set of regular divisors with normal crossings s.t. for every $m \in \text{Spec}(R)$. $N_m$ (as defined in the proof of lemma 2) $\leqslant 2$.

Using the proofs of lemma 1 and 2, one can prove :

Proposition 1 : If $\Lambda$ is a maximal order over a regular domain R in a quaternion algebra $\Sigma$, then $\Lambda$ is smooth iff :

(a). R has a regular ramification divisor with normal crossings in $\Sigma$.

(b). For every $m \in \mathrm{Spec}(R)$ one of the following cases occur :

CASE 0 : $N_m = 0$, i.e. $\Lambda_m$ is Azumaya over $R_m$

CASE 1 : $N_m = 1$ and $\dim_{R/m} (\Lambda/\Lambda m+P) = 2$

CASE 2 : $N_m = 2$ and $\dim_{R/m} (\Lambda/\Lambda m+P+Q) = 1$

where P (resp. Q) is the height one prime of $\Lambda$ lying over the ramified central prime p (resp. q).

Let us give a geometric interpretation of condidition (a) for a maximal order on a smooth surface. Let $\{p_1,\dots,p_n\}$ be the ramified height one primes, then each $p_i$ can be viewed as a curve on the surface. $N_m \leq 2$ then says that there are no three such curves which intersect at one point. Furthermore, each curve must be nonsingular and in an intersection point of two curves the tangent lines may not coincide.

As we will see later, CASE 0 (resp. 1,2) corresponds to CASE 1.1.(i) (resp. (ii), (iii)) of [ 2 ] , whereas CASE 1.1 (iv) cannot occur as a smooth maximal order.

Let us give some examples of smooth maximal orders.

Example 2: Let $\Lambda = \mathbb{C}[X,-]$ be the skew polynomial ring over $\mathbb{C}$ where - denotes conjugation. It is clear that $\Lambda$ is a maximal order with center $R = \mathbb{R}[t]$ where $X^2 = t$. It follows that $P = \{(X)\}$, so $\Lambda[P]$ is the $\mathbb{Z}$-graded ring.

$$(X^{-2})X_1^{-2} \oplus (X^{-1})X_1^{-1} \oplus \Lambda \oplus (X)X_1 \oplus (X^2)X_1^2 \oplus \dots$$

and $R[P]$ is the $\mathbb{Z}$-graded ring :

$$(t^{-1})X_1^{-2} \oplus R\, X_1^{-1} \oplus R \oplus (t)\, X_1 \oplus (t)\, X_1^2 \oplus \dots$$

For every prime ideal $q \neq (X) \in \operatorname{Spec} \mathbb{R}[t]$, $\Lambda_q$ is Azumaya over $R_q$ whence $\Lambda[P]_q$ is Azumaya over $R_q$. In the ramified prime we have :

$$\dim_{\mathbb{R}[t]/(t)}(\mathbb{C}[(X),-]/(X)) = \dim_{\mathbb{R}}(\mathbb{C}) = 2$$

i.e. case 1. So, $\Lambda$ is smooth over R.

Further, $R[P]_{(t)}/t[P]_{(t)} \cong \mathbb{R}[t_1, t_1^{-1}]$ where $t_1 = t X_1^2$ and $\Lambda[P]_{(t)}/\Lambda[P]t[P]_{(t)}$ is the $\mathbb{Z}$-graded central simple algebra $\mathbb{C}[Y_1, Y_1^{-1}, -]$ with $Y_1 = XX_1$ over $\mathbb{R}[t_1, t_1^{-1}]$.

<u>Example 3</u> : Let R be a regular local domain of dimension 2 and suppose x and y generate the maximal ideal m. Let $\Sigma$ be the quaternion-algebra $(x,y)_k$ and let $\Lambda = R[1,i,j,ij]$, i.e. $\Lambda$ is R-free with generators 1,i,j,ij and with relations :

$$i^2 = x, \quad j^2 = y \text{ and } ij = -ji$$

In [20] it was checked that $\Lambda$ is a maximal R-order. Clearly $P_c = \{(x),(y)\}$ which is a set of regular ramification indices with normal crossings.

Further, $\Lambda[P]$ is a $\mathbb{Z} \oplus \mathbb{Z}$-graded ring which can be visualized (omitting powers of $X_1$ and $X_2$) as :

$$\begin{array}{ccccccccc}
(i^{-2}j^{2}) & \oplus & (i^{-1}j^{2}) & \oplus & (j^{2}) & \oplus & (ij^{2}) & \oplus & (i^{2}j^{2}) \\
\oplus & & \oplus & & \oplus & & \oplus & & \oplus \\
(i^{-2}j) & \oplus & (i^{-1}j) & \oplus & (j) & \oplus & (ij) & \oplus & (i^{2}j) \\
\oplus & & \oplus & & \oplus & & \oplus & & \oplus \\
(i^{-2}) & \oplus & (i^{-1}) & \oplus & \Lambda & \oplus & (i) & \oplus & (i^{2}) \\
\oplus & & \oplus & & \oplus & & \oplus & & \oplus \\
(i^{-2}j^{-1}) & \oplus & (i^{-1}j^{-1}) & \oplus & (j^{-1}) & \oplus & (ij^{-1}) & \oplus & (i^{2}j^{-1}) \\
\oplus & & \oplus & & \oplus & & \oplus & & \oplus \\
(i^{-2}j^{-2}) & \oplus & (i^{-1}j^{-2}) & & (j^{-2}) & \oplus & (ij^{-2}) & \oplus & (i^{2}j^{-2})
\end{array}$$

and its center $R[\mathcal{P}]$ is the $\mathbb{Z} \oplus \mathbb{Z}$-graded ring which looks like :

$$\begin{array}{ccccccccc}
(x^{-1}y) & \oplus & (y) & \oplus & (y) & \oplus & (xy) & \oplus & (xy) \\
\oplus & & \oplus & & \oplus & & \oplus & & \oplus \\
(x^{-1}y) & \oplus & (y) & \oplus & (y) & \oplus & (xy) & \oplus & (xy) \\
\oplus & & \oplus & & \oplus & & \oplus & & \oplus \\
(x^{-1}) & \oplus & R & \oplus & R & \oplus & (x) & \oplus & (x) \\
\oplus & & \oplus & & \oplus & & \oplus & & \oplus \\
(x^{-1}) & \oplus & R & \oplus & R & \oplus & (x) & \oplus & (x) \\
\oplus & & \oplus & & \oplus & & \oplus & & \oplus \\
(x^{-1}y^{-1}) & \oplus & (y^{-1}) & \oplus & (y^{-1}) & \oplus & (xy^{-1}) & \oplus & (xy^{-1})
\end{array}$$

and $\dim_{R/m}(\Lambda/\Lambda m + (i) + (j)) = \dim_{R/m}(R/m) = 1$, so $\Lambda$ is a smooth maximal R-order of case 2.

Further, $R[\mathcal{P}]/m[\mathcal{P}] = R/m[Y_1^2, Y_1^{-2}, Y_2^2, Y_2^{-2}]$ where $Y_1 = iX_1$ and $Y_2 = jX_2$. Whereas $\Lambda[\mathcal{P}]/\Lambda[\mathcal{P}]\ m[\mathcal{P}]$ is the $\mathbb{Z} \oplus \mathbb{Z}$-graded central simple algebra: $R/m[Y_1, Y_1^{-1}, Y_2, Y_2^{-1}]$ with $Y_1Y_2 = -Y_2Y_1$. Its part of degree $(0,0)$ equals $R/m$ corresponding to the fact that $\Lambda$ is quasi-local with maximal ideal $M = (i,j)$.

<u>Example 4 :</u> Let F be any field with characteristic unequal to 2. Let $R = F[X,Y]_{(X,Y)}$ where X and Y are indeterminates over F. Then R is regular local of dimensions two and has field of fractions $K = F(X,Y)$. Let $\Sigma$ be the quaternion algebra $(X, 1+Y)_K$ and let $\Lambda$ be the R-free order $R[1,i,j,ij]$ then $\Lambda$ is a maximal order [20, p. 471]. Then $\mathcal{P} = \{(i)\}$, $\mathcal{P}_c = \{(x)\}$ and $(x) \not\subset m^2$. Because $\dim_{R/m}(\Lambda/\Lambda m + (i)) = \dim_F(F \oplus F) = 2$, $\Lambda$ is smooth. Further $R[\mathcal{P}]/m[\mathcal{P}] \cong F[Y_1^2, Y_1^{-2}]$ where $Y_1 = iX_1$ and $\Lambda[\mathcal{P}]/\Lambda[\mathcal{P}]m[\mathcal{P}]$ is the $\mathbb{Z}$-graded algebra

$$(F \oplus F\epsilon)[Y_1, Y_1^{-1}, \varphi] \overset{\alpha}{\underset{\approx}{\to}} M_2(F[Y_1^2, Y_1^{-2}])$$

where $\varphi(a \oplus b\epsilon) = a \oplus -b\epsilon$ and $\alpha$ is given by

$$\alpha(1 \oplus 0) = \begin{pmatrix} 1 & 0 \\ 0 & 1 \end{pmatrix} \qquad \alpha(0 \oplus \epsilon) = \begin{pmatrix} -1 & 0 \\ 0 & 1 \end{pmatrix}$$

$$\alpha(Y_1) = \begin{pmatrix} 0 & 1 \\ Y_1^2 & 0 \end{pmatrix}$$

Therefore, $\Lambda[P]/\Lambda[P]m[P]$ is a $\mathbb{Z}$-graded central simple algebra over $F[Y_1^2, Y_1^{-2}]$. However, the part of degree 0 of $\Lambda[P]/\Lambda[P]\ m[P]$ is semisimple $(F \oplus F\epsilon)$ corresponding to the fact that $\Lambda$ is not quasi-local. Each factor corresponds to one of the two maximal ideals of $\Lambda$ lying over $m = (X,Y)$ :

$$M_1 = \Lambda(i,j-1) \qquad M_2 = \Lambda(i,j+1)$$

Having characterized smooth maximal orders over regular domains in quaternion algebras, we will now study their Zariski local structure, i.e. the number of conjugacy classes over a regular local domain. One of the basic ingredients in this study is a result of Grothendieck [ 8 ] on descent of modules. For convenience, we state that theorem here :

<u>Theorem</u> [ 8,2.5.8 ] Let R be a Noetherian (semi) local ring, $\Lambda$ a finite R-algebra and let $M_1$, $M_2$ be finite left $\Lambda$-modules. Let $R \to S$ be a faithfully flat morphism with S. Noetherian. If $M_1 \otimes S \cong M_2 \otimes S$ as left $\Lambda \otimes S$-modules, then $M_1 \cong M_2$ as left $\Lambda$-modules.

Using the above theorem, we will have to compute the conjugacy classes of the extended orders $\Lambda \oplus_R R^{sh}$ where $\Lambda$ is smooth over R and $R^{sh}$ denotes the strict Henselization of R cfr. [ 17 ] or [ 23 ] .

<u>Case 0</u> is easy : if one maximal order $\Lambda$ over R in $\Sigma$ is Azumaya, then every other smooth order say $\Gamma$ is Azumaya too. Since $Br(R^{sh}) = 0$, $\Lambda \otimes R^{sh} \cong M_2(R^{sh}) \cong \Gamma \otimes R^{sh}$ and by descent $\Lambda \cong \Gamma$ as R-algebras, yielding that $\Lambda$ and $\Gamma$ are conjugated.

Before treating the other cases, let us recall the definition of the graded Brauer group as introduced by F. Van Oystaeyen in the $\mathbb{Z}$-graded case in [ 26 ] .

If T is any $\mathbb{Z}^{(n)}$-graded ring, then a graded Azumaya algebra over T is an Azumaya algebra over T admitting a $\mathbb{Z}^{(n)}$-gradation extending the gradation of the center. Two graded algebras $\Gamma$ and $\Omega$ are said to be gr-equivalent if there exist finitely generated graded projective T-modules P and Q such that there exists a degree preserving isomorphism

$$\Gamma \otimes_T \ \mathrm{END}_T(P) \cong \Omega \otimes_T \mathrm{END}_T(Q)$$

where the rings $\mathrm{END}_T(-)$ and the tensorproducts are equipped with the natural gradation, cfr. e.g. [ 18 ] . The set of gr-equivalence classes of graded Azumaya algebras forms a group with respect to the tensor-product, $\mathrm{Br}^g(T)$, called the graded Brauer group of T.

If T is a $\mathbb{Z}$-graded Krull domain it was shown by S. Caenepeel, M. Van den Bergh and F. Van Oystayen [ 6 ] that the natural (i.e. gradation-forgetting) morphism $\mathrm{Br}^g(T) \to \mathrm{Br}(T)$ is monomorphic. Their argument can easily be extended zo the $\mathbb{Z}^{(n)}$-graded case.

S. Caenepeel [ 5 ] calls a graded local ring R (i.e. having a unique maximal graded ideal) gr-Henselian if every finite graded commutative R-algebra B is graded decomposed, i.e. when it is the direct sum of graded local rings. In the $\mathbb{Z}$-graded case it turns out that a graded local ring is gr-Henselian iff its part of degree zero is Henselian, [ 5 ] . This result can be generalized to $\mathbb{Z}^{(n)}$-graded rings. Furthermore, if R is gr-Henselian with maximal graded ideal m then the natural map :

$$\mathrm{Br}^g R \to \mathrm{Br}^g R/m$$

is monomorphic by a similar argument as in the ungraded case.

Lemma 3 : If $\Lambda$ is a smooth maximal order over a regular local domain R in a quaternion algebra $\Sigma$, then

(1) in case 1, $R^{sh}$ splits $\Sigma$.

(2) in case 2, $R^{sh}$ does not split $\Sigma$

Proof.

$\Lambda[P] \otimes_R R^{sh}$ is a graded Azumaya algebra over $R^{sh}[P] = R[P] \otimes_R R^{sh}$. $R^{sh}[P]$ is gr-Henselian because the part of degree zero (or of $(0,0)$ is Henselian, with maximal graded ideal $m^{sh}[P]$.

(1) In this case $R^{sh}[P]/m^{sh}[P]$ is the graded field $R^{sh}/m^{sh}[Y_1^2, Y_1^{-2}]$ where $Y_1^2 = pX_1^2$. Now,

$$\overline{\Gamma} = (\Lambda[P] \otimes_R R^{sh}) / (\Lambda[P] \otimes_R R^{sh}) m^{sh}[P]$$

is a graded central simple algebra of dimension 4 over $R^{sh}/m^{sh}[Y_1^2, Y_1^{-2}]$. Calculating the part of degree zero of $\overline{\Gamma}$ it turns out that $\overline{\Gamma}_o$ need to be an algebra over $R^{sh}/m^{sh}$ of dimension two. Because $R^{sh}/m^{sh}$ is separably closed and char$(R^{sh}/m^{sh}) \neq 2$, we must have $\overline{\Gamma} \cong R^{sh}/m^{sh} \oplus R^{sh}/m^{sh}$ as algebras. hence $\overline{\Gamma}$ contains zero divisors and therefore

$$\overline{\Gamma} \cong M_2(R^{sh}/m^{sh}[Y_1^2, Y_1^{-2}])$$

(cfr. also example 4). Finally, using the injectivity of the map $Br^g R^{sh}[P] \to Br^g R^{sh}[P]/m^{sh}[P]$ it follows that

$$\Lambda[P] \otimes R^{sh} \cong END_{R^{sh}[P]}(P)$$

for some graded f.g. projective $R^{sh}[P]$ - module P. Calculating parts of degree zero on both side yields that $\Lambda \otimes_R R^{sh}$ is an order in a matrixring.

(2). In the second case, $R^{sh}[P]/m^{sh}[P]$ is a $\mathbb{Z} \oplus \mathbb{Z}$-graded field $R^{sh}/m^{sh}[Y_1^2, Y_1^{-2}, Y_2^2, Y_2^{-2}]$ where $Y_1^2 = pX_1^2$ and $Y_2^2 = q X_2^2$. Now,

$$\overline{\Gamma} = (\Lambda[P] \otimes R^{sh}) / (\Lambda[P] \otimes R^{sh}) m^{sh}[P]$$

is a $\mathbb{Z} \oplus \mathbb{Z}$ - graded central simple algebra of dimension 4 over $R^{sh}/m^{sh}[Y_1^2, Y_1^{-2}, Y_2^2, Y_2^{-2}]$. The homogeneous parts $\overline{\Gamma}_{(o,o)}, \overline{\Gamma}_{(o,1)}, \overline{\Gamma}_{(1,o)}$ and $\overline{\Gamma}_{(1,1)}$ are all non-zero and have therefore all dimension one. It follows that $\overline{\Gamma}_{(o,o)} = R^{sh}/m^{sh}$ and

$$\overline{\Gamma} \cong (aY_1^2, bY_2^2)_{R^{sh}/m^{sh}[Y_1^2, Y_1^{-2}, Y_2^2, Y_2^{-2}]}$$

for some a, b $\in (R^{sh}/m^{sh})^*$. Because $R^{sh}/m^{sh}$ is separably closed and $\mathrm{char}(R^{sh}/m^{sh}) \neq 2$, $\overline{\Gamma} \cong (Y_1^2, Y_2^2)$ so $\overline{\Gamma} \cong R^{sh}/m^{sh}[Y_1, Y_1^{-1}, Y_2, Y_2^{-1}]$ with $Y_1Y_2 = -Y_2Y_1$ (cfr. example 3). Using the norm, it is easy to check that $\overline{\Gamma}$ has no zero divisors. Therefore $\overline{\Gamma}$ is a non-trivial element in $Br^g(R^{sh}[P]/m^{sh}[P])$ and hence $\Lambda \otimes R^{sh}$ cannot be an order in a matrixring, finishing the proof.

Proposition 2 : (Zariski local structure in case 1)
All smooth maximal orders over a regular local domain in a quaternion algebra are conjugated.

proof.
For any smooth maximal order $\Lambda$ in case 1 we know by the proof of lemma 3 that

$$\Lambda[P] \otimes R^{sh} \cong \mathrm{END}_{R^{sh}[P]}(P)$$

for some f.g. graded projective $R^{sh}[P]$-module P. Because $R^{sh}[P]$ is graded local, P is graded free; i.e. of the form :

$$P \cong R^{sh}[P](\sigma_1) \oplus R^{sh}[P](\sigma_2)$$

where $\sigma_i \in \mathbb{Z}$ and $R^{sh}[P](\sigma_i)$ is the graded $R^{sh}[P]$ - module determined by taking for its homogeneous part of degree $\alpha$ : $R^{sh}[P](\sigma_i)_\alpha = R^{sh}[P]_{\sigma_i+\alpha}$. Therefore,

$$\Lambda[P] \otimes R^{sh} \cong M_2(R^{sh}[P](\sigma_1, \sigma_2)$$

where the part of degree $\alpha$ of the ring on the right side is given by the formula :

$$M_2(R^{sh}[P])(\sigma_1,\sigma_2)_\alpha = \begin{pmatrix} R^{sh}[P]_\alpha & R^{sh}[P]_{\alpha+\sigma_1-\sigma_2} \\ R^{sh}[P]_{\alpha+\sigma_2-\sigma_1} & R^{sh}[P]_\alpha \end{pmatrix}$$

A straightforward computation shows that up to a graded isomorphism the $\sigma_i$ may be chosen to be 0 or 1. Since all isomorphisms above were gradation preserving :

$$\Lambda \otimes_R R^{sh} \cong \begin{pmatrix} R^{sh} & R^{sh}[P]_{\sigma_1-\sigma_2} \\ R^{sh}[P]_{\sigma_2-\sigma_1} & R^{sh} \end{pmatrix}$$

So, we are left to show that all rings which can occur in such a way are conjugated.

| $\sigma_1 \backslash \sigma_2$ | 0 | 1 |
|---|---|---|
| 0 | $M_2(R^{sh})$ | $\begin{pmatrix} R^{sh} & R^{sh} \\ p & R^{sh} \end{pmatrix}$ |
| 1 | $\begin{pmatrix} R^{sh} & p \\ R^{sh} & R^{sh} \end{pmatrix}$ | $M_2(R^{sh})$ |

Since $\Lambda$ was supposed to be ramified, only the cases $\sigma_1 \neq \sigma_2$ can occur and clearly :

$$\begin{pmatrix} 0 & 1 \\ 1 & 0 \end{pmatrix} \begin{pmatrix} R^{sh} & p \\ R^{sh} & R^{sh} \end{pmatrix} \begin{pmatrix} 0 & 1 \\ 1 & 0 \end{pmatrix} = \begin{pmatrix} R^{sh} & R^{sh} \\ p & R^{sh} \end{pmatrix}$$

Grothendieck descent finishes the proof.

Proposition 3. (Zariski local structure in case 2)

All smooth maximal orders over a regular local domain in a quaternion algebra are conjugated.

<u>Proof.</u>

By Lemma 3 we know that $\Lambda[P]$ cannot be split by an étale extension of R. Nevertheless, since $\Lambda[P]$ is a graded Azumaya algebra over a graded local domain, $\Lambda[P]$ can be split by a gr-étale extension of $R[P]$ (where gr-étale is defined in the obvious way). If we denote by $S = R[X]/(X^2-p)$, then

$$\begin{array}{ccccccccc}
(X^2q) & \oplus & (X^{-1}q) & \oplus & (q) & \oplus & (qX) & \oplus & (qX^2) \\
\oplus & & \oplus & & \oplus & & \oplus & & \oplus \\
(X^2q) & \oplus & (X^{-1}d) & \oplus & (q) & \oplus & (qX) & \oplus & (qX^2) \\
\oplus & & \oplus & & \oplus & & \oplus & & \oplus \\
(X^{-2}) & \oplus & (X^{-1}) & \oplus & S & \oplus & (X) & \oplus & (X^2) \\
\oplus & & \oplus & & \oplus & & \oplus & & \oplus \\
(X^{-2}) & \oplus & (X^{-1}) & \oplus & S & \oplus & (X) & \oplus & (X^2) \\
\oplus & & \oplus & & \oplus & & \oplus & & \oplus \\
(X^{-2}q^{-1}) & \oplus & X^{-1}q^{-1} & \oplus & (q^{-1}) & \oplus & (Xq^{-1}) & \oplus & (X^2q^{-1})
\end{array}$$

is such a gr-étale splitting ring of $\Lambda[P]$, $S(\Phi)$. Again, it turns out that

$$\Lambda[P] \otimes S(\Phi) \cong M_2(S(\Phi))(\sigma_1,\sigma_2)$$

where $\sigma_i \in \mathbb{Z} \oplus \mathbb{Z}$ can be chosen in the set $\{(0,0),(0,1),(1,0),(1,1)\}$. We will first show that they are all graded isomorphic to $M_2(S(\Phi))$ with the usual gradation.

An easy computation shows that :

$$M_2(S(\Phi))((0,0)(0,1)) = \begin{pmatrix} 1 & 0 \\ 0 & X_2 \end{pmatrix} M_2(S(\Phi)) \begin{pmatrix} 1 & 0 \\ 0 & X_2^{-1} \end{pmatrix}$$

$$M_2(S(\Phi))(0,0),(1,0)) = \begin{pmatrix} 1 & 0 \\ 0 & X_1 \end{pmatrix} M_2(S(\Phi)) \begin{pmatrix} 1 & 0 \\ 0 & X_1^{-1} \end{pmatrix}$$

$$M_2(S(\Phi))((0,0),(1,1)) = \begin{pmatrix} 1 & 0 \\ 0 & X_1X_2 \end{pmatrix} M_2(S(\Phi)) \begin{pmatrix} 1 & 0 \\ 0 & X_1^{-1}X_2^{-1} \end{pmatrix}$$

Now, by a graded version of Grothendieck descent [ 15 ] , $\Lambda[P]$ is graded isomorphic to $\Gamma[P]$ for any other smooth maximal R-order $\Gamma$ in $\Sigma$. This isomorphism is given by conjugation with a unit $\alpha$ in $\Sigma[X_1, X_1^{-1}, X_2, X_2^{-1}]$ and because this graded ring is a domain $\alpha$ is homogeneous i.e. $\alpha = \sigma X_1^{e_1} X_2^{e_2}$ with $\sigma \in \Sigma$ . Finally, it follows that $\Gamma = \sigma \Lambda \sigma^{-1}$, finishing the proof.

<u>Corollary 1</u> : Let R be a regular domain and $\Sigma$ a quaternion algebra which contains a smooth maximal R-order $\Lambda$. If $\Gamma$ is any maximal R-order, then $S_\Gamma = \{p \in \mathrm{Spec}(R) | \Gamma_p$ is smooth$\}$ is an open set.

<u>Proof.</u>

Let $\underline{\theta}_\Lambda$ (resp. $\underline{\theta}_\Gamma$) denote the structure sheaf of the R-algebra $\Lambda$ (resp. $\Gamma$) over <u>Spec</u>(R). If $p \in \mathrm{Spec}(R)$ such that $\Gamma_p$ is smooth, then $\Lambda_p = \sigma\Gamma\sigma^{-1}$ for some $\sigma \in \Sigma^\star$. This equality carries over to a small neighbourhood of p, finishing the proof.

What can be said about the codimension of this set ? Clairly, if p is an height one prime of R, then $p \in S_\Gamma$. In view of [ 20, Th.5.4] the only height two primes of R which do not lie in $S_\Gamma$ are prime ideals p such that there are two prime ideals of $\Lambda$ lying over p. If $\Gamma$ is a projective R-order, then by [ 21, Th.2.2 ] this fact also holds for height three primes.

<u>Corollary 2</u> : If R is a regular domain and if $\Lambda$ is a smooth maximal R-order in a quaternion algebra $\Sigma$, then there is a one-to-one correspondence between conjugacy classes of smooth maximal R-orders in $\Sigma$ and elements of $H^1_{Zar}$ (Spec(R), $\underline{\mathrm{Aut}}_\Lambda$)where $\underline{\mathrm{Aut}}_\Lambda$ is the automorphism scheme of $\underline{\theta}_\Lambda$.

## 2. Brauer-Severi schemes.

Let us first sketch the general problem. Let $\Lambda$ be an order over a normal domain R in a central simple algebra $\Sigma$ of dimension $n^2$, then one can define a functor $F_\Lambda$ from the category of all commutative R-algebras to the category of sets :

$$F_\Lambda : \text{Comm Alg}_R \to \text{Sets}$$

which associates to an R-algebra A the set of all left ideals of $\Lambda \otimes_R A$ which are split projective A-modules of rank n. The main problem is now to determine whether this functor is representa ble. By this we mean the following: does there exist a scheme $\underline{BS}_\Lambda$ over $\underline{\text{Spec}}(R)$, called the Brauer-Severi scheme of $\Lambda$, such that for every commutative R-algebra A there is a natural one-to-one correspondence between elements L of $F_\Lambda(A)$ and scheme homomorphisms $\psi_L$ from $\underline{\text{Spec}}(A)$ to $\underline{BS}_\Lambda$ making the diagram below commutative :

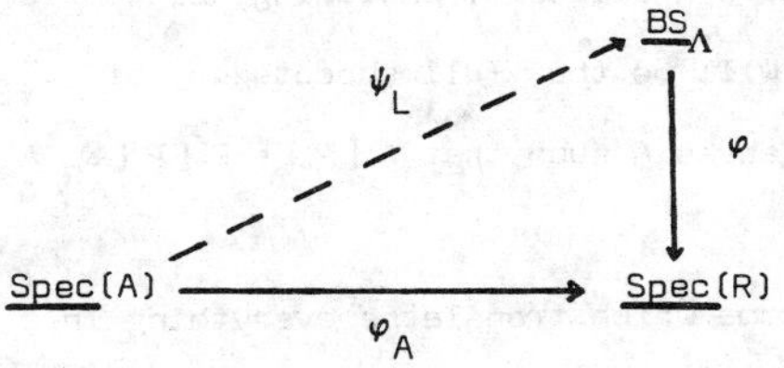


where $\varphi_A$ and $\varphi$ are the structural morphisms. Therefore, one could view the Brauer-Severi scheme of $\Lambda$ to be a scheme parametrizing the commutative R-algebras A which split $\Sigma$. A first step in this study usually consists in determining the étale local structure of this Brauer-Severi scheme, i.e. suppose that a point $x \in \text{Spec}(R)$ has an étale neighbourhood S which splits $\Sigma$, then one tries to find a representation of the functor

$$F_{\Lambda \otimes S} : \text{Comm Alg}_S \to \text{Sets}$$

This étale local structure has been determined in several cases

- Grothendieck [ 9 ] has shown that the étale local structure of the Brauer-Sévéri scheme of an Azumaya algebra is $\mathbb{P}_S^{n-1}$
- Artin and Mumford [ 1 ] calculated the Brauer-Sévéri scheme of a maximal order over a smooth surface in a ramified quaternion algebra with a regular ramification divisor.
- Recently [ 3 ] , Artin calculated the étale local structure of the Brauer Sévéri scheme for a maximal order over a Dedekind domain.

If one restricts attention in the first case to Azumaya algebras over regular domains, it turns out that all rings for which there exist Brauer-Severi schemes in the literature are smooth maximal orders. Therefore, one could ask whether the functor $F_\Lambda$ is representable for any smooth maximal order.
In this section we will prove a result which can be viewed as a first step towards this goal. We will represent $F_\Lambda$ when restricted to some nice subcategory $\underline{b}$ of all commutative R-algebras, including all étale or even smooth extensions of R. $\underline{b}$ will be the full subcategory of Comm $\text{Alg}_S$ consisting all all R-algebras A such that $A[P] = R[P] \otimes_R A$ is a regular domain.
Let us start by proving the key lemma which translates everything in a graded question. Denote by $F_\Lambda^g$ the functor

$$F_\Lambda^g : \underline{b} \otimes_R R[P] \to \text{Sets}$$

which assigns to an algebra $A[P]$ with $A \in \underline{b}$ the set of all graded left ideals of $\Lambda[P] \otimes A$ which are graded split projective $A[P]$-modules of rank n.

<u>Key Lemma :</u> If $\Lambda$ is a smooth maximal R-order and if $A \in \underline{b}$, then there is a natural one-to-one correspondence between elements of $F_\Lambda(A)$ and elements of $F_\Lambda^g(A)$.

Proof.

We claim that the maps below establish the claimed one-to-one correspondence.

$$\psi_1 : F_\Lambda(A) \to F^g_\Lambda(A) \qquad L \mapsto L \otimes_{(\Lambda \otimes A)} (\Lambda[P] \otimes A)$$
$$\psi_2 : F^g_\Lambda(A) \to F_\Lambda(A) \qquad M \mapsto M_e$$

where e is the identity element of the grading group. Because $\Lambda[P] \otimes A$ is a graded ring, $\psi_1$ and $\psi_2$ clearly give a one-to-one correspondence between left ideals of $\Lambda \otimes A$ and graded left ideals of $\Lambda[P] \otimes A$.
Because $\Lambda[P] \otimes A$ is a graded Azumaya algebra over the graded regular domain $A[P]$, gr.gldim $(\Lambda[P] \otimes A) < \infty$ whence gldim $(\Lambda \otimes A) < \infty$ using the equivalence of categories between $(\Lambda \otimes A)$-mod and $(\Lambda[P] \otimes A)$-gr. Its follows that $\Lambda \otimes A$ is a regular order over A.
Because splitting and projectivity are local conditions, we may assume from now on that A is regular local.
Now, let $L \in F_\Lambda(A)$. We claim that L is split as a left $\Lambda \otimes A$-module. Consider the exact sequence of left $\Lambda \otimes A$-modules :

$$0 \to L \to \Lambda \otimes A \to (\Lambda \otimes A) / L \to 0.$$

Because $L \in F_\Lambda(A)$, this sequence splits as a sequence of A-modules. Therefore, $(\Lambda \otimes A)/L$ is a left $(\Lambda \otimes A)$-module which is free as an A-module. Using regularity of the A-order $\Lambda \otimes A$, this entails by [20, Prop. 3.5] that $(\Lambda \otimes A)/L$ is projective as a left $(\Lambda \otimes A)$-module, finishing the proof of our claim.
But then it is clear that $\psi_1(L)$ is a graded split projective $\Lambda[P] \otimes A$-module. Finally, an easy localization argument shows that $\psi_1(L)$ has rank n.
The proof that $\psi_2$ maps elements of $F^g_\Lambda(A)$ to $F_\Lambda(A)$ is easy and is left to the reader.

Our strategy in order to represent the functor $F_\Lambda$ is now easy to grasp. First, we will represent to functor $F_\Lambda^g$ by a graded scheme. Graded schemes are defined formally in [ 14 ] , but any intelligent reader can only come up with one possible definition of them, so we will skip it here. We believe that the proofs and examples given below give a better idea what graded schemes are than any formal definition. Representing $F_\Lambda^g$ by a graded scheme is relatively easy, because $\Lambda[P]$ is a graded Azumaya algebra so we have to mimic Grothendieck's arguments [ 9 ] in the graded case.

Afterwards, we will form out of this graded scheme a usual scheme which represents $F_\Lambda$.

All graded schemes which appear in this paper have the pleasant property that their part of degree e is a usual scheme.

<u>Example 5</u> : Let $R[P]$ be the $\mathbb{Z}^{(n)}$-graded ring defined in § 1 associated to a set of height one primes $\{p_1,\dots,p_n\}$ and ramification indices $\{e_1,\dots,e_n\}$. Denote by $G = \mathbb{Z}^{(n)}$ and $H = e_1\mathbb{Z}\oplus\dots\oplus e_n\mathbb{Z}$. There is a natural one-to-one correspondence between Spec(R) and $\mathrm{Spec}_g(R[P])$, the set of all $\mathbb{Z}^{(n)}$-graded prime ideals of $R[P]$ with the induced Zariski topology.

$$\varphi_1 : \mathrm{Spec}(R) \to \mathrm{Spec}_g(R[P]);\ m \mapsto \sum_{\sigma\in} m\,R[P]_\sigma + \sum_{\tau\in G\setminus M} R[P]_\tau = m[P]$$

$$\varphi_2 : \mathrm{Spec}_g(R[P]) \to \mathrm{Spec}(R);\ M \mapsto M_e$$

where $e = (0,\dots,0)$. The maps $\varphi_i$ actually define an homeomorphism. Moreover, for any $m \in \mathrm{Spec}(R)$ we have :

$$(R[P]^g_{m[P]})_e \cong R_m$$

It follows that the part of degree e of the affine graded spectrum of $R[P]$ is isomorphic to <u>Spec</u>(R).

A : a graded representation of $F^g_\Lambda$

From now on we assume that $\Lambda$ is a smooth maximal order over a regular domain R in a quaternion-algebra $\Sigma$. We will treat the two ramified cases separatly:

CASE 1 : $P = \{p\}$ and $\dim_{R/m}(\Lambda/\Lambda m+P) = 2$

It follows from lemma 3 that there exists an étale extension S of R which splits $\Sigma$. We first represent the functor :

$$F^g_{\Lambda\otimes S} : \underline{b} \otimes S\,[P] \rightarrow \text{Sets}.$$

by a graded scheme over $\underline{\text{SPEC}^g}\,(S\,[P]\,)$, i.e. we will define a graded scheme $\underline{X^g}$ such that for any S-algebra A in $\underline{b}$ there is a natural one-to-one correspondence between elements of $F^g_{\Lambda\otimes S}(A)$ and graded scheme homomorphisms

$$\begin{array}{ccc} \underline{\text{SPEC}^g}(A\,[P]) & \dashrightarrow & \underline{X^g} \\ & {\scriptstyle \varphi_A} \searrow \quad \swarrow {\scriptstyle \varphi} & \\ & \underline{\text{SPEC}^g}(S\,[P]\,) & \end{array}$$

where $\varphi$ and $\varphi_A$ are the structural morphisms.

In § 2 we calculated the structure of $\Lambda\,[P]\otimes S$ :

$$\Lambda\,[P]\otimes S \cong M_2(S\,[P]\,(0,1)$$

Therefore, if A is any $\mathbb{Z}$-graded $S\,[P]$-algebra if we denote $\Lambda\,[P]\otimes S$ by $\Gamma\,[P]$, then

$$\Gamma\,[P]\otimes A \cong M_2(A)(0,1)$$

hence our aim is to represent the functor

$$G : \text{gr Comm Alg}_{S\,[P]} \rightarrow \text{Sets}$$

which assigns to a $\mathbb{Z}$-graded algebra A the set of all split projective graded left ideals of $M_2(A)(0,1)$ of rank 2.

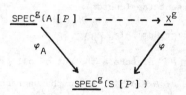

Take such a graded left ideal $L \in G(A)$, then

$$L = e_{11}L \oplus e_{22}L$$

and because all matrixelements $e_{ij}$ are homogeneous, it follows that $e_{11}L$ is a graded submodule of $A \oplus A(-1)$ which is split projective of rank 1 because

$$e_{ji}\, e_{ii}\, L = e_{jj}\, L$$

Conversely, if M is a graded split projective submodule of rank 1 of $A \oplus A(-1)$, then

$$L = M \oplus e_{21}\, M$$

is a graded split projective left ideal of rank 2 of $M_2(A)(0,1)$. Therefore it will be sufficient to represent the functor

$$\mathrm{Grass}_1^g(0,-1) : \text{gr-Comm Alg}_{S\,[P]} \rightarrow \text{Sets}.$$

which assigns to a $\mathbb{Z}$-graded $R\,[P]$ algebra A the set of all graded split projective rank one submodules of $A \oplus A(-1)$. As in the ungraded case we will do this by representing the subfunctor $Gr_i^g$, $i = 0,1$, of $\mathrm{Grass}_1^g(0,-1)$ which assigns to a $\mathbb{Z}$-graded $R\,[P]$-algebra A the set of all the elements $M \in \mathrm{Grass}_1^g(0,-1)$ such that the composite morphism (which is gradation preserving)

$$A(-i) \xrightarrow{\varphi_i} A \oplus A(-1) \xrightarrow{u} M$$

is an isomorphism. Here $\varphi_o : A \rightarrow A \oplus A(-1)$ and $\varphi_1 : A(-1) \rightarrow A \oplus A(-1)$ are the natural gradation preserving injections and u is the uniquely determined gradation preserving splitting map for M. Suppose we have a situation :

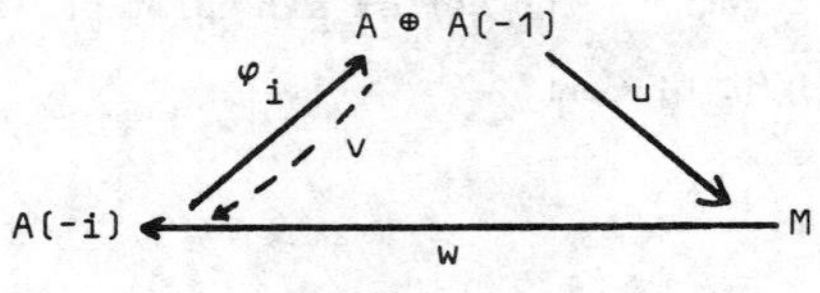

such that $v \circ \varphi_i$ is an isomorphism and let w be its inverse and $v = w \circ u$ which satisfies $v \circ \varphi_i = 1_{A(-i)}$.

Conversely, suppose we have a gradation preserving morphism v which satisfies $v \circ \varphi_i = 1_{A(-i)}$, then it is clear that

$$M = A \oplus A(-1)/\mathrm{Ker}(v)$$

is an element of $Gr_i^g(A)$. One can therefore identify $Gr_i^g(A)$ with the set of gradation preserving split morphisms of $\varphi_i$. Therefore, if one defines mappings

$$\alpha_i : HOM_A(A \oplus A(-1), A(-i))_o \to HOM_A(A(-i), A(-i))_o ; \alpha_i(v) = v \circ \varphi_i$$

$$\beta_i : HOM_A(A \oplus A(-1), A(-i))_o \to HOM_A(A(-i), A(-i))_o ; \beta_i(v) = 1_{A(-i)}$$

then $Gr_i^g(A)$ can be viewed as the kernel of the couple $(\alpha_i, \beta_i)$.

We claim that the functors :

$$A_i : \text{gr Comm Alg}_{S[P]} \to \text{Sets}; \; A_i(A) = HOM_A(A \oplus A(-1), A(-i))_o$$

$$B_i : \text{gr Comm Alg}_{S[P]} \to \text{Sets}; \; B_i(A) = HOM_A(A(-i), A(-i))_o$$

are representable by graded schemes.

In order to proves this claim, let us pause a moment and define graded vector fibres.

<u>Interludium</u> : <u>graded vector fibres.</u>

Let A be a $\mathbb{Z}^{(n)}$-graded commutative ring and let E be a graded A-module. The tensor algebra $\mathbb{T}(E) = \bigoplus_{i=1}^{\infty} (\otimes^i E)$ is given the natural $\mathbb{Z}^{(n)}$-gradation, i.e.

$$(\otimes^m E)_\gamma = \sum_{\sigma_1 + \ldots + \sigma_m = \gamma} E_{\sigma_1} \otimes \ldots \otimes E_{\sigma_m}$$

The symmetric algebra over E, S(E) is obtained $\mathbb{T}(E)$ by dividing out the homogeneous (!) twosided ideal generated by the elements $x \otimes y - y \otimes x$, x and y in E. Therefore, S(E) admits a natural $\mathbb{Z}^{(n)}$-gradation.

By the universal property of S((E) it is now fairly trivial to check that every gradation preserving A-linear morphism $E \to B$, B being a $\mathbb{Z}^{(n)}$-graded commutative A-algebra factorizes uniquely in

$$E \xrightarrow{i} S(E) \xrightarrow{g} B$$

where i is the natural map and g is a graded A-algebra morphism. Further, one verifies easily that for graded A-modules E and F, there is an isomorphism of graded A-algebras between $S(E \oplus F)$ and $S(E) \otimes S(F)$ where the tensorproduct is graded as usual.

The graded vector fibre $V^g(E)$ of the $\mathbb{Z}^{(n)}$-graded A-module E is then defined to be the graded affine spectrum $\underline{\mathrm{SPEC}}^g(S(E))$. Note that $V^g(E)$ is a graded $\underline{\mathrm{SPEC}}^g(A)$-scheme which represents the functor $\mathrm{HOM}\_(E \otimes_A -,-)_e$, e being the identity element of $\mathbb{Z}^{(n)}$. Let us give a typical example :

<u>Example 6</u> :

Let $\sigma \in \mathbb{Z}^{(n)}$, then there is an isomorphism of graded A-algebras between $S(A(\sigma))$ and $A[t]$ where t is an indeterminate with degree $-\sigma$

More generally, let $\sigma_1,\dots,\sigma_n \in \mathbb{Z}^{(n)}$, then

$$S(A(\sigma_1) \oplus \dots \oplus A(\sigma_n)) \cong A[X_1, \dots, X_n]$$

where $\deg(X_i) = -\sigma_i$.

Clearly, we are now in a position to to represent the functors $A_i$ and $B_i$ defined above. The functor $A_i$ is represented by the graded scheme $V^g(S[P](i) \oplus S[P](i-1))$ whereas the functor $B_i$ is represented by $V^g(S[P])$. More specific :

$A_0 \rightsquigarrow \underline{\mathrm{SPEC}}^g(S[P][X,Y])$ $\deg(X) = 0$ $\deg(Y) = 1$

$A_1 \rightsquigarrow \underline{\mathrm{SPEC}}^g(S[P][X,Y])$ $\deg(X) = -1$ $\deg Y = 0$

$B_0 \rightsquigarrow \underline{\mathrm{SPEC}}^g(S[P][X])$ $\deg(X) = 0$

$B_1 \rightsquigarrow \underline{\mathrm{SPEC}}^g(S[P][Y])$ $\deg(Y) = 0$

The maps $\alpha_i$ then correspond to the graded scheme morphisms :

$\alpha_o \to f_o$ : $\mathrm{SPEC}^g\ S\,[P\,][X,Y\,] \to \mathrm{SPEC}^g S\,[P\,][X\,]$

$\alpha_1 \to f_1$ : $\mathrm{SPEC}^g\ S\,[P\,][X,Y\,] \to \mathrm{SPEC}^g S\,[P\,][Y\,]$

arising from the natural algebra inclusions, whereas the maps $\beta_i$ correspond to the morphisms

$\beta_o \to g_o$ : $\mathrm{SPEC}^g\ S\,[P\,][X,Y\,] \to \mathrm{SPEC}^g S\,[P\,][X\,]$

$\beta_1 \to g_1$ : $\mathrm{SPEC}^g\ S\,[P\,][X,Y\,] \to \mathrm{SPEC}^g\ [P\,][Y\,]$

coming from the graded algebra map sending X to 1 (resp. Y to 1). Therefore, the functor $Gr^g_o$ is represented by the kernel of the following diagram of graded scheme morphisms

$$\begin{array}{ccc} \mathrm{SPEC}^g\ S\,[P\,][X,Y\,] & \xrightarrow{\ f_o\ } & \mathrm{SPEC}^g\ S\,[P\,][X\,] \\ & & \uparrow{\scriptstyle g_o} \\ & & \mathrm{SPEC}^g\ S\,[P\,][X,Y\,] \end{array}$$

It is fairly easy to verify that this kernel equals $V^g(X-1) \cong \mathrm{SPEC}^g\ S\,[P\,][Y\,]$ where deg(Y) = 1. Similarly, the functor $Gr^g_1$ is represented by the kernel of the diagram :

$$\begin{array}{ccc} \mathrm{SPEC}^G\ S\,[P\,][X,Y\,] & \xrightarrow{\ f_1\ } & \mathrm{SPEC}^g\ S\,[P\,][Y\,] \\ & & \uparrow{\scriptstyle g_1} \\ & & \mathrm{SPEC}^g\ S\,[P\,][X,Y\,] \end{array}$$

which is equal to $V^g(Y-1) \cong \mathrm{SPEC}^g\ S\,[P\,][X\,]$ where deg(X) = -1. Having that the subfunctors $Gr^g_i$ are representable by graded schemes over $\mathrm{SPEC}^g\ S\,[P\,]$, we will now glue them together in order to represent $Grass^g_1$ (0,-1).

First, we aim to compute the fundamental modules for the subfunctors $Gr^g_i$. That is an element $M_o \in Grass^g_1(0,-1)$ (S [P][Y]) and $M_1 \in Grass^g_1$ (0,-1)(S [P][X]) such that for every graded commutative

$S[\mathcal{P}]$ algebra A the natural one-to-one correspondences between $HOM(SPEC^g(A), SPEC^g(S[\mathcal{P}][Y]))$ and $Gr^g_o(A)$ and between $HOM(SPEC^g(A), SPEC^g(S[\mathcal{P}][X]))$ and $Gr^g_1(A)$ are given by assigning to a scheme morphism $\psi$, $\Gamma(SPEC^g(A), \psi^\star(\tilde{M}_o))$ resp. $\Gamma(SPEC^g(A), \psi^\star(M_1))$.

An easy computation shows that :

$$M_o \cong S[\mathcal{P}][Y]\ (0) = (S[\mathcal{P}][Y] \oplus S[\mathcal{P}][Y](-1))/(-Y,1)S[\mathcal{P}][Y]$$

$$M_1 \cong S[\mathcal{P}][X](-1) = ((S[\mathcal{P}][X](1) \oplus S[\mathcal{P}][X])/(1,-X)S[\mathcal{P}][X])(-1)$$

see [ 14 ] for a detailed computation.

The open set of $SPEC^g S[\mathcal{P}][Y]$ over which we have to glue $SPEC^g S[\mathcal{P}][Y]$ with $SPEC^g S[\mathcal{P}][X]$ is the set for which the composed map $\gamma$ is an isomorphism.

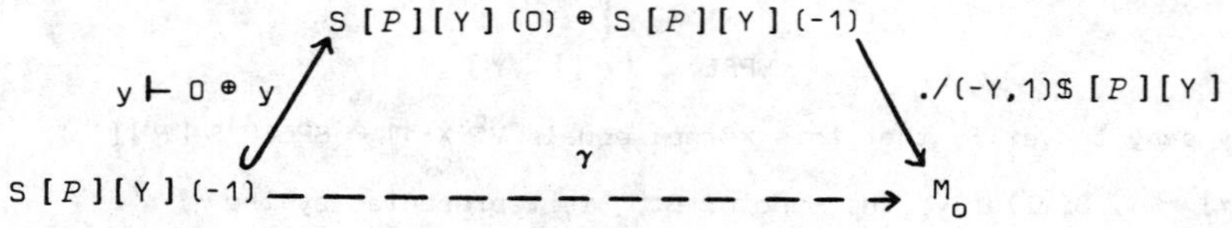


Now, $\gamma(S[\mathcal{P}][Y](-1)) = Y\,S[\mathcal{P}][Y](0)$, thus $X^g(Y)$ is the desired open set.

Similarly, $X^g(X)$ is the open set of $SPEC^g S[\mathcal{P}][X]$ for which the composed morphism $\gamma'$

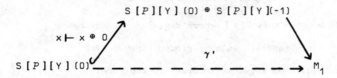

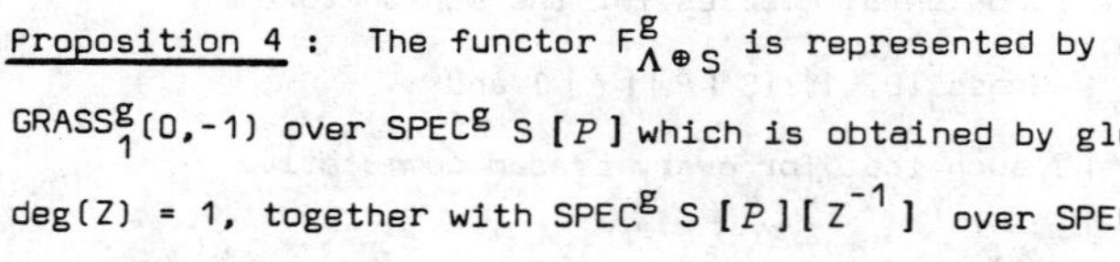


is an isomorphism. Concluding :

<u>Proposition 4</u> : The functor $F^g_{\Lambda \oplus S}$ is represented by a graded scheme $GRASS^g_1(0,-1)$ over $SPEC^g\ S[\mathcal{P}]$ which is obtained by gluing $SPEC^g S[\mathcal{P}][Z]$, $\deg(Z) = 1$, together with $SPEC^g\ S[\mathcal{P}][Z^{-1}]$ over $SPEC^g\ S[\mathcal{P}][Z,Z^{-1}]$.

The scheme $GRASS_1^g(0,-1)$ can be interpreted as the graded one-dimensional projective space over $S[P]$.

Above, we mentioned that the part of degree zero (or e) of a graded scheme is often a usual scheme. In particular, the part of degree zero of the graded scheme $GRASS_1^g(0,-1)$ is the S-scheme which is obtained by gluing $\underline{Spec}(S[P]_{\geq 0})$ with $\underline{Spec}(S[P]_{\leq 0})$ over $\underline{Spec}(S[P])$. This scheme is never regular !

Example 7 : Let $\Lambda = \mathbb{C}[X,-]$ then S can be taken to be $\mathbb{C}[t], t = X^2$. In this case, the part of degree zero of $GRASS_1^g(0,-1)$ is the scheme obtained by gluing two affine cones (i.e. Spec $\mathbb{C}[x,y,t]/(x^2-y^t)$) over the complement of a ruling.

CASE 2 : $P = \{p,q\}$ and $\dim_{R:m}(\Lambda/\Lambda m+P+Q) = 1$.

It follows from lemma 3 that there is no étale extension of R which splits $\Sigma$. However, one can find an étale extension $R_1$ of R such that $\Lambda \oplus R_1 \cong R_1 \oplus R_1 i \oplus R_1 j \oplus R_1 ij$ with $i^2 = p$ and $j^2 = q$. Moreover, there exists an extension $S = R_1[X]/(X^2-p)$ which splits $\Sigma$ and such that the ring $S(\Phi)$ defined in the proof of proposition 3 is a graded étale (and even Galois) extension of $R_1[P]$, in particular $S(\Phi)$ is graded regular. This entails that

$$(\Lambda \otimes R_1)[P] \otimes S(\Phi) \cong M_2(S(\Phi))(\sigma_1,\sigma_2)$$

Further, a small computation show that the $\sigma_i$'s may be chosen to be e or (0,1). The case that $\sigma_1 = \sigma_2 = e$ cannot occur since this would entail that

$$\Lambda \oplus R_1[P] \cong M_2(S(\Phi))(e,e)^G = M_2(R_1[P])(e,e)$$

Therefore we may may assume that $\sigma_1 = e$ and $\sigma_2 = \sigma = (0,1)$.

Our first objective will be to represent the functor :

$$F_S^g : \text{gr Comm Alg}_{S(\Phi)} \to \text{Sets}$$

which assigns to any graded commutative $S(\Phi)$-algebra A the set of all split projective graded left ideals of rank two of $M_2(A)(e,\sigma)$. As in case 1 it is readily verified that this problem is equivalent to finding a representation of

$$\text{Grass}_1^g(e,-\sigma) : \text{gr Comm Alg}_{S(\Phi)} \to \text{Sets}$$

Mimicing the arguments of case 1 in the $\mathbb{Z} \oplus \mathbb{Z}$-graded case one can prove.

<u>Proposition 5</u> : The functor $F_S^g$ (or $\text{Grass}_1^g(e,\sigma)$) is represented by a graded scheme $\text{GRASS}_1^g(e,\sigma)$ over $\text{SPEC}^g S(\Phi)$ which is obtained by gluing $\text{SPEC}^g S(\Phi)\,[Z]$, $\deg(Z) = \sigma$, together with $\text{SPEC}^g S(\Phi)\,[Z^{-1}]$ over $\text{SPEC}^g S(\Phi)\,[Z,Z^{-1}]$.

By graded Galois descent, one can then find the graded scheme over $\text{SPEC}^g R_1\,[P]$ which represents the functor

$$f^g : \text{gr Comm Alg}_{R_1[P]} \to \text{Sets}$$

which assigns to a graded $R\,[P]$-algebra A the set of all graded projective split left ideals of $\Lambda_1\,[P] \oplus A$ of rank two. A concrete computation of this graded scheme can be found in [ 14 ] .

<u>B. a representation of $F_\Lambda$.</u>

We will restrict attention here to case 1. If one has an explicit form of the graded scheme over $\text{SPEC}^g R_1\,[P]$ representing the functor $f^g$ in case 2 one can easily mimic the argument below.

If $A \in \underline{b}_S$, then it follows from the key lemma that there is a natural one-to-one correspondence between elements of $F_{\Lambda \oplus S}(A)$ and graded $\text{SPEC}^g S\,[P]$-scheme homomorphisms from $\text{SPEC}^g A\,[P]$ to $\text{GRASS}_1^g(0,-1)$.

Now, any such graded scheme morphism

$$\varphi : \mathrm{SPEC}^g A\,[P] \to \mathrm{GRASS}^g_1(0,-1)$$

is determined by graded $S\,[P]$ - algebra morphisms :

$$\varphi_1 : S\,[P][Z] \to A\,[P]$$

$$\varphi_2 : S\,[P][Z^{-1}] \to A\,[P]$$

such that their localizations $(\varphi_1)_Z$ and $(\varphi_2)_{Z^{-1}}$ yield the same graded $S\,[P]$ -algebra morphism

$$\varphi_{12} : S\,[P][Z,Z^{-1}] \to A\,[P]$$

Clearly, $\varphi_1$ is completely determined by $\varphi_1(Z) \in A\,[P]_1 = A.p$. Therefore there is a natural one-to-one correspondence between $S\,[P]$ -algebra morphisms from $S\,[P][Z]$ to $A\,[P]$ and S-algebra morphisms from $S\,[X,Y]\,/\,(X-pY)$ to A.

Similarly, $\varphi_2$ is completely determined by $y_2(Z^{-1}) \in A\,[P]_{-1} = A$. Therefore there is a one-to-one correspondence between graded $S\,[P]$ - algebra morphisms from $S\,[P][Z^{-1}]$ to $A\,[P]$ and algebra morphisms from $S\,[X^{-1}]$ to A.

Finally, there is a one-to-one correspondence between graded $S\,[P]$ - algebra homomorphisms from $S\,[P][Z,Z^{-1}]$ to $A\,[P]$ and S-algebra morphism from $S\,[X,X^{-1},Y]\,/\,(X-pY)$ to A. Therefore, if X denotes the S-scheme obtained by gluing Spec $S\,[X,Y]\,/\,(X-pY)$ with Spec $S\,[X^{-1}]$ over Spec $S\,[X,X^{-1},Y]\,/(X-pY)$, then there is a natural one-to-one correspondence between graded $\mathrm{SPEC}^g S\,[P]$ -scheme morphisms from $\mathrm{SPEC}^g A\,[P]$ to $\mathrm{GRASS}^g_1(0,-1)$ and Spec(S)-scheme morphisms from Spec(A) to X.

Theorem 1 :

In case 1 the étale local structure of the scheme representing the functor $F_\Lambda$ is the scheme obtained by gluing Spec S [ Y ] with Spec S [ Z ] over Spec S [ Y,Z ] / (p YZ-1).

The scheme $BS'_\Lambda$ over Spec(R) which is obtained by étale descent from the S-scheme X is called the (weak) Brauer-Severi scheme of $\Lambda$ and it represents the functor $F_\Lambda$.
If one can compute the full Brauer-Severi scheme of $\Lambda$, $BS_\Lambda$, we conjecture that there is always an open immersion $BS'_\Lambda \rightarrow BS_\Lambda$. Let us conclude this paper with an example :

<u>Example 8</u> : (cfr. ex 2 and ex. 7). The scheme X over Spec $\mathbb{C}$ [ t ] has fibers which can be visualized as

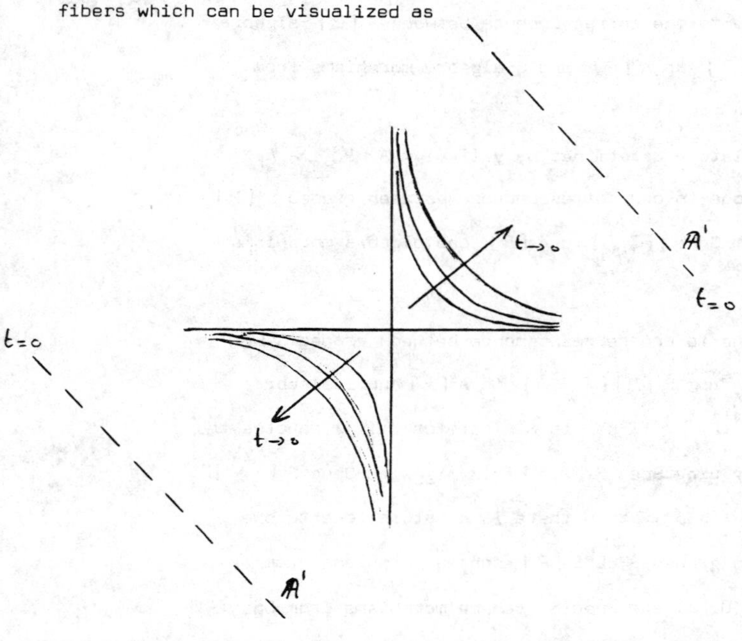


a family of conics, degenerating to a pair of distinct affine lines. The étale local structure of the full Brauer-Severi scheme was computed by Artin. Its fibers can be visualized as a family of conics, degenerating to a pair of projective lines meeting transversally at one point.

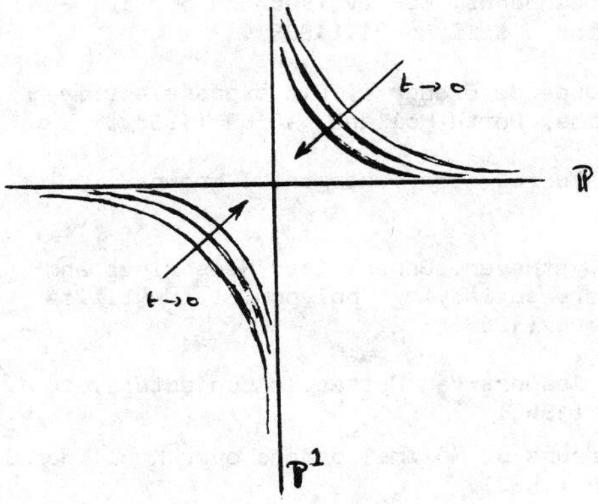


It is fairly easy to give an open immension of X in $BS_{\Lambda} \otimes A^1_{\mathbb{C}}$. The weak Brauer-Severi scheme misses one point corresponding to the left ideal of rank two L in

$$\Lambda \otimes \mathbb{C}[t]/(t) \cong \begin{pmatrix} \mathbb{C} & \mathbb{C} \\ (t)/(t^2) & \mathbb{C} \end{pmatrix}$$

where

$$L = \begin{pmatrix} 0 & \mathbb{C} \\ (t)/t^2) & 0 \end{pmatrix}$$

References :

[1] M. Artin, Munford : Some examples of unirational varieties which are not rational. Proc. London Math. Soc. 25, 3rd series, pp. 75-95 (1972).

[2] M. Artin, Local structure of maximal orders on surfaces, LNM 917, pp. 146-181 (1982).

[3] M. Artin : Left ideals in maximal orders, LNM 917, pp. 182-193 (1982).

[4] M. Auslander - O. Goldman, Maximal Orders, Trans. A.M.S. 97 (1960) 1-24.

[5] S. Caenepeel : gr-complete and gr-Henselian rings, Preprint V.U.B. (1983).

[6] S. Caenepeel, M. Van den Bergh, F. Van Oystaeyen, Generalized Crossed Products, applied to maximal orders, Brauer groups and related exact sequences, UIA-preprint 83-20.

[7] R. Fossum, Maximal orders over Krull domains, J. of Algebra 10 pp. 321-332 (1968).

[ 8 ] A. Grothendieck, J. Dieudonné, EGA IV (seconde partie) Pub. Math. Inst. Hautes Etudes Sci. No 24 (1965).

[ 9 ] A. Grothendieck : Groupe de Brauer I, Dix exposés sur la cohomologie des schémas, North Holland, 46-65 (1968).

[ 10 ] A. Grothendieck , J. Murre. Tame fundamental group ..., LNM 208, (1971).

[ 11 ] L. Le Bruyn, F. Van Oystaeyen, Generalized Rees rings and relative maximal orders satisfying polynomial identities to appear in J. Algebra (1983).

[ 12 ] L. Le Bruyn : On the Jespers-Van Oystaeyen conjecture, to appear in J. Algebra (1983).

[ 13 ] L. Le Bruyn, Class groups of maximal orders over Krull domains, Ph. D. Thesis Antwerp (1983).

[ 14 ] L. Le Bruyn : Weak Brauer-Severi schemes I, the étale local structure, in preparation.

[ 15 ] L. Le Bruyn, M. Van den Bergh : The ramification divisor of a tame order, in preparation.

[ 16 ] H. Marubayashi, E. Nauwelaerts, F. Van Oystaeyen : Graded Rings over arithmetical orders, UIA preprint (1982).

[ 17 ] J. Milne, Etale cohomology, Princeton University Press (1980).

[ 18 ] C. Nastasescu - F. Van Oysateyen : Graded ring theory, North Holland (1982).

[ 19 ] H. Orzech : Brauer groups and class groups for a Krull domain, LNM 917, pp. 66-90 (1982).

[ 20 ] M. Ramras : Maximal orders over regular local rings of dimension two, Trans. A.M.S. 142, 457-479 (1969).

[ 21 ] M. Ramras : maximal orders over regular local rings, Trans. Amer. Math. Soc. Vol 155, 345-352 (1971).

[ 22 ] M. Ramras : The center of an order with finite global dimension, Trans. Amer. Math. Soc. Vol. 20, 249-257 (1975).

[ 23 ] M. Raynaud, Anneaux Locaux Henseliens, LNM 169 (1970).

[ 24 ] J.P. Serre, Corps Locaux, Hermann, Paris (1962)

[ 25 ] L. Silver, Tame ramification etc. J. of Algebra (1966).

[ 26 ] F. Van Oystaeyen, Graded Azumaya algebras and Brauer Groups LNM 825, (1980).

[ 27 ] F. Van Oystaeyen, Some constructions of rings,UIA-preprint 82-35.

# GROUP RINGS AND DIVISION RINGS

Martin Lorenz

Max-Planck-Institut für Mathematik
Gottfried-Claren-Str. 26, D-5300 Bonn 3, FRG

*To the memory of Oberstudienrat Heinz-Joachim Dietrich*

Abstract. Continuing the work in [11],[12] we study division algebras $D = k(G)$ over a field $k$ which are generated by some polycyclic-by-finite subgroup $G$ of the multiplicative group $D^*$ of $D$. We discuss a specific class of examples of such division algebras that can be thought of as multiplicative analogs of the Weyl field. Furthermore, we show that the division algebras $D = k(G)$ always contain free subalgebras of rank $\geq 2$, provided $G$ is not abelian-by-finite. Finally, we discuss some open questions concerning commutative subfields and Lie commutators in $D = k(G)$.

## INTRODUCTION

During the past decade, a considerable amount of work has been invested in the study of prime ideals in group algebras $kG$ of polycyclic-by-finite groups $G$ over a field $k$. After the pioneering work of Zalesskii [30], Roseblade's break-through in [22], and the finishing touches by Passman and the author [14], the subject has now reached a certain state of maturity: one has a detailed recipe for constructing all primes $P$ of $kG$ starting from prime ideals of group algebras $kH$, where $H$ runs through special finite-by-abelian subquotients of $G$ (see [13] or [14] for the precise formulation). The resulting class of algebras $kG/P$ is a rich source of interesting examples of prime Noetherian rings whose fine structure is far from being well understood. For example, if $Q = Q(kG/P)$ denotes the classical ring of fractions of $kG/P$ then, by Goldie's theorem, $Q = M_n(D)$ for some integer $n$ and a suitable division k-algebra $D$ both of which are in general quite mysterious to us. In the present note, continuing the work in [11], [12], we study the Goldie field $D$ associated with a completely



*F. van Oystaeyen (ed.), Methods in Ring Theory, 265–280.*

prime ideal $P$ of $kG$. In other words, we study division k-algebras $D$ generated by some polycyclic-by-finite group $G \leq D^*$, the multiplicative group of $D$. The restriction to completely prime ideals is partly justified by the following result, due to Zalesskii [30, Theorem 4] for primitive ideals and to Brown [2] in general:

Let $P$ be a prime ideal of $kG$. Then there exists a characteristic subgroup $G_o$ of $G$ with $G/G_o$ finite and such that $P \cap kG_o$ is a finite intersection of completely prime ideals of $kG_o$ which are all conjugate under the action of $G$ on $kG_o$.

In Section 1, we study a specific class of algebras $B_\lambda$ $(\lambda \in k^*)$ and their classical division rings of fractions $E_\lambda$. Each $B_\lambda$, and $E_\lambda$, is generated by a 2-generated nilpotent group of class 2 and can be viewed as a multiplicative analog of the Weyl algebra $A_1$, resp. the Weyl field $D_1 = Q(A_1)$ (see also [10]). Although the $B_\lambda$'s are not isomorphic to $A_1$, and to each other, they share many of the well-known properties of $A_1$. The main result of Section 2 states that if $D = k(G)$ is a division algebra generated by some polycyclic-by-finite group $G \leq D^*$ and if $G$ is not abelian-by-finite, then $D$ contains a free k-subalgebra of rank at least 2. This result depends on recent work of L. Makar-Limanov [15]. Finally, in the last section, we briefly discuss commutative subfields and Lie commutators in division algebras $D = k(G)$ of the above type and mention a number of open questions.

<u>Notations and Conventions.</u> In this paper, $k$ always denotes a commutative field. If $D$ is a division k-algebra, then we use the notation $D = k(E)$ to indicate that $D$ is generated, as division k-algebra, by the subset $E$ of $D$. Furthermore, for any ring $R$ (always with 1), $R^*$ denotes the set of nonzero elements, $U(R)$ the group of units, $Z(R)$ the center, and $Q(R)$ the classical ring of fractions of $R$ (if it exists). We use square brackets to denote group theoretical and Lie commutators. Thus $[x,y] = x^{-1}y^{-1}xy$ or $[x,y] = xy-yx$, depending on the context. Otherwise the notation is standard, and follows [18].

## 1. A CLASS OF EXAMPLES

Let $\lambda \in k^*$ be given and define $B_\lambda = B_\lambda(k)$ to be the k-algebra generated by two elements $x$ and $y$ together with their inverses $x^{-1}, y^{-1}$ subject to the relation $xy = \lambda yx$. In short,

$$B_\lambda = k\{x^{\pm 1}, y^{\pm 1}\}/(xy-\lambda yx) \qquad (\lambda \in k^*).$$

Some results concerning these algebras and their tensor products have been announced in [10], even for $k$ not necessarily a field. As far as I know, the $B_\lambda$'s, or rather their power series analog, have made their first appearance in [7], with $k$ being the field of rational numbers and $\lambda = 2$.

The algebra $B_\lambda$ can be realized as the factor of the group algebra kG, with $G = \langle x,y \mid z=[x,y] \text{ central}\rangle$ the free nilpotent group of class 2 on 2 generators, modulo the ideal $(z-\lambda)kG$. $B_\lambda$ can also be viewed as a twisted group algebra of the free abelian group of rank 2,

$$B_\lambda \cong k^t[Z \oplus Z],$$

or as an iterated Ore extension,

$$B_\lambda = k[X^{\pm 1}][Y^{\pm 1};\alpha] \text{ with } X^\alpha = \lambda X.$$

In particular, $B_\lambda$ is a Noetherian domain. We denote its classical division ring of fractions by $E_\lambda = E_\lambda(k)$. Note that $E_\lambda = k(\langle x,y\rangle)$, where $\langle x,y\rangle$ is nilpotent of class 2 (if $\lambda \neq 1$).

The following lemma describes some basic properties of the $B_\lambda$'s. Some of them have also been noted in [10]. We are mostly interested in the case where $\lambda$ is not a root of unity. In this case the properties of $B_\lambda$ closely mirror those of the Weyl algebra $A_1$ in characteristic 0, whereas the case where $\lambda$ is a root of unity corresponds to the Weyl algebra in positive characteristics [21].

<u>Lemma 1.1.</u> (a) Let $\lambda \in k^*$ be of finite (multiplicative) order n. Then $B_\lambda$ is free of rank $n^2$ as a module over its center $Z(B_\lambda) = k[x^{\pm n}, y^{\pm n}]$. Moreover, for any ideal I of B , one has $I = (I \cap Z(B_\lambda))B_\lambda$.

(b) If $\lambda \in k^*$ has infinite order, then $B_\lambda$ is a central-simple k-algebra of global and Krull dimension 1 and of Gelfand-Kirillov dimension 2.

<u>Proof.</u> First suppose that $\lambda$ has finite order n and set $C = k[x^{\pm n}, y^{\pm n}] \subset B_\lambda$. Then $C \subset Z(B_\lambda)$ and $B_\lambda$ is free as a module over C, with basis $\{x^i y^j \mid 0 \leq i,j \leq n-1\}$. If $b = \Sigma_{i,j=0}^{n-1} c_{ij} x^i y^j \in Z(B_\lambda)$, with $c_{ij} \in C$, then

$$b = y^{-1}by = \Sigma_{i,j=0}^{n-1} c_{ij}\lambda^i x^i y^j$$

and so $c_{ij} = c_{ij}\lambda^i$ for all i,j. Therefore, $c_{ij} = 0$ for $i \neq 0$. Similarly, for $j \neq 0$ one has $c_{ij} = 0$ which shows that $C = Z(B_\lambda)$. Now let I be an ideal of $B_\lambda$ and set $I_1 = (I \cap C)B_\lambda \subset I$. It follows from the foregoing that $B_\lambda/I_1$ is free of rank $n^2$ over $K = C/I \cap C$, and a calculation as above shows that $K = Z(B_\lambda/I_1)$. $B_\lambda$, being an image of the group algebra of a finitely generated nilpotent group, is a polycentral ring [18, 11.3.12]. Thus, if I strictly contains $I_1$, then we must have $(I/I_1) \cap K \neq 0$ which is impossible. Therefore, $I = I_1$ and part (a) is proved.

As to part (b), the equality $Z(B_\lambda) = k$, for $\lambda$ of infinite order, follows by a straightforward calculation as above. Simplicity of $B_\lambda$ now is a consequence of polycentrality, or can be checked directly by the usual shortening trick. Finally, $\text{GK-dim}_k(B_\lambda) = 2$ follows from the fact that each monomial in the elements $x,y$ and their inverses is a scalar multiple of an "ordered" monomial $x^i y^j$ $(i,j \in Z)$, and $\text{K-dim}(B_\lambda) = \text{gl.dim}(B_\lambda) = 1$ follows from [19, Theorem 4.5], e.g. (see also [10]).

<u>Corollary 1.2.</u> (a) Let $\lambda \in k^*$ be of finite order $n$. Then $E_\lambda = Q(B_\lambda)$ is obtained by localizing $B_\lambda$ at the nonzero elements of $Z(B_\lambda)$. Thus $Z(E_\lambda) = k(x^n,y^n)$ and $E_\lambda \cong Z(E_\lambda)^t[Z/nZ \oplus Z/nZ]$ is a twisted group algebra of $Z/nZ \oplus Z/nZ$ over $Z(E_\lambda)$.

(b) If $\lambda$ is not a root of unity, then $Z(E_\lambda) = k$.

<u>Proof.</u> The assertions in (a) are immediate from Lemma 1.1(a). For (b) just note that for any simple or, more generally, polycentral ring $R$ for which $Q(R)$ exists one has $Z(Q(R)) = Q(Z(R))$. ∎

It follows from [12, Corollary 2.2] that the Gelfand-Kirillov transcendence degree of $E_\lambda$ over $k$ equals 2. If $\lambda$ has infinite order, then $E_\lambda$ contains a free k-subalgebra (see Section 2). Hence $E_\lambda$ has infinite Gelfand-Kirillov dimension in this case. On the other hand, for $\lambda$ a root of unity, $E_\lambda$ clearly has Gelfand-Kirillov dimension 2. Also, the commutative transcendence degree of $E_\lambda$ in the sense of Resco [19] is clearly 2 if $\lambda$ has finite order. For $\lambda$ of infinite order, it equals 1, by [19, Theorem 4.3] (or see Section 3.A).

Before we turn to the isomorphism question for the division algebras $E_\lambda$, let us briefly recall a few general facts concerning twisted group algebras $k^t[H]$ of ordered groups $H$. By definition, $k^t[H]$ has a k-basis $\{\dot{x} \mid x \in H\}$, and multiplication in $k^t[H]$ is determined by the rule

$$\dot{x}\cdot\dot{y} = t(x,y)\dot{(xy)} \qquad (x,y \in H),$$

where $t\colon H \times H \to k^*$ is a 2-cocycle. The basis $\{\dot{x} \mid x \in H\}$ can always be normalized so that $\dot{e} = 1$ is the identity element of $k^t[H]$. Each $\dot{x}$ is a unit of $k^t[H]$ and, using the fact that $H$ is ordered, it is easy to see that the group of units $U(k^t[H])$ of $k^t[H]$ consists precisely of the elements of the form $\alpha\dot{x}$ with $\alpha \in k^*$ and $x \in H$. Furthermore, we have a multiplicative map, the so-called lowest term map,

$$\ell\colon k^t[H] \to \{0\} \cup U(k^t[H])$$

defined by $\ell(0) = 0$ and $\ell(a) = \alpha_{x_o}\dot{x}_o$ for $0 \neq a = \Sigma_{x\in H}\ \alpha_x\dot{x}$ with $x_o = \min\{x \in H \mid \alpha_x \neq 0\}$. If $S \subset k^t[H]$ is a (right) Ore set of regular elements, then $\ell$ extends to the localization $k^t[H]S^{-1}$ by setting $\ell(ab^{-1}) = \ell(a)\ell(b)^{-1}$. It is trivial to verify that $\ell$ is well-defined and remains multiplicative on $k^t[H]S^{-1}$.

The foregoing applies conveniently to $B_\lambda$. Indeed, if $x^{\pm 1}, y^{\pm 1}$ are the canonical generators of $B_\lambda$ with $xy = \lambda yx$, then

$$B_\lambda = \bigoplus_{i,j\in Z} kx^iy^j \cong k^t[Z \oplus Z],$$

and $Z \oplus Z$ is of course an orderable group. Thus the corresponding lowest term map provides us with a multiplicative map

$$\ell\colon E^*_\mu \to U_\mu = \{\alpha x^iy^j \mid i,j \in Z,\ \ \alpha \in k^*\} = U(B_\mu)$$

which is the identity on $U_\mu$, hence on $k^* \subset U_\mu$.

<u>Proposition 1.3.</u> Let $\lambda,\mu \in k^*$ be given. Then $E_\lambda$ and $E_\mu$ are isomorphic as k-algebras if and only if $\lambda = \mu$ or $\lambda = \mu^{-1}$.

<u>Proof.</u> Since $B_\lambda = k\{x^{\pm 1},y^{\pm 1}\}/(xy-\lambda yx) = k\{y^{\pm 1},x^{\pm 1}\}/(yx-\lambda^{-1}xy) = B_{\lambda^{-1}}$, the condition is certainly sufficient. To prove necessity, let $\phi\colon E_\lambda \to E_\mu$ be a fixed k-algebra isomorphism, and let $x^{\pm 1},y^{\pm 1} \in B_\lambda$ and $u^{\pm 1},v^{\pm 1} \in B_\mu$ be the canonical generators with $xy = \lambda yx$, resp. $uv = \mu vu$. Let $\ell\colon E^*_\mu \to U_\mu = \{\alpha u^iv^j \mid i,j \in Z,\ \alpha \in k^*\}$ be the lowest term map with respect to a fixed ordering of $U_\mu/k^* \cong Z \oplus Z$. Then

$$f = \ell\circ\phi\colon E^*_\mu \to U_\mu$$

is a group homomorphism which is the identity on $k^*$. In particular, we obtain

$$\lambda = f(\lambda) = [f(x),f(y)],$$

and this belongs to the commutator subgroup $[U_\mu,U_\mu]$ of $U_\mu$. But $[U_\mu,U_\mu] = \langle\mu\rangle$ and so we have $\lambda \in \langle\mu\rangle$. By symmetry, we conclude that $\langle\lambda\rangle = \langle\mu\rangle$ in $k^*$. Therefore, if $\lambda$ has infinite order, then $\lambda = \mu$ or $\lambda = \mu^{-1}$ and we are done. Thus, in the following, we concentrate on the case where $\lambda$ and $\mu$ have finite order $n$.

We show that $f$ maps $U_\lambda$ isomorphically onto $U_\mu$. First note that $f$ induces a map $\bar{f}\colon U_\lambda/k^* \to U_\mu/k^*$. Both groups are free abelian of rank 2 generated by $\bar{x}$ and $\bar{y}$, resp. $\bar{u}$ and $\bar{v}$, where we use overbars to denote images mod $k^*$. Suppose that $\bar{f}(\bar{x}^i\bar{y}^j) = 1$ with $0 \neq |i| + |j|$ minimal. Then

$$1 = [f(x^iy^j),f(y)] = \lambda^i$$

and so $i = ni_1$ for a suitable $i_1$. Similarly, $j = nj_1$ and hence

$$1 = \bar{f}(\bar{x}^i\bar{y}^j) = \bar{f}(\bar{x}^{i_1}\bar{y}^{j_1})^n .$$

Since $U_\mu/k^*$ is torsion-free, we conclude that $\bar{f}(\bar{x}^{i_1}\bar{y}^{j_1}) = 1$ which contradicts our minimality assumption. Therefore, $\bar{f}$ is injective, and hence the same is true for $f$ on $U_\lambda$. To prove surjectivity note that, clearly, $\ell(E_\mu^*) = U_\mu$ and $\ell(E_\mu^*) = f(B_\lambda^*)f(B_\lambda^*)^{-}$. Thus it suffices to show that $f(B_\lambda^*) = f(U_\lambda)$. But every $a \in B_\lambda^*$ can be written as a finite sum $a = \Sigma_i u_i$ with $u_i \in U_\lambda$ pairwise distinct mod $k^*$. Hence, by the above, the images $\overline{f(u_i)} \in U_\mu/k^*$ are distinct, say $\overline{f(u_1)}$ is the smallest with respect to the ordering of $U_\mu/k^*$. Then we obtain that

$$f(a) = \ell(\Sigma_i \phi(u_i)) = \ell(\phi(u_1)) = f(u_1) ,$$

as required. Therefore, $f$ on $U_\lambda$ and $\bar{f}$ are isomorphisms.

We conclude that $\overline{f(x)} = \bar{u}^i\bar{v}^j$, $\overline{f(y)} = \bar{u}^r\bar{v}^s$ with $is-jr = \pm 1$. Consequently,

$$\lambda = [f(x),f(y)] = [u^iv^j,u^rv^s] = \mu^{is-jr} = \mu^{\pm 1}$$

and the proposition is proved.

Proposition 1.3 extends [10, Theorem 1] which states that (for $k$ a domain) $B_\lambda$ and $B_\mu$ are isomorphic as k-algebras iff $\lambda = \mu^{\pm 1}$. The above argument essentially follows the lines of the proof of the isomorphism theorem [11, Theorem 4.1]. In fact, for $\lambda$ of infinite order, Proposition 1.3 could have been deduced from that result.

It can be shown that in $E_\lambda$ the identity element cannot be written as a sum of Lie commutators (Section 3.B). In particular, the $E_\lambda$'s are all distinct from the Weyl field $D_1 = Q(A_1)$.

We close this section with a few facts concerning projective and injective modules for $B_\lambda$, $\lambda \in k^*$ of infinite order. The corresponding assertions for the Weyl algebra $A_1$ in characteristic 0 are well-known. We therefore restrict ourselves to a few indications and refer to the literature whenever possible.

Proposition 1.4. Let $\lambda \in k^*$ be of infinite order.
(a) For any nonzero right ideal $I$ of $B_\lambda$, $B_\lambda \oplus I \cong B_\lambda \oplus B_\lambda$. In particular, every right ideal is generated by at most 2 elements.
(b) A finitely generated projective right $B_\lambda$-module is either free or isomorphic to a right ideal of $B_\lambda$.
(c) For $n > 1$, the matrix ring $M_n(B_\lambda)$ is a principal right (and left) ideal ring, whereas $B_\lambda$ is not a principal ideal ring.
(d) $B_\lambda$ has no nonzero finitely generated injective modules.

Proof. (a) We follow Webber [29]. Write $B_\lambda$ as an Ore extension, $B_\lambda = R[Y^{\pm 1};\alpha]$ with $R = k[X^{\pm 1}]$ and $X^\alpha = \lambda X$. Then $S = R^*$ is an Ore set in $B_\lambda$ and $B_\lambda S^{-1} \cong k(X)[Y^{\pm 1};\alpha]$ is a principal ideal domain. Therefore, for any nonzero right ideal $I$ of $B_\lambda$ there exists $0 \neq d_o \in I$ such that $\bar{I} = I/d_o B_\lambda$ is R-torsion. If $\bar{I} \neq 0$ choose $0 \neq \bar{d}_1 \in \bar{I}$ such that $M_1 = \text{ann}_R(\bar{d}_1)$ is maximal among the annihilators of nonzero elements of $\bar{I}$. Then $M_1$ is a maximal ideal ideal of $R$, and this easily implies that $M_1 B_\lambda$ is a maximal right ideal of $B_\lambda$. Therefore, $\bar{d}_1 B_\lambda \cong B_\lambda / p_1 B_\lambda$ where $M_1 = (p_1)$. Continuing this way and using the fact that $B_\lambda$ is Noetherian we can write $I = \Sigma_{i=0}^m d_i B_\lambda$ with

$$\Sigma_{i=0}^{r+1} d_i B_\lambda / \Sigma_{i=0}^{r} d_i B_\lambda \cong B_\lambda / p_{r+1} B_\lambda \quad \text{for } r = 0,1,\ldots,m-1.$$

Applying Schanuel's Lemma we obtain

$$B_\lambda \oplus \Sigma_{i=0}^{r} d_i B_\lambda \cong p_{r+1} B_\lambda \oplus \Sigma_{i=0}^{r+1} d_i B_\lambda \cong B_\lambda \oplus \Sigma_{i=0}^{r+1} d_i B_\lambda$$

and so, inductively, $B_\lambda \oplus I \cong B_\lambda \oplus d_o B_\lambda \cong B_\lambda \oplus B_\lambda$.
(b) In view of part (a), this follows from [29, Theorem 1].
(c) The first assertion follows from (b) and [3, Theorem 7] and the second is immediate from [27, Corollary 1.8].
(d) This can be shown as in the case of the Weyl algebra $A_1$ [16, Theorem 5.5]. We omit the details.

## 2. FREE SUBALGEBRAS

In this section, we consider division k-algebras $D$ generated by an arbitrary polycyclic-by-finite group $G \leq D^*$. The proof of our main result (Theorem 2.3) depends upon two major ingredients which we now describe.

Let $A$ be a finitely generated free abelian group and let $H$ be a group acting on $A$. The action of $H$ on $A$ is said to be rationally irreducible if $A \otimes_Z Q$ is an irreducible module for the rational group algebra $QH$ or, equivalently, if $H$ normalizes no proper pure subgroup of $A$. The following result is due to G. Bergman [1a] (cf. also [18, 9.3.9] and [5]).

Theorem 2.1 (Bergman). Let $A$ be a finitely generated free abelian group and let $H$ be a group acting on $A$. Suppose that $H$ and all its subgroups of finite index act rationally irreducibly on $A$. If $I$ is a proper H-invariant ideal of the group algebra $kA$, then either $I = 0$ or $kA/I$ is finite dimensional over $k$.

The second result that we will need is due to L. Makar-Limanov [15]. Strictly speaking, he considers the group algebra $kH$ of the discrete Heisenberg group $H = \langle x,y \mid z=[x,y] \text{ central}\rangle$, and its division ring of fractions. His methods, however, can easily be adapted to deal with the algebras $B_\lambda(k)$ and $E_\lambda(k)$, where $\lambda \in k^*$ has infinite order (Section 1). Note that $Q(kH)$ can in fact be written as $Q(kH) = E_z(k(z))$.

Theorem 2.2 (Makar-Limanov). Let $\lambda \in k^*$ be of infinite order. Then $E_\lambda(k)$ contains a free k-subalgebra of rank 2.

The following result extends this to division algebras generated by arbitrary polycyclic-by-finite groups.

Theorem 2.3. Let $D = k(G)$ be a division k-algebra generated by some polycyclic-by-finite group $G \leq D^*$. Then $D$ contains a free k-subalgebra of rank $\geq 2$ if and only if $G$ is not abelian-by-finite.

Proof. If $G$ is abelian-by-finite, then $D$ is finite dimensional over its center, and hence $D$ does not contain free subalgebras. If $G$ is nilpotent-by-finite but not abelian-by-finite, then after dropping to a subgroup of finite index, we may assume that $G$ is non-abelian torsion-free nilpotent. Then $G$ contains elements $x$ and $y$ whose commutator $z = [x,y]$ is $\neq 1$ and commutes with $x$ and $y$. Set $K = k(z) \subset D$ and consider the K-algebra $B \subset D$ generated by $x$ and $y$ and their inverses. Clearly, $B$ is an image of $B_z(K)$. Since $z \in K^*$ has infinite order, $B_z(K)$ is simple and we do in fact have isomorphisms $B \cong B_z(K)$ and $Q(B) \cong E_z(K)$. The existence of a free subalgebra in $D$ now follows from Theorem 2.2, because the embedding $B \subset D$ extends to an embedding $Q(B) \subset D$.

Thus, in the following, assume that $G$ is not nilpotent-by-finite and that all its nilpotent subgroups are abelian-by-finite. We will proceed in three steps.

Step 1. $G$ contains a subgroup $H$ which is a semidirect product $H = A \rtimes \langle z\rangle$ with $A$ free abelian of rank at least 2 and with $z$ and all its powers acting rationally irreducibly on $A$.

Proof. After dropping to a subgroup of finite index, we may assume that the Fitting radical $B = \mathrm{Fitt}(G)$ and $G/B$ are both free abelian $\neq \langle 1\rangle$ (use [18, 12.1.5]). Fix $z \in G$, $z \notin B$ and set $V = B \otimes_Z Q$. Replacing $z$ by a suitable power if necessary, we may assume that all irreducible $Q\langle z\rangle$-submodules of $V$ remain irreducible for $Q\langle z^n\rangle$ for all $n \geq 1$ (choose $n$ so that the composition length of the $Q\langle z^n\rangle$- socle of $V$ is maximal) and, moreover, that $z$ acts trivially on the 1-dimensional $Q\langle z\rangle$-submodules of $V$ (their intersections with $B$ are infinite cyclic groups normalized by $z$ so $z^2$ acts trivially). Then $V$ must contain an irreducible $Q\langle z\rangle$-submodule $U$ of dimension at least 2, for otherwise the minimal polynomial of $z$ on $V$ would be of the form $(z-1)^r$ for some $r$

and $\langle B,z\rangle$ would be nilpotent and normal in $G$, contradicting the fact that $B = \mathrm{Fitt}(G)$. Thus we can take $A = U \cap B$ and $H = \langle A,z\rangle = A \rtimes \langle z\rangle$.

Step 2. $H$ contains a free semigroup on two generators.
Proof. This is a consequence of more general work of Rosenblatt [23]. The present special case however can quickly be dealt with using an idea of H. Bass [1]. Let $\zeta$ be the automorphism induced by $z^{-1}$ on $A$ and let $K = Q[\zeta] \subset \mathrm{End}_Q(A \otimes_Z Q)$. Then $K$ is a finite field extension of $Q$ and $\zeta$ is not a root of unity. By a theorem of Kronecker, there exists a $Q$-embedding, $\sigma$ of $K$ into the complex numbers, $C$, with $|\zeta^\sigma| > 1$ [9 , p. 215]. After replacing $z$ by a suitable power, we may assume that $|\zeta^\sigma| > 2$. We show that for any $a \in A$, $a \neq 1$, the semigroup generated by $z$ and $az$ is free. Indeed, suppose that

$$z^{i_o} a z^{i_1} \ldots a z^{i_r} = z^{j_o} a z^{j_1} \ldots a z^{j_s}$$

is a nontrivial relation, with $r,s \geq 0$, $i_1,\ldots,i_r,j_1,\ldots,j_s \geq 1$, and $i_o, j_o \geq 0$. Rewriting this relation as

$$a^{\zeta^{i_o}+\zeta^{i_o+i_1}+\ \ldots\ +\zeta^{i_o+\ldots+i_{r-1}}}\, z^{\Sigma_{l=0}^{r} i_l}$$

$$= a^{\zeta^{j_o}+\zeta^{j_o+j_1}+\ \ldots\ +\zeta^{j_o+\ldots+j_{s-1}}}\, z^{\Sigma_{l=0}^{s} j_l}$$

we see that $\Sigma_{l=0}^{r} i_l = \Sigma_{l=0}^{s} j_l$. Using the fact that $a$ generates $A \otimes_Z Q$ as a $K$-vector space, we further deduce that

$$\zeta^{i_o}+\zeta^{i_o+i_1}+\ \ldots\ +\zeta^{i_o+\ldots+i_{r-1}} = \zeta^{j_o}+\zeta^{j_o+j_1}+\ \ldots\ +\zeta^{j_o+\ldots+j_{s-1}}.$$

Since $(i_o,i_1,\ldots,i_r) \neq (j_o,j_1,\ldots,j_s)$, it follows from the above that $\zeta$ satisfies a nontrivial polynomial $f(X) = \Sigma_{l=0}^{n} r_l X^l$ with coefficients $r_l \in \{0,\pm 1\}$, $r_n \neq 0$. Therefore, in $C$ we have $0 = \Sigma_{l=0}^{n} r_l(\zeta^\sigma)^l$ and so

$$|\zeta^\sigma|^n \leq \Sigma_{i=0}^{n-1} |\zeta^\sigma|^i = \frac{|\zeta^\sigma|^n - 1}{|\zeta^\sigma| - 1}$$

which contradicts $|\zeta^\sigma| > 2$.

Step 3. The canonical $k$-algebra map $kH \to D$ given by the inclusion $H \subset D^*$ is an embedding.
Proof. Let $I$ denote the kernel of this map and suppose that $I$ is nonzero. Then $I$ is completely prime and $I \cap kA$ is also nonzero. For, every $\alpha \in kH$ can be uniquely expressed as $\alpha = \Sigma_{i=p}^{q} \alpha_i z^i$ with $\alpha_i \in kA$. Choose $0 \neq \alpha \in I$ of minimal length $q-p$ and with $p=0$. If $q \neq 0$, then $\beta^{z^{-q}} \neq \beta$ for some $\beta \in kA$, and

$$\beta\alpha - \alpha\beta = \Sigma_{i=1}^{q} \alpha_i(\beta - \beta^{z^{-i}}) z^i \in I$$

is nonzero and shorter than $\alpha$. Therefore, $q = 0$ and hence $\alpha \in I \cap kA$. Since $I \cap kA$ is z-stable, Theorem 2.1 implies that $F = kA/I \cap kA$ is a finite dimensional field extension of k, isomorphic to the k-subalgebra of D generated by A. Now z acts on F by k-automorphisms and so some power of z must act trivially on F, hence on A. However, this contradicts our construction of H so that I must be zero.

The theorem now follows, because Step 2 shows that kH contains a free k-algebra and hence so does D, by Step 3. ∎

Corollary 2.4. Let $D = k(G)$ be a division k-algebra with $G \leq D^*$ polycyclic-by-finite. Then D has finite Gelfand-Kirillov dimension over k if and only if G is abelian-by-finite.

## 3. MISCELLANY

### A) COMMUTATIVE SUBFIELDS

The following lemma is extracted from [20]. Here, Kdim denotes (Rentschler-Gabriel-) Krull dimension.

Lemma 3.1 (Resco, Small, Wadsworth). Let A be an absolutely Noetherian k-algebra (i.e., $A \otimes_k K$ is Noetherian for all field extensions K/k) and let $R = AS^{-1}$ be the localization of A with respect to a right Ore set $S \subset A$. Then every commutative subfield L of R with $k \subset L$ is finitely generated over k. Moreover, if $\mathrm{Kdim}(A \otimes_k K) \leq n$ for all field extensions K/k, then $\mathrm{tdeg}_k L \leq n$.

Proof. $R \otimes_k L$ is obtained by localizing $A \otimes_k L$ with respect to $S \otimes_k 1$, and hence $R \otimes_k L$ is Noetherian as $A \otimes_k L$ is. Moreover, $R \otimes_k L$ is free as a module over $L \otimes_k L$ and so $L \otimes_k L$ must also be Noetherian. By 28 , L is finitely generated over k. Finally, $\mathrm{tdeg}_k L = \mathrm{Kdim}(L \otimes_k L) \leq \mathrm{Kdim}(R \otimes_k L) \leq \mathrm{Kdim}(A \otimes_k L)$, where the first inequality follows from the freeness of $R \otimes_k L$ over $L \otimes_k L$ and the second holds, since $R \otimes_k L$ is a localization of $A \otimes_k L$. ∎

The lemma applies to the case where $A = kG/I$ is an image of a group algebra kG with G polycyclic-by-finite. In this case, by [25], an upper bound for the Krull dimensions of $A \otimes_k K$ is given by the Hirsch number h(G) of G. In particular, we have the following

Corollary 3.2. If $D = k(G)$ is a division k-algebra generated by some polycyclic-by-finite group G, then all commutative subfields

$L \supset k$ of $D$ are finitely generated over $k$, and $\mathrm{tdeg}_k L \leq h(G)$.

In general, $h(G)$ is a very crude bound. For example, if $\lambda \in k^*$ has infinite order, then $B_\lambda(k) \otimes_k K = B_\lambda(K)$ has Krull dimension 1, by Lemma 1.1. Therefore, Lemma 3.1 shows that commutative subfields of $E_\lambda(k)$ have transcendence degree at most 1 over $k$, whereas any division algebra $D = k(G)$ with $h(G) = 1$ is finite dimensional over its center. R. Resco has conjectured that if $G$ is torsion-free polycyclic-by-finite and $D = Q(kG)$ is the division ring of fractions of $kG$, which is a domain by [4],[6], then all commutative subfields $L \supset k$ of $D$ have transcendence degree bounded by

$$c(G) = \max \{h(A) \mid A \text{ an abelian subgroup of } G\}.$$

We cannot prove this, but the following discussion should shed some light on this problem. For the rest of this subsection, we keep the following notation:

> $G$ is torsion-free polycyclic-by-finite, and
> $D$ is the division ring of fractions of the group algebra $kG$.
> For any subgroup $H \leq G$ we set $D_G(H) = \{g \in G \mid g$ has only finitely many $H$-conjugates$\}$, and $C_R(H)$ denotes the centralizer of $H$ in $R$ (for given $R$).

The main content of the following lemma is due to M. Smith [24].

Lemma 3.3. For any subgroup $H$ of $G$, $C_D(H) = Q(C_{kG}(H)) \subset Q(kD_G(H))$ and $Q(kD_G(H))$ is finite dimensional over $C_D(H)$.

Proof. Set $R = kD_G(H)$ and note that $C_{kG}(H) \subset R$. More precisely, the action of $H$ on $R$ by conjugation factors through some finite image $\bar{H}$ of $H$, and $C_{kG}(H)$ is the fixed subring of $R$ under this action. Therefore, since $R$ is a Noetherian domain, $C_{kG}(H)$ is a Goldie domain and $Q(C_{kG}(H))$ is the fixed subring of $Q(R)$ under the action of $\bar{H}$ ([17, Theorem 5.5], e.g.). Also, $Q(R)$ is finite dimensional over $Q(C_{kG}(H))$ ([17, Lemma 2.18]). Finally, The proof of [24, Theorem 6] shows that $C_D(H) \subset Q(R)$. Hence, clearly, $C_D(H) = C_{Q(R)}(H) = Q(C_{kG}(H))$ and the lemma is proved. ∎

We will call a commutative subfield $L \supset k$ of $D$ almost maximal if $L$ is not contained in a commutative subfield of $D$ having larger transcendence degree over $k$ than $L$ or, equivalently, if $C_D(L)/L$ is algebraic. The following is immediate from Lemma 3.3.

Corollary 3.4. Let $A$ be an abelian subgroup of $G$. If $A = D_G(A)$ ($A$ has finite index in $D_G(A)$), then $k(A) = Q(kA)$ is a maximal (resp., almost maximal) commutative subfield of $D$.

We conclude this subsection by mentioning a few instances where the corollary applies.

Examples 3.5. (a) Suppose that $A$ is a maximal abelian subgroup of $G$ and $A$ is subnormal in $D_G(A)$. Then we do in fact have equality, $A = D_G(A)$, and so $k(A)$ is maximal. To see this, choose a subnormal series $A = D_n \lhd D_{n-1} \lhd \ldots \lhd D_o = D_G(A)$. Then any n-fold commutator $[a_n,[a_{n-1},\ldots,[a_1,d]]\ldots]$ with $a_i \in A$, $d \in D_o$ belongs to $A$. Consider an (n-1)-fold commutator $c = [a_{n-1},\ldots,[a_1,d]\ldots]$ . Let $a \in A$ be arbitrary and choose $m$ so that $a^m$ is central in $D_o$. Then, since $A$ is abelian and $[a,c] \in A$, we have

$$1 = [a^m,c] = [a,c]^m,$$

and hence $[a,c] = 1$. Therefore, $c \in C_G(A) = A$. By induction, we obtain $D_o = A$ , as we have claimed.

Since subnormality is automatic if $G$ is nilpotent, we recover M. Smith's result [24, Corollary 8]. In general, however, maximal abelian subgroups $A$ of $G$ need not satisfy $A = D_G(A)$ (e.g., take $G = \langle x,y \mid y^{-1}xy = x^{-1}\rangle$ and $A = \langle y\rangle$) and so $k(A)$ need not be maximal.

(b) If $A \leq G$ is abelian and satisfies $h(A) = c(G)$, then it is trivial to verify that $A$ has finite index in $D_G(A)$. Thus $k(A)$ is at least almost maximal in this case.

## B) LIE COMMUTATORS

Let $D = k(G)$ be a division k-algebra generated by some polycyclic-by-finite group $G$ and assume that char $k = 0$. It would be interesting to know whether the identity element $1 \in D$ can be written as a sum of Lie commutators in $D$. If not, then this fact would distinguish division algebras of the above type from division algebras $E$ generated by finite dimensional Lie subalgebras of $E_{[\ ,\ ]}$ (i.e., $E$ with Lie bracket $[a,b] = ab - ba$), at least if $k$ is algebraically closed. This follows from the following simple observation.

Lemma 3.6. Let $E$ be a division algebra over $k$. If $E_{[\ ,\ ]}$ contains a non-abelian nilpotent Lie algebra, or a non-abelian finite dimensional Lie algebra over $k$ and $k$ is algebraically closed, then there exist elements $a,b \in E$ with $ab - ba = 1$.

Proof. Suppose $\underline{g} \subset E_{[\ ,\ ]}$ is a nilpotent Lie algebra which is not commutative. Then there exists an element $c \in \underline{g}$ such that $[c,\underline{g}]$ is nonzero and is contained in the center of $\underline{g}$. Choose $b \in \underline{g}$ with $[c,b] = cb - bc \neq 0$ and set $a = c[c,b]^{-1} \in E$. Then

$[a,b] = 1$ in $E$.

If, on the other hand, $\underline{g} \subset E_{[\ ,\ ]}$ is finite dimensional non-nilpotent, then Engel's theorem [8, Sec. 3.2] implies that there exists $a \in \underline{g}$ such that $\mathrm{ad}(a) \in \mathrm{End}(\underline{g})$ is not nilpotent. Let $0 \neq c \in \underline{g}$ be an eigenvector for $\mathrm{ad}(a)$ with nonzero eigenvalue $\gamma \in k$ and set $b = \gamma^{-1}c \in E$. Then $[ac^{-1},b] = 1$ in $E$. ∎

Algebraic closure of $k$ is definitely required in the above. For example, the standard basis $\{1,i,j,k\}$ of the real quaternions $H$ spans a Lie subalgebra of $H$, but $1 \in H$ is not even a sum of Lie commutators (use the embedding $H \subset M_2(C)$, or the following proposition).

We now return to division algebras generated by polycyclic-by-finite groups. The following result extends, and uses, [11, Lemma 2.3].

<u>Proposition 3.7.</u> Let $G$ be a finitely generated nilpotent-by-finite group and let char $k = 0$. Let $P$ be a prime ideal of $kG$ and set $R = Q(kG/P)$. Then $1 \notin [R,R]$, the space of Lie commutators in $R$.

<u>Proof.</u> For $G$ finitely generated nilpotent, this follows from [11, Lemma 2.3]. In general, choose $G_o$ to be a nilpotent normal subgroup of finite index in $G$. Then $P \cap kG_o$ is a finite intersection of pairwise incomparable prime ideals $P_i$, $i=1,2,\ldots,l$, of $kG_o$. Moreover, it is routine to check that the Ore set $S$ of regular elements of $kG_o/P \cap kG_o$ remains Ore and regular in $kG/P$. Therefore,

$$R = (kG/P)S^{-1} = \bigoplus_x xR_o ,$$

where $R_o = (kG_o/P \cap kG_o)S^{-1}$ and $x$ runs through a transversal for $G_o$ in $G$. Now $R_o$ is the direct product of the rings $R_i = Q(kG_o/P_i)$, $i=1,2,\ldots,l$, and so

$$R \subset \mathrm{End}(R_{R_o}) \cong M_n(R_o) \cong \prod_{i=1}^{l} M_n(R_i),$$

where $n$ is the order of $G/G_o$. Thus it suffices to establish the assertion for the matrix rings $M_n(R_i)$. But we know it is true for each $R_i$. Hence the canonical map of k-spaces $T: R_i \to R_i/[R_i,R_i]$ does not vanish on $1$. $T$ can be lifted to a map $T': M_n(R_i) \to R_i/[R_i,R_i]$ by setting $T'([r_{st}]) = \Sigma_{s=1}^{n} T(r_{ss})$. Since $T'$ inherits k-linearity and the trace property $T'(AB) = T'(BA)$ from $T$ and maps the identity $1 \in M_n(R_i)$ to the nonzero element $n \cdot T(1)$, we conclude that $1 \notin [M_n(R_i), M_n(R_i)]$ as required. ∎

The proposition applies in particular to division algebras $D = k(G)$ generated by nilpotent-by-finite groups $G$. Sometimes the char 0 assumption is superfluous here. For example, if $D = k(G)$ with $G$ torsion-free nilpotent, then $1 \notin [D,D]$ holds in any characteristic. This follows from [11, Sec. 2], where explicit traces are constructed for so-called Hilbert-Neumann algebras. The same construction also applies to the division rings $E_\lambda(k)$ and, more generally, to the division rings of fractions of twisted group algebras $k^t[G]$ with $G$ an ordered group. For general polycyclic-by-finite groups, however, char 0 is definitely needed. For example, consider the Weyl algebra $A_1 = k\{x,y\}$, $xy - yx = 1$, with char k $= p > 0$. Then $D_1 = Q(A_1)$ is generated, as division algebra, by the elements $a = xy$ and $x$ which satisfy $x^{-1}ax = a - 1$. Therefore, the generating group $G = \langle a,x \rangle$ is polycyclic and is in fact isomorphic to a semidirect product of the form $Z^{(p)} \rtimes Z$. On the other hand, R. Snider [26] has shown that if char k = 0 and $G \cong Z^{(r)} \rtimes Z$ for some r, then $D = Q(kG)$ satisfies $1 \notin [D,D]$.

ACKNOWLEDGEMENT

This work was supported by the Deutsche Forschungsgemeinschaft, Grant No. Lo 261/2-1.

REFERENCES

1. H. Bass, "The degree of polynomial growth of finitely generated nilpotent groups", Proc. London Math. Soc. (3) 25 (1972) 603-614.

1a. G. M. Bergman, "The logarithmic limit set of an algebraic variety", Trans. AMS 157 (1971) 459-469.

2. K. A. Brown, "Remarks on polycyclic group algebras", preprint, Univ. of Glasgow, 1982.

3. C. Chevalley, "L'arithmétique dans les algèbres de matrices", Actualités Sci. Indust. No. 323, Paris, 1936.

4. G. H. Cliff, "Zero divisors and idempotents in group rings", Can. J. Math. 23 (1980) 596-602.

5. D. R. Farkas, "Noetherian group rings: an exercise in creating folklore and intuition", preprint, Virginia Polytechnic Inst. and State Univ., 1983.

6. D. R. Farkas and R. L. Snider, "$K_o$ and Noetherian group rings", J. Algebra 42 (1976) 192-198.

7. D. Hilbert, "Grundlagen der Geometrie", Teubner, Leipzig, 1899.

8. J. E. Humphreys, "Introduction to Lie Algebras and Representation Theory", Springer, New York, 1972.

9. K. Ireland and M. Rosen, "A Classical Introduction to Modern Number Theory", Springer, New York, 1982.

10. V. A. Jategaonkar, "Multiplicative analog of Weyl algebras", Notices AMS 23(1976)p.A-566.

11. M. Lorenz, "Division algebras generated by finitely generated nilpotent groups", to appear in J. Algebra.

12. M. Lorenz, "On the transcendence degree of group algebras of nilpotent groups", to appear in Glasgow Math. J.

13. M. Lorenz, "Prime ideals in group algebras of polycyclic-by-finite groups: vertices and sources", in: Lect. Notes in Math. No. 867, Springer, New York, 1981.

14. M. Lorenz and D. S. Passman, "Prime ideals in group algebras of polycyclic-by-finite groups", Proc. London Math. Soc. (3)43 (1981)520-543.

15. L. Makar-Limanov, "On group rings of nilpotent groups", preprint, Wayne State Univ., 1983.

16. J. C. McConnell and J. C. Robson, "Homomorphisms and extensions of modules over certain differential operator rings", J. Algebra 26(1973)319-342.

17. M. S. Montgomery, "Fixed Rings of Finite Automorphism Groups of Associative Rings", Lect. Notes in Math. No. 818, Springer, New York, 1980.

18. D. S. Passman, "The Algebraic Structure of Group Rings", Wiley-Interscience, New York, 1977.

19. R. Resco, "Transcendental division algebras and simple Noetherian rings", Israel J. Math. 32(1979)236-256.

20. R. Resco, L. W. Small, and A. Wadsworth, "Tensor products of division rings and finite generation of subfields", Proc. AMS 77(1979)7-10.

21. P. Revoy, "Algèbres de Weyl en caracteristique p", C. R. Acad. Sc. Paris (Série A) 276(1973)225-228.

22. J. E. Roseblade, "Prime ideals in group rings of polycyclic groups", Proc. London Math. Soc. (3)36(1978)385-447.

23. J. M. Rosenblatt, "Invariant measures and growth conditions", Trans AMS 193(1974)33-53.

24. M. K. Smith, "Centralizers in rings of quotients of group rings", J. Algebra 25(1973)158-164.

25. P. F. Smith, "On the dimension of group rings", Proc. London Math. Soc. (3)25(1972)288-302.

26. R. L. Snider, "The division ring of fractions of a group ring", to appear in Proc. Sém. Malliavin, Lect. Notes in Math., Springer.

27. J. T. Stafford, "Stably free, projective right ideals", preprint, Univ. of Leeds, 1982.

28. P. Vamos, "On the minimal primes of a tensor product of fields", Math. Proc. Cambridge Philos. Soc. 84(1978)25-35.

29. D. B. Webber, "Ideals and modules of simple Noetherian hereditary rings", J. Algebra 16(1970)239-242.

30. A. E. Zalesskii, "Irreducible representations of finitely generated nilpotent torsion-free groups", Math. Notes 9 (1971)117-123.

NOTE ADDED IN PROOF (Jan. 2, 1984). I learned from John McConnell that a student of his, Julian Pettit, has independently obtained Lemma 1.1 and essentially all of Proposition 1.4. These results, together with further results concerning the algebras $B_\lambda$ and generalizations thereof, are part of this thesis (Leeds). Also, V. Jategaonkar has written up her announcement [10] for publication.

# ON FREE SUBOBJECTS OF SKEW FIELDS

Leonid Makar-Limanov

Wayne State University
Department of Mathematics
Detroit, Michigan 48202

In my talk I am going to discuss several questions, which loosely can be stated as follows: does this or that free object appear in a skew field which satisfies certain conditions? Free objects which I have in mind are: free semigroup, free group and free algebra with two generators. It is not reasonable to consider a bigger number of generators, because as is well known, every free object mentioned contains the corresponding free object on a countable number of generators. On the other hand in the case of one generator every object is commutative, and these questions have been considered. It is well known that every skew field which does not coincide with its center contains a free subgroup with one generator (in its multiplicative group). The question about one-generator subalgebras constitutes the famous Kurosh problem which is still very far from the solution. However, it seems more appropriate to me to consider noncommutative subobjects in the skew field setting.

During the talk $T$ is the skew field under consideration, $Z$ is its center and it is assumed that $T \neq Z$. Any further assumptions will be stated in appropriate places.

Let us start with semigroups. Here under the assumption that $Z$ is uncountable I can prove that the multiplicative group $T^*$ of $T$ contains a free subsemigroup (with two generators). Though frankly speaking I think that this assumption is not essential for the statement, it is very essential for the proof which I am going to outline.

Lemma 1. Suppose $T$ has no such subsemigroup. Then for any pair of elements $x,y$ from $T$ there exists a semigroup



*F. van Oystaeyen (ed.), Methods in Ring Theory, 281–285.*

relation $A(a,b) = B(a,b)$ such that this relation is valid for every pair $x+c$, $y+dx$ where $c,d \in Z$ and that $\deg_a A = \deg_a B$, $\deg_b A = \deg_b B$.

Lemma 2. If the relation $A(a,b) = B(a,b)$ is nontrivial and satisfies the conclusion of Lemma 1 for a pair $x,y \in T$ then there exists some $n = n(x,y)$ such that $(ad_x)^n y = 0$, where $ad_x y = [x,y] = xy - yx$.

Lemma 3. The skew field D which is generated over Z by elements p and q such that $pq - qp = 1$ contains a free semigroup with two generators.

The main lemma here is lemma 1 in the proof of which is essential that Z is uncountable. We have only countably many possible semigroup relations. So for any fixed c (or d) there exists a relation which is valid for infinitely many values of d. But then this relation should be satisfied by any d, basically because a polynomial in a central variable over any ring can have infinitely many central roots only if it is identically zero. Then one can deduce as above that there exists a relation which is satisfied by any c and d. In the proof of lemma 2 it is sufficient to assume that Z is infinite. Finally, lemma 3 is correct for any center, though in the case of T having zero characteristic it is trivial (it is easy to see that p and pq generate free semigroup) and in the case of nonzero characteristic I have only a rather complicated proof. You will see this proof in the discussion of our next free subobject, namely the free group.

In this case I again can give proof of existence under a restriction which seems rather unnatural to me, though again this restriction is crucial for the proof I have. The restriction I am talking about is the assumption that T is finite dimensional over Z. In this setting it is true that T contains a free subgroup with two generators in its multiplicative subgroup. It is possible to prove that also every normal noncentral subgroup H of $T^*$ contains such a subgroup. Let me outline the proof that I have. It is clear that $T^*$ can be represented by finite dimensional matrices. So one can use the following well known results for linear groups: (1) Tits' alternative: every linear group either contains a free subgroup (with two generators) or is a locally finite extension of a solvable group. In the case of zero characteristic the extension must be finite. (2) The result of Zassenhaus that any solvable matrix group has bounded solvable length, where the bound depends on the dimension of the matrices only and whence any matrix group has a uniquely defined maximal solvable normal subgroup, the <u>Zassenhaus subgroup</u>. I will also use a result of W. Scott that a normal solvable subgroup of the multiplicative group of the skew field belongs to its center and a result of Kaplansky that a skew field which is locally finite over its center is a field. Now the proof that $T^*$ contains free subgroup is a simple combination of these results. Let us assume that $T^*$ does not contain a free subgroup. Then by Tits' result it contains a "big" solvable normal subgroup. But the Zassenhaus subgroup $H$ of $T^*$ is even bigger, so $T^*$ is locally finite over $H$ also. Now by Scott's result any solvable $H$ belongs to the center and according to Kaplansky $T$, which is locally finite over $Z^*$, is commutative. This brings us to a contradiction with the original assumption that $T$ is not commutative and so $T^*$ contains the desired subgroup. To obtain the result for normal subgroups one can add to the results mentioned, the theorem of Cartan-Brauer-Hua on skew subfields which are invariant under all inner automorphisms in the case of zero characteristic, the Wedderburn theorem on finite skew fields, and the Baer-Neumann theorem that a group of finite index over its center has finite commutator subgroup in the case of positive characteristic. I would like to mention that more or less all these ingredients can be extracted from the works of A. Lichtman, in Doklady 153(1963), 1425-1429 and in the Proceedings of the A.M.S. 71(1978), 174-178. Also I am rather unhappy with this proof because it depends so heavily on Tits' and Zassenhaus' results, for which it is not essential that $T$ be a skew field. Anyway, the result seems to be rather nice. Now the promised explanation about the semigroup in $D$ (from Lemma 3) for positive characteristic. In this case $D$ is finite dimensional over its center, so it contains a free group and so a semigroup.

The last free object is the free algebra. Here I have no general statements at all but only two examples of relatively small skew fields (at least they look so) which are infinite dimensional over their centers. These skew fields are the skew field $D$, and the skew field $K(G)$ of fractions of the group algebra $K[G]$ of the nilpotent group $G$ with two generators $a$ and $b$ for which $(a,b) = aba^{-1}b^{-1} = c$ and $(a,c) = (b,c) = 1$. It is possible to prove that elements $(pq)^{-1}$ and $(pq)^{-1}(1-p)^{-1}$ and elements $(1-a)^{-1}$ and $(1-a)^{-1}(1-b)^{-1}$ generate free subalgebras in the corresponding skew fields. Proofs are very similar in both cases and based on the possibility of embedding the skew fields $D$ and $K(G)$ in some bigger skew fields which are much more suitable for computations.

These skew fields consist of infinite Laurent-type series $\sum_{i=k}^{\infty} p^i r_i(t)$ where $r_i(t)$ are rational functions of $t = pq$ over $Z$ (center of $D$), and of $\sum_{i=k}^{\infty} b^i r_i(a,c)$ where $r_i$ are rational functions of $a$ and $c$ over $K$. Multiplication is given by formulas

$$\sum_i p^i r_i(t) \sum_j p^j s_j(t) = \sum_n p^n \sum_{i+j=n} r_i(t-j)s_j(t) .$$

and

$$\Sigma b^i r_i(a) \cdot \Sigma b^j s_j(a) = \Sigma b^{i+j} r_i(c^j a)s_j(a)$$

correspondingly. Now using the identities $(1-p)^{-1} = 1+p+p^2+\ldots$ and $(1-b)^{-1} = 1+b+b^2+\ldots$ in these skew fields one can check that the set of monomials spanned by the previously specified elements are linearly independent and hence the corresponding algebras are free.

Let me finish now with the following FOFS (full of free subobjects) conjecture:

(1) every noncommutative skew field contains a noncommutative free (a) semigroup, (b) group

(2) every finitely generated and infinite dimensional (over its center) skew field contains a free subalgebra with two generators.

Most probably part (2) is very difficult in such generality because it contains as very special case the Kurosh conjecture. Nevertheless one can try to investigate some concrete skew fields where it is also quite a challenge.

## Bibliography

The theorem on semigroups will appear (eventually) in the Proceedings of the AMS.

The theorem on free subalgebra in D is published in Communications in Algebra, 11(17), 2003-2006, (1983). All (p)reprints are available.

For those who are interested in free subgroups I would recommend also the work of A. Lichtman, "On normal subgroups of multiplicative group of skew fields generated by a polycyclic-by-finite group," J. Algebra 78(1982), 548-577, and his preprint "On matrix rings and linear groups over a field of fractions of enveloping algebras and group rings".

The results which are used for free subgroups appear in:

J. Tits, "Free subgroups of linear groups", J. Algebra 20(1972), 250-270.

H. Zassenhaus, "Beweis eines Satzes uber diskrete Gruppen", Abh. Math. Sem., Univ. Hamburg, 12(1938), 289-312.

(See also J. Dixon, "The solvable length of a solvable linear group", Math. Z., 107(1968), 151-158.)

W.R. Scott, "On the multiplicative group of a division ring", Proc. Amer. Math. Soc., 8(1957), 303-305.

I. Kaplansky, "A theorem on division rings", Canadian J. Math., 3(1951), 290-292.

J. Erdos, "The theory of groups with finite classes of conjugate elements", Acta Math. Acad. Sci. Hungar., 5(1954), 45-58. (Proof of the Baer-Neumann theorem.)

All other results can be found in the book of:
N. Jacobson, "Structure of Rings", Amer. Math. Soc. Coll. Publ. 37, AMS Providence, 1956.

# HERSTEIN'S LIE AND JORDAN THEORY REVISITED

W. S. Martindale, 3rd[1], and C. R. Miers[2]

[1]University of Massachusetts, Amherst, [2]University of Victoria, B. C.

## ABSTRACT

A semiprime nonassociative ring $A$ is strongly semiprime if for all $n$, given $A \supseteq U_1 \supseteq U_2 \supseteq \dots \supseteq U_n \neq 0$ with $U_i$ an ideal of $U_{i-1}$, there exists an ideal $W$ of $A$ such that $0 \neq W \subseteq U_n$. Let $R$ be a 2-torsion free semiprime associative ring with involution, $S$ be the symmetric elements, and $K$ be the skew elements. Theorem: $K/Z_K$ is a strongly semiprime Lie ring, where $Z_K = \{x \in K \mid [x,K] = 0\}$. In the course of the proof the entire Herstein Lie theory is reworked, with all questions being reduced via GPI theory to $M_n(F)$, $F$ an algebraically closed field, under transpose or symplectic involution. As a mild application, the prime radical of $T^G$ is determined, where $G$ is a finite group of certain kinds of Lie automorphisms of an arbitrary associative ring $T$. Finally, the Jordan analogue is obtained. Theorem: $S$ is a strongly semiprime Jordan ring.

The results we will describe stem from a joint effort with C. R. Miers. A complete list of theorems and full details of the proofs are to be found in a paper which (at this writing) has just been submitted for publication.

An associative ring $R$ becomes a Jordan ring $R^+$ under $x \circ y = xy + yx$ and a Lie ring $R^-$ under $[x,y] = xy - yx$. In case $R$ has an involution $*$ the set of symmetric elements $S = \{s \in R \mid s^* = s\}$ is a Jordan subring of $R^+$ and the set of skew elements $K = \{k \in R \mid k^* = -k\}$ is a Lie subring of $R^-$.



F. van Oystaeyen (ed.), Methods in Ring Theory, 287–289.

Throughout we assume that R is 2-torsion free. We are primarily interested in the situation where R is a semiprime ring with involution. Our main goals are to show that the Lie ring $\hat{K} = K/Z_K$ (where $Z_K = \{x \in K | [x,K] = 0\}$ and the Jordan ring S are <u>strongly semiprime</u> (in a sense to be made precise later on).

Part of our motivation is to redo the Herstein Lie theory from a somewhat new point of view. For R any ring with involution * and I any *-ideal of R we call $[I \cap K,K]$ a <u>standard Lie ideal</u> of K. A major object of the theory is, given a "nontrivial" Lie ideal U of K (i.e., $[U,K] \neq 0$, written $U \neq 0$), to produce a nonzero standard Lie ideal $[I_U \cap K,K]$ lying in U. In the matter of coming up with a suitable *-ideal $I_U$ we essentially use Herstein's formal construction, i.e., the ideal $I_U = R[U \circ U,U]^2R$ is such that $[I_U \cap K,K]$ always lies in U. With R now assumed to be semiprime and $0 \neq U$ a Lie ideal of K the major problem lies in showing that $I_U \neq 0$. We shall call U <u>exceptional</u> if $I_U = 0$. If $0 \neq U$ is exceptional then in some *-prime image $R_\alpha$ $U_\alpha$ remains a nontrivial exceptional ideal of K, and so without loss of generality we may assume R is *-prime. Following [1] we may embed R in $\tilde{R} = RC_* \otimes_{C_*} F$, where $C_*$ is the field of symmetric elements of the extended centroid C of R and F is the algebraic closure of $C_*$. The involution on R extends to one on $\tilde{R}$ via $r \otimes \lambda \longrightarrow r^* \otimes \lambda$, the skew elements of $\tilde{R}$ are $\tilde{K} = K \otimes F$, and $\tilde{U} = U \otimes F$ remains a nontrivial exceptional Lie ideal of $\tilde{K}$. Thus without loss of generality we may assume R is a *-prime *-closed ring over an algebraically closed field F (see [1]). The more interesting case at this point is when R is prime. We claim the socle of R is nonzero, for otherwise, picking a U such that $a^2 \notin F$, we obtain a contradiction in the form of a nontrivial GPI for K over F: $[[a,k_1]^2,[a,k_2]][[a,k_3]^2,[a,k_4]] = 0$. Now, with socle nonzero, an involution version of Litoff's Theorem reduces the problem to the classical matrix ring $M_n(F)$, where the involution is either transpose (class BD) or symplectic (class C). In easier fashion the case where R is *-prime but not prime reduces to the ring $M_n(F) \otimes M_n(F)$, with interchange involution (class A). Simple matrix calculations then show nontrivial exceptional Lie ideals of K occur precisely when $n = 2$ in class A, $n = 4$ in class BD, and $n = 2$ in class C.

A second motivating factor for our work is a desire to organize the theory in such a way that problems involving K and [K,K] can be dealt with in a more unified manner than has sometimes been the case in the past. A Lie ideal V of [K,K] is thus a <u>subideal</u> of K in the sense that $V \lhd U \lhd K$. (The symbol

$\lhd$ reads "is an ideal of"). This leads to the notion of an n-subideal $U_n$ in an arbitrary nonassociative ring A: there exists a chain $A \rhd U_1 \rhd U_2 \rhd \ldots \rhd U_n$. We then define a nonassociative ring A to be <u>strongly semiprime</u> if (1) A is semiprime and (2) every nonzero n-subideal $U_n$ of A contains a nonzero ideal W of A. It is easily shown that in (2) W may be taken to be essential in $U_n$. Of course, to say that A is <u>strongly prime</u> means changing (1) to read that A is prime.

We return now to the Lie theory in semiprime rings with involution. The analysis of nontrivial exceptional Lie ideals of K which we have outlined earlier (together with a similar treatment of exceptional subideals of K) enables us to establish our main result:

THEOREM

If R is a 2-torsion free semiprime associative ring with involution, then $\hat{K} = K/Z_K$ is a strongly semiprime Lie ring.

It is an immediate consequence that $\widehat{[K,K]} = [K,K]/Z_{[K,K]}$ is also strongly semiprime.

A similar theorem for *-prime rings also holds:

THEOREM

If R is a 2-torsion free *-prime ring with involution, then $\hat{K}$ is a strongly prime Lie ring unless $\tilde{R} = RC_* \otimes_{C_*} F$ is $M_4(F)$ with transpose involution.

Finally, turning to the Jordan theory, we are able to obtain the analogous result for S in a considerably easier fashion:

THEOREM

If R is a 2-torsion free semiprime associative ring with involution, then S is a strongly semiprime Jordan ring.

In fact we are able to show, given a nonzero n-subideal $U_n$ of S and letting $I_n(U)$ be the ideal of R generated by $\{u^{4^{2n-1}} \mid u \in U\}$, that $4^{8^{n-1}} I_n(U) \cap S$ is essential in U.

REFERENCE

[1] W. E. Baxter and W. S. Martindale, 3rd, "The extended centroid in *-prime rings," Com. in Algebra 10 (1982), pp. 847-874.

# DIVISORIALLY GRADED ORDERS IN A SIMPLE ARTINIAN RING

Hidetoshi MARUBAYASHI

Osaka University

Toyonaka, Osaka, Japan

ABSTRACT The concepts of graded Krull orders and graded v-HC orders are defined in divisorially graded orders, and it is shown that they are Krull orders and v-HC orders, respectively, as usual orders if the orders have enough graded v-invertible ideals.

## 0. INTRODUCTION

This paper is a continuation of [11]. In [11], we studied the structure of a strongly graded order $R = \Sigma \oplus R_n$ $(n \in Z)$, where $R_o$ is a prime Goldie ring and Z is the ring of integers. If $R_o$ is, for example, a Dedekind ring or a hereditary ring, then "strongly" is natural and strong enough to study the relations between $R_o$ and R. But if $R_o$ is a Krull order (in more general, a v-HC order), then it seems to me that "strongly graded" is not natural and strong. In fact, we can construct some examples of graded orders which are not strongly graded and very interesting in arithmetical ideal theory. So we introduce, in this paper, the notion of divisorially graded orders which is a Krull type generalization of strongly graded orders (see [4], for divisorially G-graded orders in which the idempotent kernel functor defining "the divisoriality" is



*F. van Oystaeyen (ed.), Methods in Ring Theory, 291–315.*

symmetric, where G is a group). After giving the definition of divisorially graded orders, we present, in §1, some elementary properties of such orders which are used in §§2 and 3.

In §§2 and 3, graded Krull orders and graded v-HC orders are defined in analogy with ungraded cases, respectively. It is shown , in §2, that graded Krull orders are Krull orders as ungraded orders. With help of this result, it is established, in §3, that a graded v-HC order with enough graded v-invertible ideals is a v-HC order with enough v-invertible ideals. It is also shown in §3 that if $R_o$ is a v-HC order with enough v-invertible ideals, then so is $R = \Sigma \oplus R_n$, a divisorially graded order. This was first proved in [11] in case R is strongly graded and $R_o$ is an HNP ring with enough invertible ideals. And, in [4], they obtained that if $R_o$ is a tame order and R is a normalizing Rees ring satisfying a polynomial identity, then R is a tame order by using the concept of relative maximal orders (note that "tame orders" are just "v-HC orders" in case of rings satisfying a polynomial identity, see [9]) . In §4, we briefly discuss on the divisor class group of a graded v-HC order with enough graded v-invertible ideals and on generalized Rees rings. In §5, we present some properties of graded v-ideals used in this paper without proofs. Concerning our terminology we refer to [7] and [12].

## 1. ELEMENTARY PROPERTIES OF DIVISORIALLY GRADED ORDERS

Let $R = \Sigma \oplus R_n$ $(n \in Z)$ be a Z-graded ring, i.e., $R_n$ is an additive group satisfying the following

(*) $$R_n R_m \subset R_{n+m} \text{ for any } n, m \in Z,$$

where Z is the ring of integers, $R_o$ is a prime Goldie ring with 1 and $Q_o$ is the quotient ring of $R_o$. Define

$F(\sigma) = \{ I_o : \text{right ideal} \mid \mathrm{Hom}_{R_o}(R_o/I_o, E(Q_o/R_o)) = 0\}$, and

$F(\sigma') = \{ J_o : \text{left ideal} \mid \mathrm{Hom}_{R_o}(R_o/J_o, E'(Q_o/R_o)) = 0\}$,

where $E(Q_o/R_o)$ $(E'(Q_o/R_o))$ is a right (left) injective hull of

$Q_o/R_o$. So $\sigma$ ($\sigma'$) stands for the idempotent kernel functor cogenerated by right (left) $R_o$-module $Q_o/R_o$. R is said to be <u>divisorially graded</u> if it satisfies the following two conditions:

(a) R is $(\sigma',\sigma)$-torsion-free.

(b) (*) induces $Q_\sigma(R_nR_m) = R_{n+m} = Q_{\sigma'}(R_nR_m)$ for any n, m $\in$ Z, where $Q_\sigma(M)$ ($Q_{\sigma'}(N)$) is the module of quotients of a right (left) module M (N) with respect to $\sigma$ ($\sigma'$). The concept "divisorially graded" was first defined in [4] in case $F(\sigma)$ contains a cofinal subset consisting of ideals. Throughout this paper, <u>R is always a divisorially graded order and $R_o$ is a prime Goldie ring with quotient ring $Q_o$</u>. In this section, we shall study some elementary properties of divisorially graded orders. We begin with the following lemma.

<u>Lemma 1.1</u>. (1) $R_{-n}R_n \in F(\sigma) \cap F(\sigma')$ for any $n \in Z$.
(2) $0 \neq r_nR_{-n}$ for any non zero $r_n \in R_n$.
(3) Any regular element in $R_o$ is regular in R.

<u>Proof</u>. (1) is clear from the fact that $Q_\sigma(R_{-n}R_n) = R_o = Q_{\sigma'}(R_{-n}R_n)$. (2) Assume that $r_nR_{-n} = 0$. Then $r_nR_{-n}R_n = 0$ and so $r_n = 0$, because $R_{-n}R_n \in F(\sigma)$ and R is $\sigma$-torsion-free. (3) Let $c_o$ be any regular element in $R_o$ and assume that $c_or_n = 0$ ($r_n \in R_n$). Then $c_or_nR_{-n} = 0$ implies that $r_nR_{-n} = 0$. Thus $r_n = 0$ by (2). Hence $c_o$ is regular in R.

For any right ideal I, we define the <u>$\sigma$-closure of I</u> as follows; $cl(I) = \{ r \in R \mid rH \subset I$ for some $H \in F(\sigma)\}$. If $I = cl(I)$, then it is called a <u>$\sigma$-closed right ideal</u>. We can define <u>$\sigma'$-closed left ideals</u> in a similar way.

<u>Lemma 1.2</u>. Let I be a graded right ideal of R. Then
(1) $cl(I) = cl(I_oR)$, where $I_o = I \cap R_o$.
(2) $cl(r_o^{-1}I) = r_o^{-1}cl(I)$ for any $r_o \in R_o$, where $r_o^{-1}I = \{ x \in R \mid$

$r_o x \in I\}$.
(3) $cl((r_o^{-1}I_o)R) = cl(r_o^{-1}(I_oR))$ for any $r_o \in R_o$, where $r_o^{-1}I_o = \{x_o \in R_o \mid r_o x_o \in I_o\}$.

Proof. (1) and (3) are easily proved by using Lemma 1.1. (2) follows from the implications: $z \in cl(r_o^{-1}I) \Longleftrightarrow r_o zH \subset I$ for some $H \in F(\sigma) \Longleftrightarrow r_o z \in cl(I) \Longleftrightarrow z \in r_o^{-1}cl(I)$.

Let $C^g$ be the set of all homogeneous regular element in R. If R satisfies the Ore condition with respect to $C^g$, then we denote by $Q^g(R)$ the ring of fractions with respect to $C^g$; $Q^g = Q^g(R) = \{ac^{-1} \mid a \in R,\ c \in C^g\} = \{d^{-1}b \mid b \in R,\ d \in C^g\}$, and is again a Z-graded ring with $(Q^g)_n = \{ac^{-1} \mid a \in R,\ c \in C^g$ and $\deg a - \deg c = n\}$.

Lemma 1.3. (1) R satisfies the Ore condition with respect to $C^g$.
(2) $Q^g$ is a strongly Z-graded ring with $(Q^g)_o = Q_o$ and is a principal ideal prime ring.

Proof. (1) Let I be any graded essential right ideal of R. Then we have $I_o = I \cap R_o$ is an essential right ideal of $R_o$. This follows from Lemmas 1.1, 1.2 and the $\sigma$-torsion-freeness of R. Hence I contains a regular element in $R_o$ which is regular in R by Lemma 1.1. From Lemmas 1.1 and 1.2, we can easily see that $\dim R_o = \dim_g R$, where $\dim R_o$ is the Goldie dimension of $R_o$ and $\dim_g R$ is the graded Goldie dimension of R. So cR is a graded essential right ideal of R for any homogeneous regular element c of R. Hence $Q^g$ exists as ungraded case and $(Q^g)_o = Q_o$.
(2) Since $R_nR_{-n}$ is a non zero ideal of $R_o$, it contains a regular element c in $R_o$ and so $1 = cc^{-1} \in R_nR_{-n}(Q^g)_o \subset (Q^g)_n(Q^g)_{-n}$. This means that $Q^g$ is strongly graded. The rest of the statements follow from [11].

Let T be any ring with the classical quotient ring. Then we denote it by Q(T). Since $Q^g$ is a principal ideal prime ring, $Q(Q^g)$ exists and it is a simple artinian ring. Thus R is a prime Goldie ring with the classical quotient ring Q (= $Q(Q^g)$). Furthermore, as it is seen from the proof of Lemma 1.3, $Q^g = \{ ac^{-1} \mid a \in R$ and c : regular element in $R_o \} = \{ d^{-1}b \mid b \in R$ and d : regular element in $R_o\}$.

Corollary 1.4. (1) $Q_\sigma(R) = R = Q_{\sigma'}(R)$.
(2) $cl(I) = Q_\sigma(I)$ for any graded right ideal I.

Proof. (1) It is clear that $Q^g$ is σ-torsion-free. Hence $Q_\sigma(Q^g) = \{ x \in E_R(Q^g) \mid xI_o \subset Q^g$ for some $I_o \in F(\sigma)\}$. Let x be any element in $Q_\sigma(Q^g)$ such that $xI_o \subset Q^g$ for some $I_o \in F(\sigma)$. Then we have $x \in xQ^g = xI_oQ^g \subset Q^g$ and so $Q_\sigma(Q^g) = Q^g$. Now, from the exact sequence $0 \longrightarrow R \longrightarrow Q^g$, we have the exact sequence $0 \longrightarrow Q_\sigma(R) \longrightarrow Q_\sigma(Q^g) = Q^g$. Let $x = x_{n_1} + \ldots + x_{n_k}$ be any element in $Q_\sigma(R)$, where $x_{n_i} \in (Q^g)_{n_i}$. Then there exists $I_o \in F(\sigma)$ such that $xI_o \subset R$ and so $x_{n_i}I_o \subset R_{n_i}$. Hence $x_{n_i} \in Q_\sigma(R_{n_i}) = R_{n_i}$ and $x \in R$. The proof of (2) is similar to one of (1).

We consider the following two kernel functors which have a close relation to F(σ);

$F(\sigma_R) = \{ I$ : right ideal $\mid Hom_R(R/I, E(Q/R)) = 0 \}$, and

$F(\sigma^g) = \{ I$ : graded right ideal $\mid HOM_R(R/I, E^g(Q^g/R)) = 0\}$ $= \{ I$ : graded right ideal $\mid (R:r^{-1}I)_l = R$ for any homogeneous element r in R$\}$, where $E^g(Q^g/R)$ is the graded injective hull of $Q^g/R$ and $(R:I)_l = \{ x \in Q \mid xI \subset R\}$, i.e., $\sigma^g$ is the rigid kernel functor cogenerated by $E^g(Q^g/R)$ (see, p. 86 of [12] for rigid kernel functors). Of course, we can define the left hand versions of $F(\sigma_R)$ and $F(\sigma^g)$, and denote them by $F(\sigma'_R)$ and $F(\sigma'^g)$, respectively. We can also define the concept of $\sigma_R$-closed ($\sigma^g$-

losed) right ideals in an obvious way.

Lemma 1.5. Let $I_o$ be any right ideal of $R_o$. Then the following are equivalent;

(1) $I_o \in F(\sigma)$.

(2) $I_oR \in F(\sigma_R)$.

(3) $I_oR \in F(\sigma^g)$.

Proof. (1) $\Longrightarrow$ (2) : This is proved by the similar way as in Theorem 3.2 of [11]. But the proof of the case (a) below is shorter and simpler than one in Theorem 3.2 of [11]. Let r be any element in R and let $q(r^{-1}(I_oR)) \subset R$ $(q \in Q)$. (a) First assume that r is a homogeneous element of degree n. Then $r^{-1}(I_oR)$ is graded and so $q \in Q^g$. Thus we may assume that q is also homogeneous of degree m. It follows that $R \supset (R_{-m}q)(r^{-1}(I_oR)) \supset (R_{-m}q)x(rx)^{-1}(I_o)$ for any $x \in R_{-n}$, and $R_o \supset R_n(R_{-m}q)x(rx)^{-1}(I_o)$. Hence $R_n(R_{-m}q)x \subset R_o$ and $R_{-n}R_n(R_{-m}qx) \subset R$. This entails that $R_{-m}qx \subset R$ by Lemma 1.1 and Corollary 1.4. By the similar discussion, we have $qx \in R$ and $qR_{-n} \subset R$. Hence $q \in R$. (b) In case r is arbitrary. Then we can prove that $q \in R$ by the exactly same way as in Theorem 3.2 of [11].
(2) $\Longrightarrow$ (3) : This is a spcial case.
(3) $\Longrightarrow$ (1) : Assume that $I = I_oR \in F(\sigma^g)$. Then, since $(r_o^{-1}I)_o = r_o^{-1}(I_o)$, we have $R = (R:r_o^{-1}I)_l \supset (R_o:r_o^{-1}I_o)_l \supset R_o$. Hence $R_o = (R_o:r_o^{-1}I_o)_l$ for any $r_o \in R_o$ and thus $I_o \in F(\sigma)$.

For any right R-ideal I, define $I_v = (R:(R:I)_l)_r$ and we say that I is a right v-ideal provided $I = I_v$. Similarly, we put ${}_vJ = (R:(R:J)_r)_l$ for any left R-ideal J, and if $J = {}_vJ$, then it is called a left v-ideal.

Lemma 1.6. Let I be a graded right R-ideal. Then
(1) $Q_{\sigma'}(R(R_o:I_o)_l) = (R:I)_l = {}_v(R(R_o:I_o)_l)$. In particular, if I

is an R-ideal, then $Q_\sigma((R_o:I_o)_l R) = Q_\sigma((R:I)_l)$.
(2) $I_v = (I_o R)_v = Q_\sigma(I_{ov} R)$.
(3) $(I_o)_v = (I_v)_o$.

Proof. First we note that $Q_\sigma(I) \subset I_v$ for any right R-ideal I. To Prove this assume that $zH \subset I$ for some $z \in Q$ and $H \in F(\sigma)$. Then we have $(R:I)_l zHR \subset R$. and hence $(R:I)_l z \subset R$ by Lemma 1.5 and $z \in I_v$. Therefore $Q_\sigma(I) \subset I_v$. In particular, if I is a right v-ideal, then we have $I = Q_\sigma(I)$.
(1) Let x be any element in $(R_o:I_o)_l$. Then we have $xI_o \subset R_o$ and $xI_o R \subset R$. Applying $Q_\sigma(\ )$ to this, we have $xI \subset xQ_\sigma(I) = xQ_\sigma(I_o R) \subset Q_\sigma(R) = R$ by Lemma 1.2 and Corollary 1.4. This entails that $x \in (R:I)_l$ and $R(R_o:I_o)_l \subset (R:I)_l$. Thus we have $(R:I)_l \supset {}_v(R(R_o:I_o)_l) \supset Q_{\sigma'}(R(R_o:I_o)_l)$. Conversely, let y be any homogeneous element in $(R:I)_l$ with deg y = n. Then $yI \subset R$ entails that $R_{-n}yI_o \subset R_o$ and so $R_n R_{-n} y \subset R(R_o:I_o)_l$. Thus $y \in Q_{\sigma'}(R(R_o:I_o)_l)$ by Lemma 1.1. If I is an R-ideal, then it is clear that $Q_\sigma((R_o:I_o)_l R) \subset Q_\sigma((R:I)_l)$. To prove the converse inclusion, let z be any homogeneous element in $Q_\sigma((R:I)_l)$ with deg z = n. Then $zH \subset (R:I)_l$ for some $H \in F(\sigma)$ and $zHR_{-n}R_n \subset (R_o:I_o)_l R$. Since $HR_{-n}R_n \in F(\sigma)$, it follows that $z \in Q_\sigma((R_o:I_o)_l R)$.
(2) Let x be any homogeneous element in $I_v$ with deg x = n. Then $(R:I)_l x \subset R$ and so, by (1), $(R_o:I_o)_l xR_{-n} \subset R_o$. It follows that $xR_{-n} \subset I_{ov}$. Thus we have $x \in Q_\sigma(I_{ov}R)$. To prove the converse inclusion, let $y_o$ be any element in $I_{ov}$, i.e., $(R_o:I_o)_l y_o \subset R_o$ and so $R(R_o:I_o)_l y_o \subset R$. Applying $Q_{\sigma'}(\ )$ to this, we have $(R:I)_l y_o = Q_{\sigma'}(R(R_o:I_o)_l)y_o = Q_{\sigma'}(R(R_o:I_o)_l y_o) \subset Q_{\sigma'}(R) = R$. So $y_o \in I_v$ and $Q_\sigma(I_{ov}R) \subset I_v$. Therefore $I_v = Q_\sigma(I_{ov}R) = (I_{ov}R)_v$. Finally, from the inclusion $I \subset Q_\sigma(I) = Q_\sigma(I_o R) \subset (I_o R)_v$, we have $I_v \subset (I_o R)_v$ and thus $I_v = (I_o R)_v$.
(3) follows from (2).

Let A be any R-ideal. Then we say that A is a <u>v-ideal</u> if ${}_vA$

$= A = A_v$. A v-ideal X is called <u>v-invertible</u> if $_v((R:X)_l X) = R = (X(R:X)_r)_v$. Following [2], a v-ideal A in R is said to be <u>v-idempotent</u> if $(A^2)_v = A = {}_v(A^2)$.

<u>Lemma 1.7</u>. Let A be a graded v-ideal. Then
(1) $O_l(A) = {}_v(RO_l(A_o))$.
(2) A is v-invertible if and only if $A_o$ is v-invertible.
(3) If A is integral, then A is v-idempotent if and only if $A_o$ is v-idempotent.

<u>Proof</u>. The proof of (1) is similar to one in Lemma 1.6. (2) and (3) are proved by combining the following three properties with (1);
(a) Let $I_o$ and $J_o$ be any right v-$R_o$-ideal with $I_o \supsetneqq J_o$. Then $I_v \supsetneqq J_v$, where $I = I_oR$ and $J = J_oR$ (this immediately follows from Lemma 1.6, (3)).
(b) A is v-invertible if and only if $O_l(A) = R = O_r(A)$ (see Lemma 1.1 of [2]).
(c) A is v-idempotent if and only if $O_l(A) = (R:A)_l$ and $O_r(A) = (R:A)_r$ (see Lemma 1.3 of [2]).

<u>Remark</u>. Let G be a torsion-free abelian group with the ascending chain condition (a.c.c.) on cyclic subgroups. Then divisorially G-graded orders satisfying a polynomial identity were studied in [4]. We do not go into divisorially G-graded orders in this paper. But, if G is a torsion-free abelian group, then all results in this section except Lemma 1.3 are true for any divisorially G-graded order $R = \Sigma \oplus R_\tau$ $(\tau \in G)$, where $R_o$ is a prime Goldie ring with the quotient ring $Q_o$. Concerning Lemma 1.3, $Q^g$ exists and is strongly G-graded which is a maximal order in $Q(Q^g)$. The detail proofs will be given in the forthcoming paper.

## 2. <u>GRADED KRULL ORDERS</u>

First we shall prove in this section that if $R_o$ is a Krull order in the sense of Chamarie, then so is R. If R is strongly graded, then this result was shown in [11] and the outlines of our proofs are similar to ones in [11]. So we shall only give proofs of some parts which are either more complicated than ones in [11] or shortened.

Proposition 2.1. If $R_o$ is a maximal order in $Q_o$, then R is a maximal order in Q.

Proof. Let T be a subring of Q containing R such that $Ta \subset R$ for some regular element a in Q. Then we have $T \subset Q^g$ by the exactly same proof as in Theorem 3.1 of [11]. Let $C_n(T) = \{q_n \in (Q^g)_n \mid t = q_n + (\text{the lower degree terms}) \in T\} \cup \{0\}$. Then $R_n \subset C_n(T)$ and $R_{-n}R_n \subset R_{-n}C_n(T) \subset Q_o$. It is clear that $R_{-n}C_n(T)$ is a ring. Write $a = a_l + (\text{the lower degree terms})$. Then we have $R_{-n}C_n(T)a_lR_{-l} \subset R_o$ and $a_lR_{-l} \neq 0$. Since $C_n(T)$ is $(R_o,R_o)$-bimodule, $A = \{r \in R_o \mid R_{-n}C_n(T)r \subset R_o\}$ is a non zero ideal. That $R_{-n}C_n(T)$ is a ring and is an $(R_o,R_o)$-bimodule implies that $(R_{-n}C_n(T) + R_o)$ is also a ring and $(R_{-n}C_n(T) + R_o)A \subset R_o$. Hence $R_{-n}C_n(T) \subset R_o$ by the maximality of $R_o$. It follows that $R_nR_{-n}C_n(T) \subset R$ and so $C_n(T) \subset (R:RR_nR_{-n})_r = R$ by Lemmas 1.1 and 1.5. Now let $t = t_n + (\text{the lower degree terms})$ be any element in T. Then $t_n \in C_n(T) \subset R$ and $t - t_n \in T$. By repeating the same argument, we finally have $t \in R$ and so $R = T$. Hence R is a maximal order in Q.

We denote $\Sigma_{n \geq o} \oplus R_n$ by $R_+$.

Lemma 2.2. Let I and J be non zero right ideals of R. Then
(1) $I \cap R_+ \neq 0$ and $Q_\sigma(C_n(I \cap R_+)R_{-n}) \subset Q_\sigma(C_m(I \cap R_+)R_{-m})$ for any $m \geq n \geq 0$.
(2) Let I and J be both $\sigma_R$-closed right ideals satisfying $J \supset I$ and $Q_\sigma(C_n(I \cap R_+)R_{-n}) = Q_\sigma(C_n(J \cap R_+)R_{-n})$ for all $n \geq 0$. Then $I = J$.

Proof. The proof of (1) is similar to one in Theorem 3.2 of [11]. (2) Assume that $J \supsetneqq I$. Pick up $j \in J - I$ and $j \in R_+$. Write $j = r$ + (the lower degree terms) ($r \in R_s$). We may assume that s (the degree of leading term of j) is minimal in $j \in (J \cap R_+) - I$. Then $r \in C_s(J \cap R_+)$ and so $rt \in C_s(J \cap R_+)R_{-s} \subset Q_\sigma(C_s(I \cap R_+)R_{-s})$ for any $t \in R_{-s}$. There exists $H \in F(\sigma)$ such that $rtH \subset C_s(I \cap R_+)R_{-s}$. This means that $rtHR_s \subset C_s(I \cap R_+)$. For any non zero element rtx ($x \in HR_s$), there is $y \in I \cap R_+$ ; $y = rtx$ + (the lower degree terms). Then we have $y - jtx \in J \cap R_+$ and $\deg (y - jtx) \lneqq s$. Hence $y - jtx \in I$ and thus $jtx \in I$. This entails that $jtHR_s \subset I$ and $jtHR_sR_{-s} \subset I$. By Lemma 1.5, $jt \in I$ for any $t \in R_{-s}$. Hence $j \in I$ by Lemma 1.5, a contradiction.

Lemma 2.3. The following three conditions are equivalent;
(1) $R_o$ satisfies the a.c.c. on $\sigma$-closed right ideals of $R_o$.
(2) R satisfies the a.c.c on $\sigma_R$-closed right ideals of R.
(3) R satisfies the a.c.c. on $\sigma^g$-closed graded right ideals of R.

Proof. (1) $\Longrightarrow$ (2) : The proof is similar to one of Theorem 3.2 of [11] by the validity of Lemma 2.2.
(2) $\Longrightarrow$ (3) : This is a spcial case.
(3) $\Longrightarrow$ (1) : This is easily proved by using Lemma 1.5.

An order T in a simple artinian ring Q(T) is called a Krull order in the sense of Chamarie, if T is a maximal order and satisfies the a.c.c. on $\sigma_T$-closed right ideals and $\sigma'_T$-closed left ideals. By Proposition 2.1 and Lemma 2.3, we have

Theorem 2.4 (Theorem 3.2 of [11]). Let R be a divisorially graded order and let $R_o$ be a Krull order. Then so is R.

Following [11], a graded order R is a graded maximal order in $Q^g$ if for each graded overring T, $R \subset T \subset Q^g$, such that $bTa \subset R$ for some regular homogeneous a, b in $Q^g$, R = T follows. We call R a graded Krull order if R is a graded maximal order and satisfies the a.c.c. on $\sigma^g$-closed graded right ideals as well as $\sigma'^g$-closed graded left ideals. In the remainder of this section we shall prove that if R is a graded Krull order, then it is a Krull order as ungraded orders. To do this, we shall study a relationship between a class of prime ideals of $Q^g$ and a class of prime ideals of $R_o$. More generally, we consider prime Goldie rings $T_o \subset T$ satisfying the condition ;

(**) T satisfies the Ore condition with respect to C and each element in C is regular in T, where C is the set of all regular elements in $T_o$.

We denote by $T_C$ the ring of fractions with respect to C. If we take $C = \{ c \in R_o \mid c : \text{regular element in } R_o\}$, then $R_o$ and R satisfy the condition (**) and $R_C = Q^g$. Put $T = T_o[X,\alpha]$, a skew polynomial ring over $T_o$ in an indeterminate X and $\alpha$ is an automorphism of $T_o$. Then $T_o$ and T satisfy the condition (**) and $T_C = Q(T_o)[X,\alpha]$.

Lemma 2.5. Under the assumption as in (**), let A be an ideal of T such that T/A is a Goldie ring. Then

(1) $M = AT_C$ is an ideal of $T_C$.

(2) If A is a prime ideal, then so is M. Furthermore if $A \cap T_o = 0$, then $M \cap T = A$.

(3) If $T_o$ is an HNP ring, then the correspondence $A \longrightarrow M = AT_C$ is one-to-one between the set of all prime ideals A of T with $A \cap T_o = 0$ and the set of all proper prime ideals M of $T_C$.

Proof. (1) If $A \cap T_o \neq 0$, then A contains an element in C. So $AT_C = T_C$, an ideal of $T_C$. Assume that $A \cap T_o = 0$. To prove that $M = AT_C$ is an ideal of $T_C$, let $x = c^{-1}a$ be any element in $T_CA$ ($c \in C$ and $a \in A$). Then there are $c_1 \in C$ and $t \in T$ such that $ct = ac_1$. Write $\bar{c} = [c + A]$ in $\bar{T} = T/A$. Then there is $n \gneq 0$ such

that $\mathrm{ann}(\bar{c}^n) = \mathrm{ann}(\bar{c}^{n+1})$. For elements $c^n$ and $t$, we have $c_2 \in C$ and $s \in T$ satisfying $c^n s = tc_2$. It follows that $c^{n+1}s = c(tc_2) = ac_1c_2 \in A$. Thus $\bar{s} \in \mathrm{ann}(\bar{c}^n)$. This means that $c^n s \in A$ and $tc_2 \in A$. Therefore $t \in tT_C = tc_2T_C \subset AT_C$ and so $x = c^{-1}a = tc_1^{-1} \in AT_C$. This entails that $AT_C$ is an ideal of $T_C$.
(2) To prove that $AT_C$ is a prime ideal, we may assume that $A \cap T_o = 0$. Put $A' = AT_C \cap T$ and assume that $A' \supsetneqq A$. Then $A'/A$ contains a regular element $a$ in $T$ ($a \in A'$), because $T$ is a prime Goldie ring. On the other hand, there is $c \in C$ such that $ac \in A$ and so $\bar{a}\bar{c} = \bar{0}$. Thus $\bar{c} = \bar{0}$ and $c \in A \cap R_o = (0)$, a contradiction. Hence $AT_C \cap T = A$ and then it immediately follows that $M = AT_C$ is also a prime ideal.
(3) From (2), the correspondence $A \longrightarrow AT_C$ is injective from the set of all prime ideals $A$ of $T$ satisfying $T/A$ is a Goldie ring and $A \cap T_O = 0$ to the set of all proper prime ideals of $T_C$. To prove the correspondence is onto, let $M$ be a proper prime ideal of $T_C$. If $M \cap T_o \neq 0$, then $M \cap T_o \cap C \neq \phi$ and so $M = (M \cap T)T_C = T_C$, a contradiction. Thus the set $\{B : \text{ideal of } T \mid B \supset M \cap T \text{ and } B \cap T_o = (0)\}$ has a maximal element by Zorn's lemma and let $A$ be one of the maximal elements in the set. It is clear that $A$ is a prime ideal. Since $M = T_C(M \cap T) = (M \cap T)T_C \subset T_CAT_C$, we have either $M = T_CAT_C$ or $T_CAT_C = T_C$, because, in HNP rings, a prime ideal is maximal. Assume that $T_CAT_C = T_C$. Then $A \cap T_o \neq 0$. This is impossible. Thus $M = T_CAT_C$ and so $M \cap T \supset T_CAT_C \cap T \supset A$. Hence $A = M \cap T$. It is now clear that $T/A$ is a prime Goldie ring, because $T_C/M$ is a simple artinian ring and is an essential extension of $T/A$.

We denote by $V(R) = \{A : \text{ideal of } R \mid A = A_v\}$ and by $V'(R) = \{A : \text{ideal of } R \mid {}_vA = A\}$.

<u>Lemma 2.6</u>. Let $R$ be a graded Krull order and let $M$ be a maximal element in $V(R)$. Then $M$ is either a maximal graded v-ideal or

a maximal v-ideal such that $M \cap R_o = 0$.

Proof. First of all we note that R satisfies the a.c.c. on $\sigma_R$-closed right ($\sigma'_R$-closed left) ideals by Lemma 2.3. Thus R/M is a right Goldie ring by Lemma 1.1 of [1]. It is clear that M is a prime ideal by Lemma 1.1 of [7]. (a) In case $M \cap R_o = 0$. Then $M = MQ^g \cap R$ and $MQ^g$ is an ideal of $Q^g$ by Lemma 2.5 and so $M = {}_vM$ by Lemma 2.3 of [7]. Hence M is a maximal v-ideal of R. (b) In case $M \cap R_o \neq 0$. Then $M_g \neq 0$ and it is a prime ideal. Since $M = M_v \supset (M_g)_v$, we have $(M_g)_v = M_g$, a v-ideal by Lemma A.1 in the appendix. Write $M_g = (M_1 \ldots M_k)_v$, where $M_i$ are maximal graded v-ideals of R (see Lemma A.7). Then $M \supset M_i$ for some i ($1 \leqq i \leqq k$). If $M \supsetneqq M_i$, then, since $M_i$ is a prime and v-invertible ideal of R, we have $C(M_i) \cap M = \phi$ by Lemma 3.1 of [3]. This contradicts to the fact that $R/M_i$ is a prime Goldie ring. Hence $M = M_i$ and it is a maximal graded v-ideal of R.

Remark. As it is seen from the proof, a maximal graded v-ideal of R is a maximal v-ideal of R.

Let A be an ideal of R. If R satisfies the Ore condition with respect to C(A), where $C(A) = \{ c \in R \mid c : \text{regular mod } A\}$, then we denote by $R_A$ the ring of fractions with respect to C(A).

Theorem 2.7. Let R be a graded Krull order. Then R is a Krull order as ungraded orders.

Proof. By Lemma 2.3, R satisfies the a.c.c. on $\sigma_R$-closed right ( $\sigma'_R$-closed left) ideals. Let M be any maximal element in V(R). Then if $M \cap R_o = 0$, then $M' = MQ^g$ is a maximal ideal of $Q^g$ and $M' \cap R = M$ by Lemma 2.5. It is well known that $(Q^g)_{M'}$ exists and it is a local Dedekind prime ring (see [5]). Then, by Proposition 1.1 of [6], $R_M$ exists and $R_M = (Q^g)_{M'}$. If $M \cap R_o \neq 0$, then it is a maximal graded v-ideal of R by Lemma 2.6. Since R is a graded Krull order, M is v-invertible. Thus it follows from Lemma 2.1 of [9] that $R_M$ exists and is a local, Dedekind prime ring. Similarly, for any maximal element N in V'(R), we have $R_N$ is a local, Dedekind prime ring. Hence R is a Krull order by Theorem 2.4 of [8].

Remarks. (1) In [13], Oystaeyen proved this theorem in case R is a graded maximal order over a commutative Krull domain.
(2) Let $R = \Sigma \oplus R_\tau$ be a divisorially graded order of type G satisfying the a.c.c. on $\sigma_R$-closed ($\sigma_R'$-closed left) ideals, where G is a torsion-free abelian group. If R is a graded maximal order, then R is a Krull order as ungraded orders.

## 3. GRADED V-HC ORDERS WITH ENOUGH GRADED V-INVERTIBLE IDEALS

In analogy with ungraded case (see [2] or [11]), a divisorially graded order R is called a graded v-H order if $_v(A(R:A)_l) = O_l(A)$ (resp. $((R:A)_r A)_v = O_r(A)$) for every non zero graded ideal A of R with $_vA = A$ (resp. $A = A_v$). We say that R is a graded v-HC order if it is a graded v-H order satisfying the a.c.c. on graded $\sigma^g$-closed right ($\sigma'^g$-closed left) ideals. R is said to be having enough graded v-invertible ideals if any non zero graded v-ideal of R contains a graded v-invertible ideal of R. In this section, we shall prove that if R is a graded v-HC order with enough graded v-invertible ideals, then it is a (real) v-HC order with enough v-invertible ideals. Note that if R is a graded v-HC order, then R satisfies the a.c.c. on $\sigma_R$-closed right ($\sigma_R'$-closed left) ideals by Lemma 2.3. We define $S^g = \cup \{X^{-1} \mid X$ : graded v-invertible ideal of R$\}$ (note that $(R:X)_l = (R:X)_r$ if X is v-invertible, and we denote it by $X^{-1}$). Clearly $R \subset S^g \subset Q^g$. Let $\kappa$ ($\kappa'$) be the idempotent kernel functor cogenerated by the right (left) R-module $Q/S^g$. Then $Q_\kappa(R) = S^g = Q_{\kappa'}(R)$ and $F(\kappa)$ contains every graded v-invertible ideal of R (see Lemma 2.9 of

[7]). Thus $S^g$ satisfies the a.c.c. on $\sigma_S g$-closed right ($\sigma'_S g$-closed left) ideals by Lemma 2.2 of [1].

Lemma 3.1. Let R be a graded v-HC order with enough graded v-invertible ideals. Then
(1) $S^g$ is a divisorially graded order.
(2) $S^g$ is a Krull order with no non-trivial graded v-ideals of $S^g$.

Proof. (1) We define $(S^g)_n = \{ q \in Q^g \mid \deg q = n$ and $qX \subset R$ for some graded v-invertible ideal X of R$\}$. In particular, $(S^g)_o = \{ q \in Q_o \mid qX_o \subset R_o$ for some v-invertible ideal $X_o$ of $R_o$ such that $(X_oR)_v = {}_v(RX_o)\}$ by Lemma 1.7. It is clear that $S^g = \Sigma \oplus (S^g)_n$ and that it is graded. It is well known, in any prime Goldie ring, that any filter corresponding to an idempotent kernel functor consists of essential right ideals or contains the zero ideal. Obviously, $F(\sigma^*)$ does not contain the zero ideal, where $\sigma^* = \sigma_{(S^g)_o}$. This entails that $S^g$ is $(\sigma^{*\prime},\sigma^*)$-torsion-free, where $\sigma^{*\prime}$ $\sigma'_{(S^g)_o}$. To prove that $S^g$ is divisorially graded, let q be any element in $(S^g)_{n+m}$. Then $qX \subset R$ for some graded v-invertible ideal X of R. It follows that $qX_o \subset R_{n+m} = Q_\sigma(R_nR_m)$. For any $x \in X_o$, there exists $H \in F(\sigma)$ such that $qxH \subset R_nR_m$. So $qxH(S^g)_o \subset (S^g)_n(S^g)_m$. But, since $H(S^g)_o \in F(\sigma^*)$ by the proof of Lemma 2.2 of [1], it follows that $qx \in Q_{\sigma^*}((S^g)_n(S^g)_m)$ and $qX_o \subset Q_{\sigma^*}((S^g)_n(S^g)_m)$. Thus $qX_oX_o^{-1}(S^g)_o \subset Q_{\sigma^*}((S^g)_n(S^g)_m)$ and $q \in Q_{\sigma^*}((S^g)_n(S^g)_m)$. Hence we have $Q_{\sigma^*}((S^g)_{n+m}) = Q_{\sigma^*}((S^g)_n(S^g)_m)$. To finish the proof, it is enough to prove that $Q_{\sigma^*}((S^g)_n) = (S^g)_n$ for any n. Let x be any element in $Q_{\sigma^*}((S^g)_n)$. Then there is $H' \in F(\sigma^*)$ such that $xH' \subset (S^g)_n$ and so $R_{-n}xH' \subset (S^g)_o$. This implies that $R_{-n}x \subset (S^g)_o$ and $R_nR_{-n}x \subset (S^g)_n$. Since $R_nR_{-n} \in F(\sigma')$, there is a finitely generated left ideal $J \in F(\sigma')$ such that $J \subset R_nR_{-n}$. Thus we have $JxY \subset R$ for some graded v-invertible ideal Y of R

and so $xY \subset Q_{\sigma'}(R) = R$. Hence $q \in Q_{\kappa}(R) = S^g$ and thus $q \in (S^g)_n$. Therefore $S^g$ is a divisorially graded order.
(2) Let I' be any graded v-ideal of $S^g$. Then $I = I' \cap R$ is a graded v-ideal of R and so it contains a graded v-invertible ideal X of R. Thus we have $S^g \supset I' = Q_{\kappa}(I) \supset Q_{\kappa}(X) = S^g$ by Lemma 2.3 of [7]. Hence $S^g$ is a graded Krull order with no non trivial graded v-ideals. Therefore it is a real Krull order by Theorem 2.7.

<u>Lemma 3.2.</u> Let R be a graded v-HC order with enough graded v-invertible ideals. Then $R = \cap R_P \cap S^g$, where P runs over all maximal graded v-invertible ideals of R, and $R_P$ is an HNP ring.

<u>Proof</u>. As in Proposition 2.11 of [7](see Lemma 2.1 of [9] and Lemma A.6).

<u>Lemma 3.3</u>. Let R be a graded v-HC order with enough graded v-invertible ideals. Then ${}_v(S^gA) = (S^gAS^g)_v = (AS^g)_v$ for any ideal A of R.

<u>Proof</u>. (a) First we prove the lemma in case $A = A_v$. To prove that $(AS^g)_v = (S^gAS^g)_v$, let q be any regular element in $S^gA$. Then there exists a graded v-invertible ideal X of R such that $qX \subset R \cap AQ^g$, Because $AQ^g = Q^gA$ by Lemma 2.5 and so $(R:A)_l qX \subset (R:A)_l \cap Q^g$. Since R satisfies the a.c.c. on left v-ideals, we have $(R:A)_l qXc \subset {}_vR = R$ for some regular element c in $R_o$ and $qXc \subset A$. Let Y be any graded v-invertible ideal of R being contained in $(XcR)_v$. Then $qY \subset A$ and so $q \in Q_{\kappa}(A) = (AS^g)_v$ by Lemma 2.3 of [7]. Thus $S^gA \subset (AS^g)_v$ by Lemma 2.2 of [6]. Hence $(S^gAS^g)_v = (AS^g)_v$.
(b) In any case; since $A_vS^g \subset Q_{\kappa}(A_v) = (AS^g)_v$, we have $(A_vS^g)_v = (AS^g)_v$, an ideal. Hence $(AS^g)_v = (S^gAS^g)_v = {}_v(S^gAS^g) = {}_v(S^gA)$, because $S^g$ is a Krull order.

<u>Lemma 3.4</u>. Let R be a divisorially graded order satisfying the a.

c.c. on $\sigma_R$-closed right ($\sigma'_R$-closed left) ideals and let A' be any maximal ideal of $Q^g$. Then

(1) $R_A = (Q^g)_{A'}$, and it is a local, Dedekind prime ring, where $A = A' \cap R$.

(2) A is a prime and v-invertible ideal of R.

Proof. (1) follows from Proposition 1.1, Lemma 5.2 of [6] and Theorem 2.7 of [5]. The proof of (2) is similar to one in Lemma 4.9 of [11].

We are now in a position to prove the following theorem.

Theorem 3.5. Let R be a divisorially graded order. If R is a graded v-HC order with enough graded v-invertible ideals, then it is a v-HC order with enough v-invertible ideals.

Proof. First note that R satisfies the a.c.c. on $\sigma_R$-closed right ($\sigma'_R$-closed left) ideals. By Lemma 3.2, we have the decomposition; $R = \cap R_P \cap S^g$, where P ranges over all maximal graded v-invertible ideals of R, $R_P$ is an HNP ring. It follows from Lemmas 3.1 and 3.3 that $S^g$ is a Krull order and $(AS^g)_v = (S^gAS^g)_v = {}_v(S^gA)$ for any ideal A of R. Hence $A = \cap AR_P \cap (AS^g)_v$ for any ideal A with $A = {}_vA$. Then we have

(a) $O_l(A) = \cap O_l(AR_P) \cap O_l((AS^g)_v)$.

(b) $O_l(AR_P) = R_PA(R_P:AR_P)_l = R_PAR_P(R:A)_l = R_PA(R:A)_l$.

(c) $S^g = O_l((AS^g)_v) = {}_v({}_v(AS^g)(S^g:AS^g)_l) = {}_v({}_v(S^gA)_v(S^g(R:A)_l)) = {}_v(S^gAS^g(R:A)_l) = {}_v(S^gA(R:A)_l)$.

Hence $O_l(A) = \cap R_PA(R:A)_l \cap {}_v(S^gA(R:A)_l) = {}_v(A(R:A)_l)$. This entails that R is a v-HC order. To prove that R has enough v-invertible ideals, let A be any v-ideal of R. In case $R_o \cap A \neq 0$, then $A \supset A_g \neq 0$, a v-ideal. Hence $A_g$ contains a graded v-invertible ideal of R. In case $A \cap R_o = 0$. Then $AQ^g \subsetneq Q^g$ and so $AQ^g = A_1'^{n_1} \cap \ldots \cap A_k'^{n_k}$ for some maximal ideals $A_i'$ of $Q^g$ and some natural numbers $n_i$. Put $B = AQ^g \cap R$ ($\supset A$). Then B contains $(A_1^{n_1} \ldots A_k^{n_k})_v$, v-invertible by Lemma 3.4, where $A_i = A_i' \cap R$. So, if $B = A$, then there is nothing to do. Assume that $B \supsetneq A$, then there is a regular element c in $R_o$ with $Bc \subset A$, because $BQ^g = AQ^g$ and R satisfies the a.c.c. on one sided v-ideals of R. By the assumption, there exists a graded v-invertible ideal X of R such that $(RcR)_v \supset X$ and so $A \supset (BRcR)_v \supset (A_1^{n_1} \ldots A_k^{n_k} X)_v$, a v-invertible ideal of R. Therefore R has enough v-invertible ideals.

Next, we shall prove that if $R_o$ is a v-HC order with enough v-invertible ideals, then so is R. This will be done by reducing the problem to Theorem 3.5. Note that if $R_o$ is a v-HC order, then R satisfies the a.c.c. on $\sigma_R$-closed right ($\sigma_R'$-closed left) ideals by Lemma 2.3.

<u>Lemma 3.6</u>. If $R_o$ is a v-H order, then R is a graded v-H order.

<u>Proof</u>. Let A be a graded ideal with ${}_vA = A$. Then it is clear that $O_l(A) \supset {}_v(A(R:A)_l)$. It follows from Lemma 1.6 that $A_o(R_o:A_o)_l \subset A(R:A)_l$. Furthermore, since ${}_vA_o = A_o$ by Lemma 1.6, we have $1 \in O_l(A_o) = {}_v(A_o(R_o:A_o)_l) \subset {}_v(A(R:A)_l)$ and so $O_l(A) \subset {}_v(A(R:A)_l)$, because ${}_v(A(R:A)_l)$ is a left $O_l(A)$-ideal. Hence $O_l(A) = {}_v(A(R:A)_l)$, which means that R is a graded v-H order.

A <u>graded cycle</u> of R is a finite set of distinct maximal graded v-ideals $M_1, \ldots, M_n$ which are v-idempotent such that $O_r(M_1) = O_l(M_2), \ldots, O_r(M_n) = O_l(M_1)$. We will also consider a maximal graded v-ideal which is v-invertible as a graded cycle, because $O_r(M) = R = O_l(M)$.

<u>Lemma 3.7</u>. Let $R_o$ be a v-HC order with enough v-invertible ideals

and let $M_1$ be a v-idempotent, maximal graded v-ideal of R. Then there exist v-idempotent, maximal graded v-ideals $M_2, \ldots, M_n$ such that $M_1, \ldots, M_n$ is a cycle. In particular, $A = M_1 \cap \ldots \cap M_n$ is v-invertible.

<u>Proof</u>. Write $M_1 = (M_{1o}R)_v = {}_v(M_{1o}R)$. Then $M_{1o}$ is v-idempotent by Lemmas 1.6 and 1.7. There is v-idempotent $M_{2o}$ of $R_o$ such that $O_r(M_{1o}) = O_l(M_{2o})$ by Corollary 1.6 and Lemma 1.7 of [2]. We shall prove that $(RM_{2o}R)_v = (M_{2o}R)_v = {}_v(RM_{2o})$. Since $(R_o:M_{1o})_r = O_r(M_{1o}) = O_l(M_{2o}) = (R_o:M_{2o})_l$, we have, by Lemma 1.6, $Q_{\sigma'}((R:M_1)_r) = Q_{\sigma'}((R(R_o:M_{1o})_r) = Q_{\sigma'}(R(R_o:M_{2o})_l) = (R:M_{2o}R)_l$. Hence $(M_{2o}R)_v$ is an R-ideal, because $Q_{\sigma'}((R:M_1)_r)$ is a right R-module and $(R:M_{2o}R)_l$ is a left R-module. Thus ${}_v(RM_{2o}R) = (RM_{2o}R)_v = (M_{2o}R)_v$ by Lemma A.1 and Lemma 3.6. We denote it by $M_2$. Since $(M_2)_o = M_{2o}$, we have $M_2 = {}_v(RM_{2o})$ by Lemma 1.6 and it is v-idempotent by Lemma 1.7. As we have just proved, $(R:M_1)_r \subset (R:M_{2o}R)_l = (R:M_2)_l$. The converse inclusion also follows from the similar method. Hence $O_r(M_1) = (R:M_1)_r = (R:M_2)_l = O_l(M_2)$. We can now check, easily, that $M_2$ is also a maximal graded v-ideal of R. We continue this method and obtain v-idempotent, maximal graded v-ideals $M_1, \ldots, M_n, \ldots$ such that $O_r(M_1) = O_l(M_2)$, $O_r(M_2) = O_l(M_3), \ldots$ . Let $M_{11}, \ldots, M_{1k}$ be the full set of maximal v-ideals of $R_o$ containing $M_{1o}$, each of which is v-idempotent. Then we have $M_{1o} = (M_{11} \ldots M_{1k})_v$ by changing suffices in a suitable way if necessary (see Lemma 1.2 of [8] and Propositions 3,4,5 of [10]). For each $M_{1j}$, there exists a v-idempotent, maximal v-ideal $M_{2j}$ of $R_o$ such that $O_r(M_{1j}) = O_l(M_{2j})$. Then $M_{21}, \ldots, M_{2k}$ is the full set of all maximal v-ideals containing $M_{2o}$ and $M_{2o} = (M_{21} \ldots M_{2k})_v$ by Propositions 3,4 and 5 of [10]. We continue this method and obtain , for any n, $M_{no} = (M_{n1} \ldots M_{nk})_v$ for some v-idempotent, maximal v-ideals $M_{nj}$. Let $n_j$ be the length of the cycle which $M_{1j}$ belongs to and let p = l.c.m. $(n_1, \ldots, n_k)$. Then $M_{1j} = M_{p+1,j}$ for each j $(1 \leqq j \leqq k)$ and so $M_{1o} = M_{p+1,o}$. This implies that $M_1 = M_{p+1}$. If

q is the smallest number m such that $M_1 = M_{m+1}$, then $M_1, \ldots, M_q$ is a graded cycle and $A = M_1 \cap \ldots \cap M_q$ is v-invertible (see Lemma 1.11 of [7] or Lemma 1.9 of [2]).

Lemma 3.8. Let $R_o$ be a v-HC order with enough v-invertible ideals. Then any graded prime v-ideal of R is a maximal v-ideal of R.

Proof. Let A be any graded prime v-ideal of R. Then there are maximal graded v-ideals $M_1, \ldots, M_n$ such that $M_1 \supset A$ and $M_1, \ldots, M_n$ is a graded cycle by Lemma 3.7. Put $P = M_1 \cap \ldots \cap M_n$. Then $AR_P$ is a prime ideal and $AR_P \subset M_1R_P$. Hence $AR_P = M_1R_P$ by Lemma 2.1 of [9] and thus $A = M_1$. That $M_1$ is a maximal v-ideal of R follows from Lemma 3.1 of [3].

Lemma 3.9. Let $R_o$ be a v-HC order with enough v-invertible ideals. Then (1) Let A be any graded v-ideal of R. Then $A = (XB)_v$, where X is a graded v-invertible ideal of R and B is a graded v-ideal of R which is not contained in any graded v-invertible ideals.
(2) Let C be any graded v-ideal of R which is not contained in any graded v-invertible ideals of R. Then it is eventually v-idempotent, i.e., $(C^m)_v$ is v-idempotent for some $m > 0$. Furthermore, let $M_1, \ldots, M_n$ be the full set of all maximal graded v-ideals of R containing C. Then $(C^n)_v = ((M_1 \cap \ldots \cap M_n)^n)_v$ and it is v-idempotent.

Proof. The proof of (1) is similar to one in HNP rings, because R is a graded v-HC order by Lemmas 2.3 and 3.6. We can define the concept of graded associated prime ideals of a graded module in analogy with ungraded case and, by Lemma 3.8, any graded associated prime ideal of R/C is a maximal graded v-ideal of R. Hence the proof of (2) is just a graded version of the proof of Proposition 1.4 of [9].

From Lemmas 3.7 and 3.9, we have

Proposition 3.10. Let $R_o$ be a v-HC order with enough v-invertible ideals. Then R has enough graded v-invertible ideals.

After all these preparations we now prove the following theorem which is the another purpose of this section.

Theorem 3.11. Let $R_o$ be a v-HC order with enough v-invertible ideals. Then so is R.

Proof. This follows from Lemmas 2.3, 3.6, Theorem 3.5 and Proposition 3.10.

Remark. Let $R = \Sigma \oplus R_\tau$ $(\tau \in G)$ be a divisorially graded order of type G, where G is a torsion-free abelian group. If R satisfies the a.c.c. on $\sigma_R$-closed right ($\sigma'_R$-closed left) ideals, then Theorems 3.5 and 3.11 hold.

## 4. SOME REMARKS ON THE GROUP OF V-INVERTIBLE IDEALS

We shall use the following notations;

$D(R) = \{ X \mid X : \text{v-R-ideal and v-invertible}\}$,

$D^g(R) = \{ Y \mid Y : \text{graded and } Y \in D(R)\}$,

$D^g(R_o) = \{ Y_o \mid Y_o : \text{v-}R_o\text{-ideal such that } (Y_oR)_v = {}_v(RY_o) \in D^g(R)\}$.

In this section, we study a relation between $D^g(R)$ and $D(R)$, and also study some properties of generalized Rees rings.

Lemma 4.1. Let R be a graded v-HC order with enough graded v-invertible ideals and let P be any maximal v-invertible ideal of R. Then P is either a maximal graded v-invertible ideal of R or $P = A' \cap R$ for some maximal ideal A' of $Q^g$.

Proof. If $P \cap R_o = 0$, then $PQ^g \subsetneqq Q^g$ and so $PQ^g \subset A'$ for some maximal ideal $A'$ of $Q^g$. Thus $P \subset PQ^g \cap R \subset A = A' \cap R$, a v-invertible ideal of R by Lemma 3.4. Therefore $P = A$. If $P \cap R_o \neq 0$, then $P_g \neq 0$ and it is a v-ideal of R. Now let M be a maximal v-ideal containing P. Then $M_g \neq 0$ and is a prime v-ideal of R. So $M_g$ is a maximal v-ideal of R by Lemma 1.2 of [8] and Theorem 3.5. Thus $M = M_g$ and, by Lemma 1.9 of [7], there are maximal v-ideals $M_1, \ldots, M_n$ of R such that $M_1 = M$, $M_1 \cap \ldots \cap M_n \supset P$ and $M_1, \ldots, M_n$ is a cycle. Hence $P = M_1 \cap \ldots \cap M_n$. It is clear that $M_i$ $(2 \leqq i \leqq n)$ are also graded. Hence P is a maximal graded v-invertible ideal of R.

Remark. As it is seen from the proof of Lemma 4.1, any maximal graded v-invertible ideal of R is a maximal v-invertible ideal of R.

Theorem 4.2. Let R be a graded v-HC order with enough graded v-invertible ideals. Then
(1) $D^g(R)$ is isomorphic to $D(R_o)$ naturally.
(2) The sequence $0 \longrightarrow D^g(R) \xrightarrow{\iota} D(R) \xrightarrow{\pi} D(Q^g) \longrightarrow 0$ is splitting exact, where $\iota$ is the inclusion map and $\pi(X) = XQ^g$ $(X \in D(R))$.

Proof. (1) follows from Lemma 1.7.
(2) It is clear that $\pi\iota = 1$. To prove the exactness of the sequence, let X be any element in D(R). Then $X = (P_1^{n_1} \ldots P_k^{n_k} A_1^{m_1} \ldots A_l^{m_l})_v$, where $P_i$ are maximal graded v-invertible ideals of R and $A_i = A_i' \cap R$ for some maximal ideals $A_i'$ of $Q^g$ by Theorem 1.13 of [7], Theorem 3.5 and Lemma 4.1. Since $XQ^g = A_1'^{m_1} \ldots A_l'^{m_l}$, if $XQ^g = Q^g$, then we have $m_1 = \ldots = m_l = 0$. Hence $X \in D^g(R)$ and the sequence is exact. To prove the sequence splits, let $A' = A_1'^{n_1} \cap \ldots \cap A_k'^{n_k}$ be any element in $D(Q^g)$, where $A_i'$ are maximal ideals of $Q^g$. Then $A = (A_1^{n_1})_v \cap \ldots \cap (A_k^{n_k})_v = (A_1^{n_1} \ldots A_k^{n_k})_v$ is v-invertible and $AQ^g = A'$. Hence the sequence splits.

Let I be a v-invertible ideal of a v-HC order $R_o$ with enough v-invertible ideals and let $\alpha$ be any automorphism of $R_o$. In case I is $\alpha$-invariant, i.e., $\alpha(I) = I$, we construct the following divisorially graded order $R = R_o[I,\alpha] = \{ a_{n_1}X^{n_1} + \cdots + a_{n_k}X^{n_k} \mid a_{n_i} \in (I^{n_i})_v,\ n_i \in Z\}$ with multiplication $a_nX^n a_mX^m = a_n\alpha^n(a_m)X^{n+m}$. We call $R_o[I,\alpha]$ a generalized Rees ring with respect to I and $\alpha$. By Theorem 3.11, $R_o[I,\alpha]$ is a v-HC order with enough v-invertible ideals. For any divisorially graded order R, let $P^g(R) = \{ A \in D^g(R) \mid A = xR = Rx$ for some homogeneous element x in $Q^g\}$ and $P^g(R_o) = \{ A_o \in D^g(R_o) \mid A_o = x_oR_o = R_ox_o$ for some $x_o \in Q_o\}$. If we define $C^g(R) = D^g(R)/P^g(R)$ and $C^g(R_o) = D^g(R_o)/P^g(R_o)$, then, for generalized Rees rings, we have

Proposition 4.3. Let $R = R_o[I,\alpha]$ be a generalized Rees ring. Then we have the following exact sequence;

$$0 \longrightarrow \mathrm{Ker}\ \bar{\Phi} \longrightarrow C^g(R_o) \xrightarrow{\bar{\Phi}} C^g(R) \longrightarrow 0,$$

where $\bar{\Phi}$ is the map induced by $\Phi : D^g(R_o) \simeq D^g(R)$ and $\mathrm{Ker}\ \bar{\Phi} = \{[(I^n)_v] \mid n \in Z\}$ ($[(I^n)_v]$ stands for the image of $(I^n)_v$ in $C^g(R_o)$).

Proof. As in [4].

## 5. APPENDIX

In this section, we shall present some properties of graded v-ideals of R, some of which were used in this paper; Mostly the proofs are formally similar to ones of ungraded case. So we do not give the proofs of them. Throughout this section, R denotes a graded v-H order satisfying the a.c.c. on graded v-ideals of R. Note that if R is a graded Krull order, then R is a graded v-HC order with enough graded v-invertible ideals.

Lemma A.1 (cf. Lemma 1.2 of [7] and Corollary 1.6 of [2]). Let A be a graded R-ideal such that $X^{-1} \supset A$ for some graded v-invertible ideal X of R. Then $A_v = {}_vA$.

Lemma A.2 (cf. Lemma 1.5 of [7]). Let M be a maximal graded v-ideal of R. Then M is either v-invertible or v-idempotent.

Lemma A.3 (cf. Lemma 1.11 of [7] and Lemma 1.9 of [2]). Let $M_1, \ldots, M_n$ be a union of graded cycles. Then $A = M_1 \cap \ldots \cap M_n$ is v-invertible.

A graded overring T of R is called a right equivalent to R if $aT \subset R$ for some regular, homogeneous a in $Q^g$.

Lemma A.4 (cf. Lemma 1.6 of [7] and Lemma 1.7 of [2]). There is a 1-1 correspondence between the set of right equivalent graded overrings T of R such that $T = T_v$ and the set of graded v-idempotent ideals of R, which is given by $T \longrightarrow (R:T)_l$ and $I \longrightarrow O_r(I)$.

Lemma A.5 (cf. Lemma 1.9 of [7]). Let M be a maximal graded v-ideal containing a graded v-invertible ideal X. Then there is a graded cycle $M_1, \ldots, M_n$ $(M = M_1)$ such that $M_1 \cap \ldots \cap M_n \supset X$.

Lemma A.6 (cf. Lemma 1.12 of [7]). Let P be a graded v-ideal of R. Then P is a maximal graded v-invertible ideal of R if and only if it is an intersection of a graded cycle.

Lemma A.7 (cf. Theorem 1.13 of [7] and Proposition 3.4 of [11]). (1) If R has enough graded v-invertible ideals, then $D^g(R) = \{ X \mid X$: graded v-ideal and v-invertible$\}$ is a free abelian group generated by the maximal graded v-invertible ideals of R.

(2) If R is a graded maximal order, then a graded v-ideal is v-invertible.

## REFERENCES

[1] M. Chamarie, Anneaux de Krull non commutatifs, J. Algebra 72 (1981), 210-222.

[2] H. Fujita, A generalization of Krull orders, preprint.

[3] H. Fujita and H. Marubayashi, Notes on v-HC orders in a simple artinian ring, preprint.

[4] L. LeBruyn and F. Van Oystaeyen, Generalized Rees rings and relative maximal orders satisfying polynomial identities, to appear in J. Algebra.

[5] J. Kuzmanovich, Localizations of Dedekind prime rings, J. Algebra 21 (1972), 378-393.

[6] H. Marubayashi, Non commutative Krull rings, Osaka J. Math. 12 (1975), 703-714.

[7] H. Marubayashi, A Krull type generalization of HNP rings with enough invertible ideals, Comm. in Algebra 11 (1983), 469-499.

[8] H. Marubayashi, Remarks on VHC orders in a simple artinian ring, to appear in the Proceedings.

[9] H. Marubayashi, A skew polynomial rings over a v-HC order with enough v-invertible ideals, to appear in Comm. in Algebra.

[10] H. Marubayashi, On v-ideals in a VHC order, to appear in Proc. Japan Academy.

[11] H. Marubayashi, E. Nauwelaerts and F. Van Oystaeyen, Graded rings over arithmetical orders, to appear in J. Algebra.

[12] C. Nastasescu and F. Van Oystaeyen, Graded and filtered rings and modules, LNM, 758, Springer Verlag, Berlin, 1979.

[13] F. Van Oystaeyen, On orders over graded Krull domains, to appear in Osaka J. Math.

# Δ-INJECTIVE MODULES AND QF-3 ENDOMORPHISM RINGS

Kanzo MASAIKE

Department of Mathematics, Tokyo Gakugei University
Koganei, Tokyo, Japan

ABSTRACT In this note we give a necessary and sufficient condition for Σ-injective modules to be Δ-injective. Further, we characterize generator modules whose endomorphism rings are QF-3 rings with ACC on annihilators.

## 0. PRELIMINARIES

Let R be a left Noetherian ring with identity. If the injective hull $I({}_R R)$ of the left R-module R is finitely generated, then R is left Artinian (cf. [1]). However, the converse does not hold in general. One of the purpose of this note is to extend this result using a notion of a hereditary torsion theory ([3], [13]). Let M be an injective left module over a ring R. Following Faith [2] we shall say M is Σ-injective (Δ-injective), if R has ACC (DCC) on annihilators of susets of M. Every Δ-injective module is Σ-injective (see [7]). In the following we denote by τ the hereditary torsion theory cogenerated by an injective module M. Then, Ann(M), the left annihilator of M, is the torsion submodule of R. A submodule N of a left R-module U is said to be τ-dense (τ-closed), if U/N is τ-torsion (τ-torsion



*F. van Oystaeyen (ed.), Methods in Ring Theory, 317–326.*

free). On the other hand, U is called τ-finitely generated if it has a finitely generated τ-dense submodule. It is to be noted that τ-closed left ideals coincide with annihilator left ideals of of subsets of M. Let U be a τ-torsion free R-module. U has DCC and ACC on τ-closed submodules, if and only if there exists a maximal chain

$$U = U_0 \supset U_1 \supset \cdots\cdots \supset U_k = 0 \qquad (*)$$

of τ-closed submodules. In this case (*) is called a τ-composition series and we shall say τ-lengtht M = k.

## 1. Δ-INJECTIVE MODULES

Throughout this note R is a ring with identity. Every homomorphism between modules will be written on the opposite side of scalars. We donote by $I({}_RU)$ the injective hull of the left R-module U.

Theorem 1.1. Assume M is a Σ-injective left R-module. Then, the following conditions are equivalent:

(1) M is Δ-injective.

(2) If N is a τ-finitely generated submodule of $I({}_RR/\mathrm{Ann}(M))$ ( = I, say) and there exists $f \in \mathrm{Hom}({}_RI, {}_RI)$ such that

$$N \subseteq Nf \subseteq Nf^2 \subseteq \cdots\cdots,$$

then, there exists a positive integer s such that $Nf^{s+i}/Nf^s$ is τ-torsion, i = 1, 2, ····.

Proof (2) → (1). Let Q be the ring of quotient with respect to τ. Then, M becomes an injective left Q-module. Let τ' be the hereditary torsion theory of the ring Q cogenerated by M. Since R satisfies ACC on τ-closed left ideals, Q satisfies ACC on τ'-closed left ideals. Then, it is easy to see that Q is a left Goldie ring. Of course, Q is the ring of quotient of itself with respect to τ'. On the other hand, $I({}_QQ)$ is eaqual to I as a left R-module, since Ann(M) is the τ-torsion submodule of R.

Assume N is a τ'-finitely generated submodule of $I({}_QQ)$ and there exists $f \in Hom({}_QI({}_QQ), {}_QI({}_QQ))$ such that $N \subsetneq Nf \subsetneq Nf^2 \subsetneq \cdots$. Since N is also τ-finitely generated R-module, there exists a positive integer s such that $Nf^s/Nf^{s+i}$ is τ-torsion and hence τ'-torsion, i = 1, 2, $\cdots$. From these facts we can deduce that Q is a semi-primary ring in view of the proof of [5, Lemma 1]. There exists a direct product of copies of the R-module M whose double centralizer is isomorphic to Q (see [9]). Let K be the direct product. Then, by [2, p.1, THEOREM] the Σ-injective module K and hence M are Δ-injective.

(1) → (2): Assume N is a τ-finitely generated submodule of I and there exists $f \in Hom({}_RI, {}_RI)$ such that $N \subsetneq Nf \subsetneq Nf^2 \subsetneq \cdots$. Since $Nf^i$ is τ-torsion free and $N/(N \cap ker\ f^i) \cong Nf^i$, $N \cap Ker\ f^i$ is a τ-closed submodule of N. As N is τ-finitely generated and R satisfies ACC on τ-closed left ideals, N also satisfies ACC on τ-closed submodules. Therefore, there exists a positive integer s such that $N \cap ker\ f^s = N \cap ker\ f^{s+1} = \cdots$. This implies that the restriction mapping $f|Nf^i : Nf^i \to Nf^{i+1}$ is an isomorphism, i = s, s+1, $\cdots$. Evidently, the τ-torsion free homomorphic image $Nf^i$ of N has a τ-composition series, since M is Δ-injective. Thus, τ-length $Nf^s$ = τ-length $Nf^{s+i}$ and this means $Nf^{s+i}/Nf^s$ is τ-torsion, i = 1, 2, $\cdots$.

<u>Remark 1.2.</u> Let R be a left Artinian ring such that $I({}_RR)$ is not finitely generated (cf. [11]). Let M be an injective left R-module which is a cogenerator (in the category of left R-modules). In this case M is Δ-injective, while $I({}_RR)$ does not have a τ-composition series. Therefore, the condition (2) in Theorem 1.1 does not imply that I satisfies ACC on τ-closed submodules.

From a result of Teply [15] an injective left R-module M is Σ-injective, if and only if there exists a cardinal number c such that every injective $\tau$-torsion free module is embedded in a direct sum of modules each of which is generated by c elements.

Corollary 1.3. Let M be an injective left R-module which is $\tau$-finitely generated. Then, the following conditions are equivalent:

(1) M is $\Delta$-injective.
(2) M is $\Sigma$-injective and the left R-module R/Ann(M) is embedded in a finite direct sum of copies of M.
(3) Every direct product of copies of M is embedded in a direct sum of coies of M.

Proof. (2) → (1): Let f be an R-monomorphism from R/Ann(M) to $\bigoplus_{i=1}^{n} M$. Then, f is extended to an R-monomorphism from I(R/Ann(M)) to $\bigoplus_{i=1}^{n} M$. Since M satisfies ACC on $\tau$-closed submodules, so does $I({}_R R/Ann(M))$. Hence M is $\Delta$-injective by Theorem 1.1.

(1) → (3): This is obvious from [7, Corollary 3.3]. A direct product of copies of M is a direct sum of injective indecomposable modules each of which has an essential submodule L such that $\tau$-length L = 1. Since L is contained in M, the consequence is immediate.

(3) → (2): Since R/Ann(M) is $\tau$-torsion free, the consequence is immediate.

Remark 1.4. Finitely generated $\Sigma$-injective modules are not necessarily $\Delta$-injective. For example, let R be a left Noetherian left QF-3 ring which is a maximal left quotient ring of itself but is not a right QF-3 ring. (For existence, see [14, p. 78]. A ring is called left QF-3, if it has a faithful injective left module which is isomorphic to a submodule of every faithful left module.) Then, a faithful injective left ideal Re is $\Sigma$-injective, where e is an idempotent. Suppose Re is $\Delta$-injective.

Then, R is semi-primary, i.e., R is left Artinian. However, this is a contradiction, since R becomes right QF-3.

## 2. QF-3 ENDOMORPHISM RINGS

Let U be a left R-module and $T = \mathrm{End}({}_RU)$. Then, U becomes a right T-module. In this section $\tau_1$ (resp. $\tau_2$) denotes a hereditary torsion theory of R (resp. T) cogenerated by $I({}_RU)$ (resp. $I(U_T)$). Let $Q_1(\ )$ (resp. $Q_2(\ )$) be the localization functor of $\tau_1$ (resp. $\tau_2$). Put $S = \mathrm{End}({}_RQ_1(U))$. Then, T is embedded in S canonically and hence $Q_1(U)$ becomes a right T-module. A left R-module K is called U-torsionless, if K is embedded in a direct product of copies of U.

Lemma 2.1. If $Q_1(U)$ is U-torsionless as a left R-module, then S is torsionless as a left T-module.
Proof. Assume $0 \neq x \in S$. Since $Q_1(U)x \neq 0$ and ${}_RQ_1(U)$ is U-torsionless, there exists $\phi \in \mathrm{Hom}({}_RQ_1(U), {}_RU)$ such that $(Q_1(U)x)\phi \neq 0$. It is evident that $\phi \in T$. On the other hand, $U\cdot S\phi \subseteq U$ implies $S\phi \subseteq T$. Dfine a T-homomorphism $\bar{\phi} : {}_TS \to {}_TT$ by the right multiplication of $\phi$. Since $(x)\bar{\phi} \neq 0$, S is torsionless as a left T-module.

The following result is proved in [6, Lemma 1.2 and Theorem 2.1].

Lemma 2.2. Assume U is a faithful left R-module such that $I({}_RU)$ is U-torsionless. If U is $\tau_1$-finitely generated and R satisfies DCC on $\tau_1$-closed left ideals, then,
(1) T satisfies dcc on $\tau_2$-closed right ideals,
(2) $Q_1(U) = Q_2(U)$ (= $\bar{U}$, say), $\mathrm{End}({}_R\bar{U}) = Q_2(T)$ and $\mathrm{End}(\bar{U}_T) = Q_1(R)$.
(3) $I(U_T)$ is $\bar{U}$-torsionless.

Now, in the following a left and right QF-3 ring is called

QF-3. If R has a QF-3 maximal two-sided quotient ring Q, then R is QF-3, if and only if Q is torsionless both as left and right R-module (see [4]). In [8] Morita has proved that if R is left Artinian and U is a finitely generated left R-module which is a generator (in the category of left R-modules), then End($_R$U) is QF-3, if and only if every injective hull of a simple submodule of U is embedded in U. In this case End($_R$U) is semi-primary and hece satisfies ACC on left (and right) annihilators.

Now, including this result we have the following

<u>Theorem 2.3.</u> If A left R-module U is a generator, the following conditions are equivalent:

(1) T = End($_R$U) is a QF-3 ring with ACC on left (and right) annihilators.

(2) (a) R has a minimal $\tau_1$-dense left ideal.

(b) U is $\tau_1$-finitely generated and every direct product copies of I($_R$U) is embedded in a direct sum of copies of U.

(3) (a) R has a minimal $\tau_1$-dense left ideal and satisfies DCC on $\tau_1$-closed left ideals.

(b) U is $\tau_1$-finitely generated and I($_R$U) is U-torsionless.

<u>Proof.</u> (2) $\leftrightarrow$ (3) is evident.

(3) $\rightarrow$ (1): Frome Lemma 2.2 $Q_1(U) = Q_2(U)$ $(= \bar{U})$ and $I(U_T)$ is $\bar{U}$-torsionless. Assume $0 \neq x \in I(U_T)$. Then, there exists f $\varepsilon$ Hom$(I(U_T)_T, \bar{U}_T)$ such that $f(x) \neq 0$. Let D be a minimal $\tau_1$-dense left ideal of R. It is evident that $D\bar{U}$ U. Since $\bar{U}$ is $\tau_1$-torsion free, there exists d $\varepsilon$ D such that $df(x) \neq 0$. Define $\bar{d} : \bar{U}_T \rightarrow U_T$ by the left multiplication of d. Then, $\bar{d}f(x) \neq 0$ and this implies $I(U_T)$ is U-torsionless. On the other hand, since $_R$U is a generator, $U_T$ is finitely generated projective and then $I(U_T)$ is torsionless. As $U_T$ is faithful, $I(T_T)$ is $I(U_T)$-torsionless and it follows that $\tau_2$ is cogenerated by $I(T_T)$. So $Q_2(T)$ is the maximal right quotient ring of T. Annihilator left ideals coincide with $\tau_2$-closed left ideals. By Lemma 2.2 T satisfies DCC (and ACC) on annihilator left ideals. Then by

[5, Theorem 2] $Q_2(T)$ is a QF-3 ring. Since $I(U_T)$ is torsionless, so is $Q_2(T)$ as a right T-module. By Lemma 2.2 $Q_2(T) = End({}_R\bar{U})$. So,by Lemma 1.1 $Q_2(T)$ is torsionless as a left T-module and then R is QF-3.

(1) → (3): As is shown above $\tau_2$ is cogenerated by $I(T_T)$, since $U_T$ is faithful torsionless. $I(T_T)$ is torsionless and then so is $I(U_T)$. It follows that $I(U_T)$ is U-torsionless. Since $U_T$ is finitely generated, from Lemma 2.2 $End(U_T) = R$ has DCC on $\tau_1$-closed left ideals, ${}_RU$ is $\tau_1$-finitely generated and $I({}_RU)$ is $\bar{U}$-torsionless. Since T is left QF-3, it has a minimal dense right ideal (see [12]), i.e., minimal $\tau_2$-dense right ideal. Therefore, by the same argument as in (3) → (1) we have that $I({}_RU)$ is U-torsionless. Now, it is sufficient to show R has a minimal $\tau_1$-dense left ideal. Since $U_T$ is finitely generated projective, $\bar{U}_{Q_2(T)}$ is a finitely generated $Q_2(T)$-module. On the other hand, Sinve T satisfies DCC on $\tau_2$-closed left ideals, $Q_2(T)$ is semi-primary. Hence $End(\bar{U}_{Q_2(T)}) = Q_1(R)$ is a semi-primary ring. Let $\tau_1'$ be the hereditary torsion theory cogenerated by the left $Q_1(R)$-module $I({}_{Q_1(R)}Q_1(R))$. Then, $Q_1(R)$ has the minimal $\tau_1'$-dense left ideal J. Put $A = \{r \in T; rQ_1(R) \subseteq R\}$. As $Q_1(R) = End({}_R\bar{U})$ is torsionless as a right R-module by the left right symmetry of Lemma 2.1, A has no non-zero right annihilator in $Q_1(R)$. Then, we can easily check that AJ is a minimal $\tau_1$-dense left ideal of R.

For a left (right) R-module K we denote by K* the right R-module $Hom({}_RK, {}_RR)$ (resp. left R-module $Hom(K_R, R_R)$). It is well known that a ring R is quasi-Frobenius, if and only if R is two-sided Artinian and K* is a simple right (resp. left) R-module for every simple left (resp. right) R-module K (see[10]). Now, considering this result we have the next

Proposition 2.4. Assume R is a maximal two-sided quotient ring of itself. Let $\tau_1$ (resp. $\tau_2$) be the hereditary torsion theory cogenerated by $I({}_RR)$ (resp. $I(R_R)$). Then, the following conditions are equivalent:

(1) R is semi-primary QF-3.

(2) (a) R satisfies DCC on $\tau_1$-closed left ideals and on $\tau_2$-closed right ideals.

(b) If K is a $\tau_1$-torsion free (resp. $\tau_2$-torsion free) R-module such that $\tau_1$-length K = 1 (resp. $\tau_2$-length K = 1), then $\tau_2$-length K* = 1 (resp. $\tau_1$-length K* = 1).

(3) (a) R satisfies DCC on $\tau_1$-closed left ideals and on $\tau_2$-closed right ideals.

(b) The class of all reflexive left R-modules coincides with the class of all $\tau_1$-finitely generated left R-modules K such that $Q_1(K) = K$.

Proof. (1) → (3): This is evident from[6, Corollary 2.4 and Theorem 3.5].

(3) → (2): A finitely generated $\tau_1$-torsion free module is embedded in a $\tau_1$-finitely generated module K such that $Q_1(K) = K$. Therefore, it is torsionless and the consequence is immediate from [6,Proposition 1.4].

(2) → (3): Let F be a finitely generated free left R-module and V a $\tau_1$-closed submodule of F. Then, there exists a $\tau_1$-composition series

$$F = F_0 \supset F_1 \supset \cdots \supset F_i = V \supset \cdots \supset F_n = 0.$$

Put $r(F_j) = \{f \in F^*; (F_j)f = 0\}$. Then, we have a sequence

$$0 = r(F_0) \subsetneq r(F_1) \subsetneq \cdots \subsetneq r(F_n) = F^* \quad (*)$$

Clearly, this is a sequence of $\tau_2$-closed submodules of F* and $r(F_{j+1})/r(F_j)$ is embedded in $(F_j/F_{j+1})^*$. Since $\tau_1$-length $F_j/F_{j+1}$ = 1, $\tau_2$-length $r(F_{j+1})/r(F_j) \leq 1$. Hence $\tau_1$-length F ≧ $\tau_2$-length F*. By the same manner we can show $\tau_2$-length F* ≧ $\tau_1$-length F. This implies (*) is a $\tau_2$-composition seris. Put $l(r(F_j)) = \{x \in F; (x)f = 0$ for every $f \in r(F_j)\}$. It follows that

$$F = l(r(F_0)) \supset l(r(F_1)) \supset \cdots \supset l(r(F_n))$$

is a $\tau_1$-composition series of F. Thus, we have $l(r(F_j)) = F_j$, $j = 0, 1, \ldots, n$. Since $l(r(V)) = V$, F/V is a torsionless left R-module. This implies every finitely generated submodule of $I({}_RR)$ is torsionless. From [5, Theorem 2] R is a semi-primary QF-3 ring.

## REFERENCES

[1] J.E. Björk, Rings satisfying certain chain condition, J.reine Angew. Math. 237 (1970), 63-73.

[2] C. Faith, Injective Modules and Injective Quotient Rings, Lecture Note in Pure and applied Math. 72 (Marcel Dekker,1982).

[3] J.S. Golan, Localizations of Noncommutative Rings, Pure and Applied Math. 30 (Marcel Dekker, 1975).

[4] K. Masaike, On embedding torsion free modules into free modules, Proc. Japan Acad. 53 (1977), 26-31.

[5] K. Masaike, Semi-primary QF-3 quotient rings, Comm. in Algebra 11 (1983), 377-389.

[6] K. Masaike, Duality for quotient modules and a characterization of reflexive modules, J. Pure and Applied Algebra 28 (1983), 265-277.

[7] R.W. Miller and M.L. Teply, The decsending chain condition relative to a torsion theory, Pacific J. Math. 83 (1979), 207-218.

[8] K. Morita, Duality in QF-3 rings, Math. Z. 108 (1968), 237-252.

[9] K. Morita, Localization in categories of modules I, Math. Z. 114 (1970), 121-144.

[10] K. Morita and H. Tachikawa, Character modules, submodules of a free modules, and quasi-Frobenius rings, Math. Z. 65 (1956), 414-428.

[11] A. Rosenberg and D. Zelinsky, On the finiteness of injective hull, Math. Z. 70 (1959), 372-380.

[12] E.A. Rutter Jr., Dominant modules and finite localizations,

Tohoku Math. J. 27 (1975) 225-239.

[13] B. Stenström, Rings of Quotients, Grund. Math. Wiss. 217 (Springer, 1975).

[14] H. Tachikawa, Quasi-Frobenius rings and Generalizations, Lecture Note in Math. 351 (Springer, 1973).

[15] M.L. Teply, Torsion free injective modules, Pacific J. Math. 28 (1969) 441-453.

# GROUP ACTIONS ON RINGS: SOME CLASSICAL PROBLEMS

S. Montgomery*

University of Southern California
Los Angeles, CA 90089-1113

A great deal of work has been done in the last ten years on actions of finite groups on non-commutative rings. However, many questions remain open, and we shall discuss here fifteen such questions. These problems are all related to a common circle of ideas which have roots in the classical theory of commutative algebra and invariant theory. In particular, our topics include integrality, prime ideals, chain conditions, and invariants of generic matrix rings. For each question, we survey the known parital results; additionally, in the section on prime ideals, we give some new proofs and even some new results. Finally, we discuss the problem of classifying the actions of a given group on a specific ring.

Throughout, R denotes an algebra with 1 over a commutative ring k, G denotes a finite group of order $|G|$, and $\alpha : G \to \mathrm{Aut}_k R$ is a group homomorphism. For simplicity, we usually suppress the particular action $\alpha$ and write $r^g$ for the image of r under $\alpha(g) \in \mathrm{Aut}\, R$. The fixed ring, or ring of invariants, is denoted by $R^G$.

## §1. INTEGRALITY OF R OVER $R^G$

If R is commutative, it is trivially true that R is integral

*Work partially supported by NSF Grant MCS 83-01393. The author also wishes to thank F. Van Oystaeyen for organizing this conference.



*F. van Oystaeyen (ed.), Methods in Ring Theory, 327–346.*

over $R^G$, since each $r \in R$ satisfies the polynomial $p_r(x) \in R^G[x]$ given by $p_r(x) = \Pi_{g\in G}(x-r^g)$. This is easily seen to be false when R is not commutative.

More generally, we consider the more general notion of integrality due to W. Schelter: that is, a ring extension $S \subset R$ is integral if each $r \in R$ satisfies some monic $p_r(x)$ in the free product (or coproduct) $S \amalg_k k[x]$. That is, $p_r(x) = x^n + q(x)$, where the x-degree of each monomial in q has degree $< n$.

However, even with this more general definition, R can fail to be integral over $R^G$, as the following example (due to G. Bergman) shows.

1.1 Example Let $R = M_2(k\langle x,y\rangle)$, the $2 \times 2$ matrices over the free algebra $k\langle x,y\rangle$, where k is a field of characteristic $p \neq 0$ containing a primitive $n^{th}$ root of 1, w, for some $n > 1$. Let G be the group generated by the three inner automorphisms which are determined by conjugation by $\begin{pmatrix} 1 & x \\ 0 & 1 \end{pmatrix}$, $\begin{pmatrix} 1 & y \\ 0 & 1 \end{pmatrix}$, and $\begin{pmatrix} 1 & 0 \\ 0 & w \end{pmatrix}$. Then $|G| = np^2$ and $R^G \cong k$. Clearly R is not integral over $R^G$.

Moreover, an example of non-integrality is given by Montgomery and Small [31] in which R is an affine, Noetherian, PI algebra. However, in both of these examples one has $|G|R = 0$.

The following fundamental question remains open:

Problem 1: If $|G|^{-1} \in k$, is R (Schelter) integral over $R^G$?

The answer is known to be yes in the following three case:

(1) R is (left) Noetherian.
For in this case, $R^G$ is also (left) Noetherian since the map $\rho_G: R \to R^G$ given by $\rho(r) = |G|^{-1}\Sigma_g r^g$ is a projection of R onto $R^G$. Now by a theorem of D. Farkas and R. Snider [10], R is a Noetherian $R^G$-module. Thus for any $r \in R$, the submodule $\Sigma_{n\geq 0} R^G r^n$ is finite over $R^G$, and so r is (classically) integral over $R^G$.

(2) R satisfies a polynomial identity.
This is proved by Montgomery and Small in [32]. It was also proved independently by Schelter (unpublished), and a partial result was obtained by J. Alev [1].

(3) G is abelian, for any R.
This is a theorem of D.S. Passman [35], who actually proves a somewhat stronger result: R is integral of bounded degree over $R^G$,

where the bound depends only on $|G|$.

We remark that Passman also gives a second proof of this result, due to Bergman, which uses group-graded rings. In particular, for any algebra A graded by a finite group G, with identity component $A_1$, the extension $A_1 \subset A$ is integral. Because of the many parallels between the extensions $R^G \subset R$ and $A_1 \subset A$, it suggests that the answer to Problem 1 is yes. As of this writing, however, it is still open for $G = S_3$.

## §2. PRIME IDEALS

In this, the main section of the paper, we are concerned with the relationship between Spec R, the set of prime ideals of R, and Spec $R^G$. We define the map

$$f : \text{Spec } R \to \text{Spec } R^G$$

by $\quad P \to \{p \mid p \text{ is minimal over } P \cap R^G\}$ .

Thus f assigns each prime P of R to a set of primes in $R^G$.

It is clear that if P, Q $\varepsilon$ Spec R with $Q = P^g$, some $g \in G$, then $Q \cap R^G = P \cap R^G$ and so f(P) = f(Q). That is, f is constant on Spec $R/_G$, the G-orbits of Spec R. This gives us an induced map

$$f' : \text{Spec } R/_G \to \text{Spec } R^G$$

Our first question is related to the injectivity of f'.

### 2A. The orbit problem

Simply stated, the orbit problem is the following: given P, Q $\varepsilon$ Spec R with $P \cap R^G = Q \cap R^G$, when are P and Q in the same G-orbit?

Equivalently, define the map

$$\tilde{f} : \text{Spec } R/_G \to \{\text{ideals of } R^G\}$$

by $\quad \{P^g\} \mapsto P \cap R^G$

The orbit problem asks when $\tilde{f}$ is injective.

This property is well-known to hold if R is commutative [4], and follows from integrality:

2.1 Lemma: If R is commutative, then $\tilde{f}$ is injective, for any group G acting on R.

Proof: Say that P, Q $\varepsilon$ Spec R with $P \cap R^G = Q \cap R^G$, and let $A = \bigcap_g P^g$, $B = \bigcap_g Q^g$. Since G is finite and the $\{P^g\}$, $\{Q^g\}$ are all primes, it will suffice to show that A = B.

Since A is integral over $A^G$, any a $\varepsilon$ A satisfies an equation of the form $a^n + b_{n-1}a^{n-1} + \ldots + b_0 = 0$, where the $b_i \varepsilon A^G = B^G \subset B$. Thus $a^n \varepsilon B$, for each a $\varepsilon$ A. Since B is an intersection of primes, a $\varepsilon$ B, and so $A \subset B$. Similarly $B \subset A$.

The proof works a bit more generally: as long as R is (Schelter) integral over $R^G$, $A + B/_B$ will be a nil ideal of the semiprime ring R/B. One still gets $A \subset B$ if either R is integral of bounded degree over $R^G$, or if R is a PI ring (so that nil ideals are locally nilpotent).

The orbit problem fails in general, as Example 1.1 shows: an infinite number of primes of R contract to (0) in $R^G$, although G is finite. It is even false if R is a PI ring [32]. However, some positive results are obtained by Montgomery and Small in [32], and it is true in the following cases:

(1) $|G|^{-1} \varepsilon R$

The argument is very close to that of Lemma 2.1, except that instead of integrality one uses the Bergman-Isaacs theorem, which says that if $(A + B/_B)^G = (0)$, then $A + B/_B = (0)$, and so $A \subseteq B$.

(2) R is semiprime PI and P,Q are identity-faithful primes.

There are several interesting cases which are open.

Problem 2: Is the map $\tilde{f}$ injective on G-orbits if either
(a) R is semiprime PI?
or (b) R is prime and G is X-outer?

(Recall that G X-outer means that any $g \neq 1$ in G remains outer when extended to the two-sided (Martindale) quotient ring of R).

Throughout the rest of §2, we assume that $|G|^{-1} \varepsilon R$. In particular, by (1) above, we know that f is always injective. In fact, much more is known about f in this case, and before proceeding with our next problem, we take time out to examine it in more detail.

First, we may define an equivalence relation on Spec $R^G$. If p, q $\varepsilon$ Spec $R^G$, we say $p \sim q$ if there exists P $\varepsilon$ Spec R so that both p and q are minimal over $P \cap R^G$. The following facts are proved by Montgomery in [27]:

2.2 Theorem (1) $\sim$ is an equivalence relation on Spec $R^G$; in particular each $p \in \operatorname{Spec} R^G$ is minimal over precisely one $P \cap R^G$, where P is unique up to its G-orbit.

(2) Each equivalence class p of $\operatorname{Spec} R^G/\sim$ contains $\leq |G|$ elements, say $\overline{p} = \{p_1, \ldots, p_m\}$, and $\cap_{i=1}^m p_i = P \cap R^G$ where $P \in \operatorname{Spec} R$ is given by (1).

(3) The map f induces a bijection

$$\overline{f} : \operatorname{Spec} R/_G \to \operatorname{Spec} R^G/\sim$$

which is a homeomorphism of the respective quotient Zariski topologies.

(4) Equivalent primes in $R^G$ have the same height; moreover ht p = ht P, where $P \in \operatorname{Spec} R$ is given by (1).

The method of proof of these results is indirect, and uses the skew group ring $R_\alpha G$. There are three ingredients. The first comes from work of Lorenz and Passman [20], who set up a correspondence between Spec R and Spec $R_\alpha G$. Since we shall need it below, we describe this correspondence in the special case when $|G|^{-1} \in R$.

2.3 <u>Theorem</u>: (1) Given $P \in \operatorname{Spec} R$, $A = \bigcap_g P^g$ is a G-prime ideal of R and $A_\alpha G = \mathcal{P}_1 \cap \ldots \cap \mathcal{P}_n$, $n \leq |G|$, where the $\mathcal{P}_i \in \operatorname{Spec} R_\alpha G$ are precisely those primes minimal over $A_\alpha G$.

Moreover, any one of the $\mathcal{P}_i$ is primitive <=> P is primitive <=> all the $\mathcal{P}_i$ are primitive.

(2) Given $\mathcal{P} \in \operatorname{Spec} R_\alpha G$, $\mathcal{P} \cap R = A$ is a G-prime ideal of R (so $A = \bigcap_g P^g$ for some $P \in \operatorname{Spec} R$), and $\mathcal{P}$ is minimal over $A_\alpha G$.

Passman has recently given fairly elementary proofs of (1) and (2), when $|G|^{-1} \in R$, in [36].

We formalize this correspondence by defining a map

$$\lambda : \operatorname{Spec} R \to \operatorname{Spec} R_\alpha G$$

by $$P \mapsto \{\mathcal{P}_1, \ldots, \mathcal{P}_n\}$$

Secondly, for any ring S with idempotent $e \neq 0$, there is a mapping $\phi$ from ideals of S to ideals of eSe given by $I^\phi = I \cap eSe = eIe$. If $\operatorname{Spec}_e S$ denotes the primes of S not containing e, then

$$\phi : \operatorname{Spec}_e S \to \operatorname{Spec} eSe$$

is a bijection.

The last ingredient is the elementary fact that for the idempotent $e = |G|^{-1} \sum_g g \in R_\alpha G$,

$$e(R_\alpha G)e = eR^G \cong R^G.$$

We will denote this isomorphism by i.

Our original map f : Spec R → Spec $R^G$ can then be factored in terms of these three maps. That is

$$f : \text{Spec } R \xrightarrow{\lambda} \text{Spec } R_\alpha G \xrightarrow{\phi} \text{Spec } e(R_\alpha G)e \xrightarrow{i} \text{Spec } R^G$$

For simplicity, we will frequently omit the last isomorphism i and write $f = \phi \circ \lambda$.

2B. Equivalence invariants

A property of prime ideals is called an equivalence invariant for G on R if, whenever some $p \in \text{Spec } R^G$ has the property and $q \sim p$, then q also has the property. The following are some known equivalence invariants:

(1) primitivity of p
This follows since the Lorenz-Passman correspondence preserves primitivity (as noted in 2.2) as does the map $\phi$.

(2) height of p
This is 2.3 (4), above.

(3) Gelfand-Kirillov (GK) dimension of $R^g/p$, when R is a PI ring.
This is proved by Montgomery and Small in [32].

(4) depth of p (that is, the classical Krull dimension of $R^G/p$) when R is an affine PI ring.
This is a lovely theorem of J. Alev [1]. We will discuss a (somewhat new) proof of his result below.

(5) depth of p when R is a Noetherian PI ring
A similar although easier argument than that given for GK-dimension in [32] shows that if $p \sim q$ in $R^G$ then $R^G/p$ and $R^G/q$ have the same Gabriel-Rentschler Krull dimension. But since R is Noetherian PI, as is $R^G$, this is the same as the classical Krull dimension.

In general, however, depth is not an equivalence invariant, even if R is affine and Noetherian. The following example is due to Montgomery and Small [32].

2.4 Example Let $A = k\langle x, y \mid xy - yx = 1\rangle$ denote the Weyl algebra over a field of characteristic 0, and let $R = \begin{pmatrix} k+Ax & A \\ Ax & A \end{pmatrix}$. Let g be conjugation by $\begin{pmatrix} 1 & 0 \\ 0 & -1 \end{pmatrix}$ and let $G = \langle g\rangle$. Then $R^G \cong (k+Ax) \oplus A$. Let $P = (0)$ in Spec R. Then in $R^G$, there are two primes minimal over (0): $p_1 = (0,A)$ and $p_2 = (k+Ax,0)$, and so $p_1 \sim p_2$. However, $p_2$ is maximal but $p_1$ is not.

In discussions with Small after this meeting, we realized that the above example can be adjusted to show that depth is not an equivalence invariant for an arbitrary PI ring. For, simply replace the Weyl algebra A by a commutative valuation domain D of rank 2, which is a k-algebra. Let the prime ideals of D be $M \supset P \supset (0)$, where M is maximal, and let $R = \begin{pmatrix} k+P & D \\ P & D \end{pmatrix}$. Since dim (k+P) = 1 and dim D = 2, the equivalent primes $p_1 = (0,D)$ and $p_2 = (k+P,0)$ have depths 1 and 2, respectively.

We now return to Alev's result, stated as (4) above. Since from (3), GK-dimension is an equivalence invariant, what Alev actually proves is the following:

2.5 <u>Theorem</u>: Let R be an affine PI ring, G a group acting on R with $|G|^{-1} \in R$, and $p \in \operatorname{Spec} R^G$. Then

$$\text{GK dim } (R^G/p) = \text{cl.K.dim } (R^G/p).$$

A new proof of Alev's theorem is given in [33]. We sketch it here to illustrate some of the ideas.

<u>proof of 2.5</u>: Since R is an affine PI k-algebra, the skew group ring $R_\alpha G$ is also an affine PI algebra. Now, for $p \in \operatorname{Spec} R^G$, let $\mathcal{P} = \phi^{-1}(p) \in \operatorname{Spec} R_\alpha G$. Then $A = R_\alpha G/\mathcal{P}$ is <u>a</u> prime affine PI algebra. Denoting the image of e in A by $\bar{e}$, it follows that

$$\bar{e}A\bar{e} \cong \bar{e}(R_\alpha G)\bar{e}/\bar{e}\mathcal{P}\bar{e} \cong R^G/p = B.$$

The theorem now follows immediately from the result of [33]: if A is any prime, affine PI algebra, and B a prime subalgebra of A, then GK dim B = cl.K.dim B.

Next, we characterize what it means for depth to be an equivalence invariant in a special case. Recall that R satisfies the <u>dimension formula</u> if for any prime P of R,

$$\text{cl.K.dim } R = \text{depth } P + \text{ht } P$$

2.6 <u>Lemma</u>: Say that R satisfies the dimension formula, and $|G|^{-1} \in R$. Then $R^G$ satisfies the dimension formula <=> depth is an equivalence invariant.

<u>proof</u>: Choose $p \in \operatorname{Spec} R^G$, and say p is minimal over $P \cap R^G$, some $P \in \operatorname{Spec} R$. By a result of Lorenz and Passman [20], R and $R^G$ have the same Krull dimension; also, ht p = ht P by 2.2(4). Thus $R^G$ satisfies the dimension formula if and only if depth p = depth P,

for all $p \varepsilon$ Spec $R^G$ and corresponding $P \varepsilon$ Spec R. Since $q \sim p \Leftrightarrow q \varepsilon f(P)$ also, clearly this is equivalent to depth being an equivalence invariant.

The lemma gives a more straightforward proof of a corollary of Alev. It is of interest since Schelter's theorem cannot be applied directly to $R^G$, as $R^G$ is not necessarily affine [33].

2.7 Corollary: Let R be an affine PI ring with $|G|^{-1} \varepsilon R$. Then $R^G$ satisfies the dimension formula.

proof: R satisfies the dimension formula by [38]. By 2.5, depth is an equivalence invariant. The result now follows by Lemma 2.6.

We raise a rather general question:

Problem 3: Study other properties of primes to determine which are or are not equivalence invariants.

## 2C. The closed map problem

Now consider Spec R and Spec $R^G$ with the usual Zariski topologies; we wish to study properties of $f$ : Spec R $\to$ Spec $R^G$ with respect to these topologies. As stated in 2.2(3), it was proved in [27] that the induced map $\bar{f}$ : Spec $R/_G \to$ Spec $R^G/\sim$ is a homeomorphism. Part of the arguments used in proving this fact show the following:

2.8 Proposition: $f$ is a closed map

proof: Consider a closed set in Spec R, say $v(I) = \{ P \varepsilon \text{ Spec } R \mid P \supseteq I \}$; we may assume $I = \bigcap_\alpha P_\alpha$. We will show that $f(v(I)) = v(I \cap R^G)$, a closed set in Spec $R^G$.

First, choose $P \varepsilon v(I)$. Then $P \supset I$ implies $P \cap R^G \supset I \cap R^G$, so $f(P)$ consists of primes containing $I \cap R^G$. Thus $f(v(I)) \subseteq v(I \cap R^G)$.

Conversely, choose $p \varepsilon v(I \cap R^G)$; then $p \subseteq \bigcap_\alpha P_\alpha \cap R^G$. By 2.3(1) there exists $p \varepsilon$ Spec R so $p$ is minimal over $P \cap R^G$. It will suffice to show that for some $h \varepsilon G$, $I = \bigcap_\alpha P_\alpha \subseteq P^h$ (for then $p \varepsilon f(P) = f(P^h) \varepsilon f(v(I)))$. We let $A = \bigcap_g P^g$, and for each $\alpha$ let $A_\alpha = \bigcap_g P_\alpha^g$ and write $A_\alpha \cap R^G = P_\alpha \cap R^G = \bigcap_i p_{\alpha,i}$. Then $I \cap R^G = \bigcap_{\alpha,i} p_{\alpha,i} \subseteq p$ implies $\bigcap_{\alpha,i} \phi^{-1}(p_{\alpha,i}) \subseteq \phi^{-1}(p)$.

Intersecting with R gives

$$\bigcap_\alpha A_\alpha = \bigcap_{\alpha,i} \phi^{-1}(p_{\alpha,i}) \cap R \subseteq \phi^{-1}(p) \cap R = A.$$

Thus $\prod_{g\varepsilon G}(\bigcap_\alpha P_\alpha)^g \subseteq \bigcap_g P^g \subseteq P$, and so for some $h \varepsilon G, (\bigcap_\alpha P_\alpha)^{h^{-1}} \subseteq P$. Then $\bigcap_\alpha P_\alpha \subseteq P^h$ and we are done.

An obvious quesion is what happens if we replace Spec R by the primitive or maximal ideals of R. We first consider the case of primitive ideals.

Let $\text{Spec}_p R$ denote the set of primitive ideas of R, and let $f_p$ denote the restriction of f to $\text{Spec}_p R$. By 2.3(1), it is clear that $\text{Im}(f_p) = \text{Spec}_p R^G$. Thus an immediate consequence of 2.8 is:

2.9 Corollary: $f_p$ is a closed map

Moreover, as in [27], one can show that

$$\bar{f}_p : \text{Spec}_p R/_G \longrightarrow \text{Spec}_p R^G/\sim$$

is a homeomorphism.

The case of maximal ideals is considerably more subtle. As above, let $\text{Spec}_m R$ denote the maximal ideals of R and let $f_m$ denote the restriction of f to $\text{Spec}_m R$. Our first observation is that $\text{Im}(f_m) \subseteq \text{Spec}_m R^G$. For if $M \varepsilon \text{Spec}_m R$, then $\bar{R} = R/\bigcap_g M^g$ has cl.k.dim 0, and thus $\bar{R}^G$ also has cl.K.dim 0 [20]. But $\bar{R}^G = R^G/\bigcap_\alpha P_\alpha$, where $\{p_\alpha\} = f(M)$. Thus each $p_\alpha$ is maximal. In particular this shows that

$$f_m(M) = f(M) = \{ m \varepsilon \text{Spec}_m R^G \mid m \supseteq M \cap R^G \}.$$

However, $f_m$ is not surjective, in general. This can be seen from Example 2.4: $p_2$ is maximal but $p_2 \varepsilon f(\text{Spec}_m R)$.

2.10 Lemma $f_m$ is surjective if and only if maximality is an equivalence invariant.

proof: Choose $m \varepsilon \text{Spec}_m R^G$. If $f_m$ is surjective, then $m \varepsilon f_m(M)$ for some $m \varepsilon \text{Spec}_m R$. By the above argument, all the primes in $\bar{m}$ are maximal; thus maximality is an equivalence invariant.

Conversely, say that $\bar{m}$ consists only of maximal ideals, say $\{ m_1 = m, m_2, \ldots, m_k \}$. Then $\bigcap_i m_i = P \cap R^G$, where $p \varepsilon$ Spec R with $f(P) = \{m_i\}$. Using $\bar{R} = R/\bigcap_g p^g$, we have $\bar{R}^G = R^G = R^G/\bigcap_i m_i$. Since $\bar{R}^G$ has cl K. dim 0, so does $\bar{R}$; thus $P = M \varepsilon \text{Spec}_m R$. But then $m \varepsilon f_m(M)$ so f is surjective.

The following question remains open:

Problem 4: If $|G|^{-1} \varepsilon R$, is $f_m$ a closed map?

It has a positive answer for PI rings.

2.11 Proposition: If R is a PI ring and $|G|^{-1} \varepsilon R$, then $f_m$ is closed.

proof: By Kaplansky's theorem, all primitive ideals are maximal. The result now follows from Corollary 2.9.

Proposition 2.11 was proved by Alev [1] for affine PI rings, with a more difficult proof. Alev's arguments, however, point out a rather interesting connection between Problem 4 and integrality of $R^G \subset R$. In a recent paper [2], Artin and Schelter prove the following theorem: if $S \subset R$ are affine PI algebras, let $f_m : \mathrm{Spec}_m R \to \mathrm{Spec}_m S$ be given by $f_m(M) = \{ m \in \mathrm{Spec}_m S \mid m \supseteq M \cap S \}$. Then the extension $S \subset R$ is integral <=> $f_m$ is "universally closed"; that is, $f_m$: $\mathrm{Spec}_m R[x] \to \mathrm{Spec}_m S[x]$ is closed for any central variable x.

Now consider the extension $R^G \subset R$. We may extend G to R[x] by letting it act trivially on x, so $(R[x])^G = R^G[x]$. Thus if we wish to show that $f_m$ is universally closed for a class of rings closed under polynomial extensions, it suffices to show $f_m$ is closed.

We therefore raise the following question.

Problem 5: What is the connection between Problem 1 and Problem 4?

## §3. CHAIN CONDITIONS

We first consider when the property that R is right or left Noetherian is inherited by $R^G$. When $|G|^{-1} \varepsilon R$, this is always true, and easy, as noted in §1. It is false in general, however, even when R is commutative. This is shown by an example of Nagarajan [34] in which $|G| = 2$ and $2R = 0$. More recently an example in characteristic 0 has been given [6]. However, these examples are not finitely-generated (affine) over a field. We ask:

Problem 6: Let R be a Noetherian domain which is affine over a field k. Is $R^G$ Noetherian?

By the Hilbert basis theorem and Noether's theorem on affine invariants, the answer to Problem 6 is "yes" if R is commutative.

The next question is closely related.

Problem 7: With the same hypotheses as Problem 6, is $R^G$ affine over k?

For a survey of what is known about affine fixed rings for non-commutative rings, see [28]. Also, the following fact may be helpful in these problems: for any finite group G acting on a domain R, there exists a non-trivial "partial trace function." That is, there exists a subset $\Lambda \subseteq G$ so that $0 \neq t_\Lambda(R) \subseteq R^G$, where $t_\Lambda(r) = \Sigma_{g\varepsilon\Lambda} r^g$ [25].

Returning to the Noetherian problem, another open case is:

Problem 8: If R is simple Noetherian and G is outer, is $R^G$ Noetherian?

In this situation, it is known that R has a "dual basis" over $R^G$; that is, there exist elements $\{x_1,\ldots,x_n;\ y_1,\ldots,y_n\}$ R so that $\Sigma_{i=1}^n x_i y_i^g = \delta_{1,g}$, all $g \in G$ [24]. Also, the usual trace function is non-trivial [26, Theorem 2.2]. On the other hand, J. Osterburg has pointed out, using an example of Zaleskii and Neroslavskii, that the hypotheses of Problem 8 do not force $R^G$ to be simple (see Example 2.8 of [26]). For this example, we do not know whether $R^G$ is Noetherian.

The last question in this section concerns the Goldie rank, or uniform dimension, of modules. Assume that $|G|^{-1} \in R$, and let $M_R$ be a (right) R-module of finite rank. Our question concerns when M has finite rank as an $R^G$-module. This is false in general, as shown by an example of Fisher and Osterburg [12]. However, in their example R is not semiprime. We ask:

Problem 9: Let R be semiprime Goldie with $|G|^{-1} \in R$, and let $M_R$ have Goldie rank n. Does $M_{R^G}$ have Goldie rank $\leqslant n|G|$?

The answer to Problem 9 is yes if either
(1) M = R; this is due to Fisher and Osterburg [12], or
(2) R is semisimple Artinian; this follows from Lorenz and Passman's result that the length of $M_{R^G}$ is $\leqslant n|G|$ if n is the length of $M_R$ [21].

Result (2) suggests the possibility of a reduced rank argument for Problem 9. Also, there has been some work recently concerning "additivity principles" which may be helpful. As of this writing, the question remains open for Ore domains.

## §4. GENERIC MATRIX RINGS

We fix the following notation. Let k be a field, and V a

vector space over k of dimension $d > 1$. Fix a basis $\{x_1,\ldots,x_d\}$ of V. We consider the following three rings:

$S = k[V] = k[x_1,\ldots,x_d]$, the commutative polynomial ring.

$F = k\langle V\rangle = k\langle x_1,\ldots,x_d\rangle$, the free algebra

$U = k\{X_1,\ldots,X_d\}$, the ring of $m \times m$ generic matrices.

That is, each $X_\ell$ is an $m \times m$ matrix whose entries $x_{ij}^{(\ell)}$ are independent commuting indeterminates over k. U can be identified with F/T, where T is the ideal of F of all identities of $m \times m$ matrices.

Let G be a finite subgroup of GL(V); that is, G acts linearly on the variables $\{x_\ell\}$, for both S and F. Since T is G-stable, there is an induced G-action on U, and G acts linearly on the $\{X_\ell\}$.

We first discuss when the fixed ring of each of these three rings is another ring of the same type. For the polynomial ring, $S^G$ is again a polynomial ring if and only if G is generated by pseudo-reflections; this is due to Shephard and Todd [39] and also Chevalley, who proved the "if" direction independently [5]. For the free algebra, $F^G$ is always free, for any G; this was proved independently by Lane [19] and by Kharchenko [17]. Very recently, the analogous problem for the generic matrices has been solved by R. M. Guralnick [13]:

4.1 <u>Theorem</u> For $|G| > 1$, $m > 1$, $U^G$ is never a generic matrix ring.

Guralnick's argument uses the classification of pseudo-reflection groups [39]. We note that Theorem 4.1 had been proved for the case of G abelian by Kharchenko [18].

Our second topic concerns when each of the fixed rings is finitely-generated. $S^G$ is always finitely-generated, even if G is not linear, by a classical result of E. Noether. For the free algebra, the situation is almost the opposite: $F^G$ is finitely-generated if and only if G consists of scalar matrices. This was proved by Dicks and Formanek [9], and independently by Kharchenko [18]. Thus, the question naturally arises as to when $U^G$ is finitely-generated. Clearly, if G consists of scalars, then $U^G$ is finitely-generated as it is the image of $F^G$. However, when G is not scalar, the answer is not simple. Details of the next two examples appear in [28].

4.2 <u>Example</u> (Formanek). Consider $U = k\{X,Y\}$, for any $m \geq 2$ and char $k \neq 2$. Let $G = \langle g\rangle_G \subseteq GL_2(V)$, where G is generated by $X^g = X$, $Y^g = -Y$. Then $U^G$ is not finitely-generated.

4.3 Example (Montgomery and Passman). Consider $U = k\{X,Y\}$ for $m = 2$ and k a field containing a primitive cube root of 1, say w. Let $G = \langle g \rangle$, where $X^g = wX$ and $Y^g = w^2Y$. Then $U^G$ is finitely-generated, with generators

$$\{XY,\ YX,\ X^3,\ Y^3,\ X(XY-YX)Y,\ X(XY-YX)X^2,\ Y(XY-YX)X,\ Y(XY-YX)Y^2\}.$$

Recent progress has been made by Fisher and Montgomery in the case of cyclic groups [11].

4.4 Theorem Let U be an $m \times m$ generic matrix ring. Consider $G = \langle g \rangle \subseteq GL(V)$, where $n = |G|$ is a unit in k and g is not scalar. Then $U^G$ is not finitely-generated whenever $m \geq n - [\sqrt{n}] + 1$.

Essentially, the theorem says that for large enough matrices, compared to $|G|$, U behaves like the free algebra. We obtain somewhat better results in special cases; for example, if g is a pseudo reflection, $U^G$ is not finitely-generated whenever $m \geq 2$.

A major difficulty in extending this theorem to arbitrary finite groups $G \subseteq GL(V)$ is Theorem 4.1. For, both known proofs of the free algebra result proceed by reducing the problem to cyclic subgroups H of G, using the fact that $F^H$ is a free algebra. Since $U^H$ is never a generic matrix ring, this approach will not work. Nevertheless, we conjecture that the answer to the following question is yes.

Problem 10: Let $G \subseteq GL(V)$ act on U, $m \times m$ generic matrices, where $|G|$ is a unit in k and G does not consist of scalar matrices. Prove that for $m \geq 3$, $U^G$ is not finitely-generated.

The case $m = 2$ may be more complicated. In particular the only known non-scalar finitely-generated example is 4.3 above; we do not know what happens if $\begin{pmatrix} w & 0 \\ 0 & w^2 \end{pmatrix}$ is replaced by $\begin{pmatrix} i & 0 \\ 0 & -i \end{pmatrix}$, when $m = 2$.

Our next problem in this section is an old one, and difficult. We mention it again in hopes someone will work on it.

Problem 11: Determine $\mathrm{Aut}_k U$, for $U = k\{X,Y\}$.

It is known that $\mathrm{Aut}\, k\langle x,y\rangle \cong \mathrm{Aut}\, k[x,y] \cong A*B$, the amalgamated free product of the group A of affine transformations and the group B of Jonquiere, or "triangular" automorphisms [8], [23], [15], [40]. One might hope that Aut $k\{X,Y\}$ would also be the same group; that is, that the automorphism group is "tame". However, Bergman has constructed a "non-tame" automorphism of $k\{X,Y\}$ [3].

On the positive side, it is known that for $U = k\{X_1,\ldots,X_d\}$,

$Aut_k U$ acts faithfully on the center of U. This follows (via the Skolem-Noether theorem) from:

4.5 Theorem: If $g \in Aut_k U$ becomes inner on the quotient division ring of U, then $g = 1$; that is, $Aut_k U$ is X-outer.

This fact was proved independently by Montgomery [29] and by Lvov and Kharchenko [22]. The proof in [29] is simpler.

The fact that $Aut_k U$ is X-outer means that Kharchenko's Galois correspondence theorem [16] applies to any finite subgroup $G \subseteq Aut_k U$. A subring S R is called an anti-ideal if whenever $a,b \in R^k$ with $ab \in S$ and $0 \neq b \in S$, then $a \in S$. One then obtains:

4.6 Proposition: Let G be a finite subgroup of $Aut_k U$, and let S be an intermediate ring, $U^G \subseteq S \subseteq U$. Then there is a one-to-one correspondence between subgroups H of G and intermediate rings S which are anti-ideals, given by

$$H \to U^H \quad \text{and} \quad S \to \{ g \in G \mid g \text{ fixes } S \}.$$

A fairly elementary proof of Kharchenko's Galois theory in the outer case is given in [30], where the problems of extending automorphisms and intermediate Galois extensions are also considered. In particular, for any intermediate ring S, any homomorphism $\phi : S \to U$ fixing $U^G$ extends to an automorphism of U.

For the free algebra F, the anti-ideals S were precisely the intermediate free algebras. Although for U, we know they cannot be generic matrix rings, we ask:

Problem 12: Find an internal characterization of the subrings S of U which are anti-ideals.

## §5. CLASSIFYING GROUP ACTIONS

In this last section we change directions and ask a different type of question. Throughout, we have been ignoring the fact that for the same ring R and group G, there can be different actions of G on R, with correspondingly different properties. A fundamental problem is to find invariants which determine the particular group action. Some very beautiful work of this type has been done in operator algebras, and we discuss here one such result.

5.1 Definition: Two actions $\alpha, \alpha'$ of G on R are conjugate if there exists $\sigma \in Aut_k R$ such that for all $g \in G$,

$$\alpha'(g) = \sigma^{-1} \alpha(g) \sigma.$$

Clearly, if we are interested in ring theoretic properties, it suffices to classify group actions up to conjugacy.

The theorem we discuss here is due to V. Jones, who in [14] classified (up to conjugacy) all finite group actions on the "hyperfinite $II_1$ factor" $\underline{R}$ (this had been done earlier for finite cyclic groups by A. Connes [7]). First, we describe the algebra $\underline{R}$.

Let $A = \varinjlim M_{n_i}(C)$, a direct limit of matrix rings over the complex numbers C, where the $\{n_i\}$ are a sequence of strictly increasing positive integers with $n_i | n_{i+1}$, all i = 1, 2,... . Now A can be imbedded in a natural way into B = B(H), the algebra of all bounded operators on a Hilbert space H. Then $\underline{R}$ is defined as the smallest von Neumann algebra in B which contains A. Equivalently, $\underline{R}$ is the double centralizer of A in B.

It is an interesting property of $\underline{R}$ that it is independent of the choice of the $\{n_i\}$. $\underline{R}$ has an involution which extends the usual Hermitian transpose on the $M_{n_i}(C)$, and it also has a trace function $t : \underline{R} \to C$ obtained by normalizing the matrix trace at each step so that t(I) = 1. We restrict our attention to those automorphisms which preserve the involution, denoted Aut*$\underline{R}$; an easy consequence of this is that any inner automorphism is induced by a unitary.

Thus, fix a finite group G, and consider an action

$\sigma : G \to \mathrm{Aut}^*\underline{R}$.

Jones defines three invariants:

(1) $N = N_\alpha = \{ g \in G \mid \alpha(g) \text{ is inner on } R \}$

(2) the "characteristic invariant"

This invariant is basically an element of a relative cohomology group. For each $h \in N$, let $v_h \in R$ induce h, and assume $v_1 = 1$. Then for all h, $k \in N$,

$v_h v_k = \mu(h,k)v_{hk}$, where $\mu : N \times N \to C^*$ is a cocycle.

Moreover, for all $g \in G$, $h \in N$,

$v_{g^{-1}hg}^{\alpha(g)} = \lambda(g,h)v_h$, where $\lambda : G \times N \to C^*$.

The pair $(\lambda,\mu)$ arising in this way is not an invariant of $\alpha$, since it depends on the choice of the $\{v_h\}$. If $\{v'_n\}$ is another choice, then for each $h \in N$,

$$v'_h = \eta(h)v_h, \text{ where } \eta : N \to C^*. \qquad (*)$$

The $\{v'_h\}$ determine a new pair $(\lambda',\mu')$. By defining pairs to be equivalent when such a relation (*) holds, the <u>characteristic invariant</u> of $\alpha$ is defined to be the equivalence class $[\lambda,\mu]$.

The set of classes $[\lambda,\mu]$ is just a relative cohomology group $\Lambda(G,N) = Z(G,N)/B(G,N)$; details appear in [14].

(3) the "trace invariant"

Given $(\lambda,\mu)$ as above, we may form the twisted group algebra $C^{\mu}N$, with C-basis $\{ z_h \mid h \in N \}$. Then there is a homomorphism $\psi : C^{\mu}N \to \underline{R}$ given by $\psi(z_h) = v_h$. Since $\underline{R}$ has a trace, $C^{\mu}N$ has an induced trace $\tau : C^{\mu}N \to C$ given by $\tau = t\circ\psi$.

Also G acts on $C^{\mu}N$ via

$$\left( \sum_{h\in N} c_h z_h \right)^g = \sum_{h\in N} \lambda(g,h) c_{h^{-1}gh} z_h .$$

Note that the fixed ring $(C^{\mu}N)^G$ is contained in the center of $C^{\mu}_GN$. The <u>trace invariant</u> is defined as the restriction of $\tau$ to $(C^{\mu}N)^G$.

We are now able to state Jones' theorem.

5.2 <u>Theorem</u>: Let $\underline{R}$ be the hyperfinite factor of type $II_1$. Then any action $\alpha : G \to \mathrm{Aut}^*\underline{R}$ is uniquely determined (up to conjugacy) by the three invariants (1), (2), and (3). Moreover, actions of G exist for any choice of the three invariants.

Jones also shows that any two outer actions of G on $\underline{R}$ are conjugate.

For more general rings, we can not expect such a theorem to hold, especially since the third invariant depends on the existence of a nice trace. However, the first two invariants certainly have analogs for any prime ring R. Let C denote the extended center of R (C is a field) and let Q(R) denote its two-sided (Martindale) quotient ring. Fix G acting on R. Then we have two invariants:

(1) The group of inners may be replaced by the group of <u>X-inners</u>: that is, $N = \{g \in G \mid \alpha(g)$ becomes inner on $Q(R)\}$.

(2) The characteristic invariant $[\lambda,\mu]$ can also be defined using the X-inners. Note now that $\lambda$, $\mu$, and $\eta$ all have values in C, the extended center.

X-inner automorphisms have already been computed for a number of examples. As a next step in classifying actions, we propose:

<u>Problem 13</u>: Compute this generalized characteristic invariant for some specific group actions.

Another invariant for group actions that has proved useful in operator algebras is the spectrum of a group action. For a finite abelian group G, this is defined as follows: given an action $\alpha : G \to \mathrm{Aut}_k(R)$, $\alpha$ extends in a natural way to an algebra homomorphism

$$\tilde{\alpha} : kG \to \mathrm{End}_k(R)$$

where kG is the usual group algebra.

5.3 Definition: $\mathrm{Spec}(\alpha) = \{ m \in \mathrm{Spec}_m kG \mid m \supseteq \ker \tilde{\alpha} \}$

If G is a locally compact abelian group, it is actually the algebra $L^1(G)$ which is used; in this case there exists a one-to-one correspondence between $\mathrm{Spec}_m L^1(G)$ and the dual group $\hat{G}$ of G. For an arbitrary discrete group G and field k, we can not define $L^1(G)$. Nevertheless, Spec($\alpha$), or its analogs for primitive or prime ideals, may prove useful.

Problem 14: Compute Spec($\alpha$) for some group actions

Finally, a special subset of Spec($\alpha$) has been used by D. Olesen and G. Pedersen in determining when a crossed product $\overline{A_\alpha G}$, for A a C*-algebra and G a locally compact abelian group, is a prime ring. This subset is called the Connes spectrum and is defined as

$$\Gamma(\alpha) = \bigcap_{B \subseteq A} \mathrm{Spec}(\alpha|_B)$$

where the intersection runs over a certain class of large G-stable subalgebras B (details appear in [37]). In particular, they prove that $\overline{A_\alpha G}$ is prime <=> A is G-prime and $\Gamma(\alpha) = \mathrm{Spec}(\alpha)$.

In view of current algebraic interest in crossed products, we ask:

Problem 15: Find an algebraic analog of the Connes spectrum.

## References


1. Alev, J., "Sur l'extension $R^G \hookrightarrow R$," Comm. in Algebra, to appear.

2. Artin, M. and Schelter, W., "Integral ring homomorphisms," Advances in Math 39 (1981), pp. 289-329.

3. Bergman, G., "Wild automorphisms of free PI algebras, and some identities," to appear.

4. Bourbaki, N., Algebré commutative, Hermann, Paris, 1964.

5. Chevalley, C., "Invariants of finite groups generated by

reflections," Amer. J. Math 77 (1955), pp. 778-782.

6. Chuang, C.L. and Lee, P.H., "Noetherian rings with involution," Chinese J. Math 5 (1977), pp. 15-19.

7. Connes, A., "Periodic automorphisms of the hyperfinite factor of type $II_1$," Acta Sci. Math. 39 (1977), pp. 39-66.

8. Czerniakiewicz, A., "Automorphisms of a free associative algebra of rank 2," Trans. AMS 160 (1971), pp. 393-401; part II, Trans. AMS 171 (1972), pp. 309-315.

9. Dicks, W. and Formanek, E., "Poincare series and a problem of S. Montgomery," Linear and Multilinear Algebra 12 (1982), pp. 21-30.

10. Farkas, D. and Snider, R., "Noetherian fixed rings," Pacific J. Math 69 (1977), pp. 347-353.

11. Fisher, J. and Montgomery, S., "Invariants of finite cyclic groups acting on generic matrices," to appear.

12. Fisher, J. and Osterburg, J., "Semiprime ideals in rings with finite group actions," J. Algebra 50 (1978), pp. 488-502.

13. Guralnick, R., "Invariants of finite linear groups on relatively free algebras," to appear.

14. Jones, V., "Actions of finite groups on the hyperfinite $II_1$ factor," Memoirs AMS, 28 (1980), no. 237.

15. Jung, H.W.E., "Uber ganze birationale Transformationen der Eben," J. Reine angew. Math 184 (1942), pp. 161-174.

16. Kharchenko, V.K., "Galois theory of semiprime rings," Algebra i Logika 16 (1977), pp. 313-363 (Russian); English translation (1978) pp. 208-258.

17. Kharchenko, V.K., "Algebras of invariants of free algebras," Algebra i Logika 17 (1978), pp. 478-487 (Russian); English translation (1979), pp. 316-321.

18. Kharchenko, V.K., "Noncommutative invariants of finite groups and nonabelian varieties," Noether 100th Birthday Volume, North Holland, to appear.

19. Lane, D.R., Ph.D. Thesis, University of London, 1975.

20. Lorenz, M. and Passman, D.S., "Prime ideals in crossed products of finite groups," Israel J. Math 33 (1979), pp. 89-132.

21. Lorenz, M. and Passman, D.S., "Observations on crossed products and fixed rings," Comm. in Algebra 8 (1980), pp. 743-779.

22. Lvov, I.V. and Kharchenko, V.K., "Normal elements of the algebra of generic matrices are central," Siberian Math J. 23 (1982), pp. 193-195 (Russian).

23. Makar-Limanov, L., "On automorphisms of free algebras with two generators," Funkc. Anal. i Ego Prilozeniya 4 (1970), pp. 107-108 (Russian).

24. Miyashita, Y., "Finite outer Galois Theory of non-commutative rings," J. Fac. Sci. Hokkaido Univ. 19 (1966), pp. 114-134.

25. Montgomery, S., "Automorphism groups of rings with no nilpotent elements," J. Algebra 60 (1979), pp. 238-248.

26. Montgomery, S., Fixed Rings of Finite Automorphism Groups of Associative Rings, Lecture Notes in Math 818, Springer-Verlag, New York, 1980.

27. Montgomery, S., "Prime ideals in fixed rings," Comm. in Algebra 9 (1981), pp. 423-449.

28. Montgomery, S., "Trace functions and affine fixed rings in non-commutative rings," Seminaire d'Algebré Paul Dubreil et Marie-Paule Malliavin, Paris, 1981, Lecture Notes in Math 924, Springer-Verlag, New York, 1982, pp. 356-374.

29. Montgomery, S., "X-inner automorphisms of filtered algebras II," Proc. AMS 87 (1983), pp. 569-575.

30. Montgomery, S. and Passman, D.S., "Outer Galois theory of prime rings," Rocky Mountain J., to appear.

31. Montgomery, S. and Small, L.W., "Fixed rings of Noetherian rings," Bull. LMS 13 (1981), pp. 33-38.

32. Montgomery, S. and Small, L.W., "Integrality and prime ideals in fixed rings of PI rings," Noether 100th Birthday Volume, North Holland, to appear.

33. Montgomery, S. and Small, L.W., "Some remarks on affine rings," to appear.

34. Nagarajan, K., "Groups acting on Noetherian rings," Nieuw Archief voor Wiskunde 16 (1968), pp. 25-29.

35. Passman, D.S., "Fixed rings and integrality," J. Algebra 68 (1981), pp. 510-519.

36. Passman, D.S., "Its essentially Maschke's theorem," Rocky Mountain J. Math 13 (1983), pp. 37-54.

37. Pedersen, G., C*-algebras and their automorphism groups, Academic Press, London, 1979.

38. Schelter, W., "Non-commutative affine PI rings are catenary," J. Algebra 51 (1978), pp. 12-18.

39. Shephard, G.C. and Todd, J.A., "Finite unitary reflection groups," Canad. J. Math 6 (1954), pp. 274-304.

40. Van der Kulk, W., "On polynomial rings in two variables," Nieuw Archief voor Wiskunde 1 (1953), pp. 33-41.

LINKS BETWEEN MAXIMAL IDEALS IN BOUNDED NOETHERIAN PRIME RINGS OF KRULL DIMENSION ONE

Bruno J. Müller

McMaster University, Hamilton, Ontario, L8S 4K1, Canada

ABSTRACT. We study the rings of the title via their (artinian) factor rings. If these factors are serial, then the ring has enough clans, and we obtain conceptual proofs of theorems due to Lenagan and Singh. If they are quasi-Frobenius, then the ring has injective dimension one, and we develop a technique for the construction of examples with pre-assigned properties.

1. <u>Summary and Introduction</u>. The paper initiates a detailed study of the directed graph of links between prime ideals of a noetherian ring. It seemed appropriate to confine attention initially to the simplest case, bounded noetherian prime rings of Krull dimension one, where all non-zero prime ideals are maximal.

Recall that, by definition [18], a <u>link</u> $P \rightsquigarrow Q$ between two maximal ideals of an FBN-ring exists if and only if $QP \subsetneq P \cap Q$ holds. The crucial relationship between links and localization lies in the fact that the finite link connectivity components are the <u>clans</u>, i.e. the minimal classically localizable sets of prime ideals [18].



*F. van Oystaeyen (ed.), Methods in Ring Theory, 347–377.*

The present paper has two parts: The first one starts off with the fundamental observation (Theorem 3) that every link component of maximal ideals, in an indecomposable non-artinian FBN-ring, contains a circuit. As an immediate consequence, if all proper factor rings are serial, then all link components are finite. We deduce a simple conceptual proof for Lenagan's theorem (Theorem 5) that every bounded HNP-ring has enough invertible ideals - attempts to find such a proof were the original motivation for our work. We also obtain a fairly direct proof for Singh's theorem (Theorem 6) that a bounded noetherian prime ring all whose proper factor rings are serial, is HNP. Motivated by the localization-globalization technique used in the argument, we show a general globalization theorem (Theorem 7), which overcomes the difficulty that the familiar (commutative) technique can only deal with ideals and bi-modules.

The second part studies prime FBN-rings of injective dimension one. Such rings have automatically Krull dimension one (Proposition 9), and constitute the proper setting for the definition of the <u>cycles</u> of maximal ideals, either via duality [5], or via orders ([7], and Proposition 8). Our interest was aroused by the two facts, that the cycle map is an automorphism of the link structure (Section 7), and that cycle map and links coincide for HNP-rings ([24], and Proposition 11). We show that in the presence of enough invertible ideals, a prime noetherian ring has injective dimension one if and only if all factor rings modulo invertible ideals are quasi-Frobenius (Theorem 14). Then, the cycle map can be identified with the Nakayama permutations of these factor rings (end of Section 7). Theorem 14 is combined with the description of the structure of quasi-Frobenius rings in [16], to develop a construction technique for noetherian prime rings of injective dimension one, which yields examples with a great variety of pre-assigned properties (Section 9). We conclude with a list of open questions.

Our rings have identity elements, and our modules are unital right modules. $|M|$ denotes the Krull dimension of the module M. For a discussion of Krull series and Krull primes, see [22]. Noetherian means left- and right-noetherian, and ideal means two-sided ideal.

A set of prime ideals is called (co)hereditary if it is closed with respect to set-theoretically smaller (larger) prime ideals. For the general background on torsion theories, see [28]; for terminology and the relationship between torsion theories and sets of prime ideals, compare [21], and Section 1 of [20]; for a study of links, consult [18].

2. A Fundamental Observation. Richards [24] (cf. also Golan [8]) introduced the concept of a direct decomposition of a torsion theory $\underline{T}$ : $\underline{T} = \underline{T}_0 \oplus \underline{T}_1$ holds if and only if $\underline{T} \geq \underline{T}_0$, $\underline{T}_1$ and every $\underline{T}$-torsion module is the direct sum of its $\underline{T}_0$- and $\underline{T}_1$-torsion radicals. He obtained the following application, which is fundamental to our work: if $\underline{T} = \underline{T}_0 \oplus \underline{T}_1$, and if $\underline{T}$ is faithful, then it is perfect if and only if $\underline{T}_0$ and $\underline{T}_1$ are perfect; and then for every $\underline{T}$-dense ideal the $\underline{T}_0$-quotient is an ideal of the quotient ring. As a special case, one obtains:

LEMMA 1. *Let R be a prime FBN-ring. Then any direct summand of the Goldie Torsion Theory is perfect, and the corresponding quotient of any ideal is again an ideal. Consequently, the quotient ring is right-FBN, and its spectrum consists precisely of the quotients of the closed prime ideals.*

PROOF. The Goldie Torsion Theory is obviously faithful and perfect, and each ideal is either zero, or dense. Therefore,

Richards' results apply. Since the quotients of ideals are again ideals, one sees easily that the quotients of closed prime ideals are again prime ideals. (Note that, due to the perfectness of the torsion theory, the quotient of an ideal I is the product, IT, with the quotient ring T.) ∎

The next lemma provides a simple criterion for when a torsion theory decomposes directly. We recall that, over a right-FBN ring, the torsion theories are in 1-1 correspondence with the hereditary, as well as with the cohereditary sets of prime ideals. Explicitly, this correspondence associates with a torsion theory $\underline{T}$, the hereditary set $\underline{P}$ of the closed prime ideals, and the cohereditary set $\underline{D}$ of the dense prime ideals.

LEMMA 2. *Let* R *be an* FBN-*ring*. *Then* $\underline{T} = \underline{T}_0 \oplus \underline{T}_1$ *holds if and only if* $\underline{D}$ *is the disjoint union of* $\underline{D}_0$ *and* $\underline{D}_1$, *and there is no link between any prime ideals* $P \in \underline{D}_0$ and $Q \in \underline{D}_1$.

REMARK. The lemma puts into evidence that $\underline{T} = \underline{T}_0 \oplus \underline{T}_1$ is a stronger requirement than $\underline{T} = \underline{T}_0 \vee \underline{T}_1$, $\underline{T}_0 \wedge \underline{T}_1 = 0$. In fact, for FBN-rings, the last condition just means that $\underline{D}$ is the disjoint union of $\underline{D}_0$ and $\underline{D}_1$.

PROOF. The necessity of the conditions is easy to verify. In particular, if a link $\underline{D}_0 \ni P \rightsquigarrow Q \in \underline{D}_1$ exists, we have an exact sequence $0 \to X \to Y \to Z \to 0$, with Y uniform, and X and Z critical with annihilators P and Q, respectively. We conclude $X \in \underline{T}_0$, $Z \in \underline{T}_1$, and $Y \in \underline{T}$. From $\underline{T} = \underline{T}_0 \oplus \underline{T}_1$, we obtain $Y = \rho_0 Y \oplus \rho_0 Y$. As Y is uniform, and as $X \in \underline{T}_0$, we get $Y = \rho_0 Y \in \underline{T}_0$, and consequently $Z \in \underline{T}_0$. This yields the contradiction $Z \in \underline{T}_0 \cap \underline{T}_1 = 0$.

We proceed to show sufficiency. $\underline{D}_0 \cap \underline{D}_1 = \emptyset$ implies immediately $\rho_0 M \cap \rho_1 M = 0$, for any module $M \in \underline{T}$, and therefore

$M \supset \rho_0 M \oplus \rho_1 M = N$. If $M \supsetneq N$, then, replacing M by an intermediate module if necessary, we may assume that $0 \neq M/N \in \underline{T}_0$ (or $\in \underline{T}_1$). We write $\bar{M}$ for $M/\rho_0 M$, and we observe that the inclusion $\bar{M} \supset \bar{N}$ is essential: otherwise, we would obtain a module $M \supset B \supsetneq \rho_0 M$ with $B \cap N = \rho_0 M$, and deduce $\bar{B} = B/(B \cap N) \cong (B+N)/N \subset M/N \in \underline{T}_0$, hence $B \in \underline{T}_0$, a contradiction.

We claim that all the Krull primes of $\bar{M}$ belong to $\underline{D}_1$; this implies then that $\bar{M}$ is annihilated by a product of primes in $\underline{D}_1$, and consequently that $\bar{M} \in \underline{T}_1$ holds. We deduce the contradiction $M/N \in \underline{T}_1 \cap \underline{T}_0 = 0$.

Assume, contrary to the claim, the existence of a Krull prime P of $\bar{M}$ with $P \notin \underline{D}_1$. As $\bar{M} \in \underline{T}$ holds, we have $P \in \underline{D}$ hence $P \in \underline{D}_0$. Among all such P, we select one which appears in a Krull series of shortest length. Then, it is associated with the top factor of this Krull series, while all the other factors have associated primes which belong to $\underline{D}_1$; and at least one of these latter ones occurs since $\bar{N} \cong \rho_1 M \in \underline{T}_1$. Thus, we obtain a link $\underline{D}_1 \ni Q \rightsquigarrow P \in \underline{D}_0$, contrary to our conditions. ∎

We combine now these observations to prove a fundamental result. To put into perspective the restriction that it imposes upon the link structure in non-artinian FBN-rings, we recall that an arbitrary finite (connected) graph can be realized as the link graph of a (ring-directly indecomposable) artinian ring [17].

THEOREM 3. _If a ring-directly indecomposable FBN-ring is not artinian, then every link component of maximal ideals contains a circuit._

PROOF. The exceptional case arises, if there is a link component $\underline{C}$ of maximal ideals, which consists entirely of minimal primes. Then the sets $\underline{D}_0 = \underline{C}$ and $\underline{D}_1 = \text{spec } R - \underline{C}$ are both link closed and cohereditary, and their disjoint union is all of spec R. Therefore, by Lemma 2, the largest torsion theory $\underline{T} = \text{mod } R$ decomposes: $\text{mod } R = \underline{T}_0 \oplus \underline{T}_1$. In particular, one obtains the ring-direct decomposition $R = \rho_0 R \oplus \rho_1 R$. By indecomposability of R, one concludes $R = \rho_0 R \in \underline{T}_0$, and therefore $R/P \in \underline{T}_0$ hence $P \in \underline{D}_0$ for all $P \in \text{spec } R$. Thus, spec $R = \underline{C}$, and R is an FBN-ring of Krull dimension zero, hence artinian.

In the complementary case, for any link component $\underline{C}$ of maximal ideals, we can find prime ideals $Q \subsetneq P \in \underline{C}$. We select one such Q of minimal Krull dimension dim Q, and consider the factor ring $\bar{R} = R/Q$. Since no links are created by passage to a factor ring, the component comp($\bar{P}$) of $\bar{P}$ in $\bar{R}$ is a subset of $\underline{C}$. By the minimality of dim Q, all $\bar{P} \in$ comp($\bar{P}$) are of height one in $\bar{R}$.

The sets $\underline{D}_1 = \text{comp}(\bar{P})$ and $\underline{D}_0 = \text{spec } \bar{R} - \{\bar{0}\} - \underline{D}_1$ are both link closed and cohereditary, and their disjoint union corresponds to the Goldie Torsion Theory $\underline{T}$ for $\bar{R}$. By Lemma 2, we obtain a direct decomposition $\underline{T} = \underline{T}_0 \oplus \underline{T}_1$. By Lemma 1, the $\underline{T}_0$-quotient ring T is a right-FBN-ring, whose spectrum is identified with comp($\bar{P}$) $\cup \{\bar{0}\}$. We conclude that T has right-Krull dimension one.

Next, we consider an arbitrary regular element $c \in \bar{R}$. Clearly, c is regular in T, and therefore T/cT has Krull dimension zero, hence is artinian, as T-module. Thus, it is annihilated by a product of maximal ideals, $P_i T$, where $P_i \in$ comp($\bar{P}$). We deduce $cT \supset (\Pi P_i)T$, and more generally $c^n T \supset (\Pi P_i)^n T$, for every n.

Now we assume, contrary to our claim, that C contains no circuit. Then we can reindex the $P_i$ in such a manner that $i \leq j$ implies $P_i \not\rightsquigarrow P_j$; and this means $P_j P_i = P_j \cap P_i \supset P_i P_j$. We conclude $c^n T \supset P_1 \ldots P_s T$. But $T/P_1 \ldots P_s T$ is again artinian, and we obtain $c^n T = c^{n+1} T$ for some n, and therefore $c^{-1} \in T$. This shows that T coincides with the simple Goldie quotient ring of $\bar{R}$, in contradiction with spec T = comp($\bar{P}$) $\cup \{\bar{0}\}$. ∎

If we restrict attention to prime FBN-rings of Krull dimension one, then the above proof simplifies considerably: the first case does not occur, and in the second one the passage to the factor ring R/Q is unnecessary. We observe that we arrive at the same final contradiction if we admit self-links $P \rightsquigarrow P$, as long as all the maximal ideals P are eventually idempotent, that is $P^n = P^{n+1}$ for some n. Note that this is equivalent to $\bigcap_{n=1}^{\infty} P^n \neq 0$, in our restricted situation. Thus, we obtain:

COROLLARY 4. <u>Any link component of maximal ideals, of a bounded prime noetherian ring of Krull dimension one, contains a proper circuit, or a maximal ideal P with</u> $\bigcap_{n=1}^{\infty} P^n = 0$. ∎

3. <u>A Question of Warfield</u>. In the notes of the 1979 Durham Conference [30], R.B. Warfield asks: "<u>Find an</u> FBN-<u>ring with an infinite set of simple modules</u>, $S_i$, $-\infty < i < \infty$, <u>all nonisomorphic, such that for all</u> i, $\mathrm{Ext}^1(S_i, S_{i+1}) \neq 0$."

Obviously, this is equivalent to finding a set of distinct maximal ideals, $P_i$, with links $P_{i+1} \rightsquigarrow P_i$.

If we strengthen the question, by requiring that these links should be the only ones in the component, then such a ring cannot exist. Indeed, we would require a component which is an infinite chain, in contradiction to Theorem 3.

If we take the question at face value, then examples are easy to find. Our counterexample 1 in [18] is a (non-prime) noetherian PI (hence FBN-) ring of Krull dimension one, whose components of maximal ideals are all infinite chains, but with self-links everywhere:

$$\rightsquigarrow o \rightsquigarrow o \rightsquigarrow o \rightsquigarrow$$

An example which is a prime PI-ring, cannot exist in Krull dimension one, since for every such ring all the link components are finite [18]. Such an example in Krull dimension two is the matrix ring $R = \begin{pmatrix} A & X \\ Y & B \end{pmatrix}$, where $C = k[x,y]$ is the polynomial ring in two variables over an (algebraically closed) field R of characteristic zero, $X = Y = yC$, $A = k[x + ax^2] + yC$ and $B = k[x + bx^2] + yC$, for distinct, $a,b \in k$. Here, every PI-degree-two maximal ideal constitutes a singleton component, while the PI-degree-one maximal ideals form infinite chains, with links in both directions, and self-links everywhere (except for two components which extend only in one direction):

$$\leftrightsquigarrow o \leftrightsquigarrow o \leftrightsquigarrow o \leftrightsquigarrow$$
$$o \leftrightsquigarrow o \leftrightsquigarrow o \leftrightsquigarrow$$
$$\leftrightsquigarrow o \leftrightsquigarrow o \leftrightsquigarrow o$$

(The techniques for an anlysis of this example have been developed in [20]).

4. Lenagan's Theorem. The original motivation for our work, was to "understand" and to generalize the following result of

Lenagan [13]:

THEOREM 5. Every bounded HNP-ring has enough invertible ideals.

Though its previous proofs, particularly the (privately circulated) versions due to A. Heinicke and P.F. Smith, are not difficult, they rely on details of the ideal theory which are only available for HNP-rings [7]. We outline here a new proof, which is based on Corollary 4 and uses no ideal theory.

The well known fact that every proper factor ring of an HNP-ring is serial, is proved with the basic machinery of module theory [6]. It implies immediately that for every maximal ideal P, there can be at most one incoming link $Q \rightsquigarrow P$ and one outgoing link $P \rightsquigarrow Q'$. It follows that the link components are chains or circuits. Since Corollary 4 forbids the first possibility, they are circuits and therefore finite. This proves that any bounded HNP-rings has enough clans.

It is again well known that in any HNP-ring, the intersection $N = \cap_{i=1}^{s} P_i$ of a clan is invertible [17]. For the convenience of the reader, we supply a simple torsion theoretical proof of this fact:

Any overring of an HNP-ring, and in particular the right-order $O_r(N)$, is the quotient ring for a perfect torsion theory, whose quotient functor we denote by $\underline{T}$. We observe $\underline{T}(N) = NO_r(N) = N$, and $\underline{T}(R/N) = \overline{R/N}$, the latter since any quotient of any semi-simple artinian ring is a factor ring. We obtain the diagram

$$\begin{array}{ccccccccc} 0 & \to & N & \to & R & \to & R/N & \to & 0 \\ & & \| & & \downarrow & & \downarrow & & \\ 0 & \to & N & \to & O_r(N) & \to & \overline{R/N} & \to & 0 \end{array}$$

with exact rows and natural maps. Diagram chasing shows $R = O_r(N)$. We deduce $*N\,N \subset \mathrm{endo}({}_RN) = O_r(N) = R$; on the other hand projectivity of ${}_RN$ yields $*N\,N \supset R$. ∎

5. <u>Singh's Theorem</u>. Surjeet Singh [27] proved the following characterization of bounded HNP-rings, by a fairly lengthy argument. We shall give a quite different proof, based on localization techniques.

THEOREM 6. <u>A bounded noetherian prime ring of Krull dimension one is</u> HNP <u>if and only if all its proper factor rings are serial</u>.

PROOF. The "only if" part is the result of Eisenbud-Griffith [6] quoted earlier. Conversely, if all proper factor rings of R are serial, then the argument of the preceding section shows that the link components are finite circuits, hence clans. The localization, T, at any such clan is a semilocal noetherian prime ring, all whose proper factor rings are again serial. We denote the Jacobson radical of T by N.

We index the maximal ideals of T such that $P_i \rightsquigarrow P_{i+1}$ (i modulo s), where s is the size of the clan in question. For a fixed but arbitrary n, we have corresponding indecomposable idempotents $e_1 \in \bar{T} = T/N^{n+s}$. Nakayama's Lemma implies that the Loewy-length of $T/N^{n+s}$ is precisely n+s. Therefore this serial ring has a uniserial module of length n+s, which in turn possesses subfactors of length n+1 with arbitrary top composition factors $e_iT/N$.

It follows that all the modules $e_i T/N^{n+1}$ have length n+1.

Then, all the $e_i N/N^{n+1}$ have length n, and are consequently isomorphic to $e_{i-1} T/N^n$ hence projective as $T/N^n$-modules. We conclude that $N/N^{n+1}$ is projective as $T/N^n$-module. It follows readily that N is X-projective for any T-module X with $XN^n = 0$.

As this holds true for every n, N is X-projective for any finitely generated Goldie-torsion module X. By the fundamental results of Azumaya [1], the class $C^p(N)$ of all modules X with respect to which N is X-projective, is closed under submodules and direct sums (since N is finitely generated), and therefore under colimits. We conclude that $C^p(N)$ contains all Goldie-torsion modules, and in particular the injective hull E for any simple module S. This implies immediately that $\mathrm{Ext}^1_T(N,S) = 0$, and we obtain that N is projective [3]. It follows readily that each localization T of R at a clan is hereditary.

To show now that R is hereditary, we use a globalization argument (cf. Corollary 8 in the next section): the familiar commutative technique yields immediately that every ideal of R is projective as an R-right-module. It follows from boundedness that every simple module, and therefore every module of finite length, has homological dimension at most one. Consequently, every (essential) right-ideal is projective, and R is right-hereditary. ∎

REMARK. The considerations of the last two sections suggest the problem of characterizing those bounded noetherian prime rings of Krull dimension one - as a generalization of bounded HNP-rings - for which each maximal ideal has at most one incoming and one outgoing link. Obviously, these rings have enough clans. A non-hereditary example is $R = \begin{pmatrix} A & A \\ M & A \end{pmatrix}$, for any non-regular commutative noetherian local domain A of Krull dimension one, with maximal ideal M; for instance $A = k[\xi,\eta]_{\langle\xi,\eta\rangle}$ with $\xi^3 = \eta^2$.

6. Globalization. The known non-commutative globalization theorems [26], [4] (except for noetherian prime rings of Krull dimension one where ad hoc arguments are possible) seem to require the assumption that the homological dimension in question is finite, an assumption which severely limits their usefulness. Our next theorem avoids this shortcoming.

THEOREM 7. *Let R be an FBN-ring all whose maximal ideals belong to clans. Then* whd $M = \sup_T$ whd $M_T$ *holds for every module M, where the supremum is taken over the localizations T at the clans of maximal ideals.*

PROOF. Let $\sup_T$ whd $M_T = N \leq \infty$. Since the quotient functors $- \otimes_R T$ are exact, we have immediately whd $M_T \leq$ whd M, hence $n \leq$ whd M.

To obtain the reverse inequality (for finite n), we show first $\mathrm{Tor}_m^R(M,R/I) = 0$ for every ideal I and every $m > n$, using the standard commutative technique: if F denotes a free resolution of the right-module R/I, then the homology of $F \otimes_R R/I$ is $\mathrm{Tor}^R(M,R/I)$. By exactness of $- \otimes_R T$, the homology of $F \otimes_R R/I \otimes_R T$ is $\mathrm{Tor}^R(M,R/I) \otimes_R T = \mathrm{Tor}^R(M,R/I)_T$. On the other hand, as $IT = TIT$ holds, we have $R/I \otimes_R T \cong T/IT = T/TIT$. Since $F \otimes_R R/I \otimes_R T \cong F \otimes_R T/TIT \cong F \otimes_R T \otimes_T T/TIT$ is true, and since $F \otimes_R T$ is a free resolution of $M_T$, the homology is also $\mathrm{Tor}^T(M_T,T/TIT)$. Therefore we have $\mathrm{Tor}_m^R(M,R/I)_T \cong \mathrm{Tor}_m^T(M_T,T/TIT) = 0$ for all T, and consequently $\mathrm{Tor}_m^R(M,R/I) = 0$.

The desired inequality whd M $\leqslant$ n will follow once we show $\mathrm{Tor}_m^R(M,R/L) = 0$ for every left-ideal L. We do this by induction over $|R/L| = \alpha$.

Let $L = L_0 \subsetneq L_1 \subsetneq \ldots \subsetneq L_s = R$ be a Krull series of R/L, with associated Krull primes $P_i$. Each $L_i/L_{i-1}$ is a critical non-singular $R/P_i$-module, and hence isomorphic to a left-ideal of $R/P_i$. Therefore, for suitable $t_i$, there is an essential monomorphism $(L_i/L_{i-1})^{t_i} \to R/P_i$, whose image we denote by $K_i/P_i$. We obtain $\alpha = |R/L| \geqslant |L_i/L_{i-1}| = |R/P_i| > |R/K_i|$, and consequently $\mathrm{Tor}_m^R(M,R/K_i) = 0$ by induction hypothesis. As we know already $\mathrm{Tor}_m^R(M,R/P_i) = 0$ from the previous consideration, we deduce $\mathrm{Tor}_m^R(M,L_i/L_{i-1}) = 0$, for every i. The required conclusion $\mathrm{Tor}_m^R(M,R/L) = 0$ follows. ∎

REMARK. Our proof goes through under the weaker assumption that R is an FBN-ring with a family of perfect torsion theories $\underline{T}_i$ for right-modules, with quotient rings $T_i$, for which the quotients of ideals are again ideals, and with $\cap\, \underline{T}_i = o$. As discussed in Section 2, such a family exists for every bounded noetherian prime ring of Krull dimension one.

COROLLARY 8. <u>Under the same assumptions,</u>

$$\mathrm{gl.dim.}R = \sup_T \mathrm{gl.dim.}R_T. \qquad ∎$$

<u>7. Injective Dimension One.</u> Throughout this section, R will be a prime FBN-ring of left- and right-injective dimension one,

and with Goldie quotient ring K.

Whether, in this setting, rt-id R = 1 already implies lt-id R = 1, appears to be an open question. Zaks [29] proves it, under the assumption lt-id $R < \infty$. It also follows from our Theorem 13, provided R has enough invertible ideals.

R is a reflexive ring, in the sense of Jans [10], ie. every finitely generated torsionless module is reflexive, with respect to $\hom_R(-,R)$. In particular, every finitely generated submodule X of K is reflexive. We denote the dual of X, identified with the appropriate subset of K, by *X or X*, depending on whether X is a left- or right-module. We remark that, due to the boundedness of R, any bi-submodule of K which is finitely generated on one side, is also finitely generated on the other.

We are grateful to F. Sandomierski for communicating to us the proof of the following fact (for another proof, see [30], Theorem 4.6):

PROPOSITION 9. A prime FBN-ring of right-injective dimension one has Krull dimension one.

PROOF. We have to show that every non-zero prime ideal P is maximal. We write $\bar{R} = R/P$ and $\bar{K} = K/R$. Then, $\hom_R(\bar{R},\bar{K})$ is an injective $\bar{R}$-right-module, since rt-id R = 1. It can be identified with P*/R, via the evaluation of homomorphisms at 1. As observed above, P* and hence P*/R are finitely generated on both sides.

For each maximal ideal M containing P, we obtain $P^* \supseteq M^* \supsetneq R$. Therefore, P*/R contains a copy of the simple right-module annihilated by M, and is an injective cogenerator for $\bar{R}$. Consequently, the minimal cogenerator is finitely generated, and the number of simple $\bar{R}$-right-modules is finite. This implies that the Jacobson radical $J(\bar{R})$ is semi-maximal. Since the minimal cogenerator, the direct sum of the injective hulls of the simple modules, is 0-homogeneous [12] and finitely generated, it is annihilated by a certain power of $J(\bar{R})$; and since it is faithful, this power of $J(\bar{R})$ is zero. Primeness of $\bar{R}$ then yields $J(\bar{R}) = 0$; and $\bar{R}$ is artinian, as required. ∎

We return to the reflexivity of the finitely generated bi-submodules of K, and observe that it creates a bijection of this set with itself, via $X \Rightarrow Y = {}^{**}X$ (or ${}^{*}X = Y^{*}$). This bijection is a lattice automorphism, and restricts to bijections of the sets of non-zero ideals and of maximal ideals. We call it the <u>cycle map</u>, and the orbits of the maximal ideals the <u>cycles</u>, of R (cf. [5]).

The cycle map is also compatible with products. Indeed, if $X_i \Rightarrow Y_i$ and $q \in K$, then

$$q \in {}^{*}(X_1X_2) \Leftrightarrow X_1X_2q \subset R \Leftrightarrow X_2q \subset {}^{*}X_1 = Y_1{}^{*} \Leftrightarrow X_2qY_1 \subset R \Leftrightarrow q \in (Y_1Y_2)^{*}.$$

This observation has a number of consequences:

First of all, <u>the cycle map is an automorphism of the link structure</u>. Indeed, a link $P_1 \rightsquigarrow P_2$ between maximal ideals exists if and only if $P_2P_1 \subsetneq P_2 \cap P_1$; if $P_i \Rightarrow Q_i$, then we deduce $Q_2Q_1 \subsetneq Q_2 \cap Q_1$, hence $Q_1 \rightsquigarrow Q_2$.

Secondly, <u>any overring which is a quotient ring for a left-torsion theory, is also a quotient ring for a right-torsion theory</u>, namely the one corresponding under the cycle map. Indeed, the quotient ring of the left-torsion theory determined by the set <u>D</u> of maximal ideals is $\cup_{P \in \underline{D}} {}^{*}(\Pi P)$, and equals $\cup_{Q \in \underline{D}'} (\Pi Q)^{*}$, where $\underline{D} \Rightarrow \underline{D}'$. It is also easily checked that the correspondence between torsion theories and quotient rings is one-to-one.

Finally, <u>any clan is closed under the cycle map</u>. Indeed, by the preceding observation, $\underline{D}$ is closed under the cycle map whenever the left- and right-quotient rings at $\underline{D}$ coincide; and this holds true in particular if $\underline{D}$ is the complement of a clan.

It follows that the cycles are finite (ie. periodic) if R has enough clans. Whether infinite cycles ever occur, is an open question. Now we show that our cycles are the same as those defined in [7] for HNP-rings via orders.

PROPOSITION 10. <u>A maximal ideal</u> P <u>is either invertible, in which case</u> $P \Rightarrow P$ <u>and</u> $O_r(P) = O_\ell(P) = R$ <u>hold; or it is not invertible, in which case</u> $P \Rightarrow Q$ <u>is equivalent to</u> $O_r(P) = O_\ell(Q)$.

PROOF. If P is invertible, then $*P = P*$ hence $P \Rightarrow P$, and $PO_r(P) = P$ hence $O_r(P) = P^* P O_r(P) = P^* P = R$ are true.

If P is not invertible, then $P *P = P$ holds. Indeed, otherwise, $P *P = R$ would hold. This implies $O_r(P) \subsetneq *P$ hence $O_r(P) = R$, and also that ${}_RP$ is a generator hence $P_R = P_{O_r(P)}$ is projective. Consequently, $P *P$, which equals either R or P, is idempotent. But $P *P = R$ is impossible since P is not invertible, and $P *P = P$ implies $P = P^2$ hence the contradiction $R = P *P = P(P *P) = P$. Now, $P *P$ yields $O_r(P) = *P$; and similarly one gets $O_\ell(Q) = Q^*$. ∎

Next, we give a simple proof of the fact, observed in [24], that for bounded HNP-rings, links and the cycle map are the same thing.

PROPOSITION 11. <u>For a bounded</u> HNP-<u>ring</u>, $P \rightsquigarrow Q$ <u>holds if and only if</u> $P \Rightarrow Q$ <u>holds</u>.

PROOF. Let $P \Rightarrow Q$. If $P = Q$, then $P$ is invertible, since $*PP \supset R$ holds by projectivity, and $P*P \subset R$ is always true. Therefore $P$ is not idempotent, and we obtain a self link $P \rightsquigarrow P$.

If $P \neq Q$, then the idealizer of $P$ in $T = *P = Q* = O_r(P) = O_\ell(Q)$ is $\mathbb{I}_T(P) = *P \cap P* = R$. We consider a maximal T-right-ideal $M$ containing $P$. Since $TP = T$ holds, by projectivity of ${}_RP$, we have $M \not\supset TP$. As in the proof of Proposition 1.4 in [25], we obtain a unique maximal intermediate R-module $T \supset C \supset M$, namely $C = \{t \in T: \ tP \subset M\}$. And as in the same proof, we see that $C/M$ is isomorphic to $\hom_T(T/P,T/M)$, as module over $\mathrm{endo}_T(T/P) \cong R/P$. It is therefore a simple R-module annihilated by $P$. On the other hand, the inclusions $TQP \subset RP \subset M$ show that $T/C$ is annihilated by $Q$. Thus, the R-module $T/M$ defines a link $P \rightsquigarrow Q$.

So far, we have found a link $P \rightsquigarrow Q$ whenever $P \Rightarrow Q$ holds. There cannot be any other links, since proper factor rings of $R$ are serial. ∎

Next, we will show that every proper factor ring of $R$ has Morita duality [15].

PROPOSITION 12. <u>For all non-zero ideals</u> $X \Rightarrow Y$, <u>the bi-module</u> $*X/R = Y*/R$ <u>induces a Morita duality between</u> $R/X$ and $R/Y$.

PROOF. As in the proof of Proposition 8, with $Y$ taking the place of $P$, we see that $*X/R = Y*/R$ is an injective cogenerator as $R/Y$-right-module (and by symmetry also as $R/X$-left-module). It remains to show that its endomorphism ring is $R/X$.

We pick a regular element $c \in X$, and consider any $\phi \in \mathrm{endo}(X/R_{R/Y})$. The map $c{*}X \cong {*}X \to {*}X/R \overset{\phi}{\to} {*}X/R \subset \bar{K}$ is given, due to the injectivity of $\bar{K}$, by left-multiplication with an element $\bar{q} \in \bar{K}$; this means that $\phi(\bar{a}) = \bar{q}ca$ holds for all $a \in {*}X \subset K$. Since $\phi(\bar{1}) = 0$ is true, we conclude $\overline{qc} = 0$, or $qc = r \in R$. Therefore $\phi$ is given by left-multiplication with the element $\bar{r} \in R/X$, and the claim follows. ∎

It appears to be an open question whether these Morita dualities are always self-dualities, that is whether R/X and R/Y are always Morita equivalent. This is certainly true if X is a maximal ideal; and then the Morita equivalence of the simple artinian rings R/X and R/Y means that they are matrix rings over the same division ring. We infer the following generalization of a result proved in [25] for HNP-rings:

COROLLARY 13. <u>The simple modules annihilated by the maximal ideals of a cycle have isomorphic endomorphism rings.</u> ∎

Finally, we utilize invertible ideals. For any invertible ideal Y, there exists another one, X, which is contained in the same maximal ideals as Y, and commutes with each of them. Indeed, mapping P to $YPY^{-1}$ defines a permutation on the set of maximal ideals containing Y; and if t is its order, then $X = Y^t$ has the desired property. (Note that Y itself already works, if it is centrally generated.)

If $P \Rightarrow Q$ and P  X, then $R \supset XQ^* \supset X$ holds, and therefore the bi-module $XQ^*/X$ defines a bi-module link between its left-annihilator $XPX^{-1} = P$, and its right-annihilator Q. By Corollary 16 of [20], such a bi-module link $P \sim Q$ is equivalent to a "long link"

$Q \rightsquigarrow\rightsquigarrow P$; and by [18], the long link implies a chain of links $Q \rightsquigarrow Q' \rightsquigarrow \dots \rightsquigarrow P$. Thus, we have shown a result which should be compared with Proposition 11: if $P$ contains an invertible ideal, then $P \Rrightarrow Q$ implies $Q \rightsquigarrow \dots \rightsquigarrow P$.

The factor ring $R/X$ modulo any invertible ideal $X$ is quasi-Frobenius ([5], or Theorem 14 below). The Nakayama permutation $\pi$ of an arbitrary quasi-Frobenius ring $A$ is defined ([23], [9]) as a permutation $\pi$ on the set of isomorphism types of indecomposable idempotents $e$, by $\mathrm{socle}(eA) \cong \mathrm{top}(e^{\pi}A)$, or equivalently by $e\,\mathrm{socle}(A)\,e^{\pi} \neq 0$. For the maximal ideal $\bar{P} = A(1-e)A + J(A)$ corresponding to $e$, this definition transcribes as $\bar{P}^{\pi} = \text{rt-ann}(\text{rt-ann}(\bar{P}))$. If $P \Rightarrow Q$, and if $X$ is an invertible ideal as above, then the calculation

$$\bar{P}a = 0 \Leftrightarrow Pa \subset X \Leftrightarrow PaX^{-1} \subset R \Leftrightarrow ax^{-1} \subset {}^*P = Q^* \Leftrightarrow a \in Q^*X$$

shows $\text{rt-ann}(\bar{P}) = Q^*X/X$, and

$$Q^*Xb \subset X \Leftrightarrow Q^*XbX^{-1} \subset R \Leftrightarrow XbX^{-1} \subset Q \Leftrightarrow b \in X^{-1}QX = Q$$

demonstrates $\text{rt-ann}(\text{rt-ann}(\bar{P})) = \bar{Q}$. Thus, we have established an intimate connection between the cycle map and the Nakayama permutation: the restriction of the cycle map, to the set of maximal ideals containing a given invertible ideal $X$ which commutes with each of them, coincides with the Nakayama permutation of the quasi-Frobenius ring $R/X$.

8. Enough Invertible Ideals. A semi-prime noetherian ring $R$ has enough invertible ideals if every primitive ideals contains an invertible ideal. For instance, this condition is fulfilled if $R$ is prime, and satisfies a polynomial identity or is integral over the centre, because in both cases every non-zero ideal contains a

non-zero central element, which generates an invertible ideal. For rings with enough invertible ideals, we give a criterion for injective dimension one, which generalizes a result of Bass [2]:

THEOREM 14. _Let_ R _be a semi-prime noetherian ring with enough invertible ideals. Then,_ $\ell$t- _and/or_ rt-id R = 1 _hold if and only if_ R/X _is quasi-Frobenius, for all/enough invertible ideals_ X.

PROOF. As the Goldie quotient ring K is injective as R-right-module, rt-id R = 1 holds if and only if $\bar{K} = K/R$ is injective as $\bar{R}$-right-module.

For an invertible ideal X, $\bar{R} = R/X$ is quasi-Frobenius if and only if $X^{-1}/R$ is injective as $\bar{R}$-right-module. Indeed, if $X^{-1}/R$ is $\bar{R}$-injective, then the splitting epimorphism $\bigoplus_X X^{-1}/R \to \bar{R}$ induced by the natural epimorphism $\bigoplus_X X^{-1} \to XX^{-1} = R$ shows that $\bar{R}$ is self-injective hence quasi-Frobenius. Conversely, a free extension $\bar{R}^n \to X^{-1}/R$ splits, since $X^{-1}$ is projective as R-module hence $X^{-1}/R$ is projective as $\bar{R}$-module, and therefore if $\bar{R}$ is self-injective, then $X^{-1}/R$ is injective as $\bar{R}$-module.

After these preliminary observations, we consider an injective hull $0 \to \bar{K} \overset{\alpha}{\to} E \overset{\beta}{\to} C \to 0$ of $\bar{K}$ as R-right-module. Applying $\mathrm{hom}_R(\bar{R},-)$, we obtain the exact sequence

$$0 \to \mathrm{hom}_R(\bar{R},\bar{K}) \to \mathrm{hom}_R(\bar{R},E) \to \mathrm{hom}_R(\bar{R},C) \to \mathrm{Ext}^1_R(\bar{R},\bar{K}) \to 0.$$

Making the appropriate identifications, and observing

$$\mathrm{Ext}^1_R(\bar{R},\bar{K}) \cong \mathrm{Ext}^2_R(\bar{R},R) \cong \mathrm{Ext}^1_R(X,R) = 0$$

since X is invertible hence projective, we deduce the exact sequence $0 \to X^{-1}/R \xrightarrow{\alpha'} \mathrm{ann}_E(X) \xrightarrow{\beta'} \mathrm{ann}_C(X) \to 0$ of $\bar{R}$-modules.

The easy direction is the one where rt-id R = 1 is given (cf. [5]). Then, $\bar{K}$ is R-injective, and therefore C = 0. Thus, $X^{-1}/R \cong \mathrm{ann}_E(X)$, which is $\bar{R}$-injective. Our preliminary considerations show that $\bar{R}$ is quasi-Frobenius.

In the opposite direction, we are given that $\bar{R} = R/X$ is quasi-Frobenius for enough invertible ideals X. We know then that $X^{-1}/R$ is $\bar{R}$-injective, and our exact sequence splits, producing a decomposition $\mathrm{ann}_E(X) = U \oplus V$, where $U = \mathrm{im}(\alpha') = \ker(\beta')$.

For the moment, we make the additional assumption that X is contained in the Jacobson radical J(R). It follows that $\mathrm{ann}_C(X)$ is essential in C. Indeed, by Theorem 2.4 of [14], $|R| = |R/X| + 1 = 1$ holds. Therefore, and since $\bar{K}$ and hence E are Goldie-torsion, every finitely generated submodule of E and hence of C is artinian. We conclude that C is essential over its socle. This socle is contained in $\mathrm{ann}_E(X)$, due to $X \subset J(R)$.

$0 = \ker(\beta') \cap V$ implies $0 = E(\ker \beta) \cap E(V)$, where E(-) denotes the injective hull as R-module. In particular, $\beta|E(V)$ is a monomorphism, and therefore $\beta(E(V))$ is an injective submodule of C. It contains the essential submodule $\beta(V) = \beta(U \oplus V) = \mathrm{ann}_C(X)$. We conclude $\beta(E(V)) = C$.

We claim that ker $\beta$ is R-injective. Indeed, if there is an element $e \in E(\mathrm{Ker}\ \beta) - \ker \beta$, then $0 \neq \beta(e) \in C = \beta(E(V))$ hence $\beta(e) = \beta(v)$ for some $v \in E(V)$. Consequently we have $v - e \in \ker \beta$ and $v \in E(\ker \beta) \cap E(V) = 0$, a contradiction. We deduce that

$\bar{K} \cong \text{im } \alpha = \ker \beta$ is injective, and therefore rt-id R = 1.

Finally, we remove the auxilliary assumption $X \subset J(R)$. As every invertible ideal X has the AR-property, we can localize at the semi-maximal ideal $\sqrt{X}$, and the localization $R_X$ fulfills the condition $J(R_X) = \sqrt{X}\, R_X \supset XR_X$. We obtain rt-id $R_X = 1$, for every X.

We complete the proof with a standard globalization argument: If F denotes a finitely generated free resolution for an arbitrary cyclic R-right-module M, then $F_X$ is a similar resolution for $M_X$, and

$$\hom_{R_X}(F_X, R_X) \cong \hom_R(F, R_X) \cong R_X \otimes {}_R\hom_R(F,R)$$

holds. We deduce

$$R_X \otimes {}_R\text{Ext}^2_R(M,R) \cong \text{Ext}^2_{R_X}(M_X, R_X) = 0.$$

The localizations $R_X$ are jointly faithful, since they involve all primitive ideals of R. We arrive at the conclusion $\text{Ext}^2_R(M,R) = 0$, which means rt-id R = 1. ∎

Theorem 14 applies, in particular, to the integral group ring $R = \mathbb{Z}G$ of a finite group G. As $p\,\mathbb{Z}G$ is invertible, and $(\mathbb{Z}/p\,\mathbb{Z})G$ is quasi-Frobenius, for every prime number p, we obtain the well known fact id $\mathbb{Z}G = 1$. - We shall use Theorem 14 in the next section, to construct additional examples.

<u>9. Examples</u>. The following constructions are based (besides

on the preceding considerations) on the description of the structure of quasi-Frobenius rings which we gave in [16]. In Theorem 4 of that paper, we obtained a matrix representation for an arbitrary basic quasi-Frobenius ring whose Nakayama permutation is just one cycle of length n, along the following lines (The details are too involved to be reproduced here, and we refer to [16]. Unfortunately, in that paper, we worked with the inverse of what is here, and everywhere else, called the Nakayama permutation. We remove this discrepancy by transposing all matrices.)

A basic cyclic quasi-Frobenius ring is isomorphic to a $n \times n$-matrix ring $(X_{ik})$, where the $X_{ii}$ are local artinian rings which constitute a "cycle of Morita dualities", the $X_{i,i+1}$ are bi-modules inducing these dualities, and the remaining $X_{ik}$ are bi-modules obtained from a small number ($m - 1$ if $n = 2m$ or $n = 2m-1$) of left-and right-finitely generated bi-modules subject to certain additional conditions. The multiplication maps $X_{ij} \otimes X_{jk} \to X_{ik}$ are given by the ring and bi-module structure, or by evaluation, wherever this makes sense, and are otherwise arbitrary except for associativity and certain other conditions.

We mimic this description here, to construct prime noetherian rings R, whose factors modulo invertible ideals have the described structure and are therefore quasi-Frobenius. It follows that the rings R have injective dimension one (Theorem 14) and possess clans which are cycles.

Specifically, and to overcome the technical difficulties hidden in the additional conditions which were suppressed above, we start with a commutative noetherian domain A of injective dimension one, and quotient field F. Localizing at a maximal ideal M, we may (and will) assume without loss of generality that A is local.

For the rings $X_{ii}$, and the duality inducing bi-modules $X_{i,i+1}$, we always take A. For the remaining m-1 bi-modules, we take non-zero finitely generated submodules of F; then they are torsionless hence reflexive [10], and their duals can be identified with similar submodules of F. The multiplication is determined by selecting a system of non-zero elements $\alpha_{ik}$ of F, and using the natural multiplication on $(\alpha_{ik}X_{ik})$ as a subset of $M_n(F)$; this guarantees associativity. The system $\alpha_{ik}$ has to satisfy the following three sets of conditions: $\alpha_{ii} = 1$; $\alpha_{ij}\alpha_{jk}X_{ij}X_{jk} \subset \alpha_{ik}X_{ik}$; $\alpha_{k,k+1} = \sigma\alpha_{kj}\alpha_{j,k+1}$ $(k \neq j \neq k+1)$. The last one expresses the requirement that the multiplication is evaluation whenever appropriate; the quantity $\sigma$ equals 1 if n is even, and satisfies $\sigma Z^* = Z$ if n is odd, for the exceptional bi-module Z with $Z^* \cong Z$ which occurs in this case (cf. [16]).

The complete solution of these conditions can be found, but is rather complicated, and will be used below only in the manageable cases n = 3 and n = 4. The arising rings R are prime, since all $X_{ik}$ are non-zero. They are finitely generated as modules over the centre A, and therefore noetherian and bounded. The clans of maximal ideals are just the fibres of the maximal ideals of A [17]. Since we assume (A,M) to be local, there is only one such clan. We fix a particular non-zero element $\mu \in M$.

We observe that id.A = 1 implies $\mathrm{Ext}^1_A(X,A) = 0$, for every finitely generated submodule X of F ([10]). This allows us to identify $\hom_{A/\mu A}(X/\mu X, A/\mu A)$ with $X^*/\mu X^*$, via the exact sequence $0 \to A \xrightarrow{\mu} A \to A/\mu A \to 0$. It follows that the factor ring $R/\mu R$, modulo the invertible ideal $\mu R$, has precisely the structure of a cyclic quasi-Frobenius ring, provided it is basic. The links in R can be determined quite easily, by comparing J(R) and $J(R)^2$, in matrix form.

Without presenting the details of the arguments, we summarize their results, for the smallest interesting values of n:

EXAMPLE 1; n = 3: *The most general ring obtained from our procedure is isomorphic to the subring* $R = \begin{pmatrix} A & A & Z \\ Z & A & \sigma A \\ A\sigma^{-1} & Z & Z \end{pmatrix}$ *of* $M_3(F)$, *where Z is a non-zero ideal and* $\sigma$ *a non-zero element of A, subject to* $\sigma Z^* = Z$ *and* $Z^2 \subset \sigma A$. (We note that these conditions are fairly restrictive, as they imply $Z^* \cong Z$ and determine $\sigma$ up to a factor which is a unit in O(Z).) There arise two cases:

(1) $Z = A$: Here, R is isomorphic to $M_3(A)$, and the clan is a singleton set equipped with a self-link.

(2) $Z \subset M$: Here, $R/\mu R$ is basic cyclic quasi-Frobenius, with three maximal ideals, corresponding to the three diagonal positions of the matrix. Therefore, the clan comprises three maximal ideals $P_1$, $P_2$ and $P_3$; and the cycle map, being the same as the Nakayama permutation, is $P_i \Rightarrow P_{i+1}$ (i modulo 3). The link structure - for which the cycle map is an automorphism - is determined as follows: $P_i \rightsquigarrow P_i$ iff $Z \subsetneq M$; $P_i \rightsquigarrow P_{i+1}$ iff $\sigma A \subsetneq Z$; $P_i \rightsquigarrow P_{i+2}$ iff $Z^2 \subsetneq \sigma A$.

The middle conditions, $\sigma A \subsetneq Z$, is always satisfied, since $\sigma A = Z$ together with $\sigma Z^* = Z$ implies the contradiction $Z = A$. Therefore, for these examples, as for HNP-rings (Proposition 11), $P \Rightarrow Q$ always implies $P \rightsquigarrow Q$.

Concrete realizations for the four remaining possibilities are the following ones:

| A | Z | σ | |
|---|---|---|---|
| $\mathbb{Z}_p$ | M | $p^2$ |  |
| $k[\xi,\eta]_{\langle\xi,\eta\rangle}$ with $\xi^3=\eta^2$ | M | $\xi^2$ |  |
| | $\langle\xi\rangle$ | $\xi^2$ |  |
| | $M^2$ | $\xi^3$ |  |

(Header of last column: )

EXAMPLE 2; n = 4: The general ring obtained here is isomorphic to the subring $R = \begin{pmatrix} A & A & \alpha Y^* & Y \\ Y & A & \alpha A & \alpha Y \\ Y^*\alpha^{-1}Y & & A & A \\ A & Y^* & Y & A \end{pmatrix}$ of $M_4(F)$, where Y is a non-zero ideal and α is a non-zero element of A, subject to $\alpha Y^{*2} \subset A$ and $Y^3 \subset \alpha A$. (We note that these conditions, though similar to the ones in Example 1, are far less restrictive.) There arise three cases:

(1) Y = A: Then, R is isomorphic to $M_4(A)$, and the clan is a singleton with a self-link.

(2) $Y \subset M$ but $Y^{*2} = A$: One deduces $Y^* \cong A$, hence $Y = \beta A$, for some element $\beta \in M$ with $\alpha A = \beta^2 A$. Then, R is isomorphic to the matrix ring $\begin{pmatrix} A & A & A & A \\ A & A & A & A \\ \beta A & \beta A & A & A \\ \beta A & \beta A & A & A \end{pmatrix}$, which is Morita equivalent to $\begin{pmatrix} A & A \\ \beta A & A \end{pmatrix}$. The factor ring $R/\mu R$ is quasi-Frobenius, and the clan contains two prime ideals $P_1$ and $P_2$. There are always links $P_i \rightsquigarrow P_{i+1}$; and self-links $P_i \rightsquigarrow P_i$ exist if and only if $\beta A \subsetneq M$.

(3) $Y \subset M$ and $\alpha Y^{*2} \subset M$: Then, $R/\mu R$ is basic cyclic quasi-

Frobenius, with four maximal ideals. Therefore, the clan consists of four maximal ideals $P_1$, $P_2$, $P_3$ and $P_4$, and the cycle map is $P_i \Rightarrow P_{i+1}$ (i modulo 4). The link structure looks as follows: $P_i \rightsquigarrow P_i$ iff $Y + \alpha Y^{*2} \subsetneq M$; $P_1 \rightsquigarrow P_{i+1}$ iff $\alpha Y^* \subsetneq Y$; $P_i \rightsquigarrow P_{i+2}$ iff $\alpha^{-1} Y^2 + A \subsetneq Y^*$; $P_i \rightsquigarrow P_{i+3}$ iff $YY^* \subsetneq A$.

We are particularly interested in the case which could not arise for $n = 3$, namely $P \Rightarrow Q$ and $P \not\rightsquigarrow Q$. In the present situation, this means $\alpha Y^* = Y$. We use this condition to eliminate $Y^*$, and to simplify the link criteria: $P_i \rightsquigarrow P_i$ iff $Y^2 \subsetneq \alpha M$; $P_i \rightsquigarrow P_{i+2}$ iff $\alpha A \subsetneq Y$; $P_i \rightsquigarrow P_{i+3}$ iff $Y^2 \subsetneq \alpha A$. The two last conditions are easily seen to be always satisfied, and therefore the cycle and the link graph look like this:

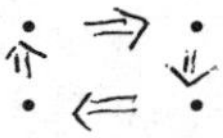

Concrete realizations for the two remaining possibilities are the following:

$A_1 = k[\xi,\eta]_{\langle\xi,\eta\rangle}$ with $\xi^3 = \eta^2$, $Y = \langle \xi^2, \xi\eta \rangle$, $\alpha = \xi^3$, has no self-link;

$A_2 = k[\xi,\eta]_{\langle\xi,\eta\rangle}$ with $\xi^5 = \eta^2$, $Y = \langle \xi^2, \eta \rangle$, $\alpha = \xi^2$, has self-links.

EXAMPLE 3. So far, we have only dealt with clans which are just one cycle. For the general case, rather than going through the same elaborate procedure, on the basis of [(16], Theorem 6), we confine ourselves to listing the simplest example:

Let $R = \begin{pmatrix} A & X \\ X^* & A \end{pmatrix}$, where X is an arbitrary non-zero finitely generated submodule of F. There arise two cases:

(1) $XX^* = A$: Then, R is isomorphic to $M_2(A)$, and the clan is a singleton set with a self-link.

(2) $XX^* \subset M$: Then, $R/\mu R$ is a basic quasi-Frobenius ring whose Nakayama permutation is the identity. Therefore, the clan consists of two maximal ideals $P_1$ and $P_2$, and the cycle map is the identity $P_i \Rightarrow P_i$. Links $P_i \rightsquigarrow P_{i+1}$ always exist; and self-links $P_i \rightsquigarrow P_i$ occur if and only if $XX^* \subsetneq M$.

The following are concrete examples:

$A_1$ (as in Example 2 above), $X = \langle \xi, \eta \rangle$, has no self-link;
$A_2$, $X = \langle \xi^2, \eta \rangle$, has self-links.

10. Open Questions. Bounded noetherian prime rings of Krull dimension one are the simplest non-artinian noetherian rings, and we have occasionally heard the opinion that everything interesting is known about them. We find it therefore worthwhile to list a few open problems:

Do they have enough invertible ideals?, enough clans? Is the link graph locally finite? Is each left-quotient ring at a link component also the right-quotient ring at a link component? (this is true if id.R = 1 or gl.d.R $\leq$ 2.) Does lt-id-R = rt-id.R hold? If lt-id.R = rt-id.R = 1 holds, is every cycle contained in a link component, and is every cycle finite?

All of this is true if the ring has enough invertible ideals, and in particular if it satisfies a polynomial identity or is integral over its centre. What are the clans, explicitly?

If the ring is finitely generated as module over its centre, we know that the clans of maximal ideals are the fibres of the maximal ideals of the centre, but we know little about the link graph (for some additional information, cf. [11]). Does every maximal ideal belong to a circuit? Does every link belong to a circuit? What are the rings for which every maximal ideal has at most one incoming and one outgoing link?

REFERENCES

1. G. Azumaya, M-projective and M-injectivite modules (unpublished).

2. H. Bass, Injective dimension in noetherian rings, Trans. Amer. Math. Soc. 102 (1962), 18-29.

3. M. Boratyński, A change of rings theorem and the Artin-Rees property, Proc. Amer. Math. Soc. 53 (1975), 307-310.

4. K.A. Brown, C.R. Hajarnavis and A.B. MacEacharn, Noetherian rings of finite global dimension, Proc. London Math. Soc. 44(1982), 349-371.

5. J.H. Cozzens and F.L. Sandomierski, preprint.

6. D. Eisenbud and P. Griffith, Serial rings, J. Algebra 17 (1971), 389-400.

7. D. Eisenbud and J.C. Robson, Hereditary noetherian prime rings, J. Algebra 16 (1970), 86-104.

8. J.S. Golan, Localization of Noncommutative Rings, M. Dekker 1975.

9. T.A. Hannula, On the construction of quasi-Frobenius rings, J. Algebra 25 (1973), 403-414.

10. J.P. Jans, Rings and Homology, Holt, Rhinehart and Winston 1964.

11. W. Jansen, Bounded noetherian prime rings of injective dimension one, Ph.D. thesis, McMaster University 1983.

12. A.V. Jategaonkar, Jacobson's conjecture and modules over fully bounded noetherian rings, J. Algebra 30 (1974), 103-121.

13. T.H. Lenagan, Bounded hereditary noetherian prime rings, J. London Math. Soc. 6 (1973), 241-246.

14. T.H. Lenagan, Krull dimension and invertible ideals in noetherian rings, Proc. Edinburgh Math. Soc. 20(1976), 81-86.

15. K. Morita, Duality for modules and its applications to the theory of rings with minimum condition, Sci. Report Tokyo Kyoiku Daigaku 6 (1958), 83-142.

16. B.J. Müller, The structure of quasi-Frobenius rings, Canad. J. Math. 26 (1974), 1141-1151.

17. B.J. Müller, Localization in non-commutative noetherian rings, Canad. J. Math. 28 (1976), 600-610.

18. B.J. Müller, Localization in fully bounded noetherian rings, Pacific J. Math. 67 (1976), 233-245.

19. B.J. Müller, Noncommutative localization and invariant theory, Comm. Algebra 6 (1978), 839-862.

20. B.J. Müller, Ideal invariance and localization, Comm. Algebra 7 (1979), 415-441.

21. B.J. Müller, Two-sided localization in noetherian PI-rings, J. Algebra 63 (1980), 359-373.

22. B.J. Müller, The quotient problem for noetherian rings, Canad. J. Math. 33(1981), 734-748.

23. T. Nakayama, On Frobenusean algebras II, Ann. Math. 42 (1941), 1-21.

24. R. Richards, Noetherian prime rings of Krull dimension one, Comm. Algebra 7 (1979), 845-873.

25. J.C. Robson, Idealizers and hereditary noetherian prime rings, J. Algebra 22 (1972), 45-81.

26. J.E. Roos, Détermination de la dimension homologique globale des algebres de Weyl, C.R. Acad. Sci. Paris 274 (1972), A 23-26.

27. Surjeet Singh, Modules over hereditary noetherian prime rings, Canad. J. Math. 27 (1975), 867-883.

28. B. Stenström, Rings and Modules of Quotients, Lecture Notes in Math 237, Springer 1971.

29. A. Zaks, Injective dimension of semi-primary rings, J. Algebra 13 (1969) 73-86.

30. A Zaks, Hereditary noetherian rings, J. Algebra 29 (1974), 513-529.

31. Noetherian Rings and Rings with Polynomial Identities, Proc. Conf. Univ. Durham 1979.

# NOETHERIAN SUBRINGS OF QUOTIENT RINGS

Ian M. Musson

University of Wisconsin-Madison

Let $R$ be a right Noetherian ring which has a classical right quotient ring $Q$. We study subrings $S$ of $Q$ which are generated by $R$ and finitely many elements of $Q$. We shall be interested in conditions under which the ring $S$ is right Noetherian. This is a Morita invariant property for right orders in right Artinian rings.

To make further progress we study the case where $S = \langle R, x^{-1}\rangle$ and $x$ is a semisimple element of $R$. If $R$ is a ring with center $Z$ and $x \in R$, $z \in Z$ we set $D_x(z) = \{r \in R \mid [x,r] = zr\}$ where $[a,b] = ab - ba$. The element $x$ will be called <u>semisimple</u> if $R = \oplus D_x(z)$, where the sum is over $z \in Z$. In the case where $R$ is a right Noetherian domain, we show that there is a certain right Ore set $\mathcal{C}'(x)$ containing $x$ such that $S$ is right Noetherian if and only if $S = R_{\mathcal{C}'(x)}$ the localization at $\mathcal{C}'(x)$.



*F. van Oystaeyen (ed.), Methods in Ring Theory, 379–389.*

Semisimple elements exist in abundance in enveloping algebras and certain matrix rings and we give several examples.

Lemma 1. Consider the following property of a ring $R$. (P) $R$ is a right order in a right Artinian ring $Q$ and for every finite set of elements $\{x_1, \ldots, x_n\}$ of $Q$ the ring $\langle R, x_1, \ldots, x_n\rangle$ is right Noetherian. Then (P) is a Morita invariant property.

Proof. By (4), Page 177 the ring $S$ is Morita equivalent to $R$ if and only if there is an idempotent $e$ in $M_n(R)$, the $n \times n$ matrix ring over $R$ (for some $n$) such that $S \cong eM_n(R)e$. By a result of Small (9) if $R$ is a right order in an Artinian ring $Q$, then $eM_n(R)e$ is a right order in $eM_n(Q)e$. First note that if $R$ has (P) and $x = (xij) \in M_n(Q)$, then $\langle M_n(R), x\rangle = M_n(S)$ where $S$ is the subring of $Q$ generated by $R$ and all the $xij$, so $M_n(R)$ has (P). Finally if $R$ has (P) and $x \in eQe$, then it is easily checked that $\langle eRe, x\rangle = e\langle R, x\rangle e$ so that $eRe$ has (P).

We note that if $R$ is commutative, Noetherian and has an Artinian quotient ring, then $R$ has (P) by the Hilbert basis theorem.

Now suppose that $x$ is semisimple in $R$ and set

$$A'(x) = \{z \in Z \mid D_x(z) \neq 0\}$$
$$'A(x) = \{z \in Z \mid D_x(-z) \neq 0\}$$

and let $B'(x)$ (resp. $'B(x)$) be the additive subsemigroups of $Z$ generated by $A'(x)$ (resp. $'A(x)$). Note that $B'(x) = A'(x)$ if $R$ is a domain. Finally let $\mathcal{C}'(x)$ (resp. $'\mathcal{C}(x)$) be the multiplicatively closed set in $R$ generated by the elements $x + z$ for $z \in B'(x)$ (resp. $'B(x)$).

<u>Theorem 2</u>. Let $x$ be a semisimple element of $R$

i) $\mathcal{C}'(x)$ is a right Ore set

ii) If $R$ is a right Noetherian domain then $R_{\mathcal{C}'(x)}$ is the smallest right Noetherian subring of $Q$ containing $\langle R, x^{-1}\rangle$.

<u>Proof</u>. (i) By (2), Lemma 1.2, it is enough to show that for $c = (x + z) \in \mathcal{C}'(x)$, $r \in R$ there exist $r_1 \in R$, $c_1 \in \mathcal{C}'(x)$ with $rc_1 = cr_1$. Since $x$ is semisimple we may assume $r \in D_x(z')$ for some $z' \in Z$. Then $[xr] = z'r$ so $xr = r(z' + x)$ and $cr = (x + z)r = r(x + z + z') = rc_1$ with $c_1 = x + z + z' \in \mathcal{C}'(x)$.

ii) Let $T$ be a right Noetherian subring of $Q$ containing $\langle R, x^{-1}\rangle$. We wish to show that $(x + z)^{-1} \in T$ for all $z \in A'(x)$. Choose $r \in D_x(z)$, $r \neq 0$. Then $[x,r] = zr$ so $xr = r(z + x)$.

Since the right ideal $\sum_{t\geq 0} x^{-t}rT$ is finitely generated there exists $n \geq 0$ such that

$$x^{-(n+1)}r \in \sum_{t=0}^{n} x^{-t}rT$$

Hence $r \in \sum_{t=0}^{n} x^{n+1-t}\, rT = r(x+z) \sum_{t=0}^{n} (x+z)^t T$.

Since $R$ is a domain this shows that $(x + z)$ is a unit in $T$ as required.

Given an Ore set $\mathcal{C}$ in a ring $R$, we would like to know whether $\mathcal{C}$ consists of regular elements. If $R$ is prime Noetherian, then $\mathcal{C}$ consists of regular elements provided $0 \notin \mathcal{C}$ by (1), Lemma 2.11. Now clearly $0 \in \mathcal{C}'(x)$ if and only if $x$ satisfies an equation of the form

$$(x + z_1)(x + z_2) \dots (x + z_n) = 0$$

with $z_i \in A'(x)$. In particular if $0 \in \mathcal{C}'(x)$ then $x$ satisfies a monic polynomial with coefficients in $Z$.

However, even if $\mathcal{C}'(x)$ consists of regular elements, Theorem 2 ii) may fail if $R$ is not a domain.

Example 3. If $A$ is a commutative ring and $R = M_n(A)$ then any diagonal matrix is semisimple. For example take $R = M_2(\mathbb{Z})$ and $x = \text{diag}(3,1)$. Then $\langle R, x^{-1}\rangle = M_2(\mathbb{Z})[1/3]$ which is Noetherian, by Lemma 1. Also $A'(x) = {}'A(x) = 2\mathbb{Z}I$ where $I$ is the identity matrix, $\mathcal{C}'(x)$ consists of regular elements and $R_{\mathcal{C}'(x)}$ strictly contains $\langle R, x^{-1}\rangle$.

If $R$ is a right Noetherian domain, then Theorem 2 says that $S = \langle R, x^{-1}\rangle$ is right Noetherian if and only if $S = R_{\mathcal{C}'(x)}$. In a sense this explains <u>why</u> $S$ is right Noetherian, if it is, but gives little information about <u>when</u> this happens. However it seems that $S$ will not be right Noetherian unless the semigroup of eigenvalues $A'(x)$ is a group, and we prove this under an additional assumption (see Theorem 5).

We need to work inside the ring $T = R_{C'(x)}$ where $C'(x)$ is the right Ore set introduced above. Note that $x$ is a semisimple element of $T$ also. In arguments involving elements of $D_x(z)$ it will be clear from the context which ring they belong to so we write $T = \oplus D_x(z)$. An element of $T$ is <u>homogeneous</u> if it belongs to some $D_x(z)$. Clearly the centralizer of $x$ in $T$, $C_T(x)$ is the subring $D_x(0)$.

We need to know that $R$ has a certain factorization property for homogeneous elements.

<u>Lemma 4</u>. Suppose $\alpha, \beta \in R$ and $\alpha\beta$ is homogeneous

i) if $R$ is a domain of characteristic zero, then $\alpha$ and $\beta$ are homogeneous

ii) if $\alpha$ is homogeneous and a non-zero divisor in $R$, then $\beta$ is homogeneous

<u>Proof</u>. i) Since char $R = 0$, the semigroup $A'(x)$ is torsion free abelian and so can be ordered. Suppose $\alpha = a_1 + a_2 + \ldots a_t$ and $\beta = b_1 + \ldots b_s$ where $a_i \in D_x(z_i)$, $b_i \in D_x(z_i')$ and $z_1 < \ldots < z_t$, $z_1' < \ldots < z_s'$. By looking at the coefficient of $\alpha\beta$ in $D_x(z_1 z_1')$ and $D_x(z_t z_s')$ we see that $t = s = 1$ and $\alpha, \beta$ are homogeneous.

The proof of ii) is similar and we omit it. Note however, that if char $R = 2$ and $[x,r] = zr \neq 0$ with $z$ central then $[x,r^2] = 0$ and $(x + r)(x + r + z) = x^2 + r^2 + xz \in C_R(x)$ and neither factor is homogeneous.

Now assume that $R$ is a domain and that the eigenvalues $A'(x)$ do not form a group. Then for some $z_0 \in Z$ $D_x(z_0) \neq 0$ and $D_x(-z_0) = 0$. Let $I$ be the ideal of $T$ given by $I = \oplus D_x(z)$ where the sum is over elements $z \in Z$ with $z - z_0 \in A'(x)$. Then $I$ is a proper ideal of $T$, in fact $I \cap C_T(x) = 0$ since $D_x(-z_0) = 0$. Let $\bar{\ } : T \to T/I = \bar{T}$ be the natural map. It is easy to see that elements of $C_T(x)$, and in particular elements of $Z$ are regular mod $I$. It follows that the sum $\bar{T} = \sum \overline{D_x(z)}$ is direct so $\bar{x}$ is a semisimple element of $\bar{T}$. We note that homogeneous elements of $\bar{T}$ may be lifted mod $I$. That is if $r \in \overline{D_x(z)}$ there is $s \in D_x(z)$ with $\bar{s} = r$. Set $A' = \{z \in Z \mid \overline{D_x(z)} \neq 0\}$ and let $\mathcal{C}'$ be the multiplicatively closed subset of $\bar{R}$ generated by the elements $\overline{(x + z)}$ where $z \in A'$. Then $\mathcal{C}'$ is a right Ore set in $\bar{R}$ by Theorem 2 i). Note that $z_0 \notin A'$. Also $\bar{R}$ is a subring of $\bar{T}$ with the same 1 and elements of $\mathcal{C}'$ are units in $\bar{T}$ and so are regular in $\bar{R}$ and $\bar{R}_{\mathcal{C}'}$ exists and is (isomorphic to) a subring of $\bar{T}$, by the universal property of localization.

Now suppose in addition that $T = R_{\mathcal{C}'(x)} = \langle R, x^{-1} \rangle$. Then $x + z_0$ is a unit in $T$ so $\overline{(x + z_0)}$ is a unit in $\bar{T}$ and $\bar{T}$ is generated by the subring $\bar{R}$ and $\bar{x}^{-1}$. Therefore $\bar{T} \subseteq \bar{R}_{\mathcal{C}'}$

and we can write

$$\overline{(x + z_0)}^{-1} = \bar{\alpha}\, \overline{(x + z_1)}^{-1} \dots \overline{(x + z_n)}^{-1}$$

where $z_1, \dots, z_n \in A'$ and $\bar{\alpha} \in \bar{R}$. Then

$$\overline{(x + z_n)} \dots \overline{(x + z_1)} = \overline{(x + z_0)}\,\bar{\alpha}.$$

Now by Lemma 4 ii) $\bar{\alpha} \in \overline{C_R(x)}$, since $\overline{x + z_0}$ is a regular homogeneous element of $\bar{R}$. By the lifting property for homogeneous elements there is an $\alpha \in C_R(x)$, $\beta \in I$ with $(x + z_n) \dots (x + z_1) = (x + z_0)\alpha + \beta$, but then $\beta \in I \cap C_T(x) = 0$. Hence

$$(x + z_n) \dots (x + z_1) = (x + z_0)\alpha.$$

We seem to require some additional assumption to obtain a contradiction from this last equation. For example if we assume that $C_R(x) = Z[x]$ is a polynomial ring over $Z$ then $(x + z_0)$ generates a prime ideal, and so divides $(x + z_i)$ for some $i$. Then comparing degrees in $x$, we have $x + z_i = x + z_0$ so $z_0 = z_i \in A'$ a contradiction. In conclusion we have proved.

<u>Theorem 5</u>. Suppose $R$ is a domain and $x$ a semisimple element of $R$ such that $C_R(x) = Z[x]$ a polynomial ring over the center $Z$, and that the eigenvalues $A'(x)$ do not form a group. Then $\langle R, x^{-1} \rangle \subsetneq R_{C'(x)}$.

<u>Example 6</u>. Let $k$ be a field of characteristic

zero and $\underline{g}$ the 2-dimensional soluble Lie algebra with basis $x,y$ such that $[x,y] = y$. If $R$ is the universal enveloping algebra of $\underline{g}$, then $x$ is a semisimple element of $R$ and $R = \bigoplus_{n \in \mathbb{N}} D_x(n)$ where $D_x(n) = k[x]y^n$. The eigenvalues $\mathbb{N}$ do not form a group and the centralizer $C_R(x) = k[x]$ is a polynomial ring. Hence $\langle R,x^{-1}\rangle \subsetneq R_{\mathcal{C}'(x)}$ and $\langle R,x^{-1}\rangle$ is not right Noetherian.

We remark that the ring $R_{\mathcal{C}'(x)}$ is an example of a right Noetherian right and left Ore domain which is not left Noetherian. The first example of such a ring was given by D. A. Jordan (6).

Clearly we can obtain analogous results to those given above using the left Ore sets $'\mathcal{C}(x)$. Also the condition on eigenvalues in Theorem 6 is right-left symmetric, since $A'(x)$ is a group if and only if $A'(x) = 'A(x)$.

<u>Example 7</u>. Let $k$ be a field of characteristic zero and $R = A_n(k)$ the nth Weyl algebra over $k$ with generators $p_i$, $q_j$, $1 \leq i,j \leq n$ such that $[p_i, q_j] = \delta_{ij}$. If $\lambda_1, \ldots, \lambda_n$, $\mu$ are elements of $k$ then $x = \sum_{i=1}^{n} \lambda_i p_i q_i + \mu$ is a semisimple element of $R$. The eigenvalues satisfy $A'(x) = 'A(x) = \sum_{i=1}^{n} \mathbb{Z}\lambda_i$. In general, we have been unable to determine whether the ring $\langle R, x^{-1}\rangle$ is Noetherian. However if $n = 1$, then $R$ is an HNP ring and every ring contained between $R$ and its quotient ring is Noetherian (7), Proposition 1.5 or

(5), Proposition 2.

In connection with this last result we mention that if $R$ is a prime Noetherian ring of Krull dimension one which is finitely generated as a module over its center, then every ring contained between $R$ and its quotient ring is Noetherian of Krull dimension at most one. This follows from a result of Schelter (8), Theorem 1. Another proof has been obtained by the author based on the proof of the commutative Krull-Akizuki theorem given in (3) Ch. VII, p. 29. However if $R$ is merely assumed to have an Artinian quotient ring the corresponding result may fail.

<u>Example 8</u>. Let $R = \begin{pmatrix} \mathbb{Z} & \mathbb{Z} \\ 0 & \mathbb{Z} \end{pmatrix}$ the ring of $2 \times 2$ upper triangular matrices over $\mathbb{Z}$. Then $R$ is Noetherian of Krull dimension one, and finitely generated as a module over its center and has an Artinian quotient ring $\begin{pmatrix} \mathbb{Q} & \mathbb{Q} \\ 0 & \mathbb{Q} \end{pmatrix}$. If $x = \mathrm{diag}(m,n)$, then $x$ is a semisimple element of $R$, $\langle R, x^{-1}\rangle = \begin{pmatrix} \mathbb{Z}[1/m] & \mathbb{Z}[1/mn] \\ 0 & \mathbb{Z}[1/n] \end{pmatrix}$ and

i) $\langle R,x^{-1}\rangle$ is right Noetherian if and only if every prime factor of $m$ is a prime factor of $n$

ii) $\langle R,x^{-1}\rangle$ is left Noetherian if and only if every prime factor of $n$ is a prime factor of $m$.

Varying our choice of $m,n$ we obtain examples where $\langle R, x^{-1}\rangle$ is Noetherian on both sides, one side only, or neither side.

Finally we prove a reduction result modulo a nilpotent ideal.

Theorem 9. Let $R$ be a right Noetherian ring which has a quotient ring. Let $c$ be a regular element of $R$, $S = \langle R, c^{-1}\rangle$ and let $N$ be the nilpotent radical of $R$ and $N^* = NS$. Then $S$ is right Noetherian if and only if $SN^* = N^*$ and $S/N^*$ is right Noetherian.

Proof. Suppose that $S$ is right Noetherian. To show $SN^* \subseteq N^*$ it suffices to show $c^{-1}N^* \subseteq N^*$. Since the right ideal $\sum_{t\geq 0} c^{-t}N^*$ is finitely generated we have $c^{-(t+1)}N^* \subseteq \sum_{n=0}^{t} c^{-n}N^*$ for some $t$ and $c^{-1}N^* \subseteq \sum_{n=0}^{t} c^{t-n}N^* \subseteq N^*$. Clearly $S/N^*$ is right Noetherian.

Conversely if $S/N^*$ is right Noetherian and $SN^* = N^*$, then $N^*$ is a nilpotent ideal of $S$. Also $N^*$ is finitely generated as a right ideal of $S$, since $N$ is a finitely generated right ideal of $R$. It follows that each $(N^*)^i/(N^*)^{i+1}$ is a finitely generated right $S/N^*$-module and hence $S$ is Noetherian.

REFERENCES


(1) Borho, W., Gabriel, P. and Rentschler, R., Primideale in Einhüllenden auflösbarer Lie-Algebren, Springer, Berlin, Heidelberg, New York, 1973.

(2) Borho, W., and Rentschler, R., Oresche Teilmengen in Einhüllenden Algebren, Math. Ann 217, (1975), pp. 201-210.

(3) Bourbaki, N., Algèbre Commutative, Hermann, Paris, 1961-65.

(4) Chatters, A. W. and Hajarnavis, C. R., Rings with Chain Conditions, Pitman, Boston, London, Melbourne, 1980.

(5) Goodearl, K. R., Localization and splitting in heriditary Noetherian prime rings, Pacific J. Math, 53(1974), pp. 137-151.

(6) Jordan, D. A., A left Noetherian right Ore domain which is not right Noetherian, Bull. London Math. Soc., 12 (1980), pp. 202-204.

(7) Kuzmanovich, J., Localizations of Dedekind prime rings, J. Algebra, 21(1972), pp. 378-393.

(8) Schelter, W., On the Krull-Akizuki theorem, J. London Math. Soc. (2), 13, (1976), pp. 263-264.

(9) Small, L. W., Orders in Artinian rings II, J. Algebra 9 (1968), pp. 266-273.

# STABILITY CONDITIONS FOR COMMUTATIVE RINGS WITH KRULL DIMENSION

Constantin Năstăsescu and Şerban Raianu

Facultatea de Matematică
Str. Academiei 14
R 70109 Bucharest 1, Romania

As Gordon and Robson show in their work "Krull dimension" [4], rings with Krull dimension have a lot of noetherian-like properties, such as : a ring with Krull dimension has ACC on prime ideals, every ideal in such a ring is a finite intersection of irreducible ideals, every ideal contains a power of its radical, etc.

However, rings with Krull dimension must not have DCC on prime ideals. Also, there are rings with Krull dimension in which the Principal Ideal Theorem does not hold, and the intersection of the powers of their Jacobson radical is not equal to zero [4].

The aim of this paper is to investigate other possible similarities (or differences) between noetherian rings and rings with Krull dimension. We restrict our attention to commutative rings only.

The main result of this paper is Theorem 2.3. Loosely, it says that under a certain stability assumption, a commutative ring with Krull dimension is noetherian with respect to a given Gabriel topology. One of the most interesting consequences of this result is that there is a close connection between Krull domains and domains with Krull dimension (see Corollary 2,10 and Remark 2.11).



*F. van Oystaeyen (ed.), Methods in Ring Theory, 391–402.*

## 1. NOTATION AND PRELIMINARIES

Throughout this paper, the word "ring" will mean "commutative and unitary ring". R will always denote such a ring and Mod-R will denote the category of all R-modules.

For the definition of Krull dimension see [5]. The Krull dimension of a module M will be denoted by K.dim(M).

Let $\mathcal{C}$ be a Grothendieck category which has simple objects. An object of $\mathcal{C}$ is said to be <u>semiartinian</u> iff it belongs to the smallest localizing subcategory of $\mathcal{C}$ containing all simple objects.

<u>PROPOSITION 1.1</u> [4] If X is a semiartinian object of $\mathcal{C}$ and has Krull dimension, then X is artinian.

We will let Spec(R) (resp. Max(R)) denote the set of all prime (resp. maximal) ideals of R. If M is an R-module, we will denote $\mathrm{Ass}(M) = \{ \underline{p} \in \mathrm{Spec}(R) \mid \exists\, x \in M,\ x \neq 0,\ \underline{p} = \mathrm{Ann}(x) \}$.
If R has Krull dimension and $M \in$ Mod-R, $M \neq 0$, then $\mathrm{Ass}(M) \neq \emptyset$ [4]. E(M) will denote the injective envelope of the R-module M. Recall that Ass(M) = Ass(E(M)) [6].

Let F be a Gabriel topology on the ring R [9; p. 146], and let $(\mathcal{T}, \mathcal{F})$ be the corresponding hereditary torsion theory. If the injective envelope of any module in $\mathcal{T}$ is again in $\mathcal{T}$ we will say that R is <u>F-stable</u>, or that $\mathcal{T}$ is <u>stable</u>.

An R-module is said to be <u>F-noetherian</u> (resp. <u>F-artinian</u>) iff its image in the quotient category Mod-R/$\mathcal{T}$ is a noetherian (resp. artinian) object.

We will let $\mathrm{Spec}_F(R) = \mathrm{Spec}(R) \setminus F$ and $\mathrm{Max}_F(R)$ will denote the set of all maximal members of $\mathrm{Spec}_F(R)$.

If $\underline{p} \in \mathrm{Spec}(R)$ we let $F_{\underline{p}} = \{ I \leqslant R \mid I \not\subset \underline{p} \}$, which is a Gabriel topology on R. For any $n \in \mathbb{N}$, we put $F_n = \bigcap_{ht(\underline{p}) \leqslant n} F_{\underline{p}}$,

We will say that R is n-noetherian (resp. n-stable), instead of $F_n$-noetherian (resp. $F_n$-stable).

LEMMA 1.2 Let R be a ring with Krull dimension and F a Gabriel topology on R. Let ( $\mathcal{T}$ , $\mathcal{F}$ ) be the corresponding torsion theory and t its torsion radical. Then:

1) $M \in \mathcal{F} \Longleftrightarrow \mathrm{Ass}(M) \subseteq \mathrm{Spec}_F(R)$.

2) $F = \bigcap_{\underline{p} \in \mathrm{Spec}_F(R)} F_{\underline{p}} = \bigcap_{\underline{p} \in \mathrm{Max}_F(R)} F_{\underline{p}}$.

Proof: 1) Let $M \in$ Mod-R, $M \neq 0$, such that $\mathrm{Ass}(M) \subseteq \mathrm{Spec}_F(R)$. If $M \notin \mathcal{F}$, then $t(M) \neq 0$, and hence $\mathrm{Ass}(t(M)) \neq \emptyset$. But $\mathrm{Ass}(t(M)) \subseteq$ $\subseteq F$, and $\mathrm{Ass}(t(M)) \subseteq \mathrm{Ass}(M)$, hence $\emptyset \neq \mathrm{Ass}(t(M)) \subseteq F \cap \mathrm{Spec}_F(R)$, a contradiction.

2) We will show only that $\bigcap_{\underline{p} \in \mathrm{Spec}_F(R)} F_{\underline{p}} \subseteq F$. Let $\underline{a}$ be an ideal of R, $\underline{a} \notin F$. Then $A/\underline{a} \notin \mathcal{T}$ and hence $\underline{b}/\underline{a} = t(A/\underline{a}) \neq A/\underline{a}$. Now $0 \neq A/\underline{b} \in \mathcal{F}$ and thus $\emptyset \neq \mathrm{Ass}(A/\underline{b}) \subseteq \mathrm{Spec}_F(R)$. Hence there exists $\underline{p} \in \mathrm{Spec}_F(R)$ such that $\underline{a} \subseteq \underline{b} \subseteq \underline{p}$, and so $\underline{a} \notin \bigcap_{\underline{p} \in \mathrm{Spec}_F(R)} F_{\underline{p}}$.

PROPOSITION 1.3 Let M be a module with Krull dimension, $\mathcal{T}$ a localizing subcategory of Mod-R, and T : Mod-R $\longrightarrow$ Mod-R/$\mathcal{T}$ the canonical functor. Then T(M) has Krull dimension.

Proof: We may clearly suppose that $\varphi : M \longrightarrow ST(M)$ (S is the right adjoint of T) is a monomorphism. Then the mapping $X \longmapsto \varphi^{-1}(S(X))$ (from the lattice of all subobjects of T(M) to the lattice of all submodules of M) is injective and preserves order.

All unexplained facts concerning torsion theory (resp. category theory) used in this paper may be found in [1] and [9] (resp. in [3] and [8]).

## 2. MAIN RESULTS

PROPOSITION 2.1 Let $R$ be a ring with Krull dimension and $\mathcal{T}$ a localizing subcategory of Mod-R. Denote by $\mathcal{C}$ the quotient category Mod-R/$\mathcal{T}$ and by $T : \text{Mod-R} \longrightarrow \mathcal{C}$ the canonical functor. Suppose that $Q \in \mathcal{C}$ is a cogenerator which is $\sum$-injective (i.e. any direct sum of copies of $Q$ is injective) and semiartinian. Then $T(R)$ is noetherian.

Proof: Put $U = T(R)$. Since $R$ has Krull dimension, it follows from Proposition 1.3 that $U$ has Krull dimension too. We assume that

$$U_1 \subset U_2 \subset \ldots \subset U_{n-1} \subset U_n \subset \ldots \quad \subset U$$

is a strictly ascending chain of subobjects of $U$, and look for a contradiction. Since $U_n/U_{n-1} \neq 0$ for all $n \geqslant 2$, we have that $\text{Hom}_{\mathcal{C}}(U_n/U_{n-1}, Q) \neq 0$, and so there exists $\overline{f}_n : U_n \longrightarrow Q$, $\overline{f}_n \neq 0$, and $\overline{f}_n(U_{n-1}) = 0$. Since $Q$ is injective, there exists $f_n : U \longrightarrow Q$ extending $\overline{f}_n$, and hence $f_n(U_n) \neq 0$, $f_n(U_{n-1}) = 0$. We denote by $Q^{\mathbb{N}}$ (resp. $Q^{(\mathbb{N})}$) the direct product (resp. sum) of a countable set of copies of $Q$, and by $\pi_k : Q^{\mathbb{N}} \longrightarrow Q$ the k-th projection. There exists $f : U \longrightarrow Q^{\mathbb{N}}$ such that $\pi_n \circ f = f_n$ for all $n \geqslant 2$. Put $K = \bigcup_{n \in \mathbb{N}} U_n$ and denote by $f_1 : K \longrightarrow Q^{\mathbb{N}}$ the restriction of $f$ to $K$. We claim that $f_1$ factors through the inclusion $Q^{(\mathbb{N})} \subset Q^{\mathbb{N}}$. To see this, it is enough to show that $f_1(U_n) = f(U_n) \subset Q^{(\mathbb{N})}$ for all $n \in \mathbb{N}$, and this is obvious, since $(\pi_k \circ f)(U_n) = f_k(U_n) = 0$ for $k > n$. We have the commutative diagram

$$\begin{array}{ccc} U & \xrightarrow{\quad f \quad} & Q^{\mathbb{N}} \\ \cup & \nearrow^{f_1} & \cup \\ K & \xrightarrow[g_1]{- - - -} & Q^{(\mathbb{N})} \end{array}$$

Since $Q^{(\mathbb{N})}$ is injective, there exists $g : U \longrightarrow Q^{(\mathbb{N})}$ extending $g_1$

$$\begin{array}{ccc} K & \xrightarrow{g_1} & Q^{(\mathbb{N})} \\ \cap & \nearrow_{g} & \\ U & & \end{array}$$

Now $g(U) \subset Q^{(\mathbb{N})}$ has Krull dimension and is semiartinian, and hence $g(U)$ is artinian by Proposition 1.1. From the exact sequence $U \longrightarrow g(U) \longrightarrow 0$, we deduce that there exists an ideal $I$ of $R$ such that $g(U) \simeq T(R/I)$. Let $F$ be the Gabriel topology on $R$ such that $\mathcal{T}$ is the class of all F-torsion modules, and let $\varphi : R \longrightarrow R/I$ be the natural surjection. Since $T(R/I)$ is artinian, it follows that $R/I$ is an F-artinian module, and so $R/I$ is a $\varphi(F)$-artinian ring [1; 1.10]. By [1; 4.7] we get that $R/I$ is a $\varphi(F)$-noetherian ring, and so $R/I$ is an F-noetherian module, i.e. $T(R/I)$ is a noetherian object of $\mathcal{C}$. This means that $T(R/I) \simeq g(U)$ is an object of finite length in $\mathcal{C}$. If we put $A_n = Q^n$ for $n \in \mathbb{N}$, we have that $Q^{(\mathbb{N})} = \sum_{n \in \mathbb{N}} A_n$ and $g(U) = (\sum_{n \in \mathbb{N}} A_n) \cap g(U) = \sum_{n \in \mathbb{N}} (A_n \cap g(U))$. It follows that there exists $n \in \mathbb{N}$ such that $g(U) = A_n \cap g(U)$, i.e. $g(U) \subset A_n$, contradicting the fact that $g(U_n) = f_n(U_n) \neq 0$ for all $n \in \mathbb{N}$.

<u>REMARK 2.2</u> 1) Proposition 2.1 fails to be true when $R$ is an arbitrary commutative ring. In fact, if $\mathcal{C}$ is a Grothendieck category with a generator $U$, and $\mathcal{C}$ has a cogenerator which is $\sum$-injective, it does not follow that $U$ is noetherian. As it is shown in [7], there exists a ring $A$, and a localizing subcategory $\mathcal{T}$ of Mod-$A$, such that any direct sum of injective objects of Mod-$A/\mathcal{T}$ is injective, but Mod-$A/\mathcal{T}$ has no simple objects, and hence it is not a locally noetherian category.

2) From the proof of Proposition 2.1 it follows at once the following result : if $\mathcal{C}$ is a Grothendieck category having a generator U which is finitely generated [9; p.121] and $\mathcal{C}$ has a $\Sigma$-injective cogenerator, then U is noetherian.

We are now in a position to state and prove the main result of this paper.

<u>THEOREM 2.3</u> Let R be a ring with Krull dimension and F a Gabriel topology on R. Let $X = \text{Spec}_F(R) \setminus \text{Max}_F(R)$ and put $F' = \bigoplus_{\underline{p} \in X} F_{\underline{p}}$. Assume that R is F'-stable. Then R is F-noetherian.

<u>Proof</u>: Denote by $(\mathcal{T}, \mathcal{F})$ (resp. $(\mathcal{T}', \mathcal{F}')$) the hereditary torsion theory corresponding to F (resp. F') and by t (resp. t') its torsion radical. Let us remark first that if $X = \emptyset$, then $\mathcal{T}' = \text{Mod-}R$ is stable.

Let $Q = E(\bigoplus_{\underline{p} \in \text{Max}_F(R)} E(R/\underline{p}))$ and $F_Q$ the Gabriel topology cogenerated by Q (i.e. $I \in F_Q \Longleftrightarrow \text{Hom}_R(R/I, Q) = 0$). We claim that $F = F_Q$. The inclusion $F_Q \subseteq F$ being obvious, let $I \in F$. We assume that $\text{Hom}_R(R/I, Q) \neq 0$ and look for a contradiction. Let $x \in Q$ such that $x \neq 0$ and $I \subseteq \text{Ann}(x)$. It follows that there exists an $a \in R$ such that $0 \neq ax \in \bigoplus_{\underline{p} \in \text{Max}_F(R)} E(R/\underline{p})$. Let $\underline{q} \in \text{Max}_F(R)$ such that the component of $ax$ in the direct summand $E(R/\underline{q})$ is non-zero. Then $I \subseteq \text{Ann}(x) \subseteq \text{Ann}(ax) \subseteq \underline{q}$, and hence $\underline{q} \in F$, a contradiction. Thus $F \subseteq F_Q$, and so $F = F_Q$.

We denote by $T : \text{Mod-}R \longrightarrow \text{Mod-}R/\mathcal{T}$ the canonical functor, and by S its right adjoint. It is well-known that T is exact and S is left exact.

It is clear that T(Q) is a cogenerator of $\text{Mod-}R/\mathcal{T}$. Indeed, let $X \in \text{Mod-}R/\mathcal{T}$, $X \neq 0$. Then S(X) is F-torsion free, $S(X) \neq 0$, and so there exists $f \in Hom_R(S(X), Q)$, $f \neq 0$. If $T(f) = 0$, it follows that $f(S(X)) \in \mathcal{T}$. On the other hand, $f(S(X)) \subseteq Q \in \mathcal{F}$, and hence $f(S(X)) = 0$, a contradiction. Hence $T(f) \neq 0$ and $T(f) \in Hom_{Mod\text{-}R/\mathcal{T}}(X, T(Q))$.

We will show that $T(R)$ is noetherian by applying Proposition 2.1, and so the only thing to prove is that $T(Q)$ is a $\Sigma$-injective semiartinian object of $Mod\text{-}R/\mathcal{T}$. We denote by $\mathcal{A}$ the localizing subcategory of all semiartinian objects of $Mod\text{-}R/\mathcal{T}$. Since $T$ is exact and commutes with direct sums, it follows that

$$\mathcal{C} = \{ M \in Mod\text{-}R \mid T(M) \in \mathcal{A} \}$$

is a localizing subcategory of $Mod\text{-}R$.

We will show that $\mathcal{C} = \mathcal{T}'$. Let $M \in \mathcal{C}$, $M \neq 0$. If $M \notin \mathcal{T}'$, it follows that $0 \neq M/t'(M) \in \mathcal{F}'$ and $M/t'(M) \in \mathcal{C}$. Thus, we assume that $M \in \mathcal{C}$, $M \in \mathcal{F}'$, $M \neq 0$, and look for a contradiction. Now $T(M) \in \mathcal{A}$, and so $T(M)$ contains a simple object of $Mod\text{-}R/\mathcal{T}$, say $T(R/\underline{p}) \subseteq T(M)$, $\underline{p} \in Max_F(R)$. Then $ST(R/\underline{p}) \subseteq ST(M)$. Since $\mathcal{F}' \subseteq \mathcal{F}$, we have that $M \subseteq ST(M)$ and the inclusion is essential. It follows that $M \cap R/\underline{p} \neq 0$, and so $R/\underline{p} \subseteq M$. Hence $\underline{p} \in Ass(M)$, and since $\underline{p} \in Max_F(R)$, this contradicts the fact that $M \in \mathcal{F}'$. Thus $\mathcal{C} \subseteq \mathcal{T}'$. Conversely, let $M \in \mathcal{T}'$, $M \neq 0$. Clearly we may assume that $M \in \mathcal{F}$. In order to prove that $M \in \mathcal{C}$, it is enough to show that $T(M)$ contains a simple object of $Mod\text{-}R/\mathcal{T}$. Now $Ass(M) \neq \emptyset$ and let $\underline{p} \in Ass(M)$. Clearly $\underline{p} \in Max_F(R)$ and $R/\underline{p} \subseteq M$. Thus $T(R/\underline{p}) \subseteq T(M)$. Hence $\mathcal{T}' \subseteq \mathcal{C}$ and so $\mathcal{C} = \mathcal{T}'$.

Now, since $\mathcal{C}$ is stable, it follows that $\mathcal{A}$ is stable. Indeed, let $X \in \mathcal{A}$ and let $E(X)$ be the injective envelope of $X$ in $Mod\text{-}R/\mathcal{T}$. It follows that $S(E(X))$ is an essential extension of $S(X)$. Since $T$ is exact, $S(E(X))$ is injective and $S(X) \in \mathcal{C}$ since $X = TS(X)$. It follows that $S(E(X)) \in \mathcal{C}$, and so $E(X) = TS(E(X)) \in \mathcal{A}$. Hence $\mathcal{A}$ is stable and so an injective object of $\mathcal{A}$ is injective in Mod-$R/\mathcal{T}$.

The next step is to prove that $T(\bigoplus_{\underline{p}\in Max_F(R)} E(R/\underline{p}))$ is a $\Sigma$-injective semiartinian object of Mod-$R/\mathcal{T}$.

It is obvious that the family $\{R/I \mid I \leqslant R,\ R/I \in \mathcal{C}\}$, is a family of generators for $\mathcal{C}$, and so $\{T(R/I) \mid R/I \in \mathcal{C}\}$ is a family of generators for $\mathcal{A}$. Since for each ideal $I$ of $R$ such that $R/I \in \mathcal{C}$, $T(R/I)$ has Krull dimension and is semiartinian, it follows that $T(R/I)$ is noetherian (see the proof of Proposition 2.1). Hence $\mathcal{A}$ is a locally noetherian category, and, consequently, any direct sum of injective objects of $\mathcal{A}$ is injective [8].

Now for every $\underline{p} \in Max_F(R)$, $E(R/\underline{p})$ is F-closed, and so $T(E(R/\underline{p}))$ is injective, and $T(E(R/\underline{p})) \in \mathcal{A}$, because $\mathcal{A}$ is stable. Hence $T(\bigoplus_{\underline{p}\in Max_F(R)} E(R/\underline{p})) = \bigoplus_{\underline{p}\in Max_F(R)} T(E(R/\underline{p})) \in \mathcal{A}$ and is $\Sigma$-injective.

We end the proof by showing that $T(\bigoplus_{\underline{p}\in Max_F(R)} E(R/\underline{p})) = T(Q)$. Clearly $\bigoplus_{\underline{p}\in Max_F(R)} E(R/\underline{p})$ is F-torsion free, and thus we have that $ST(\bigoplus_{\underline{p}\in Max_F(R)} E(R/\underline{p}))$ is an essental extension of $\bigoplus_{\underline{p}\in Max_F(R)} E(R/\underline{p})$. Since $T$ is exact, $ST(\bigoplus_{\underline{p}\in Max_F(R)} E(R/\underline{p}))$ is injective and hence $ST(\bigoplus_{\underline{p}\in Max_F(R)} E(R/\underline{p})) = E(\bigoplus_{\underline{p}\in Max_F(R)} E(R/\underline{p})) = Q$. Thus we have that $TST(\bigoplus_{\underline{p}\in Max_F(R)} E(R/\underline{p})) = T(E(\bigoplus_{\underline{p}\in Max_F(R)} E(R/\underline{p}))) = T(Q)$ and the proof is complete.

REMARK 2.4 Theorem 2.3 is not true for arbitrary rings. As it is shown in [7], if $R$ is a noetherian ring and $\Lambda$ is an infinite set, then $A = R[X_\alpha]_{\alpha\in\Lambda}$ has the following properties:

1) A is n-noetherian for every $n \in \mathbb{N}$.

2) A is n-stable for every $n \in \mathbb{N}$.

3) A is not $F_W$-noetherian, where $F_W = \bigcap_{n \geq 0} F_n$.

<u>COROLLARY 2.5</u> Let R be a ring with Krull dimension. Then the following assertions are equivalent:

1) R is noetherian.

2) The localizing subcategory of all semiartinian R-modules is stable.

<u>Proof</u>: 1)$\Longrightarrow$2). It follows from [6; 4.3]. Take $F = \{R\}$ in Theorem 2.3 to prove 2)$\Longrightarrow$1).

<u>REMARK 2.6</u> Corollary 2.5 improves a result of Damiano and Papp [2], where it is proved that a commutative ring with Krull dimension, having the property that every localizing subcategory of Mod-R is stable, is noetherian.

<u>COROLLARY 2.7</u> Let R be a ring with Krull dimension and let $X = \{\underline{p} \in \mathrm{Spec}(R) \mid \forall \underline{q} \in \mathrm{Spec}(R),\ \underline{q} \subseteq \underline{p},\ \underline{q}$ is finitely generated$\}$. Let $F_X = \bigcap_{\underline{p} \in X} F_{\underline{p}}$. Then R is $F_X$-noetherian. Moreover, if R/I is noetherian for each $I \in F_X$, then R is noetherian.

<u>Proof</u>: For each $\underline{p} \in X$, R is $F_{\underline{p}}$-stable by [6; 4.1], and hence R is $F_X$-noetherian by Theorem 2.3. The second assertion follows from the following general lemma:

<u>LEMMA 2.8</u> Let R be a ring and F a Gabriel topology on R such that R is F-noetherian and F-stable. Assume that R/I is noetherian for each $I \in F$. Then R is noetherian.

<u>Proof</u>: Denote by $(\mathcal{T}, \mathcal{F})$ the hereditary torsion theory corresponding to F, and by t its torsion radical. We will

show that any injective R-module is a direct sum of injective indecomposable R-modules. Let $Q$ be an injective R-module. Then clearly $t(Q)$ is injective in $\mathcal{T}$, and since $\mathcal{T}$ is stable, $t(Q)$ is injective in Mod-R. Thus $Q = t(Q) \oplus Q/t(Q)$. Since $\mathcal{T}$ is locally noetherian, $t(Q)$ is a direct sum of injective and indecomposable R-modules, and so is $Q/t(Q)$ (because R is F-noetherian and $Q/t(Q)$ is injective and F-torsion free [6]). So $Q$ is a direct sum of injective indecomposable R-modules, and hence R is noetherian.

<u>COROLLARY 2.9</u> Let R be a ring with Krull dimension and $n \in \mathbb{N}$. Suppose that for each $\underline{p} \in \operatorname{Spec}(R)$, $ht(\underline{p}) < n$, R is $F_{\underline{p}}$-stable. Then R is n-noetherian.

<u>COROLLARY 2.10</u> A domain with Krull dimension is 1-noetherian.

<u>Proof</u>: Any domain is 0-stable, sice clearly the injective envelope of each torsion module (in the classical sense) is a torsion module.

<u>REMARK 2.11</u> Corollary 2.10 shows that domains with Krull dimension are"very close" to Krull domains, since Krull domains are 1-noetherian, and this is an essential fact in their characterisation [9].

<u>COROLLARY 2.12</u> A Krull domain having Krull dimension at most two is noetherian.

<u>Proof</u>: Krull domains are 1-stable [9].

<u>COROLLARY 2.13</u> Let R be a ring with Krull dimension such that for each $\underline{p} \in \operatorname{Spec}(R)$, $ht(\underline{p}) < \infty$, R is $F_{\underline{p}}$-stable. Then any direct sum of $\Sigma$-injective R-modules is $\Sigma$-injective.

Proof: Let $\{Q_\alpha\}_{\alpha\in\Lambda}$ be a family of $\Sigma$-injective R- modules. It follows that R is $F_{Q_\alpha}$-noetherian for each $\alpha\in\Lambda$, and hence $R_{\underline{p}}$ is noetherian for each $\underline{p}\in \mathrm{Spec}_{F_{Q_\alpha}}(R)$. Clearly we have that $\mathrm{Spec}_{(\bigcap_{\alpha\in\Lambda} F_Q)}(R) = \bigcap_{\alpha\in\Lambda} \mathrm{Spec}_{F_{Q_\alpha}}(R)$, and hence R is $(\bigcap_{\alpha\in\Lambda} F_Q)$-noetherian by Theorem 2.3. Let $Q = \bigoplus_{\alpha\in\Lambda} Q_\alpha$. Since all $Q_\alpha$ are clearly $(\bigcap_{\alpha\in\Lambda} F_{Q_\alpha})$-torsion free, it follows that Q is injective [6; 1.6]. Now $F_Q = \bigcap_{\alpha\in\Lambda} F_{Q_\alpha}$ and so R is $F_Q$-noetherian, i.e. Q is $\Sigma$-injective.

REMARK 2.14 Corollary 2.13 does not hold when R does not have Krull dimension. With the notation of Remark 2.4, we have that $Q_n = \bigoplus_{\substack{\underline{p}\in \mathrm{Spec}(A)\\ ht(\underline{p})=n}} E(R/\underline{p})$ is $\Sigma$-injective for each $n\in\mathbb{N}$, but $\bigoplus_{n\in\mathbb{N}} Q_n$ is not $\Sigma$-injective.

EXAMPLES 2.15 1) Let $\alpha\neq 0$ be an ordinal and let R be a commutative noetherian domain having Krull dimension $\alpha$ [4; 9.8]. Let $\underline{m}\in \mathrm{Max}(R)$ and $Q = E(R/\underline{m})$. It is well-known that Q is artinian. Q is not noetherian, because if Q would be finitely generated, then it would be finitely generated as an $R_{\underline{m}}$-module [6; 3.7], and hence $K.\dim(R_{\underline{m}})$ would be equal to zero. Let

$$S = \left\{ \begin{pmatrix} a & b \\ 0 & a \end{pmatrix} \mid a\in R,\ b\in Q \right\}.$$

Then S is commutative, S is not noetherian, and $K.\dim(S) = \alpha$.

Proof: Same as the proof of [4; 10.6].

2) (See [6]) Let R be a valuation ring whose value group is $\mathbb{Z}\times\mathbb{Z}$, ordered lexicographically. Clearly R is not noetherian. Let $0\subset \underline{p}\subset \underline{m}$ be the prime ideals of R. Then $E(R/\underline{m})$ is not semiartinian.

Proof: We suppose that $E(R/\underline{m})$ is semiartinian and look for a contradiction. Let $x \in E(R/\underline{m})$, $x \neq 0$, and $I = \mathrm{Ann}(x)$. Then $R/I$ is semiartinian and has Krull dimension, hence it is artinian by Proposition 1.1. Thus $I$ is $\underline{m}$-primary. It follows that $E(R/\underline{m}) = \bigcup_{i\geq 1}\{x \in E(R/\underline{m}) \mid \underline{m}^i x = 0\}$, and so $\bigcap_{k\geq 1}\underline{m}^k$ annihilates $E(R/\underline{m})$. Since $E(R/\underline{m})$ is a cogenerator for Mod-R, it follows that $\bigcap_{k\geq 1}\underline{m}^k$ annihilates $R$, and so $\bigcap_{k\geq 1}\underline{m}^k = 0$, contradicting the fact that $0 \neq \underline{p} \subset \bigcap_{k\geq 1}\underline{m}^k$.

## REFERENCES

1. T. ALBU and C. NĂSTĂSESCU, "Décompositions primaires dans les catégories de Grothendieck commutatives", I, II, J. Reine Angew. Math. 280, 282 (1976), 172-194.

2. R. F. DAMIANO and Z. PAPP, "On consequences of stability", Comm. Algebra 9(1981), nr.7, 747-764.

3. P. GABRIEL, "Des catégories abéliennes", Bull. Soc. Math. France 90(1962), 323-448.

4. R. GORDON and J. C. ROBSON, "Krull dimension", Mem, Amer. Math. Soc. 133(1973).

5. B. LEMONNIER, "Déviation ordinale des ensembles ordonnés et groupes abéliens totalement ordonnés", C. R. Acad. Sci. Paris 273(1971), 1013-1016.

6. C. NĂSTĂSESCU, "La structure des modules par rapport à une topologie additive", Tôhoku Math. J. 26(1974), 173-201.

7. C. NĂSTĂSESCU, "Modules $\Sigma$-injectifs" in "Ring Theory", Proc. of the 1978 Antwerp Conference, Lecture Notes in Pure and Applied Math. 51(1979), Marcel Dekker, 729-740.

8. N. POPESCU, "Abelian categories with applications to rings and modules", L.M.S. Monographs nr.3 (1973), Academic Press.

9. B. STENSTRÖM, "Rings of quotients", Grundlehren 217(1975), Springer-Verlag.

CANCELLATIVE GROUP-GRADED RINGS

D. S. Passman

University of Wisconsin - Madison
Madison, Wisconsin 53706

In (2) we obtained necessary and sufficient conditions for a strongly group-graded ring to be prime or semiprime. In this note, we point out how the same methods can apply to yield analogous results for the somewhat larger class of cancellative group-graded rings. Since a complete proof of these new results would require totally rewriting (2) and since these results can probably be further generalized, we do not offer a detailed proof here. Rather we indicate the few new ideas involved, list a few pertinent lemmas and then refer the reader to (2).

## §1. MIDDLE CANCELLATION

If $R$ is a ring and $I,J \lhd R$, we write $I \sim J$ if and only if $UIV = 0$ is equivalent to $UJV = 0$ for all $U,V \subseteq R$.

Lemma 1.1. With the above notation, $\sim$ is an equivalence relation and we denote the equivalence class of the ideal $I$ by $[I]$. Moreover

i. $[0] = \{0\}$.

ii. $I_1 \sim J_1$ and $I_2 \sim J_2$ implies $I_1I_2 \sim J_1J_2$.



F. van Oystaeyen (ed.), Methods in Ring Theory, 403–414.

iii. $I_i \sim J_i$ implies $\sum_i I_i \sim \sum_i J_i$.

iv. Each class has a unique largest member.

v. If $I$ is an annihilator ideal (left annihilator, right or middle) then $I$ is the largest member of its class.

Part (iv) is immediate from (iii). Note that $I \sim R$ if and only if $UIV = 0$ implies $UV = 0$. We call such ideals "middle cancellable". We call the classes of this equivalence relation the "cancellation classes" of $R$.

Let $G$ be a multiplicative group. An associative ring $S$ is $G$-graded if $S = \sum_{x \in G} R_x$, the direct sum of the additive subgroups $R_x$ indexed by the elements of $G$, with $R_x R_y \subseteq R_{xy}$ for all $x, y \in G$. Then $R = R_1$ is a subring of $S$ and we assume throughout that $1 \in R$ is the unit element of $S$. By abuse of notation, we write $S = R*G$ to merely indicate the base ring $R = R_1$ and the group $G$. $R*G$ is strongly $G$-graded if $R_x R_y = R_{xy}$ for all $x, y \in G$.

We say that $R*G$ is "cancellable" if and only if $UR_x R_y V = 0$ implies $UR_{xy}V = 0$ for all homogeneous sets $U, V \subseteq R*G$. Note that, since $R_x R_y \subseteq R_{xy}$, the reverse implication always holds. It is clear that strongly $G$-graded rings are cancellable. An alternate characterization is

<u>Lemma 1.2</u>. $R*G$ is cancellative if and only if for all $x \in G$

i. $R_x R_{x^{-1}}$ is a middle cancellable ideal of $R$.

ii. $\underline{r}_{R*G}(R_x) = 0$ (or $\underline{\ell}_{R*G}(R_x) = 0$).

<u>Proof</u>. Assume (i) and (ii) and let $x \in G$. We first show $\underline{\ell}_{R*G}(R_x) = 0$. Indeed if $U_y R_x = 0$ (where $U_y$ is homogeneous of grade $y$) then $0 = (R_{y^{-1}} U_y) R_x R_{x^{-1}}(R)$ implies $(R_{y^{-1}} U_y) R = 0$ and hence $U_y = 0$ since $\underline{r}_{R*G}(R_{y^{-1}}) = 0$. Next it follows immediately from (i) and the fact that

$\underline{r}_{R*G}(R_y) = \underline{\ell}_{R*G}(R_y) = 0$ that for all homogeneous $U,V$ if $UR_xR_{x^{-1}}V = 0$ then $UV = 0$. Finally suppose $UR_xR_yV = 0$ with $U,V$ homogeneous. Since $R_{y^{-1}}R_{x^{-1}}R_{xy} \subseteq R$ this yields $UR_xR_yR_{y^{-1}}R_{x^{-1}}R_{xy}V = 0$. Now cancel the $R_yR_{y^{-1}}$ term and then the $R_xR_{x^{-1}}$ term.

We remark that (i) above is not sufficient to imply cancellable since it does not imply (ii). For example let $S = Z \oplus Z_4$ and $S[G] = S \oplus Sx$ with $x^2 = 1$. Let $R = Z1 \subseteq S$ so that $R = \{(z,\bar{z}) \mid z \in Z$ and $\bar{z}$ is $z \bmod 4\}$. Take

$$R*G = R \oplus (2S)x \subseteq S[G].$$

Since $(2S)^2 = 4S \subseteq R$, $R*G$ is a subring with $R \simeq Z$. Note that $R_x = (2S)x$ so $R_xR_{x^{-1}} = 4S$ ess $R$. But $\underline{r}_{R*G}(R_x) \supseteq (2Z_4)x \neq 0$.

If $R$ is semiprime and (ii) above holds then clearly $R_xR_{x^{-1}}$ ess $R$ and hence is middle cancellable. Thus we have

<u>Corollary 1.3</u>. Suppose $R$ is semiprime. Then $R*G$ is cancellable if and only if $\underline{r}_{R*G}(R_x) = 0$ for all $x$.

For the remainder of this section we assume that $R*G$ is cancellable. If $I \lhd R$ and $x \in G$ we define $I^x = R_{x^{-1}}IR_x$.

<u>Lemma 1.4</u>. Let $I,J \lhd R$ and $x \in G$. Then

i. $I^1 = I$, $I \subseteq J$ implies $I^x \subseteq J^x$.

ii. $(\sum_i I_i)^x = \sum_i I_i{}^x$.

iii. $I \sim J$ implies $I^x \sim J^x$ and $\underline{r}_{R*G}(I) = \underline{r}_{R*G}(J)$.

iv. $(I^x)^y \sim I^{xy}$, $(I^x)^y \subseteq I^{xy}$.

v. $I^xJ^x \sim (IJ)^x$, $I^xJ^x \subseteq (IJ)^x$.

<u>Proof</u>. (iii). If $UI^xV = 0$, then $(R_xUR_{x^{-1}})I(R_xVR_{x^{-1}}) = 0$ so $(R_xUR_{x^{-1}})J(R_xVR_{x^{-1}}) = 0$. Cancelling the outside $R_x$ and $R_{x^{-1}}$ factors then yields $UJ^xV = 0$. Finally $I \sim J$ implies

that $\underline{r}_R(I) = \underline{r}_R(J)$. Since $\underline{r}_{R*G}(I)$ is graded and $\underline{\ell}_{R*G}(R_x) = 0$ for all $x \in G$, this part is proved.

It follows from (i),(iii) and (iv) that $G$ permutes the cancellation classes of $R$. Furthermore, by Lemma 1.1(ii)(iii), products and arbitrary sums of classes are well defined and (ii), (v) above imply that $G$ preserves these sums and products.

Let $H$ be a subgroup of $G$ and $I \triangleleft R$. We say that $H$ "stabilizes" $I$ if and only if $I^x \subseteq I$ for all $x \in H$.

Lemma 1.5. Let $H \subseteq G$ and $I \triangleleft R$.

i. If each $I_i$ is H-stable, then so is $\sum_i I_i$ and $\bigcap_i I_i$.

ii. $H$ stabilizes $I^H = \sum_{x \in H} I^x$.

iii. If $H$ stabilizes $I$, then it stabilizes $[I]$.

iv. If $H$ stabilizes $[I]$, then $H$ stabilizes the unique maximal member of that class. Also $I^H \sim I$ and $H$ stabilizes $I^H$.

v. If $H$ stabilizes $I$, then it stabilizes $\underline{r}_R(I)$ and $\underline{\ell}_R(I)$.

Proof. Part (ii) follows from $(I^x)^y \subseteq I^{xy}$. For (iii) it suffices to show that $I^x \sim I$ for all $x \in H$. Since $I^x \subseteq I$ we know that $UIV = 0$ implies $UI^xV = 0$. Conversely since $I^{x^{-1}} \subseteq I$ we have $(R_{x^{-1}}R_x)I(R_{x^{-1}}R_x) \subseteq I^x$ and hence $UI^xV = 0$ implies, by cancelling the $R_{x^{-1}}R_x$ factors that $UIV = 0$. Finally for (v) suppose $K = \underline{r}_R(I)$. Then $I^x \subseteq I$ yields $I^xK = 0$ so $IR_xK = 0$ and $IK^{x^{-1}} = 0$. Thus $K^{x^{-1}} \subseteq K$ and $K$ is H-stable.

## §2. CANCELLATIVE GROUP-GRADED RINGS

The proof of the main theorem of (2) now carries over quite easily to the more general cancellative group-graded rings. The idea is to replace the action of $G$ on the ideals of $R$, used in the original proof, by the action of $G$ on the cancellation classes of $R$. For example if $R*G$ is cancellative, we have the following useful generalization of Lemma 1.5(v).

Lemma 2.1. Let $H \subseteq G$ and let $[I]$ and $[J]$ be H-stable classes. Then $[IJ]$ is H-stable and $K = \{r \in R | IrJ = 0\}$ is an H-stable ideal of $R$.

Proof. Let $x \in H$. Since $I^x \sim I$ and $J^x \sim J$ the equation $IKJ = 0$ yields $I^xKJ^x = 0$. Cancelling the outside $R_x$ and $R_{x^{-1}}$ factors then yields $0 = IK^{x^{-1}}J = 0$ so $K^{x^{-1}} \subseteq K$.

Recall that $I \lhd R$ is H-nilpotent free if it is H-stable and contains no nonzero H-stable nilpotent ideal.

Lemma 2.2. Let $I$ be an H-nilpotent free ideal.

i. $\underline{r}_R(I) = \underline{r}_R(I^2)$ and $\underline{r}_{R*G}(I) = \underline{r}_{R*G}(I^2)$.

ii. If $J$ is H-stable, then $IJ \cap \underline{r}_R(IJ) = 0$.

Proof. (i). $\underline{r}_R(I) \subseteq \underline{r}_R(I^2) = J$. Since $[I^2]$ is H-stable so is $J$, by Lemma 2.1, and hence so is $[IJ]$. But $IJ \cdot IJ \subseteq I^2J = 0$ so $(IJ)^H$ is an H-stable ideal of square zero. Since $(IJ)^H \subseteq I$, we have $IJ = 0$ and $J \subseteq \underline{r}_R(I)$.

(ii). Since $[IJ]$ is H-stable we have $\underline{r}_R(IJ)$ H-stable and also $(IJ)^H \cdot \underline{r}_R(IJ) = 0$. Then $(IJ)^H \cap \underline{r}_R(IJ)$ is an H-stable nilpotent ideal contained in $I$ so $(IJ)^H \cap \underline{r}_R(IJ) = 0$.

We leave the remaining verification to the reader. For the most part, the translation of the proof in (2) to cancellative rings is now routine. Once in awhile, minor ad hoc arguments are required. In any case, the main result is

Theorem 2.3. Let $R*G$ be a cancellative G-graded ring with base ring $R$. Then $R*G$ contains nonzero ideals $A,B$ with $AB = 0$ if and only if there exist:

i. Subgroups $N \lhd H \subseteq G$ with $N$ finite.

ii. An H-stable ideal $I$ of $R$ with $I^xI = 0$ for all $x \in G\setminus H$.

iii. Nonzero ideals $\tilde{A},\tilde{B}$ of $R*N$ with $\tilde{A},\tilde{B} \subseteq I(R*N)$ and $\tilde{A}(R*H)\tilde{B} = 0$.

Furthermore $A = B$ if and only if $\tilde{A} = \tilde{B}$.

Recall that $R$ is a G-prime ring if the product of any two nonzero G-stable ideals of $R$ is nonzero. Similarly $R$ is G-semiprime if the square of any nonzero G-stable ideal is nonzero. Furthermore an ideal $I$ of $R$ is a T.I. (trivial intersection) ideal if for all $x \in G$ either $I^x \subseteq I$ or $I^x \cap I = 0$.

<u>Lemma 2.4.</u> Let $R*G$ be cancellative with $R$ a G-semiprime ring and let $I$ be an ideal of $R$.

i. If $I$ is a T.I. ideal then $G_I = \{x \in G \mid I^x \subseteq I\}$ is a subgroup of $G$, the stabilizer of $I$. Moreover $I^xI = 0$ for all $x \in G\setminus G_I$.

ii. If $H$ is a subgroup of $G$ stabilizing $I \neq 0$ and $I^xI = 0$ for all $x \in G\setminus H$, then $I$ is a T.I. ideal and $G_I = H$.

<u>Proof</u>. (i). We may assume that $I \neq 0$ and observe that for any nonzero ideal $J$ of $R$ and $x \in G$ we have $J^x \neq 0$. Let $x,y \in G_I$. Then $((I^x)^{x^{-1}})^y \neq 0$ and since $(I^x)^{x^{-1}} \subseteq I^1 = I$ we have $((I^x)^{x^{-1}})^y \subseteq I^y \subseteq I$ since $y \in G_I$. On the other hand, $I^x \subseteq I$ so $((I^x)^{x^{-1}})^y \subseteq (I^{x^{-1}})^y \subseteq I^{x^{-1}y}$. Thus $I \cap I^{x^{-1}y} \neq 0$ and $x^{-1}y \in G_I$.

(ii). Since $I^xI = 0$ for all $x \in G\setminus H$ and $R$ is G-semiprime, it follows that $I$ is H-nilpotent free. Thus by Lemma 2.4(ii) with $J = R$ we see that $I \cap \underline{r}_R(I) = 0$ and similarly $I \cap \ell_R(I) = 0$. Therefore $I \cap I^x = 0$ for all $x \in G\setminus H$. Since $0 \neq I^x \subseteq I \cap I^x$ for all $x \in H$, the result follows.

With the above notation we have

Corollary 2.5. Let $R*G$ be a cancellative G-graded ring with the base ring $R$ being G-prime. Assume that for every T.I. ideal $I \neq 0$ of $R$ we have $\Delta^+(G_I) = 1$. Then $R*G$ is prime.

In particular if $R$ is prime and $\Delta^+(G) = 1$ or if $R$ is G-prime and $G$ is torsion free, then $R*G$ is prime. For the corresponding semiprime corollary we require yet another version of Maschke's theorem.

## §3. THE QUINTESSENTIAL MASCHKE'S THEOREM

Let $R$ be a ring and $W_R \subseteq V_R$ right R-modules. We say $W_R$ is "quintessential" in $V_R$, and write $W_R$ qess $V_R$, if and only if for all subsets $\Omega$ of $V$ with $\Omega \not\subseteq 0$ there exists $r \in R$ with $0 \neq \Omega r \subseteq W$.

Lemma 3.1. Let $W_R \subseteq V_R$. Then $W$ qess $V$ if and only if $\Pi W$ ess $\Pi V$ for all (strong) direct products $\Pi V$.

Some examples which occur naturally are

Lemma 3.2. Let $I \triangleleft R$.
i. $I + \ell_R(I)$ qess $R_R$.
ii. If $\ell_R(I) = 0$ then $I$ qess $R_R$.
iii. If $I^2 = 0$ then $\ell_R(I)$ qess $R_R$.

Proof. Let $\Omega \not\subseteq 0$ be a subset of $R$. If $\Omega \subseteq \ell_R(I)$ then $0 \neq \Omega 1 \subseteq I + \ell_R(I)$. Otherwise there exists $r \in I$ with $\Omega r \neq 0$. But then $\Omega r \subseteq I$ so $0 \neq \Omega r \subseteq I + \ell_R(I)$. This yields (i) and parts (ii) and (iii) are now immediate.

Let $R*G = \sum_{x\in G} R_x$ be a G-graded ring and let $V$ be a unital right $R*G$-module. Then $V$ is "pointed" if there exists

$t \in \bigcap_x R_x R_{x^{-1}}$ which acts regularly on $V$, that is $vt = 0$ implies $v = 0$. We say that $V$ is "component regular" if each $R_x$ acts regularly on $V$, that is $vR_x = 0$ implies $v = 0$.

If $R*G$ is strongly G-graded, then any unital module $V$ is pointed with $t=1$. Furthermore any pointed module is certainly component regular. The goal here is to obtain a Maschke theorem for component regular modules. We start by considering pointed modules. Here, of course, $G$ is finite.

Lemma 3.3. Let $V$ be a pointed $R*G$-module with no $|G|$-torsion and let $W$ be an essential $R*G$-submodule. If $W_R | V_R$ then $W = V$.

Proof. Write $V_R = W_R \oplus U_R$ and let $\pi: V \to W$ be the natural R-projection. Let $t \in \bigcap_x R_x R_{x^{-1}} \subseteq R$ be the element given by the definition of pointed module. Since $t \in R_x R_{x^{-1}}$ we can write $t = \sum_i a_{xi} b_{x^{-1}i}$ with $a_{xi} \in R_x$, $b_{x^{-1}i} \in R_{x^{-1}}$. Define $\lambda: V \to W$ by

$$v^\lambda = \sum_{x,i} (va_{xi})^\pi b_{x^{-1}i}.$$

It is clear that $\lambda$ is additive and, since $W$ is an $R*G$-submodule, that $v^\lambda \in W$. It is not true however that $\lambda$ is an $R*G$-homomorphism.

If $w \in W$ the $w^\lambda = wt|G|$ so we deduce that $W \cap (\ker \lambda) = 0$. Next for any $v \in V$ and $\alpha \in R*G$ we observe by the usual computation that $(vt\alpha)^\lambda = v^\lambda \alpha t$. In particular, when $\alpha = 1$ we have $(vt)^\lambda = v^\lambda t$.

Now suppose $v \in \ker \lambda$. Then $(vt\alpha)^\lambda = v^\lambda \alpha t = 0$ so $vt(R*G) \subseteq \ker \lambda$. Observe that $vt(R*G)$ is an $R*G$-submodule, $W \cap \ker \lambda = 0$ and $W$ ess $V_{R*G}$. It follows that $vt(R*G) = 0$ and hence that $v = 0$ since $t$ acts regularly. Thus $\ker \lambda = 0$.

Finally let $u \in U$ and set $w = u^\lambda \in W$. Then

$$(ut|G|)^\lambda = (ut)^\lambda |G| = u^\lambda t|G| = wt|G| = w^\lambda.$$

Thus since $\lambda$ is one-to-one this yields $ut|G| = w \in U \cap W = 0$. Again we conclude that $u = 0$ so $U = 0$ and $V = W+U = W$.

The next two results are now routine.

Lemma 3.4. Let $V$ be an $R*G$-module and $W$ an $R$-submodule.

i. $WR_x$ is an $R$-submodule of $V$.

ii. $\bigcap_{x\in G} WR_x$ is an $R*G$-submodule of $V$.

iii. If $W$ ess $V_R$ and $V$ is component regular, then $WR_x$ ess $V_R$.

Theorem 3.5. Let $V$ be a pointed $R*G$-module with no $|G|$-torsion and let $W$ be an $R*G$-submodule. Then $W$ ess $V_R$ if and only if $W$ ess $V_{R*G}$.

The above is of course the essential version of Maschke's theorem. The following observation is useful.

Lemma 3.6. Let $V$ be an $R*G$-module with $G$ finite and suppose each $R_x$ contains an element acting regularly on $V$. Then $V$ is a pointed $R*G$-module.

Proof. Let $t(x) \in R_x$ be the given element and set $t = \Pi_{x\in G} t(x)t(x^{-1})$, a product in some order. Clearly $t$ acts regularly on $V$. Moreover since $t(x)t(x^{-1}) \in R_xR_{x^{-1}} \lhd R$, it follows that $t \in \bigcap_x R_xR_{x^{-1}}$.

If $R$ is a ring and $Z$ a set of noncommuting variables, then we let $R\langle\langle Z\rangle\rangle$ denote the ring of formal power series over $R$ in these variables. The uniqueness of expression in the semigroup generated by $Z$ guarantees that multiplication is well defined. If $V$ is an $R$-module, then $V\langle\langle Z\rangle\rangle$ becomes an $R\langle\langle Z\rangle\rangle$-module.

Observe that, if $G$ is finite, then

$(R*G)\langle\langle Z\rangle\rangle = R\langle\langle Z\rangle\rangle * G$

with $R\langle\langle Z\rangle\rangle_x = R_x\langle\langle Z\rangle\rangle$.

Lemma 3.7. Let $V$ be a component regular $R*G$-module with no $|G|$-torsion. If $W$ qess $V_{R*G}$, then $W$ ess $V_R$.

Proof. Let $Z$ be a set of noncommuting variables of size at least $|R*G|$, the cardinality of $R*G$. Form the G-graded ring $R'*G = (R*G)\langle\langle Z\rangle\rangle$ with modules

$$V' = V\langle\langle Z\rangle\rangle \supseteq W\langle\langle Z\rangle\rangle = W'.$$

We show that $V'$ is a pointed $R'*G$-module. To this end, let $x \in G$ and since $|Z| \geq |R*G| \geq |R_x|$, we can label some of the elements of $Z$ by the elements of $R_x$. Thus $Z \supseteq \{z_\alpha | \alpha \in R_x\}$ and we form

$$t'(x) = \sum_{\alpha\in R_x} \alpha \cdot z_\alpha \in R_x\langle\langle Z\rangle\rangle = R'_x.$$

If $0 \neq v' \in V'$, write $v' = \sum_\mu v_\mu \cdot \mu$ where the elements $\mu$ are monomials in the variables of $Z$ and $v_\mu \in V$. Suppose $\tau$ has minimal length with $v_\tau \neq 0$. Since $V$ is component regular, $v_\tau R_x \neq 0$ so we can choose $\beta \in R_x$ with $v_\tau\beta \neq 0$. It now follows from uniqueness of expression that the $\tau z_\beta$ coefficient of $v't'(x)$ is precisely $v_\tau\beta \neq 0$. Hence $v't'(x) \neq 0$ so $t'(x)$ acts regularly on $V'$. Lemma 3.6 now implies that $V'$ is a pointed $R'*G$-module.

Clearly $V'$ has no $|G|$-torsion. Furthermore since $W$ qess $V_{R*G}$, it follows from Lemma 3.1 that $W'$ ess $V'_{R*G}$ and hence $W'$ ess $V'_{R'*G}$. Theorem 3.5 now implies that $W'$ ess $V'_{R'}$ and from this we conclude immediately that $W$ ess $V_R$.

Now the quintessential Maschke theorem.

Theorem 3.8. Let $V$ be a component regular $R*G$-module with no $|G|$-torsion and let $W$ be an $R*G$-submodule of $V$. Then $W$ qess $V_R$ if and only if $W$ qess $V_{R*G}$.

Proof. Clearly $W$ qess $V_R$ implies $W$ qess $V_{R*G}$. Assume conversely that $W$ qess $V_{R*G}$. Then it follows easily from Lemma 3.1 that $\Pi W$ qess $\Pi V_{R*G}$. Lemma 3.7 now yields $\Pi W$ ess $\Pi V_R$ and hence by Lemma 3.1 again we have $W$ qess $V_R$.

## §4. SEMIPRIME GROUP-GRADED RINGS

Let $R*G$ be a group-graded ring with $R$ semiprime. In view of the left analog of Corollary 1.3, $R*G$ is cancellative if and only if $R*G$, as a right $R*G$-module, is component regular. As a consequence of Theorem 3.7 and Lemma 3.2(iii) we obtain

Corollary 4.1. Let $R*G$ be a cancellative group-graded ring and assume that $R$ is semiprime with no $|G|$-torsion. Then $R*G$ is semiprime. Furthermore if $I \triangleleft R*G$ and $I$ is not contained in any minimal prime, then $(I \cap R)$ ess $R_R$.

For infinite groups, the argument of (2) and Theorem 2.3 now yield

Corollary 4.2. Let $R*G$ be a cancellative group-graded ring whose base ring $R$ is $G$-semiprime. Suppose that for every trivial intersection ideal $I$ of $R$, $I$ has no $|\Delta^+(G_I)|$-torsion. Then $R*G$ is semiprime.

As in (2), if $G$ is infinite, then a module $V$ has no $|G|$-torsion if, for all finite subgroups $N$ of $G$, $V$ has no $|N|$-torsion. We close with a few comments and questions.

1. It appears that the appropriate hypothesis for studying infinite group-graded rings is that $\ell_{R*G}(R_x) = 0$ for all $x$ (or perhaps the right annihilator condition or perhaps both). In any

case, the results here should be extendible to this larger class of rings.

2. The concept of quintessential submodule was of course introduced because Theorem 3.8 could be proved and because the study of these submodules suffices in dealing with ideals of R*G. On the other hand, it is certainly of interest to ask whether Theorem 3.8 actually holds for essential submodules.

3. The best result on semiprime group-graded rings of finite groups concerns nondegenerate rings and is contained in (1). One wonders whether the Maschke techniques can be extended to obtain that result.

REFERENCES

1. M. Cohen and S. Montgomery, Group-graded rings, smash products and group actions, Trans. AMS.

2. D. S. Passman, Infinite crossed products and group-graded rings, Trans. AMS.

# THE AUTOMORPHISM GROUP OF A POLYNOMIAL ALGEBRA

Marilena Pittaluga

Columbia University, New York

## 1. INTRODUCTION

This is a summary of my Ph.D. thesis [8] on which Hyman Bass reported at the Antwerp meeting in one of his two lectures, the other one of which was devoted to recent developments on the Jacobian conjecture (see [2]).

The object of study of this work is the group $\underline{GA_n(k)}$ of k-algebra automorphisms of a polynomial ring $\underline{k^{[n]}} = k[X_1,\ldots,X_n]$ in n variables over a commutative ring k. This group can also be thought of as the group of automorphisms of the scheme $\mathbb{A}_k^n$ ( = Spec $(k^{[n]})$), and may be considered as a non-linear analogue of its subgroup $GL_n(k)$. If F is an element of $GA_n(k)$, its jacobian determinant, det $(\partial(F(X_i))/\partial X_j)$, is a unit in $k^{[n]}$, hence a unit in k whenever k is a reduced ring (i.e. k does not have non-zero nilpotent elements). Since this fact turns out to have interesting consequences, that do not depend on the ring being reduced, we will often consider, instead of the whole group, the subgroup $\underline{SA_n(k)}$ of automorphisms with jacobian determinant 1. We remark that, if k is reduced, the jacobian



*F. van Oystaeyen (ed.), Methods in Ring Theory, 415–432.*

determinant is a homomorphism onto $k^\times$ with kernel $SA_n(k)$.

The structure of the group $GA_n(k)$ is well known in case $n = 2$ and k is a field. Precisely, let $BA_2(k)$ be the group of automorphisms F of the following form:

$$F: \begin{array}{l} X \to \alpha X + \gamma \\ Y \to \beta Y + f(X) \end{array}$$

where $\alpha, \beta \in k^\times$, $\gamma \in k$, $f(X) \in k[X]$; and let $Af_2(k)$ be the group of affine automorphisms:

$$F: \begin{array}{l} X \to \alpha X + \beta Y + \gamma \\ Y \to \alpha' X + \beta' Y + \gamma' \end{array}$$

where $\alpha, \beta, \gamma, \alpha', \beta', \gamma' \in k$ and $\alpha\beta' - \beta\alpha' \in k^\times$.
A well known theorem by Jung-Van der Kulk (see [6] and [10]) asserts that, if k is a field, $GA_2(k)$ is the free product of $BA_2(k)$ and $Af_2(k)$, amalgamated over their intersection.

For $n > 2$ the group $GA_n(k)$ is an almost complete mystery, for we have no precise information about what its elements, its algebraic subgroups or its normal subgroups look like. For example, let <u>$EA_n(k)$</u> be the group generated by <u>elementary automorphisms</u>, i.e. automorphisms F of the form:

$$F(X_j) = X_j + f, \qquad F(X_i) = X_i \quad \text{for } i \neq j,$$

where $f \in k^{[n]}$ is independent of $X_j$. As an obvious generalization of the above result for $n = 2$, the following question arises naturally: is $GA_n(k)$ generated by $EA_n(k)$ and the linear group when k is a field? This question is so far unanswered, and it seems not even known whether $EA_n(k)$ is a normal subgroup of $GA_n(k)$ (see [7] and [11] for a discussion of this problem).

We will be mostly concerned with the problem of determining the normal subgroups of $GA_n(k)$.
In what follows we allow k to be an arbitrary $\mathbb{Q}$-algebra. This generalization, beside the intrinsec interest, might also be useful in studying the case of a field. For example $GA_n(k)$ naturally contains a copy of the group $GA_{n-1}(k[X_n])$, so

information on this second group might help in answering questions about the first. Expectedly, when k is no longer a field, the ideal structure of k will play a role in the determination of the normal subgroups because ideals of k give rise to congruence subgroups.

The proofs of results announced here may be found in [8].

## 3. THE CONGRUENCE SUBGROUPS

From now on k is assumed to be a commutative $\mathbb{Q}$-algebra.

A natural way to produce congruence subgroups is to take kernels of actions of $GA_n(k)$ on $k^{[n]}/\mathscr{I}$ where $\mathscr{I}$ is an invariant ideal of $k^{[n]}$.

The full group $GA_n(k)$, since it contains the translations, tends to leave very little invariant and to have few interesting normal subgroups. Instead we focus on $\underline{GA_n^o(k)}$, the subgroup of automorphisms that leave invariant the ideal $\underline{(X)}$ generated by $X_1, X_2, \ldots, X_n$; these correspond precisely to those automorphisms of $A_k^n(k)$ which fix the origin.

(2.1) Theorem. Every ideal of $k^{[n]}$ invariant under $GA_n^o(k)$ has the form:

$$I^{[n]} = \bigoplus_{d \geq 0} I_d k_d^{[n]} = \sum_{d \geq 0} I_d k^{[n]} ,$$

where $I = (I_0, I_1, I_2, \ldots)$ is an increasing sequence of ideals of k (i.e. $I_0 \subset I_1 \subset I_2 \subset \ldots$). We call I an ideal level of k.

If, in theorem (2.1), invariance under the whole group $GA_n(k)$ is required, the ideal level must be constant (i.e. $I_d = I_0$ for all $d \geq 0$).

Now let I be an ideal level, and let $\underline{I(1)}$ be the "shifted" ideal level: $I(1) = (0, I_0, I_1, I_2, \ldots)$. We define the congruence

subgroup of level I, denoted by $GA_n^I(k)$, as the subgroup of those automorphisms in $GA_n^o(k)$ that act trivially on $k^{[n]}/\mathbb{I}(1)^{[n]}$. (The shifting is introduced for technical purposes).
We set $SA_n^I(k) = SA_n(k) \cap GA_n^I(k)$. On the other hand, while putting $EA_n^o(k) = EA_n(k) \cap GA_n^o(k)$, we define $EA_n^I(k)$ in general to be the normal subgroup of $EA_n^o(k)$ generated by all elementary automorphisms lying in $GA_n^I(k)$.

(2.2) Proposition. Let I, J be two ideal levels. Then:

$$(GA_n^I(k),\ GA_n^J(k)) \subset GA_n^{IJ}(k), \quad \text{and}$$

$$EA_n^{IJ}(k) \subset (EA_n^I(k),\ EA_n^J(k)) \quad \text{if} \quad n \geqslant 3.$$

We consider the special ideal levels $\underline{(d)} = (0,..,0,\overset{d}{\downarrow}{k},k,...)$ defined so that:

$$(d)^{[n]} = (X)^d \qquad (\text{all } d \geqslant 0).$$

The corresponding subgroups $GA_n^d(k)$ form a descending central filtration of $GA_n^o(k)$. When k is a reduced ring, $GA_n^d(k) = SA_n^d(k)$.

## 3. THE NORMAL LEVEL CONJECTURE

We will describe several approximations to the following conjecture; which we state first in a preliminary form:

(3.1) Normal Level Conjecture. Let k be a nöetherian Q-algebra of finite dimension. There is a constant $n_0(k) \geqslant 3$ (possibly $n_0(k) = 3$ always) such that, for $n \geqslant n_0(k)$, if N is a subgroup of $SA_n^1(k)$ normalized by $EA^o(k)$, then, for some (uniquely determined) ideal level I:

$$EA_n^I(k) \subset N \subset SA_n^I(k).$$

Remarks.

(3.2) This conjecture is inspired by the well known description of normal subgroups of $GL_n(k)$, (see [1] or [12]).

(3.3) There is a more definitive form of the Normal Level Conjecture predicting a similar constraint on any subgroup of $GA_n^o(k)$ normalized by $GL_n(k)$ and $EA_n^o(k)$. However, this is somewhat more technical in that it involves a related pair (I, J) of ideal levels. It is best understood in the setting of the group $\hat{GA}_n(k)$ of power series automorphisms described below. For the group $GA_n(k)$ the secondary ideal level J plays only a marginal role when k is reduced, but if k has nilpotent elements there are significant new families of normal subgroups that the parameter J is needed to account for.

(3.4) Suppose that k is a field (characteristic zero). Then we expect that $n_0(k)$ can be taken to be 3. That, when n = 2, no such conclusions can be drawn for $GA_2(k)$ is known (see [4]), and can be also easily deduced from the Jung-Van der Kulk theorem described above.

(3.5) When k is a field, every non-zero ideal level is of the form (d) for some $d \geqslant 0$. Confirmation of the Normal Level Conjecture would imply that a non trivial subgroup N of $GA_n^1(k)$ normalized by $EA_n^o(k)$ satisfies:

$$EA_n^d(k) \subset N \subset GA_n^d(k) \qquad \text{for some } d \geqslant 1.$$

(3.6) It has been conjectured (see [7] and [11]) that when k is a field one has $SA_n^d(k) = EA_n^d(k)$ for all $d \geqslant 0$. If true this would simplify and make quite precise the above conjectural form of normal subgroups of $GA_n^o(k)$ and of $GA_n(k)$.

(3.7) When k is a field, Šafarevič has shown how to interpret $GA_n(k)$ as an infinite dimensional algebraic group (see [9]). Using results announced below, in combination with the methods of Šafarevič we can show that every Zariski-closed, normal subgroup N of $GA_n^o(k)$ has the form:

$$N = SA_n^d(k) \qquad \text{or} \qquad N = \Gamma \cdot SA_n^1(k)$$

where $\Gamma = N \cap GL_n(k)$ is either scalar or contains $SL_n(k)$.

In addition to (3.7) we establish the following approximations to the Normal Level Conjecture:

- it is valid "stably" for $GA_\infty(k)$ (section 4);
- it is valid "infinitesimally" for the Lie algebra $\mathfrak{ga}_n(k)$ (section 5);
- it is valid "topologically" for the power series automorphism group $\hat{GA}_n^o(k)$, which can be viewed in part as a completion of $GA_n^o(k)$ (section 6).

While the conjecture itself is not yet proved, these approximations render it quite plausible.

## 4. THE STABLE APPROXIMATION

We embed $GA_n(k)$ in $GA_{n+1}(k)$ in the obvious way and we take the union:

$$GA_\infty(k) = \bigcup_{n\geq 1} GA_n(k) \quad .$$

For an ideal level I we define:

$$GA_\infty^I(k) = \bigcup_{n\geq 1} GA_n^I(k) \quad , \qquad EA_\infty^I(k) = \bigcup_{n\geq 1} EA_n^I(k)$$

A stronger form of conjecture (3.1) then holds:

(4.1) Theorem. Let k be any $\mathbb{Q}$-algebra and N a subgroup of $GA_\infty^o(k)$ normalized by $EA_\infty^o(k)$. Then:

$$EA^{I}_{\infty}(k) \subset N \subset GA^{I}_{\infty}(k) \quad ,$$

for some ideal level I.

If $N \subset GA_{\infty}(k)$ and is normalized by $EA_{\infty}(k)$, the level must be constant and the congruence subgroups must be appropriately enlarged to allow for translations.

## 5. THE INFINITESIMAL APPROXIMATION

We study the Lie algebra $\mathfrak{ga}_n(k)$ of k-derivations of the polynomial ring $k^{[n]}$. These correspond to polynomial vector fields on $\mathbb{A}^n_k$. The connection between $\mathfrak{ga}_n(k)$ and $GA_n(k)$ is two-fold.
On one side $\mathfrak{ga}_n(k)$ is the Lie algebra of the group valued functor on commutative rings $k \to GA_n(k)$ (see [5], Chapter II, § 4). Considering the functor $k \to SA_n(k)$, we obtain the subalgebra $\mathfrak{sa}_n(k)$ of derivations with zero divergence. (The divergence map is defined by:

$$\text{Div:}\quad \mathfrak{ga}_n(k) \longrightarrow k^{[n]}$$
$$D \longmapsto \sum_{i=1}^{n} \frac{\partial(DX_i)}{\partial X_i} \quad ).$$

In case k is a field, $SA_n(k)$ is an infinite dimensional algebraic group, and $\mathfrak{sa}_n(k)$ is also the Lie algebra of $SA_n(k)$ in the sense defined by Šafarevič in [9].

$\mathfrak{ga}_n(k)$ admits a grading:

$$\mathfrak{ga}_n(k) = \bigoplus_{d\geq -1} \mathfrak{ga}_n(k)_d$$

which makes it into a graded Lie algebra and into a free graded $k^{[n]}$-module on the basis $\partial/\partial X_i$, which are homogeneous derivations of degree -1. We put:

$$\mathfrak{ga}^d_n(k) = \bigoplus_{e\geq d} \mathfrak{ga}_n(k)_e \quad , \quad \text{and}$$

$$\underline{sa_n^d(k)} = sa_n(k) \cap ga_n^d(k) \quad .$$

The $ga_n^d(k)$ (respectively $sa_n^d(k)$) form a descending central filtration of ideals of $ga_n^o(k)$ (resp. of $sa_n^o(k)$), and can be also described as the Lie algebras of the subgroups $GA_n^d(k)$ (resp. $SA_n^d(k)$).

The other connection of $ga_n(k)$ with the group $GA_n(k)$ is provided by the following result (due to D. Anick and H. Bass):

There exists a Lie algebra isomorphism between the associated graded of $SA_n^1(k)$ and $sa_n^1(k)$:

$$\mu: \quad gr(SA_n^1(k)) \xrightarrow{\sim} sa_n^1(k) \quad .$$

(We recall that the associated graded of a group G with a central filtration $G = G^1 \triangleright G^2 \triangleright G^3 \triangleright \cdots$ is defined by

$$gr(G) = \bigoplus_{d\geq 1} G^d/G^{d+1}$$

with Lie bracket $[\overline{x}, \overline{y}] = \overline{xyx^{-1}y^{-1}}$ for $x \in G^d$, $y \in G^e$, where for $z \in G^f$, $\overline{z}$ denotes its class in $G^f/G^{f+1}$ ).

$GA_n(k)$ acts naturally on $ga_n(k)$ via the adjoint action:

$$Ad(F)D = FDF^{-1} \qquad (F \in GA_n(k),\ D \in ga_n(k)\ ) \ .$$

In particular the subgroup $GL_n(k)$ of $GA_n^o(k)$ acts on $ga_n(k)$. We also remark that the isomorphism $\mu$ introduced above is equivariant for the action of $GL_n(k)$ (the action on $gr(SA_n^1(k))$ being induced by conjugation in $GA_n(k)$).

The following theorem provides the basic clue for analizing the $GL_n(k)$-structure of $ga_n(k)$.

(5.1) Theorem. There is a $GL_n(k)$-invariant decomposition:

$$ga_n(k) = k^{[n]}\Delta \oplus sa_n(k) \quad .$$

Here $\Delta$ is the Euler derivation defined by $\Delta X_i = X_i$, and though $\mathfrak{sa}_n(k)$ is not an ideal of $\mathfrak{ga}_n(k)$, the complement $k^{[n]}\Delta$ is a Lie subalgebra of $\mathfrak{ga}_n(k)$ isomorphic to $k^{[n]}$ as a graded $GL_n(k)$-module.

The isomorphism $\mu\colon \mathrm{gr}(SA_n^1(k)) \to \mathfrak{sa}_n^1(k)$ quoted above can be extended to $\mathrm{gr}(GA_n^1(k))$, and it produces a Lie algebra isomorphism of $GL_n(k)$-modules:

$$\mu\colon \ \mathrm{gr}(GA_n^1(k)) \xrightarrow{\sim} \mathcal{N}\cdot(X)\Delta \oplus \mathfrak{sa}_n^1(k)$$

where $\mathcal{N}$ is the nil radical of k.

Let now I be an ideal level of k, then $I(1)^{[n]}$ is an ideal of $k^{[n]}$, and we consider the congruence subalgebra $\underline{\mathfrak{ga}_n^I(k)}$ (resp. $\underline{\mathfrak{sa}_n^I(k)}$) of those derivations in $\mathfrak{ga}_n(k)$ (resp. $\mathfrak{sa}_n(k)$) that vanish on $k^{[n]}/I(1)^{[n]}$. If J is another ideal level, we set

$$\underline{\mathfrak{ga}_n^{(I,J)}(k)} = J^{[n]}\Delta \oplus \mathfrak{sa}_n^I(k) .$$

We remark that in case $I_0 = 0$,

$$\mathfrak{sa}_n^I(k) = \mu(\mathrm{gr}(SA_n^I(k)) = \mu(\mathrm{gr}(EA_n^I(k))$$

where $\mu$ is the isomorphism introduced above.

We have the following approximation to the Normal Level Conjecture:

(5.2) Theorem. Let M be an additive subroup of $\mathfrak{ga}_n^o(k)$, invariant under $EA_n^o(k)$. Assume that one of the following additional hypothesis holds:

a) $n \geqslant 3$ ,

b) $M \subset \mathfrak{sa}_n^o(k)$ ,

c) M is invariant under $GL_n(\mathbb{Z})$

Then there is an ideal level I and a quasi ideal level J satisfying $J(1) \subset I$ such that:

$$M = \mathfrak{ga}_n^{(I,J)}(k) .$$

(A quasi ideal level is a sequence of additive subgroups of k $J = (J_0, J_1, \ldots)$ such that $\tilde{J} = (0, J_1, J_2, \ldots)$ is an ideal level).

If M is invariant under $GA_n^o(k)$, the following additional conditions must hold: (i) $\mathcal{N} J_0 \subset J_1$, (ii) $\mathcal{N} I(1) \subset J$.

Under the same hypothesis a similar result holds for subgroups M of $\mathfrak{ga}_n(k)$; in this case M has the form:

$$M = \mathfrak{ga}_n^{\mathscr{I}}(k)_{-1} \oplus \mathfrak{ga}_n^{(I,J)}(k) ,$$

where $\mathscr{I}$ is an ideal of k, $\mathscr{I} \subset I_0$, and $\mathfrak{ga}_n^{\mathscr{I}}(k)_{-1}$ is the set of derivations of degree -1 that vanish modulo $\mathscr{I}$.
Finally, if we assume that M is invariant under $EA_n(k)$, the ideal levels I and J have to be constant, $J \subset I$ and $\mathscr{I} = I$; if M is actually invariant under $GA_n(k)$, then, in addition, $\mathcal{N} I \subset J$.

A more direct connection with the Normal Level Conjecture lies in the following corollary:
(5.3) Corollary. Let N be a subgroup of $SA_n^1(k)$, normalized by $EA_n^o(k)$. Then there exists an ideal level I of k with $I_0 = 0$, such that:

$$gr(N) = gr(SA_n^I(k)) = gr(EA_n^I(k))$$

Theorem (5.2) implies that, if k is a field (of characteristic zero), the only non-zero $EA_n^o(k)$-invariant subalgebras of $\mathfrak{sa}_n(k)$ are the $\mathfrak{sa}_n^d(k)$. In view of the correspondance between connected subgroups of $GA_n(k)$ and their Lie algebras given by Safarevic in [9], we have:
(5.4) Corollary. Let N be a non trivial, Zariski closed subgroup of $SA_n^o(k)$, normalized by $EA_n^o(k)$. Then N has the form:

$N = SA_n^d(k)$ or $N = \Gamma \cdot SA_n^1(k)$ ,

where $\Gamma = N \cap GL_n(k)$ is either scalar or contains $SL_n(k)$.

We should also remark at this point that the subalgebras $ga_n^{(I,J)}(k)$ (where I and J satisfy the conditions stated in theorem (5.2)) are invariant under $sa_n^o(k)$ and that they are ideals of $ga_n^o(k)$ if and only if, in addition, the following conditions hold: (i) $J_0 \subset J_1$ , (ii) $I(1) \subset J$ .

## 6. THE TOPOLOGICAL APPROXIMATION

We study the group $\underline{\hat{GA}_n^o(k)}$ of k-algebra automorphisms of the power series ring $\underline{k^{[[n]]}} = k[[X_1,X_2,\ldots,X_n]]$ preserving the ideal $\underline{(\hat{X})}$ generated by $X_1,\ldots,X_n$. Inside $\hat{GA}_n^o(k)$ we have the congruence subgroups $\underline{\hat{GA}_n^d(k)}$ and $\underline{\hat{GA}_n^I(k)}$ (defined in the natural way), and we give $\hat{GA}_n^o(k)$ the topology induced by the filtration $(\hat{GA}_n^d(k))_{d\geq 0}$ . We remark that $GA_n^o(k)$ is contained in $\hat{GA}_n^o(k)$ and, more precisely:

(6.1) Theorem. (Anick, Bass)

(a) The closure of $GA_n^o(k)$ in $\hat{GA}_n^o(k)$ is the subgroup consisting of automorphisms $F \in \hat{GA}^o(k)$ whose jacobian determinant is congruent to a unit in k modulo the nil radical of k.

(b) The closure of $SA_n^o(k)$ as well as the closure of $EA_n^o(k)$ in $\hat{GA}_n^o(k)$ is the group $\underline{\hat{SA}_n^o(k)}$ of automorphisms $F \in \hat{GA}_n^o(k)$ of jacobian determinant 1.

The Lie algebra of the functor $k \to \hat{GA}_n^o(k)$ is $\underline{\hat{ga}_n^o(k)}$, the Lie algebra of derivations of $k^{[[n]]}$, which we can also view as the completion of $ga_n^o(k)$ with respect to the filtration $(ga_n^d(k))_{d\geq 0}$. Note that both $\hat{GA}_n^o(k)$ and $\hat{ga}_n^o(k)$ live inside the associative algebra of k-linear endomorphisms of $k^{[[n]]}$, and there, for endomorphisms L that map $(X)^d$ to $(X)^{d+1}$ (all $d \geq 0$), we can define the exponential map by:

$$\exp(L) = \sum_{k\geq 0} \frac{1}{k!} L^k .$$

(6.2) Theorem. Under the exponential, the Lie algebra $\hat{ga}_n^1(k)$ is mapped bijectively to $\hat{GA}_n^1(k)$, and the subalgebra $\hat{sa}_n^1(k)$ is mapped bijectively to $\hat{SA}_n^1(k)$.

The second part of the theorem asserts that for a derivation $D \in \hat{ga}_n^1(k)$, the jacobian determinant of exp(D) is 1 if and only if the divergence of D is zero, but, contrary to what one may expect from the linear theory, the jacobian determinant of exp(D) does not coincide with exp(Div(D)), as one can easily see using $D = X_1\Delta$.

The usefulness of the above theorem lies in the following:
(6.3) Proposition. The exponential map establishes a one-to-one correspondance between closed Lie subalgebras of $\hat{ga}_n^1(k)$ and closed subgroup of $\hat{GA}_n^1(k)$ (in the filtration topology). Closed ideals correspond to closed normal subgroups. (See also [3], Chapter II, § 6).

We now analyse the exponentials of the particular Lie subalgebras described in section 5.
First of all we notice that the exponential of a derivation of the form $f\Delta$, $f \in (\hat{X})$, is a "pseudo-homothety", i.e. an automorphism of the following form:

$$\Lambda_h : X_i \to hX_i \qquad i = 1,\dots,n;\ h \in 1 + (\hat{X}) .$$

The pseudo-homotheties (we can allow h to be any power series with invertible constant term) form a subgroup of $\hat{GA}_n^{\circ}(k)$, denoted by $\hat{\Delta A}_n(k)$. We define, for an ideal level J, $\hat{\Delta A}_n^J(k)$ as the intersection of $\hat{\Delta A}_n(k)$ with $\hat{GA}_n^J(k)$. We remark that $\hat{\Delta A}_n^J(k)$ is a normal subgroup of $\hat{\Delta A}_n(k)$, normalized also by $GL_n(k)$.

The decomposition $ga_n(k) = k^{[n]}\Delta \oplus sa_n(k)$ has the following group theoretic analogue:
(6.4) Proposition. $\hat{GA}_n^1(k)$ is the product of the subgroups $\hat{\Delta A}_n^1(k)$

and $\hat{SA}_n^1(k)$, in the sense that every automorphism $F \varepsilon \hat{GA}_n^1(k)$ has a unique decomposition:

$$F = \Lambda_f \cdot S \quad , \quad (\Lambda_f \varepsilon \hat{\Delta A}_n^1(k), \quad S \varepsilon \hat{SA}_n^1(k)).$$

Let now $(I,J)$ be a pair of ideal levels, we define

$$\underline{\hat{GA}_n^{(I,J)}(k)} = \hat{\Delta A}_n^J(k) \cdot \hat{SA}_n^I(k) .$$

These turn out to be subgroups of $\hat{GA}_n^o(k)$, normalized by $GL_n(k)$, and moreover,

$$\hat{GA}_n^{(I,J)}(k) = \exp(\hat{ga}_n^{(I.J)}(k)) \quad \text{if} \quad I_0 = J_0 = 0 .$$

We also remark that the closure of $GA_n^1(k)$ in $\hat{GA}_n^1(k)$ can be described as the subgroup $\hat{GA}_n^{(1, \mathcal{N}(1))}(k)$, where, by abuse of notation, here $\mathcal{N}$ denotes the constant ideal level of value $\mathcal{N}$, the nil radical of k.

Thanks to our classification of invariant subalgebras of $ga_n(k)$ and the correspondance given by the exponential map, we conclude that:

(6.5) <u>Theorem</u>. Every closed subgroup N of $\hat{GA}_n^1(k)$, normalized by $EA_n^o(k)$, has the form:

$$N = \hat{GA}_n^{(I,J)}(k)$$

where I and J are ideal levels with $I_0 = J_0 = 0$ and $J(1) \subset I$. If N is normalized by $GA_n^o(k)$, furthermore we have $\mathcal{N}I(1) \subset J$, and if N is normalized by $\hat{GA}_n^o(k)$, then $I(1) \subset J$.

Now we must take care of subgroups of the whole group $\hat{GA}_n^o(k)$. For this we can use the description of the normal subgroups of GL (k) given in [1], Chapter V, § 4, or [12], theorem 3.

We assume henceforth that $n \geqslant 3$.

(6.6) Theorem. Let $N$ be a subgroup of $\hat{G}A_n^o(k)$ normalized by $E_n(k)$. Assume furthermore that one of the following conditions holds:

(i) $N$ is normalized by $GL_n(k)$,

(ii) $N \subset \hat{S}A_n^o(k)$.

Then:

$$N = N^1 \rtimes \Gamma$$

where $N^1 = N \cap GA_n^1(k)$, and $\Gamma = N \cap GL_n(k)$.

This result permits us to make the description of the normal subgroups of $\hat{G}A_n^o(k)$ quite precise.
Recall that, under the stable range conditions, every subgroup $\Gamma$ of $GL_n(k)$ normalized by $E_n(k)$ satisfies a relation:

$$E_n^{I_0}(k) \subset \Gamma \subset G'L_n^{I_0}(k) \quad ,$$

where $I_0$ is an ideal of $k$, $GL_n^{I_0}(k) = \ker(GL_n(k) \to GL_n(k/I_0))$, $G'L_n^{I_0}(k)$ is the group of matrices that are scalar mod $I_0$, and $E_n^{I_0}(k)$ is the normal subgroup of $E_n(k)$ generated by the elementary matrices which are trivial mod $I_0$.
If we have two ideals $I_0$ and $I_1$, we define $\underline{G'L_n^{(I_0,I_1)}(k)}$ as the inverse image under the map $GL_n(k) \to GL_n(k/I_0)$ of the group of scalar matrices congruent to the identity mod $I_1+I_0/I_0$.

Let now $I$, $J$ be two ideal levels, and let $\tilde{I}$, $\tilde{J}$ be the ideal levels obtained as $\tilde{I} = (0, I_1, I_2, \ldots)$; $\tilde{J} = (0, J_1, J_2, \ldots)$.
We define:

$$\underline{G'\hat{A}_n^{(I,J)}(k)} = \hat{G}A_n^{(\tilde{I},\tilde{J})}(k) \rtimes G'L_n^{(I_0,I_1)}(k) \quad ;$$

$$\underline{G''\hat{A}_n^{(I,J)}(k)} = \hat{G}A_n^{(\tilde{I},\tilde{J})}(k) \rtimes G'L_n^{(I_0,I_1 \cap J_1)}(k) \quad ;$$

$$\underline{\hat{E}A_n^{(I,J)}(k)} = \hat{G}A_n^{(\tilde{I},\tilde{J})}(k) \rtimes E_n^{I_0}(k)$$

(6.7) Proposition.

(a) The subgroups defined above are normalized by $GL_n(k)$.

(b) If $J(1) \subset I$, they are normalized by $\hat{SA}_n^o(k)$.

(c) If $J(1) \subset I$ and $\mathcal{N}I(1) \subset J$, $G''\hat{A}_n^{(I,J)}(k)$ and $\hat{EA}_n^{(I,J)}(k)$ are normalized by $\hat{GA}_n^{(0,\mathcal{N})}(k)$.

(d) If $J(1) \subset I$ and $I(1) \subset J$, $G''\hat{A}_n^{(I,J)}(k)$ and $\hat{EA}_n^{(I,J)}(k)$ are also normalized by $\hat{GA}_n^o(k)$.

We are now able to completely describe the normalized subgroups of $\hat{GA}_n^o(k)$.

(6.8) Theorem. Let N be a closed subgroup of $\hat{GA}_n^o(k)$, normalized by $GL_n(k)$. Assume that, respectively,

(i) N is normalized by $EA_n^o(k)$;

(ii) N is normalized by $GA_n^o(k)$;

(iii) N is normalized by $\hat{GA}_n^o(k)$.

Then there is a pair of ideal levels $(I, J)$ with $J_0 = 0$ such that, respectively:

(i) $\hat{EA}_n^{(I,J)}(k) \subset N \subset G'\hat{A}_n^{(I,J)}(k)$, and $J(1) \subset I$;

(ii) $\hat{EA}_n^{(I,J)}(k) \subset N \subset G''\hat{A}_n^{(I,J)}(k)$, and $J(1) \subset I$, $\mathcal{N}I(1) \subset J$;

(iii) $\hat{EA}^{(I,J)}(k) \subset N \subset G''\hat{A}^{(I,J)}(k)$, and $J(1) \subset I$, $I(1) \subset J$.

The pair $(I, J)$ is called the level of N.

In particular, if k is a field, the subgroups of $\hat{GA}_n^o(k)$ normalized by $GL_n(k)$ and by $EA_n^o(k)$ are:

$\hat{GA}_n^{(p,q)}(k)$ where $p \geqslant 0$, $q \geqslant 1$, $p \leqslant q + 1$, and

$\hat{GA}_n^{(1,q)}(k) \cdot \Gamma$ where $q \geqslant 1$ and $\Gamma$ is a subgroup of $GL_n(k)$ which is either scalar or contains $SL_n(k)$.

The normal subgroups of $\hat{GA}_n^o(k)$ are:

$\hat{GA}_n^{(p,q)}(k)$ where $p \geqslant 1$, $q \geqslant 1$, $q - 1 \leqslant p \leqslant q + 1$ ;

$\hat{GA}_n^{(1,q)}(k) \cdot \Gamma$ where $q = 1$ or $2$, and $\Gamma$ is subgroup of $GL_n(k)$ containing $SL_n(k)$ ;

$\hat{GA}_n^{(1,1)}(k) \cdot \Gamma$ where $\Gamma$ is a subgroup of the scalars in $GL_n(k)$ .

## 7. CONSEQUENCES

We now go back to the group $GA_n(k)$.
Let $\Delta A_n(k)$ be the group of pseudo-homotheties in $GA_n(k)$, then $\Delta A_n(k)$ is the group of those $\Lambda_h$, with $h \in k^\times + \mathcal{N}\cdot(X)$ ($\mathcal{N}$ denotes the nil radical of $k$).
We again have a decomposition of $GA_n^1(k)$ as a product $\Delta A_n^1(k) \cdot SA_n^1(k)$.

For a pair of ideal levels $(I, J)$ with $J_0 = 0$, $J_d \subset \mathcal{N}$ for all $d \geqslant 1$, we can still define subgroups:

$$GA_n^{(I,J)}(k) = \Delta A_n^J(k) \cdot SA_n^I(k) ;$$

$$G'A_n^{(I,J)}(k) = GA_n^{(I,J)}(k) \rtimes G'L_n^{(I_0, I_1)}(k) ;$$

$$G''A_n^{(I,J)}(k) = GA_n^{(I,J)}(k) \rtimes G'L_n^{(I_0, I_1 \cap J_1)}(k) ;$$

$$EA_n^{(I,J)}(k) = \Delta A_n^J(k) \cdot EA_n^I(k) .$$

We remark that these subgroups are dense in the corresponding ones in the power series group, and that, except for $EA_n^{(I,J)}(k)$, they can be described as the intersections of the corresponding subgroups of $\hat{GA}_n^o(k)$ with $GA_n^o(k)$.

We define the level of a normalized subgroup $N$ of $GA_n^o(k)$ as the level of its closure in $\hat{GA}_n^o(k)$; thus the level of $N$ is a pair of ideal levels $(I, J)$ where $J$ is a nil ideal level with $J_0 = 0$, and $J(1) \subset I$. If N is normal in $GA_n^o(k)$ then, in addition $\mathcal{N}I(1) \subset J$.

Theorem (6.8) shows that a subgroup of $GA_n^o(k)$ of level (I, J) is contained in $G'A_n^{(I,J)}(k)$ and even in $G''A_n^{(I,J)}(k)$ if the subgroup is normal in $GA_n^o(k)$.

Another consequence of (6.8) is that, for a subgroup N of level (I, J) with $I_0 = J_0 = 0$,

$$gr(N) = \mathfrak{ga}_n^{(I,J)}(k) = gr(GA_n^{(I,J)}(k)) = gr(EA_n^{(I,J)}(k)).$$

We conclude by stating a more definitive form of the Normal Level Conjecture:

(7.1) Normal Level Conjecture (definitive form). Let k be a noetherian Q-algebra of finite dimension. There exists an integer $n_0(k) \geqslant 3$ such that, for $n \geqslant n_0(k)$, if N is a subgroup of $GA_n^o(k)$, normalized by $EA_n^o(k) \cdot GL_n(k)$, of level (I, J),

$$EA_n^{(I,J)}(k) \subset N \subset G'A_n^{(I,J)}(k).$$

## BIBLIOGRAFY

[1] H. Bass, *Algebraic K-theory*, Benjamin, New York, 1968.

[2] H. Bass, E.H. Connell and D.L. Wright, *The Jacobian conjecture: reduction of degree and formal expansion of the inverse*, Bull. Amer. Math. Soc. 7 (1982), pp. 287-329.

[3] N. Bourbaki, *Groupes et algebres de Lie*, Chap. II, Hermann, Paris.

[4] V.I. Danilov, *Non-simplicity of the group of unimodular automorphisms of an affine plane*, Mat. Zametki 15 (1974), pp. 289-293. English transl. in Math. Notes 15 (1974).

[5] M. Demazure and P. Gabriel, *Groupes algebriques*, Masso & CIE, Paris, 1970.

[6] H.W.E. Jung, *Uber ganze birationale Transformatione der Ebene*, J. Reine Angew. Math. 194 (1942), pp. 161-174.

[7] M. Nagata, *On the automorphism group of k[x,y]*, Lectures in Math., Kyoto University N° 5 (1972), Kinokuniya, Tokyo.

[8] M. Pittaluga, *On the automorphism group of a polynomial algebra*, thesis, Columbia University, 1983.

[9] I.R. Safarevic, *On some infinite dimensional algebraic groups*, Math. U.S.S.R. Izvestija, Vol. 18 (1982), N° 1.

[10] W. Van der Kulk, *On polynomial rings in two variables*, New Archief voor Wiskunde, 3 I, (1953), pp. 33-41.

[11] D.L. Wright, *Algebras which resemble symmetric algebras*, thesis, Columbia University, 1975.

[12] L.N. Vaserstein, *On the normal subgroups of $GL_n$ over a ring*, appeared in "Algebraic K-theory", Evanston 1980, (Proc. Conf. Northwestern Univ. Evanston), Lecture Notes in Mathematics 854, Springer, Berlin, 1981, pp. 456-465.

# AUSLANDER-REITEN QUIVERS FOR SOME ARTINIAN TORSION THEORIES AND INTEGRAL REPRESENTATIONS

Klaus W.Roggenkamp

Universität Stuttgart (West Germany)

Almost split sequences and Auslander-Reiten quivers (species) have turned out to be an extremely useful tool in the representation theory of artinian algebras and integral representations of complete orders. Whereas the theory is well developed in the artinian situations, it is far behind for integral representations. The main reason is that the powerful tool of universal coverings and covering techniques - as developed by Riedtmann [14], Bongartz - Gabriel [4] - has not yet found its integral analogue. The main difficulty is that it is unclear to me, when a representation quiver - in the sense of C.Riedtmann [14] - is the Auslander-Reiten quiver of an order of finite lattice type; in particular, one obstacle is the multiplication with the parameter of the ground ring. A little progress has been made in this direction by Wiedemann. Apart from my general philosophy, that most of the results in the representation theory of artinian algebras have more or less obvious analoga in integral representation theory, there is a more solid connection between integral representations and some categories of torsionfree modules over artinian algebras as observed by C.M.Ringel and the author [15], such that in some instances integral representation theory might also influence the representation theory of artinian algebras - e.g. that some torsionfree categories of artinian modules have almost split sequences was - before its general treatment by M.Auslander and S.O.Smalø [1] - in secial cases - influenced by the results in integral representation theory - noted by the author [17] and independently by R.Bautista and R.Martinez [3].



*F. van Oystaeyen (ed.), Methods in Ring Theory, 433–449.*

In this lecture I would like to point out the structure of the Auslander-Reiten quivers of the category of the torsionfree modules with respect to the socle of a hereditary algebra. A necessary and sufficient condition for this category to be of finite representation type was already given by C.M.Ringel and the author [16]. At the same time this is used to describe the Auslander-Reiten quiver of generalized Bäckström orders - these are the analoga of hereditary algebras - thus extending the corresponding results of Bäckström orders [19], which correspond to radical-square-zero algebras.

In § 1 I shall talk about the connection between integral representations and torsion theories.

In § 2 the Auslander-Reiten quiver of the above mentioned torsion theories is described.

In § 3 these results are applied to complete orders.

In § 4 I mention some more results on orders; in particular on Gorenstein orders - the artinian analogue of this are the quasi Frobenius algebras, and on Schurian orders.

For the terminologiy and the definitions in this note I refer to [18], and most of the results will be found with detailed proofs in [20].

## § 1 Integral representations and torsion theories.

We first fix our notation: Let $R$ be a complete Dedekind domain with field of fractions $K$ and residue field $\mathfrak{k} = R/\mathrm{rad}\, R$ where $\mathrm{rad}\, R = R\pi$ . $A$ is a finite dimensional separable K-algebra and $\Lambda$ is an R-order in $A$ . ${}_{\Lambda}\mathfrak{M}^{o}$ stands for the left $\Lambda$-lattices and $\mathrm{ind}^{o}(\Lambda)$ denotes the isomorphism classes of indecomposable objects in ${}_{\Lambda}\mathfrak{M}^{o}$ - sometimes no distinction is made between modules and their isomorphism classes.

The connection between orders and artinian algebras is best demonstrated in the situation of subhereditary orders.

(1.1) Definition: An R-order $\Lambda$ is said to be subhereditary provided there exists a hereditary R-order $\Gamma$ in $A$ with

$$\mathrm{rad}\,\Gamma \subset \Lambda \subset \Gamma .$$

(1.2) Remarks: (i) Special subhereditary orders are the Bäckström orders where in addition $\operatorname{rad}\Gamma = \operatorname{rad}\Lambda$. In this case the representation theory of finite type is well understood [2,16,19].

(ii) There is some evidence that by using covering techniques it will be possible to reduce the study of arbitrary orders of finite type to the study of subhereditary orders of finite type. This is for example the case when $\Lambda$ contains a complete set of primitive orthogonal idempotents of $A$ – these are the so called Schurian orders, which were studied in [24, 22, 21]. For these orders there are coverings, which reduce the study of general Schurian orders to that of subhereditary Schurian orders.

(iii) We recall that an order $\Lambda$ is said to be hereditary, if all its lattices are projective. The structure of hereditary orders is well understood [10, 6, 11]. Their Auslander-Reiten quiver is a union of oriented cycles.

(iv) It should be noted that an order $\Lambda$ can be subhereditary with respect to two different hereditary orders. If we talk about a subhereditary order $\Lambda$ we think of a fixed hereditary order $\Gamma$; i.e. a pair $(\Lambda,\Gamma)$ satisfying (1.1).

(1.3) Definition: Let $(\Lambda,\Gamma)$ be a subhereditary order. We put $\mathfrak{B} = \Gamma/\operatorname{rad}\Gamma$ and $\mathfrak{A} = \Lambda/\operatorname{rad}\Gamma$. Then $\mathfrak{A}$ and $\mathfrak{B}$ are finite dimensional $\mathfrak{k}$-algebras, and $\mathfrak{B}$ is semisimple.

$$\mathfrak{D} = \begin{pmatrix} \mathfrak{B} & {}_{\mathfrak{B}}\mathfrak{B}_{\mathfrak{A}} \\ 0 & \mathfrak{A} \end{pmatrix} ,$$

where ${}_{\mathfrak{B}}\mathfrak{B}_{\mathfrak{A}}$ is $\mathfrak{B}$ as $(\mathfrak{B},\mathfrak{A})$-bimodule, is called the $\mathfrak{k}$-algebra of the subhereditary order $(\Lambda,\Gamma)$.

(1.4) Lemma: (i) Let $\mathfrak{D}$ be the $\mathfrak{k}$-algebra of the subhereditary order $(\Lambda,\Gamma)$, then $\operatorname{soc}({}_{\mathfrak{D}}\mathfrak{D})$, the socle of ${}_{\mathfrak{D}}\mathfrak{D}$ as left module is projective.

(ii) Let $\mathfrak{D}$ be a finite dimensional $\mathfrak{k}$-algebra such that $\operatorname{soc}({}_{\mathfrak{D}}\mathfrak{D})$ is projective. Then – up to Morita equivalence – $\mathfrak{D}$ is the $\mathfrak{k}$-algebra of a subhereditary R-order $(\Lambda,\Gamma)$.

Proof: (i) Since $\mathfrak{B}$ is semisimple and $\mathfrak{A}$ can be viewed as a subalgebra of $\mathfrak{B}$, it is obvious, that

$$\operatorname{Soc}({}_{\mathfrak{D}}\mathfrak{D}) = \begin{pmatrix} \mathfrak{B} & \mathfrak{B}^{\mathfrak{B}}{}_{\mathfrak{A}} \\ 0 & 0 \end{pmatrix}$$

and hence it is projective.

(ii) There is no loss of generality if we assume that $\mathfrak{D}$ is indecomposable as ring. We decompose the left module ${}_{\mathfrak{D}}\mathfrak{D}$ as

$$ {}_{\mathfrak{D}}\mathfrak{D} = S' \oplus T $$

where $S'$ is semisimple and $T$ does not have a simple direct summand. Let $S = \mathrm{Soc}_{\mathfrak{D}}(T)$. Since $\mathfrak{D}$ is indecomposable as ring, and since it has a projective left socle, $\mathfrak{D}$ is Morita equivalent to

$$ \mathfrak{D}_o = \begin{bmatrix} \mathrm{End}_{\mathfrak{D}}(S) & \mathrm{Hom}_{\mathfrak{D}}(S,T) \\ O & \mathrm{End}_{\mathfrak{D}}(T) \end{bmatrix}, $$

noting $\mathrm{Hom}_{\mathfrak{D}}(T,S) = O$. Moreover, $\mathrm{Hom}_{\mathfrak{D}}(S,T) = \mathrm{Hom}_{\mathfrak{D}}(S,S)$ since $S = \mathrm{Soc}_{\mathfrak{D}}(T)$. Hence, if we put $\mathfrak{B} = \mathrm{End}_{\mathfrak{D}}(S)$ and $\mathfrak{A} = \mathrm{Hom}_{\mathfrak{D}}(T,T)$, then

$$ \mathfrak{D}_o = \begin{bmatrix} \mathfrak{B} & \mathfrak{B}^{\mathfrak{B}}_{\mathfrak{A}}\mathfrak{A} \\ O & \end{bmatrix} $$

and $\mathfrak{B}$ is semisimple. We next note that the map

$$ \alpha: \quad \mathfrak{A} \to \mathfrak{B} \quad \text{defined as} $$

$$ \varphi \in \mathrm{End}_{\mathfrak{D}}(T) \qquad \varphi|_S : \mathrm{Soc}_{\mathfrak{D}}(T) \to \mathrm{Soc}_{\mathfrak{D}}(T) $$

is injective. In fact, if $\varphi|_S = O$, then we have - thanks to the ker-coker-lemma - the situation:

$$
\begin{array}{ccccccccc}
 & & O & & & & & & \\
 & & \downarrow & & & & & & \\
 & & \mathrm{Soc}_{\mathfrak{D}}(T) & & O & & O & & \\
 & & \downarrow & & \downarrow & & \downarrow & & \\
O & \to & \mathrm{Soc}_{\mathfrak{D}}(T) & \to & T & \to & T/\mathrm{Soc}\,T & \to & O \\
 & & \downarrow \varphi|_S & & \downarrow \varphi & & \downarrow & & \\
 & & O & \to & T & \to & T & \to & O \\
 & & \downarrow & & \downarrow & & \downarrow & & \\
 & & O & \to & \mathrm{Coker}\,\varphi & \to & \mathrm{Soc}_{\mathfrak{D}}(T) & & \\
 & & & & & & \downarrow & & \\
 & & & & & & O & &
\end{array}
$$

and $\mathrm{Soc}_{\mathfrak{D}}(T)$ being projective, $T$ would have a simple direct summand, a contradiction.

We now choose a hereditary R-order $\Gamma$ such that

$\Gamma/\mathrm{rad}\,\Gamma \xrightarrow{\sim} \mathfrak{B}$ - note that $\Gamma$ is by no means unique - and we define $\Lambda$ as the pullback

$$\begin{array}{ccc} \Lambda & \to & \Gamma \\ \downarrow & & \downarrow \beta \\ \mathfrak{A} & \xrightarrow{\alpha} & \mathfrak{B} \end{array}$$

Then it is easily seen that $(\Lambda,\Gamma)$ is a subhereditary order which has $\mathfrak{D}_o$ as associated $\mathfrak{k}$-algebra. #

(1.5) Remark: In general there are more than one subhereditary orders $(\Lambda,\Gamma)$ which have the same $\mathfrak{k}$-algebra $\mathfrak{D}$ (cf.proof of (1.4)); however, once $\Gamma$ is fixed, $\Lambda$ is uniquely determined by $\mathfrak{D}$.

(1.6) Definition: Let $\mathfrak{D}$ be a finite dimensional $\mathfrak{k}$-algebra with $\mathrm{Soc}({}_{\mathfrak{D}}\mathfrak{D})$ projective. Then we put

$$\mathfrak{C}(\mathfrak{D}) = \{X \in \mathrm{mod}({}_{\mathfrak{D}}\mathfrak{D}) : \mathrm{Soc}_{\mathfrak{D}}(X) \text{ is projective}\}$$

and

$$\mathfrak{C}^o(\mathfrak{D}) = \{X \in \mathfrak{C}(\mathfrak{D}) : X \text{ has no simple projective direct summand}\}$$

(1.7) Theorem [8,15]: Let $(\Lambda,\Gamma)$ be a subhereditary R-order in A with $\mathfrak{k}$-algebra $\mathfrak{D}$. Then the functor

$$\mathbb{F}: {}_{\Lambda}\mathfrak{M}^o \to \mathfrak{C}^o(\mathfrak{D}) \quad \text{induced by}$$

$$M \to \begin{pmatrix} \Gamma M/(\mathrm{rad}\,\Gamma)M \\ M/(\mathrm{rad}\,\Gamma)M \end{pmatrix}$$

is a representation equivalence.

If we denote for a "decent" category $\mathfrak{C}$ by $n(\mathfrak{C})$ the number of indecomposable object of $\mathfrak{C}$, we have

(1.8) Corollary: If $(\Lambda,\Gamma)$ is a subhereditary order with algebra $\mathfrak{D}$, then $n(\Lambda) = n(\mathfrak{C}^o(\mathfrak{D})) - t$, where t is the number of non-isomorphic simple projective left $\mathfrak{D}$-modules. Moreover, $\mathfrak{C}(\mathfrak{D})$ is the category of torsionfree modules in a hereditary torsion theory on $\mathrm{mod}({}_{\mathfrak{D}}\mathfrak{D})$.

(1.9) Remark: The above functor is not only a representation equivalence but also induces a map from the Auslander-Reiten species of $\mathrm{ind}(\mathfrak{C}(\mathfrak{D}))$ to that of $\mathrm{ind}^o(\Lambda)$. Though the results hold in more general situations, we shall restrict ourselves here to the simple situation - namely to the case where $\mathfrak{D}$ is hereditary (most of our results hold in case $\mathfrak{D}$ is simply connected).

(1.10) <u>Definition:</u> An R-order $\Lambda$ in A is called a <u>generalized Bäckström order</u> if

(i) $\Lambda$ is a subhereditary order $(\Lambda, \Gamma)$, and

(ii)

$$\mathfrak{D} = \begin{pmatrix} \Gamma/\mathrm{rad}\,\Gamma & \Gamma/\mathrm{rad}\,\Gamma \\ 0 & \Lambda/\mathrm{rad}\,\Gamma \end{pmatrix}$$

is a hereditary artinian $\mathfrak{k}$-algebra.

We next give an <u>internal characterization of generalized Bäckström orders</u>.

(1.11) <u>Proposition:</u> An R-order $\Lambda$ in A is a generalized Bäckström order if and only if

(i) $\Lambda$ is a subhereditary order $(\Lambda, \Gamma)$ and

(ii) if P is an indecomposable projective left $\Lambda$-lattice, then $\mathrm{rad}_\Lambda(P) = X \oplus Y$, where X is a projective $\Lambda$-lattice.

(1.12) <u>Remark:</u> We recall that $\Lambda$ is said to be a Bäckström order if (i) is satisfied and in (ii), $Y = 0$ i.e.

$$\mathrm{rad}\,\Lambda = \mathrm{rad}\,\Gamma \subset \Lambda \subset \Gamma .$$

In that case $\mathfrak{D}$ is hereditary with radical square zero.

For the <u>proof of (1.11)</u> we refer to [20, I, Theorem I §1].

In studying the Auslander-Reiten species of generalized Bäckström orders - which were originally defined via (1.11) - we were lead to study the Auslander-Reiten species of $\mathfrak{S}(\mathfrak{D})$, where $\mathfrak{D}$ is a hereditary tensor algebra. We shall present the results for $\mathfrak{S}(\mathfrak{D})$ in § 2. § 3 gives the translation to the Auslander-Reiten species of generalized Bäckström orders.

## § 2 <u>Auslander-Reiten species of "socle categories" for hereditary tensor algebras</u>.

We keep the notation of § 1. In addition we assume that $\mathfrak{D}$ is the $\mathfrak{k}$-tensor algebra of a valued oriented graph T with modulation over $\mathfrak{k}$. A necessary and sufficient criterion that the socle category $\mathfrak{S}(\mathfrak{D})$ of $\mathfrak{D}$ is of finite representation type was given in [16]. Before we can state that result we have to introduce some more <u>notation</u>.

(2.1) <u>Definition:</u> A valued oriented graph T is said to be <u>contractible</u>, if there exists a pair of vertices a,b linked by an arrow $b \xrightarrow{(1,1)} a$ with valuation (1,1) in such a way that the graph obtained from T by removing the edge between a and b is a disjoint union of a graph $T_1$ and a graph $T_2$, such that a is a source in $T_1$ and $T_2$ is a Dynkin graph of type $A_m$ such that b is a source in $T_2$. If $T_{a=b}$ is obtained from T by identifying a and b in T. We say that T is <u>1-step-contractible</u> to $T_{a=b}$. We say that T is <u>contractible to $T_1$</u>, if $T_1$ can be reached from T by a finite number of 1-step contractions.

(2.2) <u>Theorem:</u> Let $\mathfrak{D}$ be the $\mathfrak{k}$-tensor algebra of a valued oriented graph T and let $\mathfrak{C}(\mathfrak{D})$ be the category of finitely generated left $\mathfrak{D}$-modules with a projective socle.

(i) [16] $\mathfrak{C}(\mathfrak{D})$ has a finite number of non-isomorphic indecomposable objects if and only if T can be contracted to a finite disjoint union of Dynkin graphs.

(ii) [20] If T is a Dynkin graph, then the Auslander-Reiten quiver of $\mathfrak{C}(\mathfrak{D})$ is simply connected. Moreover, every indecomposable is obtained as iterated Auslander-Reiten translate - to the right - starting with the indecomposable projective $\mathfrak{D}$-modules (these lie in $\mathfrak{C}(\mathfrak{D})$).

Before we go on, let us demonstrate the result by means of an

(2.3) <u>Example:</u> Let

T = 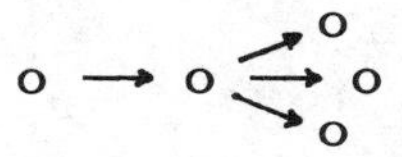 ,

where the valuation is trivial. Then

$$\mathfrak{D} = \begin{pmatrix} \mathfrak{k} & \mathfrak{k} & \mathfrak{k} & \mathfrak{k} & \mathfrak{k} \\ 0 & \mathfrak{k} & \mathfrak{k} & \mathfrak{k} & \mathfrak{k} \\ 0 & 0 & \mathfrak{k} & 0 & 0 \\ 0 & 0 & 0 & \mathfrak{k} & 0 \\ 0 & 0 & 0 & 0 & \mathfrak{k} \end{pmatrix} \quad \text{is surely of infinite}$$

representation type; by (2.1) T reduces to the Dynkin diagram $D_4$ and so $\mathfrak{C}(\mathfrak{D})$ has a finite number of non-isomorphic indecomposables. The section of irreducible maps between the indecomposable projectives is given as

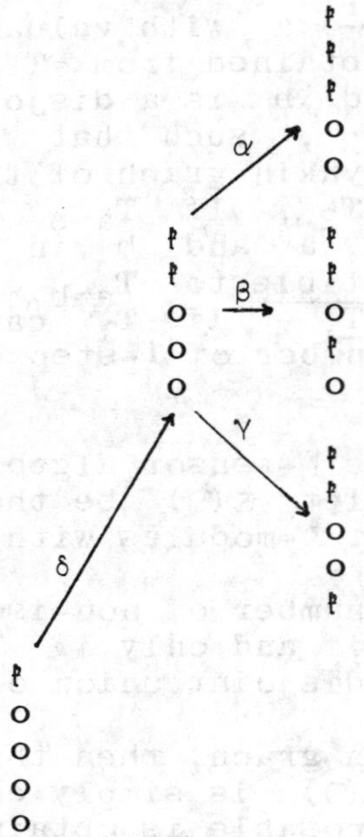


We now construct the Auslander-Reiten quiver of $\mathfrak{C}(\mathfrak{D})$ by iteratively forming the cokernel, if this lies in $\mathfrak{C}(\mathfrak{D})$ and stopping otherwise e.g. $\mathrm{Coker}(\alpha \oplus \beta \oplus \gamma) \in \mathfrak{C}(\mathfrak{D})$, $\mathrm{Coker}\,\delta \notin \mathfrak{C}(\mathfrak{D})$. Continuing this way we get the Auslander-Reiten quiver.

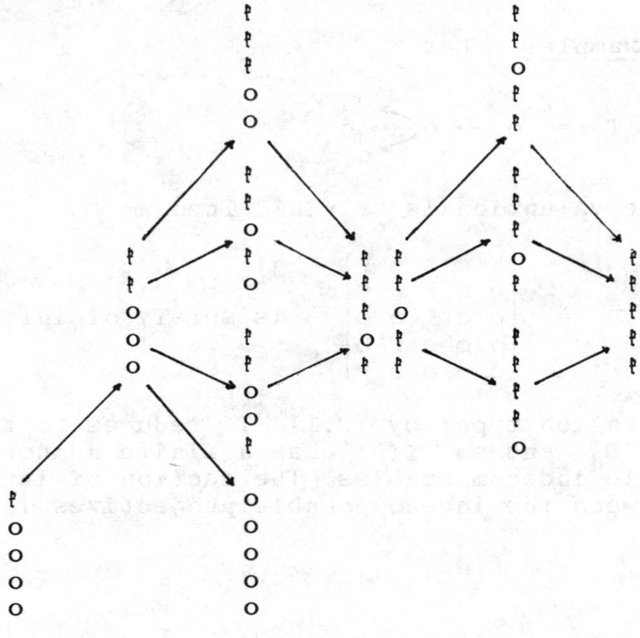

The $\operatorname{ext}_{\mathfrak{C}(\mathfrak{D})}$-injectives are

$$\begin{pmatrix} \mathfrak{k} \\ o \\ o \\ o \\ o \end{pmatrix}, \quad \begin{pmatrix} \mathfrak{k} \\ \mathfrak{k} \\ o \\ \mathfrak{k} \\ \mathfrak{k} \end{pmatrix}, \quad \begin{pmatrix} \mathfrak{k} \\ \mathfrak{k} \\ \mathfrak{k} \\ o \\ \mathfrak{k} \end{pmatrix}, \quad \begin{pmatrix} \mathfrak{k} \\ \mathfrak{k} \\ \mathfrak{k} \\ \mathfrak{k} \\ o \end{pmatrix} \text{ and } \begin{pmatrix} \mathfrak{k} \\ \mathfrak{k} \\ \mathfrak{k} \\ \mathfrak{k} \\ \mathfrak{k} \end{pmatrix}$$

(2.4) We next turn to a <u>purely combinatorical description of the Auslander-Reiten quiver</u> of $\mathfrak{C}(\mathfrak{D})$ provided there are only finitely many non-isomorphic indecomposables in $\mathfrak{C}(\mathfrak{D})$. Let $T$ be a <u>valued oriented tree</u>, then we define the <u>disturbed additive function</u> of $T$ as follows.

Let $\mathbb{Z}T$ be the <u>translation quiver</u> of $T$ in the sense of Riedtmann [14]. The vertices of $\mathbb{Z}T$ are labelled $(\nu, i)$, $\nu \in \mathbb{Z}$, $i$ a vertex of $T$. We view $T$ as embedded in $\mathbb{Z}T$ via $(o,i)$. By $\tau$ we denote the translation on $\mathbb{Z}T$,

$$\tau: \quad (\nu, i) \to (\nu+1, i) \quad .$$

Let $i_1, \ldots, i_s$ be the sources of $T$. On $T$ let $f$ have the value which is the sum of the generating functions of the sources $i_1, \ldots, i_s$. Then continue $f$ additively as long as the values are positive. If we reach a non-positive value - by the additive generation - say at $(\nu, i)$ then put $f((\nu,i)^{\tau^n}) = 0$ for all $n \in \mathbb{N}$.

We then call $f$ the <u>disturbed additive function</u> on $\mathbb{Z}T$.

(2.3) <u>Example continued:</u> For $T =$ 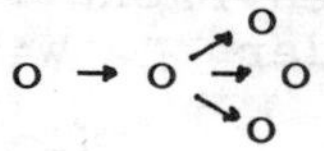 we get the following values

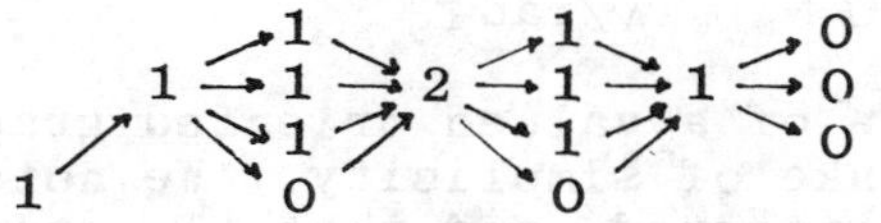

and one sees that $\{(\nu, i) : f(\nu, i) \neq 0\}$ gives the Auslander-Reiten quiver; moreover, the values give the size of the corresponding module.

(2.5) <u>Theorem</u> [20]: Let $T$ be a connected valued oriented tree, $\mathfrak{D}$ and $\mathfrak{C}(\mathfrak{D})$ as in (2.2). Then the following are equivalent:

(i) $\mathfrak{C}(\mathfrak{D})$ has a finite number of indecomposables.

(ii) $T$ can be contracted to a Dynkin diagram.

(iii) The disturbed additive function $f$ of $T$ has finite support on $\mathbb{Z}T$.

Moreover, if one of these equivalent conditions is satisfied, then $\{(\nu,i): f(\nu,i) \neq 0\}$ is the Auslander-Reiten quiver of $\mathfrak{C}(\mathfrak{D})$ and if $M \in \mathfrak{C}(\mathfrak{D})$ corresponds to $(\nu,i)$, then $f(\nu,i)$ is the number of simple modules in the socle of $M$.

(2.6) Remarks: The proof is done by showing the equivalence of (i) and (iii). The equivalence with (ii) follows then from (2.2). The proof requires a careful analysis of almost split sequences in $\mathfrak{C}(\mathfrak{D})$.

(i) How they are obtained from the ordinary almost split sequences.

(ii) If $\varphi: X \to Y$ is irreducible in $\mathfrak{C}(\mathfrak{D})$, $X, Y$ indecomposable, then either

α.) $\varphi$ is surjective or

β.) $\operatorname{Coker}\varphi \in \mathfrak{C}(\mathfrak{D})$ or

γ.) $\operatorname{Im}\varphi$ is a maximal submodule.

(iii) A construction of almost split sequences in $\mathfrak{C}(\mathfrak{D})$ by using the transpose and a duality on $\mathfrak{C}(\mathfrak{D})$, which is influenced from the construction of almost split sequences for orders.

(iv) A determination of the extinjective objects in $\mathfrak{C}(\mathfrak{D})$.

(For details we refer to [20], where most of the results are proved for more general torsion theories.)

## § 3 Auslander-Reiten quivers for generalized Bäckström orders of finite type.

With the notation of § 1 we assume that $\Lambda$ is a basic generalized Bäckström order (1.10) with associated hereditary order $\Gamma$ with $\operatorname{rad}\Gamma \subset \Lambda \subset \Gamma$, and that

$$\mathfrak{D} = \begin{pmatrix} \Gamma/\operatorname{rad}\Gamma & \Gamma/\operatorname{rad}\Gamma \\ 0 & \Lambda/\operatorname{rad}\Gamma \end{pmatrix}$$

is the tensor algebra of a valued oriented graph $T$. (This only for the sake of simplicity.) We note that in general $T$ is not connected, and that $T$ can be determined entirely from $(\Lambda,\Gamma)$ [20].

(3.1) Example: Let $\mathfrak{p} = \operatorname{rad} R$

$$\Lambda = \begin{pmatrix} R & R & R & R \\ \mathfrak{p} & R & R & R \\ \mathfrak{p} & \mathfrak{p} & R & R \\ \mathfrak{p} & \mathfrak{p} & \mathfrak{p} & R \end{pmatrix} \qquad \Gamma = \begin{pmatrix} R & R & R & R \\ \mathfrak{p} & R & R & R \\ \mathfrak{p} & \mathfrak{p} & R & R \\ \mathfrak{p} & \mathfrak{p} & \mathfrak{p} & R \end{pmatrix}$$

$$P_1 \; P_2 \qquad\qquad X_1 X_2 X_3 X_4$$

where R - R means that the corresponding elements are congruent modulo $\mathfrak{p}$ .

Then

$$\mathfrak{D} = \begin{pmatrix} \mathfrak{p} & o & o & o & \mathfrak{p} & o & o & o \\ o & \mathfrak{p} & o & o & o & \mathfrak{p} & o & o \\ o & o & \mathfrak{p} & o & o & o & \mathfrak{p} & o \\ o & o & o & \mathfrak{p} & o & o & o & \mathfrak{p} \\ o & o & o & o & \mathfrak{p} & o & o & o \\ o & o & o & o & o & \mathfrak{p} & o & o \\ o & o & o & o & o & o & \mathfrak{p} & o \\ o & o & o & o & o & o & o & \mathfrak{p} \end{pmatrix}$$

$\overline{X}_1 \overline{X}_2 \overline{X}_3 \overline{X}_4 \quad \overline{P}_1 \quad \overline{P}_2$

is the tensor algebra of the graph

T = 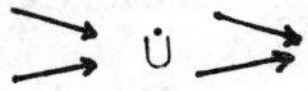 .

So by the results of § 2, the Auslander-Reiten quiver of $\mathfrak{S}(\mathfrak{D})$ has the form - with the function f -

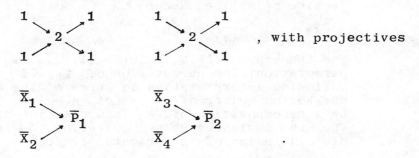

, with projectives

.

The Auslander-Reiten quiver of ${}_{\Lambda}\mathfrak{M}^o$ , the $\Lambda$-lattices, has the form:

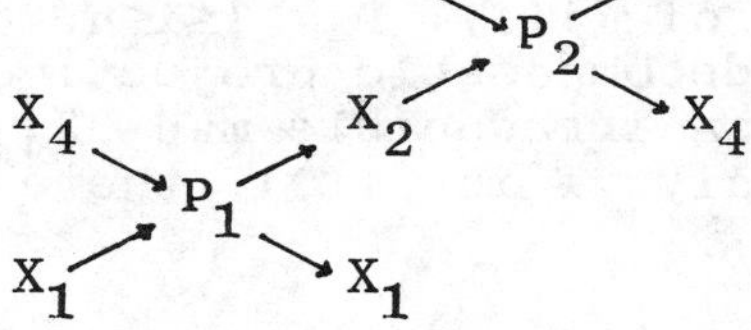


identifying the modules which have the same labels.

One sees in this example, that the Auslander-Reiten quiver of $\mathfrak{S}(\mathfrak{D})$ maps onto the Auslander-Reiten quiver of ${}_{\Lambda}\mathfrak{M}^o$ .

Before we can state this as a general result, we report on some

(3.2) Observations: (i) Let $\varphi: M \to N$ be a homomorphism of indecomposable $\Lambda$-lattices. If $\mathbb{F}(\varphi) \neq 0$ ($\mathbb{F}$ the functor from (1.7)), then $\varphi$ is irreducible if and only if $\mathbb{F}(\varphi)$ is irreducible. If $\mathbb{F}(\varphi)$ is zero, then $N$ is projective and $M$ is a $\Gamma$-lattice.

(ii) For every irreducible homomorphism $\bar{\varphi}$ in $\mathfrak{S}^o(\mathfrak{D})$ there exists an irreducible homomorphism $\varphi$ of lattices with

$$\mathbb{F}(\varphi) = \bar{\varphi} \text{ . (Note } \mathfrak{S}^o(\mathfrak{D}) = \mathrm{Im}(F) = \mathfrak{S}(\mathfrak{D}) \setminus \{\text{semi simple projectives}\}\text{ .}$$

(iii) The permutation of $\Gamma$. Let $\{X_i\}_{1 \leq i \leq m}$ be the indecomposable $\Gamma$-lattices. Since $\Gamma$ is hereditary, every projective $\Lambda$-lattice $X_i$ is at the same time an injective lattice, it has a unique minimal overlattice $S(X_i)$, which is again a projective $\Gamma$-lattice. Hence

$$S(X_i) = X_{\sigma(i)} ,$$

and the map $\sigma: \{1,2,\ldots,m\} \to \{1,2,\ldots,m\}$ is a permutation, the permutation of $\Gamma$. It has the following interpretation in terms of the Auslander-Reiten quiver of $\Gamma$. Let $\sigma = \sigma_1,\ldots,\sigma_s$ be a decomposition of $\sigma$ into disjoint cycles. Then the Auslander-Reiten quiver of $\Gamma$ is the disjoint union of $s$ oriented cycles

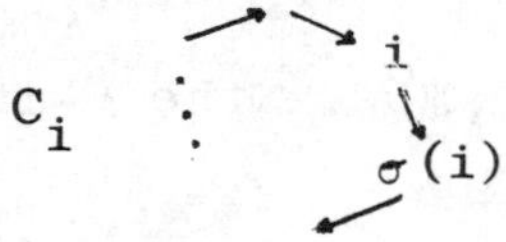


.

(iv) With the notation of (iii) $P_i$ $1 \leq i \leq n$ being the non-isomorphic indecomposable projective $\Lambda$-lattices: There is an irreducible map $X_j \to P_i$ in ${}_\Lambda\mathfrak{M}^o$ if and only if in $\mathfrak{S}(\mathfrak{D})$ there is an irreducible map

$$(X_{\sigma(j)} / \operatorname{rad} \Gamma X_{\sigma(j)})_0 \to \mathbb{F}(P_i) .$$

(3.3) Theorem: Let the notation be as in (3.2). Assume that $\Lambda$ is of finite lattice type.

(i) Then the Auslander-Reiten quiver of $\Lambda$ is obtained from the Auslander-Reiten quiver of $\mathfrak{S}(\mathfrak{D})$ in the following way: The sinks in the Auslander-Reiten quiver of $\mathfrak{S}(\mathfrak{D})$ correspond to the

$\mathrm{ext}_{\mathfrak{C}(\mathfrak{D})}$-injective modules

$$E_j = \begin{pmatrix} X_j/\mathrm{rad}\,\Gamma X_j \\ X_j/\mathrm{rad}\,\Gamma X_j \end{pmatrix},$$

and the sources correspond to the simple projective $\mathfrak{D}$-modules

$$S_j = \begin{pmatrix} X_j/\mathrm{rad}\,\Gamma X_j \\ 0 \end{pmatrix}.$$

We now identify $E_j$ with $S_{\sigma(j)}$ to obtain the Auslander-Reiten quiver of ${}_{\Lambda}\mathfrak{M}^{o}$, and via $\mathbb{F}^{-1}$ we obtain the lattices.

(ii) All indecomposable $\Lambda$-lattices can be constructed from the projective one and from the $\Gamma$-lattices by forming almost split sequences.

(3.1) Example continued: In (3.1) the permutation is (1 2 3 4). So with the above notation of $E_j, S_j$, we have for the Auslander-Reiten quiver of $\mathfrak{C}(\mathfrak{D})$

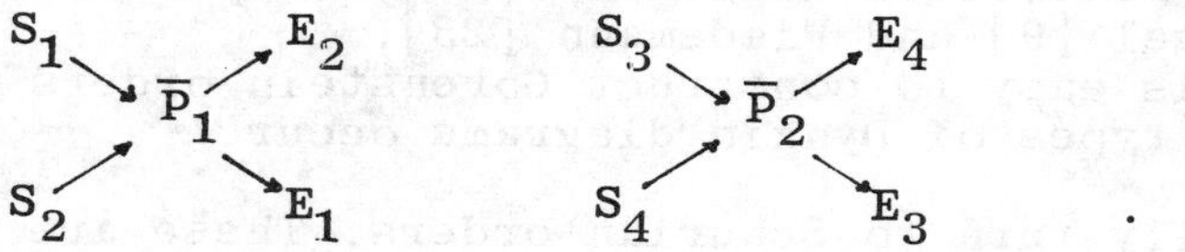


According to the recipe we have to identify $E_2 = S_3$, $E_4 = S_1$, $E_1 = S_2$, $E_3 = S_4$ and obtain - as predicted - the Auslander-Reiten quiver of $\Lambda$.

## § 4 Gorenstein orders and Schurian orders of finite lattice type.

We recall that $\Lambda$ is said to be a Gorenstein order provided $\Lambda$ is an injective $\Lambda$-lattice or equivalently $\mathrm{Hom}_R(\Lambda, R)$ is a projective $\Lambda$-lattice.

(4.1) Proposition [5]: Let $\Lambda$ be a Gorenstein order indecomposable as ring, and assume that one of the following holds.

(i) There exists an indecomposable projective left $\Lambda$-lattice $P$ such that $\mathrm{rad}\, P$ decomposes non-trivially.

(ii) There exists an indecomposable left $\Lambda$-lattice X such that $X \oplus X$ is a direct summand of $\mathrm{rad}_\Lambda P_1 \oplus \mathrm{rad}_\Lambda P_2$, $P_i$ indecomposable projective, $P_1 \not\simeq P_2$.

Then $\Lambda$ is a Bäckström order with associated valued graph a disjoint union of Dynkin graphs of type $A_2$, $A_3$ and $B_2$. (In particular, the Auslander-Reiten quiver of $\Lambda$ is well understood.)

In the next result let $\mathfrak{A}_s(\Lambda)$ be the stable Auslander-Reiten quiver of $\Lambda$.

(4.2) Theorem: Let $\Lambda$ be a Gorenstein order, indecomposable as ring.

(i) $\mathfrak{A}_s(\Lambda)$ has an isolated point if and only if $\Lambda$ is a Bäckström order with graph a disjoint union of $A_3$, $A_2$ and $B_2$.

(ii) Assume that $\mathfrak{A}_s(\Lambda)$ has no isolated point. Then $\Lambda$ is of finite lattice type if and only if the universal cover of $\mathfrak{A}_s(\Lambda)$ is of the form $\mathbb{Z}D$, $D$ a Dynkin diagram in the sense of Riedtmann [14] or $\mathbb{Z}L_t$, where $L_t = \bullet\!-\!\bullet\!-\!\bullet\!-\!\cdots\bullet\!\circlearrowleft$.

Remarks: 1.) This is an easy consequence of (4.1) using the results of Riedtmann [14], Happel-Preiser-Ringel [9] and Wiedemann [23].

2.) It is easy to construct Gorenstein orders such that all types of Dynkin diagrams occur.

We finally turn to Schurian orders. These are basic orders:

$$\Lambda = (\mathfrak{p}^{n_{ij}})_n \subset (R)_n \quad , \quad \mathfrak{p} = \mathrm{rad}\, R \ .$$

(4.3) The partially ordered set of a Schurian order: The vertices of $P(\Lambda)$ are the points $(\nu, i) \in \mathbb{Z} \times \{1, \dots, n\}$, and the partial order is induced by

(i) $(\nu, i) \leq (\nu+1, i)$,

(ii) $(\nu, i) \leq (\nu+n_{ij}, j)$.

We denote this partially ordered set by $P(\Lambda)$.

Let $W$ be the path algebra of $P(\Lambda)$ over $\mathfrak{k}$. Then $W$ is a $\mathfrak{k}$-algebra, whose indecomposable projectives are in bijection to the vertices of $P(\Lambda)$. Let $\mathfrak{C}$ be the category of W-modules which can be embedded into a finite number of indecomposables. Then $\mathfrak{C}$ is the category of torsionfree objects in a suitable torsion theory.

Let R = [[ t ]] .

(4.4) <u>Theorem:</u> Let $\Lambda$ be a basic Schurian order with associated partially ordered set.

(i) [24] $\Lambda$ is of finite lattice type if and only if $P(\Lambda)$ does not contain a critical subset in the sense of [13] (this was not constructive).

(ii) [21] $\mathfrak{C}$ has a simply connected Auslander-Reiten quiver $\mathfrak{A}(\mathfrak{C})$ (infinite) which can algorithmically be constructed if and only if $P(\Lambda)$ does not contain a critical set. $\mathfrak{A}(\mathfrak{C})$ is a covering of the Auslander-Reiten quiver of $\Lambda$, $\mathfrak{A}(\Lambda)$. (This allows to construct the indecomposable $\Lambda$-lattices by an algorithm.

<u>Remark:</u> 1.) To get results in the non-graded case we have to employ model theory [12].

2.) Using techniques developed by Rump [22] one has finite covering techniques in the sense of Bongartz-Gabriel [7] available.

<u>References:</u>

[1] Auslander,M. - S.O.Smalø: "Preprojective modules over artin algebras", Journal of Algebra <u>66</u>, (1980), 61 - 122.

[2] Bäckström,K.J.: "Orders with finitely many indecomposable lattices", Ph.D.thesis, Göteborg (1972).

[3] Bautista,R. - R.Martinez: "Representations of partially ordered sets and 1-Gorenstein Artin algebras", Proc.Conf.Ring Theory Antwerpen 1978, Marcel Dekker (1979).

[4] Bongartz,K. - P.Gabriel: "Covering spaces in representation theory", Invent.math. <u>65</u>, (1982), 331-378.

[5] Bongartz,K. - K.W.Roggenkamp: "Stable Auslander-Reiten quivers of Gorenstein orders", to be published.

[6] Brumer,A.: "Structure of hereditary orders", Bull.Amer.Math.Soc. <u>69</u>, (1963), 721-729.

[7] Gabriel,P.: "The universal cover of a representation-finite algebra", Representations of Algebras, Springer Lecture Notes in Math. 903 (1981).

[8] Green,E.L. - I.Reiner: " Integral representations and diagrams", Mich.J.Math. 25 (1978), 53-84.

[9] Happel,D. - U.Preiser - C.M.Ringel: "Vinberg's characterization of Dynkin diagrams using subadditive functions - with application to DTr-periodic modules", Proc.ICRA II, Ottawa 1979, Springer Lecture Notes in Math. 832 (1980).

[10] Harada,M.: "Structure of hereditary orders over local rings", Osaka J.Math. 14 (1963), 1-22.

[11] Jacobinski,H.: Sur les ordres commutatifs avec un nombre fini de réseaux indécomposable", Acta Math. 118 (1967), 1-31.

[12] Jensen,C.U. - H.Lenzing: "Model theory and representations of algebras", Representation Theory II, Springer Lecture Notes in Math. 832 (1980), 302-308.

[13] Kleiner,M.M.: "On explicit representations of partially ordered sets of finite type", Sem.Leningrad Mat.Inst.Steklov 28 (1972), 32-42.

[14] Riedtmann, Chr.: "Algebren, Darstellungsköcher, Überlagerungen und zurück", Comm.Math.Helv. 55 (1980), 199-224.

[15] Ringel,C.M. - K.W.Roggenkamp: "Diagrammatic methods in representation theory", Journal of Algebra 60 (1979), 11-42.

[16] Ringel,C.M. - K.W.Roggenkamp: "Socle determined categories of representations of artinian hereditary tensor algebras", Journal of Algebra 64, 1 (1980), 249-269.

[17] Roggenkamp,K.W.: "Auslander-Reiten sequences for 'nice' torsion theories of artinian algebras", Canad.Math.Bull. 23 (1) (1980), 61-65.

[18] Roggenkamp,K.W.: "The lattice type of orders I,II", Ring theory Antwerpen 1980, Springer Lecture Notes in Math. 825 (1980), 104-129 and Integral representations and applications, Springer Lecture Notes in Math. 882 (1981), 430-477.

[19] Roggenkamp,K.W.: "Auslander-Reiten species of Bäckström orders",to appear in Journal of Algebra.

[20] Roggenkamp,K.W.: "Auslander-Reiten species for socle determined categories of hereditary algebras and for generalized Bäckström orders" in memoriam Hermann Boerner, Mitteilungen Math. Seminar Giessen, Heft 159 (1983).

[21] Roggenkamp,K.W. - A.Wiedemann: "Auslander-Reiten quivers of Schurian orders", to be published.

[22] Rump,W.: "Systems of lattices in vector spaces and their invariants", Comm.Alg.9 (1981), 893-932.

[23] Wiedemann,A.: "Orders with loops in their Auslander-Reiten graph", Comm.Alg.9 (1981),641-656.

[24] Zavadskij,A.G. - V.V.Kirichenko: "Semimaximal rings of finite type", Mat.Sbornik 103 (145) (1977), 323-345, Math.USSR Sbornik 32 (1977), 273-291.

# AUTOMORPHISMS AND ISOMORPHISMS OF INTEGRAL GROUP RINGS OF FINITE GROUPS

Klaus W.Roggenkamp

Universität Stuttgart (West Germany)

This talk presented a preliminary version of joint work with L.L.Scott on the isomorphism problem for finite groups. Since a detailed version of the results is going to appear in Proceedings Groups - Korea 1983, I shall here only state the main results.

The <u>isomorphism problem asks</u>:

(1) Does $\mathbb{Z}G \simeq \mathbb{Z}H$ as augmented rings imply $G \simeq H$, $G, H$ finite groups.

The <u>Zassenhaus conjecture</u> predicts

(2) $\mathbb{Z}G = \mathbb{Z}H$ augmented, $G,H$ finite implies the existence of a unit $a \in \mathbb{Q}G$ with $aGa^{-1} = H$.

During our struggle with the isomorphism problem we found that some group rings have an even stronger property. But first some <u>notation</u>:
If $G$ is finite we put $\pi(G) = \{p \text{ prime}, p \mid |G|\}$, and for a finite set of primes $\pi$ we let $\mathbb{Z}_\pi$ be the semilocalization of $\mathbb{Z}$ at $\pi$.

We say that a finite group $G$ has the <u>trivial automorphism property</u> TAP, if every normalized automorphism of $\mathbb{Z}_{\pi(G)}G$ is a group automorphism of $G$ followed by conjugation with a unit in $\mathbb{Z}_{\pi(G)}G$. (If $V(\mathbb{Z}_{\pi(G)}G)$ are the normalized units, then TAP is equivalent to maximal finite subgroups of $V(\mathbb{Z}_{\pi(G)}G)$ being conjugate; a result which does not hold in general for local arithmetic groups.)



*F. van Oystaeyen (ed.), Methods in Ring Theory, 451–454.*

The importance of TAP for the isomorphism problem comes from the following

Proposition 1 [2]:

Let $1 \to A \to E \to G \to 1$ be a group extension with $A$ finite abelian and $G$ having TAP. Then the isomorphism problem has a positive solution for $E$.

The main general result we have in this direction is

Theorem I:

Let $G$ be a finite nilpotent group of class 2, then $G$ has TAP.

Remarks: 1.) In view of Proposition 1 this settles the isomorphism problem for nilpotent groups of class at most 5.
2.) There are other special classes of groups - except dihedral 2-groups which were dealt with by [1] and certain metacyclic p-groups [3] - namely $C_p \wr C_p$, and the quaternion 2-groups where we have verified TAP. The proof is based on various reductions. From now on we restrict our attention to p-groups $G$ and put $R = \hat{\mathbb{Z}}_p$, the p-adic integers.

Reduction 1: Once one has TAP for p-groups we can prove TAP for nilpotent groups.

Reduction 2: Let $C = \langle c : c^p = 1\rangle$ be a central subgroup of $G$, and put $\varepsilon = 1 - (\frac{1}{p}\sum_{i=0}^{p-1} c^i)$ and $\Lambda = RG\varepsilon$.
Then $\Lambda$ is an $R[\zeta]$-order, $\zeta$ a primitive p-th root of unity, and we put $\tau = (\zeta - 1)$. To prove TAP it is enough by induction to show.

Let $\alpha$ be a central automorphism of $\Lambda$ with

$$\alpha \equiv id_\Lambda \mod \tau \operatorname{rad} \Lambda .$$

Then $\alpha$ is an inner automorphism.

A careful analysis of $\Lambda$ allows to almost prove that:

Theorem II:

Let $\alpha$ be a central automorphism of $\Lambda$ such that

$$\alpha \equiv \mathrm{id}_\Lambda \mod \tau\ \mathrm{rad}^2\Lambda .$$

Then $\alpha$ is an inner automorphism.

So one needs a device to modify $\alpha \equiv \mathrm{id}_\Lambda \mod \tau\ \mathrm{rad}\,\Lambda$ by an inner automorphism in order to make it trivial modulo $\tau\ \mathrm{rad}^2\Lambda$. A possible reduction comes from the observation, that

$$\mathrm{rad}\,\Lambda/(\tau R + \mathrm{rad}^2\Lambda) \simeq G/\mathrm{Fr}(G)\cdot C$$

and that conjugation by $\alpha$ induces a cocycle $\gamma$. If $\gamma$ were a coboundary we could modify $\alpha$ in the desired term.

Again analyzing carefully the structure of $\Lambda$ one becomes convince the obstruction for $\gamma$ to be inner comes from the following class of groups:

<u>Definition 1</u>:

A finite p-group $G$ is said to be a PC-group (possible counter-example) provided for every central subgroup $C$ of order $p$ there are elements $y_1,\dots,y_n \in G$ such that

(i) $G \neq C_G(y_i) = C_G(y_i C)$,

(ii) there are cocycles $y_i \in H^1(C_G(y_i)/Z, \mathbb{F}_p)$ such that the transferred cocycles

$$\gamma_i\!\uparrow^{G/Z}(y_{i+1}) = 0 ,\quad y_{n+1} = y_1 ,\quad Z = \text{centre of } G.$$

The first group we could find with PC has order $2^{15}$.

Using Theorem II and properties of PC-groups we can prove

<u>Theorem III</u>:

Assume that $G$ is not a PC-group with respect to $C$. If $G/C$ has TAP, so does $G$. Theorem I follows from this by noting that no nilpotent class 2-group is a PC.

References:

[1] Endo,S. - T.Miyata - K.Sekiguchi: "Picard groups and automorphism groups of dihedral group rings of metacyclic groups", J.of Algebra 77 (1982), 286-310.

[2] Gruenberg,K.W. - K.W.Roggenkamp: "Extension categories of groups and modules, I: Essential covers", "..., II: Stem extensions", J.of Algebra 49, Nr.2 (1977), 564-594, 67, Nr.2 (1980), 342-368.

[3] Sekiguchi,K.: "On the automorphism group of the p-adic group ring of a metacyclic p-group", Preprint.

SELF-INJECTIVE DIMENSION OF SERIAL RINGS

Hideo SATO

Wakayama University

Following Eisenbud and Griffith [3,4], we say that a ring R is serial if both $_{R}R$ and $R_{R}$ are direct sums of uniserial modules of finite length. Serial rings were introduced by Nakayama[14], who established the well-known theorem that each module such a ring is a direct sum of factor module of indecomposable projective modules and conversely. With each indecomposable serial ring R, Kupisch [10] associated a series $P_1,\dots,P_n$ of left indecomposable projective modules, which is called a left Kupisch series for R. Let c(i) be the composition length of $P_i$. Then the series (c(1),...,c(n)) is called the left admissible sequence corresponding to a left Kupisch series $P_1,\dots,P_n$.

Fuller [5] showed that the global dimension of serial ring is determined by its left admissible sequence. Under the above notation we let c(R) be the least integer among the set {c(1),...,c(n)}. Murase [11,12,13] called a serial ring R with c(R) = 1 of the first category. It was shown in Murase [11], Eisenbud and Griffith



F. van Oystaeyen (ed.), Methods in Ring Theory, 455–481.

[4] that any serial ring of the first category has a finite global dimension.

A noetherian ring R is called left n-Gorenstein if ${}_R R$ is of finite self-injective dimension at most n. A left and right n-Gorenstein ring is called n-Gorenstein, and Gorenstein means n-Gorenstein for some n. In the case of commutative rings, our definition of Gorenstein rings coincides with that of Gorenstein rings with finite Krull dimension originally defined by Bass [1], who described the distribution of prime ideals in terms of the injective resolution of ${}_R R$.

On the other hand, it is natural to consider that injective indecomposable modules replace prime ideals in the theory of non-commutative rings. From the point of the view above, Iwanaga [7,8] tried to describe the distribution of injective indecomposable modules in the minimal injective resolution of ${}_R R$ especially in the case where R is a serial ring. But the situation is rather complicated even if R is a serial Gorenstein ring. Therefore, at first, we are interested in the self-injective dimension of serial rings, which can be infinite or finite.

Let R be an indecomposable serial ring with left admissible sequence (c(1),...,c(n)). Let n=n(R) and c(R)=min{c(1),...,c(n)}. In a similar manner to Fuller [5], we shall show in §1 that the self-injective dimension of R is determined by (c(1),...,c(n)). We shall show in §2 that the self-injective dimension of a serial ring $\tilde{R}$ with left admissible sequence (c(1)+n,...,c(n)+n) can be calculated by those of projective indecomposable R-modules. Moreover we shall show that gl.dim $R < \infty$ implies $c(R) \leqq n(R)$. In §3 we shall treat serial rings of the first category and give a direct and simple proof of the theorem that the global dimension of such a ring does not exceed n-1.

In 4 we shall treat serial rings with $c(R) \geqq n(R) = n$. At first, with such a given serial ring R we shall associate two oriented graphs G(R) and G*(R) which are called the river and the coriver respectively. Let $P_1, \ldots, P_n$ be a left Kupisch series for such a ring R and let $X_i = P_i/N^n P_i$ where N denotes the radical of R. For a non-injective $X_i$, let $\Omega^2(X_i)$ be the cokernel of $E^0(X_i) \to E^1(X_i)$ where $(E^k(X_i))$ is the minimal injective resolution of $X_i$. Then $\Omega^2(X_i) \cong X_{\sigma(i)}$ for some $1 \leqq \sigma(i) \leqq n$. The river G(R) is defined by letting $i \to \sigma(i)$. The coriver G*(R) is defined dually. In §5 we shall study the properties of G(R) anf G*(R) for the case $c(R) = n(R)$ and establish the transformation theorem between G(R) and G*(R). As their application, we shall show in §6 that there exists an integer s, $0 \leqq s \leqq n-1$, such that $\text{inj.dim}_R R = \text{inj.dim}\, R_R = 2s$ for a serial ring R with $c(R) = n(R) = n$.

We shall present some examples and some remarks in §7. Example 7.2 is found in Iwanaga's thesis [9]. The author wishes to thank Dr. Iwanaga for his generosity of permission to write the example in this paper.

## 1. PRELIMINARIES

In this section we list up notations used throughout this paper and we shall show that the self-injective dimension of a serial ring is determined by its left admissible sequence. A ring R with radical N is an indecomposable serial ring with left admissible sequence (c(1),...,c(n)) corresponding to a left Kupisch series $P_1,\ldots,P_n$. Let n = n(R) and c(R) = min{c(1),...,c(n)}. For our purpose we can assume that R is not quasi-Frobenius and hence c(1) = c(R) and c(n) > c(R) (Cf. [5] and [10]). On the contrary to our assumption that R is indecomposable, we assume that R may be decomposable if c(R) = 1.

Unless otherwise specified, we use the following notations for a left R-module M.

Soc(M) = the socle of M

top(M) = M/NM

pd(M) = the projective dimension of M

id(M) = the injective dimension of M

|M| = the copmosition length of M

P(M) = the projective cover of M

E(M) = the injective hull of M

Following Fuller [5], let $d(k) = |E(top(P_k)|$ and for each integer z we shall let [z] be the least strictly positive remainder of z modulo n.

For an R-module M, we denote its minimal injective resolution by $0 \to M \to (E^k(M))$ and define $\Omega^k(M)$ by letting $\Omega^0(M) = M$ and $\Omega^k(M)$ = Cokernel of $\Omega^{k-1}(M) \to E^{k-1}(M)$ for $k \geq 1$. Dually we denote the minimal projective resolution of M by $(P^k(M)) \to M \to 0$ and define $\Lambda^k(M)$ by letting $\Lambda^0(M) = M$ and $\Lambda^k(M)$ = Kernel of $P^{k-1}(M) \to \Lambda^{k-1}(M)$ for $k \geqq 1$. If M is indecomposable, then it follows from [14, Theorem 17] that all of $E^k(M)$, $\Omega^k(M)$, $P^k(M)$ and $\Lambda^k(M)$ are uniserial or zero. Unless otherwise specified, all modules that we shall consider are (finitely generated) indecomposable. Now let $\omega_k(M) = |\Omega^k(M)|$ and $\lambda_k(M) = |\Lambda^k(M)|$. In the following discussion, we write $\omega_k(M)$ (resp. $\lambda_k(M)$) only when $\omega_{k-1}(M) \neq 0$ (resp. $\lambda_{k-1}(M) \neq 0$), and we say that $\omega_k(M)$ (resp. $\lambda_k(M)$) can be defined.

In a similar manner to [5,Theorem 3.1], we have the following inductively.

<u>Lemma 1.1</u>. For an indecomposable left R-module M, with $\mathrm{top}(M) \cong \mathrm{top}(P_k)$, we have

(1) $\omega_1(M) = d([k+1-\omega_0(M)]) - \omega_0(M)$

$\omega_2(M) = d([k+1]) - \omega_1(M)$

$\omega_j(M) = d([k+1+\omega_1(M)+\cdots+\omega_{j-2}(M)]) - \omega_{j-1}(M) \qquad (j \geqq 3)$

$\mathrm{Soc}(\Omega^1(M)) \cong \mathrm{top}(P_{[k+1]})$

$\mathrm{Soc}(\Omega^j(M)) \cong \mathrm{top}(P_{[k+1+\omega_1(M)+\cdots+\omega_{j-1}(M)]}) \qquad (j \geqq 2)$

(2) $\lambda_1(M) = c(k) - \lambda_0(M)$

$\lambda_j(M) = c([k-\lambda_0(M)-\cdots-\lambda_{j-2}(M)]) - \lambda_{j-1}(M) \qquad (j \geqq 2)$

$\mathrm{top}(\Lambda^j(M)) \cong \mathrm{top}(P_{[k-\lambda_0(M)-\cdots-\lambda_{j-1}(M)]}) \qquad (j \geqq 1)$

<u>Corollary 1.2</u>. Let M be an indecomposable left R-module with $|M| = n$.

(1) If M is not injective, then $E^0(M) \cong E^1(M)$ and $|\Omega^2(M)| = n$.

(2) If M is not projective, then $P^0(M) \cong P^1(M)$ and $|\Lambda^2(M)| = n$.

Proof. Assume that M is not injective and $top(M) \cong top(P_k)$. By Lemma 1.1, we have $Soc(M) \cong Soc(\Omega^1(M)) \cong top(P_{[k+1]})$. Therefore $E^0(M) \cong E^1(M)$ and hence $|\Omega^2(M)| = n$. The second half of our statement can be verified similarly.

Theorem 1.3. For an indecomposable left R-module M, both id(M) and pd(M) are determined only by the left admissible sequence of R, $|M|$ and top(M). In particular, both $id({}_RR)$ and $id(R_R)$ are determined by the left admissible sequence of R.

Proof. It follows from [5,Theorem 2.2] that the right admissible sequence (d(n),...,d(1)) is determined by (c(1),...,c(n)). Therefore our statement follows from Lemma 1.1 and its right hand version.

## 2. PERIODICITY OF SELF-INJECTIVE DIMENSION

In this section we shall show that our problem to calculate self-injective dimension of a serial ring R can be reduced to the case $c(R) \leq n(R)$.

Lemma 2.1. Let R be an indecomposable serial ring with $c(R) > n(R) = n$. Let M be an indecomposable left R-module with M = n. Then

$\mathrm{id}(M) = \infty$ and $\mathrm{pd}(M) = \infty$.

Proof. It follows from [5,Theorem 2.2] that M is not injective. By Corollary 1.2, $\Omega^2(M)$ is not injective and $|\Omega^2(M)| = n$. This shows $\mathrm{id}(M) = \infty$. Similarly we have $\mathrm{pd}(M) = \infty$.

Theorem 2.2. If a serial ring R is of finite global dimension, then $c(R) \leq n(R)$.

Proof. Obvious by Lemma 2.1.

Recall that Kupisch [10] constructed an indecomposable self-basic serial algebra R over a field K whose left admissible sequence coincides with the given one $(c(1),\ldots,c(n))$, as follows: Let $\pi = (n,n-1,\ldots,1)$, the cycle permutation. Let $S = \{a_i^s \mid 1\leq i\leq n, 0\leq s\leq c(i)-1\}\cup\{0\}$. Define the multiplication on S as

$$a_i^s \cdot a_j^k = \delta_{i,\pi^k(j)} \cdot a_j^{s+k} \quad \text{if } s+k < c(j)$$

$$= 0 \quad \text{if } s+k \geq c(j)$$

Then S is easily verified to be a Kupisch semigroup in the terminology of [15]. Let $R = K[S]$, the semigroup algebra of S over K, as is the desired algebra.

In the following discussion, we shall let $\tilde{R}$ be an indecomposable serial ring with left admissible sequence $(c(1)+n,\ldots,c(n)+n)$ whose corresponding left Kupisch series $\tilde{P}_1,\ldots,\tilde{P}_n$.

Lemma 2.3. Let $(\widetilde{d}(n),\ldots,\widetilde{d}(1))$ be the right admissible sequence of $\widetilde{R}$. Then the following statements hold.

(1) $\widetilde{d}(i) = d(i) + n$ for each $1 \leqq i \leqq n$.

(2) If $\omega_1(P_k),\ldots,\omega_p(P_k)$ can be defined, then $\omega_1(\widetilde{P_k}),\ldots,\omega_p(\widetilde{P_k})$ can be defined. Moreover for each $q \leqq p$, we have

$$\begin{aligned} \omega_q(\widetilde{P_k}) &= \omega_q(P_k) && \text{if } q \text{ is odd} \\ &= \omega_q(P_k) + n && \text{if } q \text{ is even.} \end{aligned}$$

Proof. The statement (1) is clear by [5,Theorem 2.2]. The statement (2) is then easily verified inductively.

Theorem 2.4. The following statements hold for each k, $1 \leqq k \leqq n$.

(1) If $id(P_k) = \infty$, then $id(\widetilde{P_k}) = \infty$.

(2) Assume $id(P_k) = m < \infty$. (i) If m is even, then $id(\widetilde{P_k}) = m$. (ii) If m is odd, then $id(\widetilde{P_k}) = \infty$.

Proof. The statement (1) is clear by Lemma 2.3. Assume $id(P_k) = m < \infty$. Let $\omega_j = \omega_j(P_k)$ and $\widetilde{\omega}_j = \omega_j(\widetilde{P_k})$ for each $1 \leqq j \leqq m+1$. Then $\omega_1,\ldots,\omega_m \neq 0$ and $\omega_{m+1} = 0$. These yields $\widetilde{\omega}_1,\ldots,\widetilde{\omega}_m \neq 0$ by Lemma 2.3. If m is even, then $\widetilde{\omega}_{m+1} = \omega_{m+1} = 0$ by Lemma 2.3 and hence $id(\widetilde{P_k}) = m$. If m is odd, then we have $\widetilde{\omega}_{m+1} = n$ by Lemma 2.3. By Lemma 2.1, we have $id(\Omega^{m+1}(\widetilde{P_k})) = \infty$ and hence $id(\widetilde{P_k}) = \infty$.

## 3. SELF-INJECTIVE DIMENSION FOR THE CASE c(R) = 1

Let R be an indecomposable serial ring with c(R) = 1, that is, of the first category in the terminology of Murase [11]. Eisenbud and Griffith [4] applied Chase's result [2] and Murase's result [11] to show that R is of finite global dimension. We give here a prrof for it by showing the existence of upper bound depending only on n(R).

Theorem 3.1. Let R be an indecomposable serial ring with c(R) = 1. Then gl.dim R $\leqq$ n - 1 where n = n(R).

Proof. Let (c(1),...,c(n)) be the left admissible sequence of R and c(1) = 1. Let M be an indecomposable left R-module with top(M) $\cong$ top($P_k$). Define t(M) = k. Let M' be a non-trivial submodule of M. From the assumption c(1) = 1, we have t(X) < t(Y) for any composition factor X of M' and for any composition factor Y of M/M'. If $\Omega^q(M)$ is not injective for each q, $0 \leqq q < p$, then we have an ineqaulity

$$1 \leqq t(M) < t(\Omega^1(M)) < t(\Omega^2(M)) < \cdots < t(\Omega^p(M)) \leqq n.$$

Therefore $\Omega^p(M)$ is injective for some p, $0 \leqq p \leqq n-1$. This shows gl.dim R $\leqq$ n-1. Therefore $id({}_RR) = id(R_R)$ = gl.dim R.

## 4. RIVERS AND CORIVERS

Let R be an indecomposable serial ring with c(R) $\geqq$ n(R) = n and $P_1,\ldots,P_n$ its left Kupisch series. Let $X_k = P_k/N^nP_k$ where N is the radical of R.

Lemma 4.1. (1) $X_k$ is injective if and only if $\Omega^2(X_k) = 0$.

(2) If $X_k$ is not injective, then there exists $1 \leqq \tau(k) \leqq n$ such that $\Omega^2(X_k) \cong X_{\tau(k)}$.

Proof. (1) Assume $\Omega^2(X_k) = 0$. Then we have an exact sequence :

$$0 \to X_k \to E^0(X_k) \to E^1(X_k) \to 0.$$

If $X_k$ is not injective, then $E^1(X_k)$ is nonzero and hence $|E^1(X_k)| \geqq n$ because $c(R) \geqq n$. On the other hand, we have $|E^1(X_k)| = |E^0(X_k)| - |X_k| < n$. This is a contradiction. The converse is trivial. The statement (2) is nothing but Corollary 1.2.

With each indecomposable serial ring R with $c(R) \geqq n(R)$, we associate the following oriented graph G(R), which we call the river of R.

The set of verteces of $G(R) = \{1,...,n\}$ where $n = n(R)$.

The arrow of i to j exsists if and only if $\Omega^2(X_i) \cong X_j$.

A vertex $i \in G(R)$ is said to be an initial point if there does not exist a vertex j and arrow $j \to i$. Dually a final point is defined. We remark that a river has not necessarily any final point. A final point is said to be of height 0, and a non-final point k is said to be of height p if there exist verteces $k = k_0, k_1, ..., k_p$ such that $k_p$ is a final point and there exist arrows $k_{i-1} \to k_i$ for $1 \leqq i \leqq p$. Then we let $h(k) = p$. If each vertex has height, we let $h(R) = \sup_{k \in G(R)} h(k)$, which we call the height of G(R). Otherwise we let $h(R) = \infty$.

Examples. Let R be an indecomposable serial ring with left

admissible sequence $(c(1),\ldots,c(n))$. Then we denote R by

$_RR = (c(1),\ldots,c(n))$.

(1) $_RR = (8,9,10,11,11,10,10,11)$.

$$G(R) \quad : \quad 1 \to 4 \to 6 \to 8 \qquad ; \; h(R) = 3.$$
$$2 \to 5 \qquad 7$$
$$3$$

(2) $_RR = (9,9,10,11,11)$.

$$G(R) : 2 \to 1 \to 5 \leftarrow 4 \leftarrow 3 \qquad ; \; h(R) = \infty.$$

(3) $_RR = (6,6,7,7,7)$.

$$G(R) : 2 \to 4 \to 5 \to 1 \qquad ; \; h(R) = \infty.$$
$$3$$

(4) $_RR = (6,7,8,6,7,8)$.

$$G(R) : 1 \to 3 \leftarrow 2 \; , \; 4 \to 6 \leftarrow 5 \; ; \; h(R) = 1.$$

Lemma 4.2. The following conditions are equivalent for $X_k$.

 (1) There exists $X_h$ such that $\Omega^2(X_h) \cong X_k$.

 (2) The projective cover of $X_k$ is injective.

Proof. Assume $\Omega^2(X_h) \cong X_k$. By definition, we have a canonical exact

sequence : $0 \to X_h \to E^0(X_h) \to E^1(X_h) \to X_k \to 0$. Let $\pi:P(X_k) \to X_k$

be the projective cover. Then there exists a map g of $P(X_k)$ into

$E^1(X_h)$ which makes the following diagram commutative.

$$P(X_k)$$
$$g \swarrow \qquad \downarrow \pi$$
$$E^1(X_h) \to X_k \to 0$$

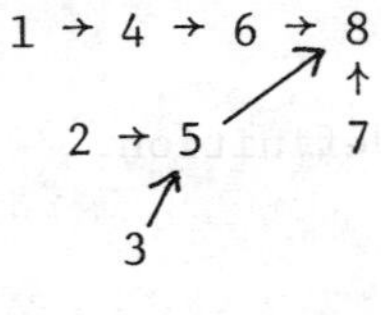

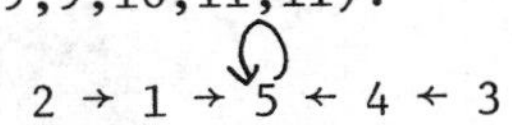

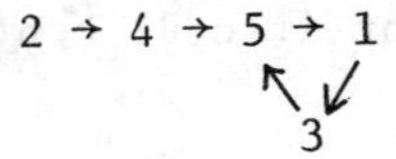

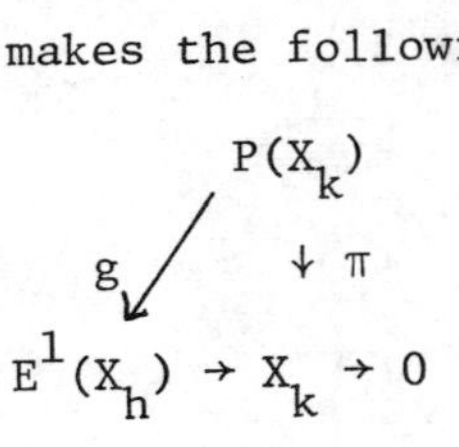

Since $E^1(X_h)$ is uniserial, g is an epimorphism. Since R is serial, $P(X_k)$ must be injective. Conversely assume that $P(X_k)$ is injective. Then $\mathrm{top}(P(X_k)) \cong \mathrm{top}(\ker\pi)$ because $|X_k| = n$. Therefore we have the following exact sequence :

$$0 \to Y \to P(X_k) \to P(X_k) \to X_k \to 0.$$

Then we have $|Y| = n$ and $\Omega^2(Y) \cong X_k$ by definition.

The following is trivial.

<u>Lemma 4.3</u>. A vertex $k \in G(R)$ is a final point if and only if $X_k$ is injective. Hence G(R) has no final point if $c(R) > n(R)$.

We can easily show the dual statements of Lemmas 4.1 and 4.2. Therefore, for an indecomposable serial ring R with $c(R) \geqq n(R) = n$, we can obtain the dual notion of the river G(R), which we call the coriver of R and denote by G*(R). More precisely,

The set of verteces of G*(R) = {1,...,n}.

The arrow of j to i exists if and only if $\Lambda^2(X_i) \cong X_j$.

Also we define initial points and final points in G*(R) similarly, and we define coheights h*(k) and h*(R) dually.

<u>Example</u>. Let ${}_RR = (8,9,10,11,11,10,10,11)$. Then

```
G*(R) :      2 → 5 → 7        ; h*(R) = 3
            ↗         ↘
          1 → 3         8
            ↘
             4 → 6
```

## 5. RIVERS AND CORIVERS FOR THE CASE $c(R) = n(R)$

Let R be an indecomposable serial ring with $c(R) = n(R)$. Then we can obtain precise informations about the river $G(R)$ and the coriver $G^*(R)$. We retain the notations in the preceding sections. In the sequel we assume $|P_1| = n = n(R)$. We begin with the river.

Lemma 5.1. If $X_k$ is not injective, then we have

$$k < \tau(k) = k + d([k+1]) - n.$$

Proof. Since $Soc(X_k) \cong top(P_{[k+1]})$, it follows from the assumption that $d([k+1]) = |E(X_k)| > |X_k| = n$. Hence $d([k+1]) - n \geqq 1$. Since $|P_1| = n$, we have $d([k+1]) \leqq 2n - k$ for each $1 \leqq k \leqq n$. Therefore $2 \leqq k + d([k+1]) - n \leqq n$. Since $E^0(X_k) \cong E^1(X_k)$ and $X_{\tau(k)}$ is a factor module of $E^1(X_k)$, we have $\tau(k) = [k + d([k+1])]$ and hence $\tau(k) = k + d([k+1]) - n$. Thus $\tau(k) - k = d([k+1]) - n > 0$, as is the desired conclusion.

Lemma 5.2. If neither $X_k$ nor $X_{k'}$ is injective, then $k < k'$ implies $\tau(k) \leqq \tau(k')$.

Proof. Let $k' = k + h$ for $1 \leqq h \leqq n - 1$. From the above lemma, we have $\tau(k') - \tau(k) = h + d([k'+1]) - d([k+1])$. On the other hand, we have $e_p N^h \neq 0$ for each p because $1 \leqq h \leqq n - 1$ and $c(R) = n$. Hence we have $d([k'+1]) = d([k+1+h]) \geqq d([k+1]) - h$ by [5,Lemma (2.1)'].

This shows $\tau(k') \geqq \tau(k)$.

From the preceding lemmas, we have immediately

Proposition 5.3.

(1) $G(R)$ contains no oriented cycle.

(2) Each connected component contains exactly one final point.

(3) Each connected component contains one point $k$ such that $X_k$ is projective, and then $k$ is necessarily an initial point.

(4) The maximal height in each component is given by an initial point $k$ such that $X_k$ is projective.

(5) If $k$ and $p$ such that $X_k$ is injective and $X_p$ is projective belong to the same connected component, then $p \leqq k$ and each $j$ $(p \leqq j \leqq k)$ belongs to that component.

(6) The number of connected components is equal to that of the final points.

Corollary 5.4. The following statements hold for an indecomposable serial ring $R$ with $c(R) = n(R)$.

(1) $id(X_k) = 2 \cdot h(k)$ for each $k$.

(2) $h(R) \leqq n(R) - 1$.

(3) $\sup\{id(X_k)\} = 2 \cdot h(R)$.

Next we treat the coriver.

Lemma 5.1'. If $X_k$ is not projective, then we have

$k > \sigma(k) = k - c(k) + n.$

Proof. We have easily $\sigma(k) = [k - c(k)]$. By assumption, we have $|P(X_k)| > |X_k|$ and hence $c(k) > n$. This shows $k - c(k) + n < k$. Since we assume $|P_1| = n$, we have $c(k) \leqq n + (k-1)$ for $1 \leqq k \leqq n$. This shows $1 \leqq k - c(k) + n$. Therefore we have $\sigma(k) = k - c(k) + n$ and hence $\sigma(k) < k$.

A dual argument to Lemma 5.2 shows

Lemma 5.2'. If neither $X_k$ nor $X_{k'}$ is projective, then $k < k'$ implies $\sigma(k) \leqq \sigma(k')$.

Therefore we can obtain the dual statements of Proposition 5.3 and Lemma 5.4, which are omitted. Now we can directly obtain the coriver $G^*(R)$ from the river $G(R)$ and conversely.

Theorem 5.5. (Transformation Theorem)

(1) Assume that k is not an initial point in G(R). Let $p_1 < \cdots < p_t$ be all the points such that $\tau(p_1) = \cdots = \tau(p_t) = k$. Then $\sigma(k) = p_1$.

(2) Assume that k is an initial point in G(R) such that $X_k$ is not projective. Then there exists at least one point k' (> k) which is not initial point in G(R). Let p be the least number among them. Then $\sigma(k) = \sigma(p)$.

<u>Proof</u>. Let $k = \tau(p)$. Then $X_k$ is not projective and $X_p$ is not injective. By Lemma 5.1 we have $p < n$. On the other hand, $P(X_k)$ is injective by Lemma 4.2. Now we can find maps $\phi = \phi_0, \phi_1, \phi_2$ which make the following diagram commutative.

$$\begin{array}{ccccccccc} 0 \to & X_{\sigma(k)} & \to & P(X_k) & \to & P(X_k) & \to & X_k & \to 0 \quad \text{(exact)} \\ & \downarrow \phi_2 & & \downarrow \phi_1 & & \downarrow \phi_0 & & \| & \\ 0 \to & X_p & \to & E(X_p) & \to & E(X_p) & \to & X_k & \to 0 \quad \text{(exact)} \end{array}$$

As is clearly seen, $\sigma(k) < \tau(\sigma(k)) = k$ and $\phi$ is an epimorphism. Therefore there exists $p_i$ such that $\sigma(k) = p_i$. On the other hand, $\mathrm{Soc}(E(X_p)) \cong \mathrm{Soc}(X_p) \cong \mathrm{top}(P_{p+1})$ and $\mathrm{Soc}(P(X_k)) \cong \mathrm{Soc}(X_{\sigma(k)}) \cong \mathrm{top}(P_{\sigma(k)+1})$. Since $|P_1| = n$ by assumption, we have $2n - 1 \geqq |P(X_k)| \geqq |E(X_p)| > n$ and hence $\sigma(k) + 1 \leqq p + 1$. This shows $\sigma(k) \leqq p_1$ and hence $\sigma(k) = p_1$.

Next, let k be an initial point in G(R) such that $X_k$ is not projective. Suppose that $P_k = P(X_k)$ is injective. Then we have $\tau(\sigma(k)) = k$, which is a contradiction because k is an initial point. Thus $P_k$ is not injective. Now $E(P_k)$ is projective and let $P_p \cong E(P_k)$. Then $p = \tau(\sigma(p))$ and hence p is not an initial point. Also $p > k$ is clear. $X_{\sigma(p)}$ is the submodule of $P_p$ of length n, while $X_{\sigma(k)}$ is the submodule of $P_k$ of length n. Since $P_p$ is uniserial and $P_k \subset P_p$, we have $\sigma(p) = \sigma(k)$. Take q such that $k < q < p$. Since $E(P_k) = P_p$, we have $E(P_q) = P_p$ and hence $P_q$ is not injective. By Lemma 4.2, q is an initial point. This completes the proof.

The following is clear by Proposition 5.3 and its dual.

Lemma 5.5. Let $\{k,\ldots,p\}$ be a subset of $\{1,\ldots,n\}$. Then it forms a connected component of $G(R)$ if and only if it forms a connected component of $G^*(R)$.

Theorem 5.6. Let $R$ be an indecomposable serial ring. Then we have

(1) $h(R) = h^*(R)$.

(2) $\sup\{\mathrm{id}(X_k)\} = \sup\{\mathrm{pd}(X_k)\}$.

Proof. The statement (2) is an immediate consequence of the statement (1) by Corollary 5.4 and its dual. We have only to show $h^*(R) \leqq h(R)$ because the inverse inequality is its dual. Without loss of generality, we can assume that $G(R)$ (and hence $G^*(R)$) is connected by Lemma 5.5. Then $X_1$ is the only projective module and $X_n$ is the only injective module in the set $\{X_1,\ldots,X_n\}$. By Lemma 5.2 and its dual, we have $h(R) = h(1)$ and $h^*(R) = h^*(n)$. Let $h = h(R)$ and $h^* = h^*(R)$. By Proposition 5.3 and its dual, we have $\tau^h(1) = n$ and $1 = \sigma^{h^*}(n)$. In order to show $h^* \leqq h$, it is sufficient to show that $1 \neq j \leqq i = \tau(k)$ implies $\sigma(j) \leqq k$. If $j = i$, then we have $\sigma(j) \leqq k$ by Theorem 5.5. If $j \leqq k < i$, then $j > \sigma(j)$ by Lemma 5.1' and hence $\sigma(j) < k$. Assume $k < j < i$. Then we have $n + \ell - 1 \geqq c(\ell) \geqq n + (\ell - k)$ for each $\ell$, $k$ $\ell$ $i$. Therefore $\sigma(j) = j - (c(j)-n) \leqq k$.

## 6. SELF-INJECTIVE DIMENSION FOR THE CASE $c(R) = n(R)$

Throughout this section, $R$ is assumed to be an indecomposable serial ring with $c(R) = n(R) = n$.

Lemma 6.1. Let X and Y be nonzero indecomposable R-modules, both of which are non-injective. Consider the following commutative diagram with exact rows.

$$\begin{array}{lccccccccc} & 0 & \to & Y & \to & E(Y) & \to & \Omega(Y) & \to & 0 \\ (Y,X:\eta,\varepsilon,\eta') : & & & \downarrow\eta & & \downarrow\varepsilon & & \downarrow\eta' & & \\ & 0 & \to & X & \to & E(X) & \to & \Omega(X) & \to & 0 \end{array}$$

(1) If $\eta$ is a monomorphism, then $\eta'$ is an epimorphism.

(2) If $\eta$ is an epimorphism, then $\eta'$ is a monomorphism.

Proof. (1) It is clear that $\varepsilon$ is an isomorphism. Therefore $\eta'$ is an epimorphism.

(2) By the snake lemma, $0 \to \operatorname{Ker}\eta \to \operatorname{Ker}\varepsilon \to \operatorname{Ker}\eta' \to 0$ is an exact sequence. Since E(X) is uniserial, we have $|\operatorname{Im}(\varepsilon)| - |\operatorname{Im}(\eta)| = |E(Y)| - |Y|$ and hence $|Y| - |\operatorname{Im}(\eta)| = |E(Y)| - |\operatorname{Im}(\varepsilon)|$. Hence we have $|\operatorname{Ker}(\eta)| = |\operatorname{Ker}(\varepsilon)|$. This shows $\operatorname{Ker}(\eta') = 0$.

Corollary 6.2. Let Y be an indecomposable non-injective R-module with $|Y| \geqq n$. Then $0 < |\Omega^1(Y)| < n$ and $|\Omega^2(Y)| \geqq n$.

Proof. Let $X = Y/N^nY$ and let $\eta_0$ be the canonical map of Y onto X. Then X is also non-injective, and $|X| = n$. Let $(E^i(Y))$ and $(E^i(X))$ be the minimal injective resolutions of Y and X respectively. Let $(\varepsilon_i)$ be any complex morphism of $(E^i(Y))$ into $(E^i(X))$ over $\eta_0$. For each $i \geqq 1$, let $\eta_i$ be the map of $\Omega^i(Y)$ into $\Omega^i(X)$ induced by $(\varepsilon_i)$. Applying Lemma 6.1 to the diagram $(Y,X;\eta_0,\varepsilon_0,\eta_1)$, we see that $\eta_1$ is a monomorphism. Therefore $|\Omega^1(Y)| \leqq |\Omega^1(X)| = |E^0(X)| - |X| < n$. Since Y is non-injective, we have $0 < |\Omega^1(Y)| < n$. Therefore both $\Omega^1(Y)$ and $\Omega^1(X)$ are non-injective. Hence we can apply Lemma 6.1 to the diagram $(\Omega^1(Y),\Omega^1(X);\eta_1,\varepsilon_1,\eta_2)$ and we see that $\eta_2$ is an epimorphism. Hence we have $|\Omega^2(Y)| \geqq |\Omega^2(X)| = n$ by Corollary 1.2.

Theorem 6.3. Let R be an indecomposable serial ring with $c(R) = n(R) = n$, and Y an indecomposable R-module with $|Y| \geqq n$. Let $X_k = Y/N^nY$. Then there exists an integer s such that $id(Y) = 2 \cdot s$ and $0 \leqq s \leqq h(k)$.

Proof. We have shown $id(X_k) = 2 \cdot h(k)$ in Corollary 5.4. Let $h = h(k)$. Then $\Omega^{2h}(X_k)$ is injective. Assume $id(Y) > 2(h-1)$. Then $\Omega^{2(h-1)}(Y)$ is nonzero noninjective. Applying Lemma 6.1 repeatedly, we see that $\eta_{2h} : \Omega^{2h}(Y) \to \Omega^{2h}(X)$ is an epimorphism. Hence $\Omega^{2h}(Y)$ is injective and hence $id(Y) = 2h$. Therefore we see $id(Y) \leqq 2h$ in general. Furthermore it follows from Corollary 6.2 that id(Y) is not odd. This completes the proof.

We can easily prove the dual statement of Lemma 6.1. Thus we can obtain the followings.

Corollary 6.2'. Let Y be an indecomposable non-injective R-module with $|Y| \geqq n$. Then $0 < |\Omega^1(Y)| < n$ and $|\Omega^2(Y)| \geqq n$.

Theorem 6.3'. Let R be an indecomposable serial ring with $c(R) = n(R) = n$, and Y an indecomposable R-module with $|Y| \geqq n$. Let $X_k$ be the submodule of Y with length n. Then there exists an integer s such that $pd(Y) = 2\cdot s$ and $0 \leqq s \leqq h^*(k)$.

Theorem 6.4. Let R be an indecomposable serial ring with $c(R) = n(R) = n$. Then $id(_RR) = id(R_R) = 2\cdot h(R)$.

Proof. By Theorem 6.3 and Corollary 5.4, we have $id(P_k) \leqq id(X_k) = 2\cdot h(k) \leqq 2\cdot h(R)$. On the other hand it follows from Proposition 5.3 that there exists a point $k \in G(R)$ such that $X_k$ is projective and $h(k) = h(R)$. This shows $id(_RR) = 2\cdot h(R)$. Since $Soc(X_i) \cong top(P_{[i+1]})$ for each i, $\{E(X_1),\ldots,E(X_n)\}$ forms a complete set of non-isomorphic injective indecomposable modules. By the duals of Theorem 6.3 and Corollary 5.4, we have $pd(E(X_k)) \leqq pd(X_k) = 2\cdot h^*(k) \leqq 2\cdot h^*(R)$. On the other hand, it follows from the dual of Proposition 5.3 that there exists a point $k \in G^*(R)$ such that $X_k$ is injective and $h^*(k) = h^*(R)$. This shows $\sup\{pd(E(X_k))\} = 2\cdot h^*(R)$. By [8,Propostion 1], we have $id(R_R) = 2\cdot h^*(R)$. Thus we have $id(_RR) = id(R_R) = 2\cdot h(R)$ by Theorem 5.6.

We are interested in the last term of the minimal injective resolution of $_RR$ for an indecomposable serial ring R.

Theorem 6.5. Let R be an indecomposable serial ring with c(R) = n(R) = n. Let $id({}_RR) = 2h$. Then an indecomposable injective R-module Y is a direct summand of $E^{2h}({}_RR)$ if and only if $|Y| = n$ and $pd(Y) = 2h$.

Proof. Since $id(P_k) \leqq id(X_k)$ by Theorem 6.3, each direct summand Y of $E^{2h}({}_RR)$ is isomorphic to $\Omega^{2h}(P_k)$ for some k such that $h(k) = h = h(R)$. Let G' be the connected component which contains k, and $p \in G'$ the point such that $X_p$ is projective (see Proposition 5.3). Then $k \geqq p$ and k is an initial point. Furthermore any point k', $p \leqq k' \leqq k$, is also an initial point by Lemma 5.2 and by the assumption $h(k) = h$. Therefore $X_p = P_p \subset P_k$. Then $\Omega^{2h}(X_p)$ is a nonzero injective module of length n. Applying Lemma 6.1 repeatedly, we see that a naturally induced map of $\Omega^{2h}(X_p)$ into $\Omega^{2h}(P_k)$ is a monomorphism. Since $\Omega^{2h}(P_k)$ is indecomposable, we have $\Omega^{2h}(X_p) \cong \Omega^{2h}(P_k) \cong Y$ and hence $|Y| = n$. Since $h(R) = h^*(R)$ by Theorem 5.6, we have $pd(Y) = 2h$ from Lemma 5.5, Proposition 5.3 and its dual. The converse is also deduced from Lemma 5.5, Proposition 5.3 and its dual.

## 7. EXAMPLES AND REMARK

Remark 7.1. We have shown in Theorem 6.3 that $id(P_k) \leqq id(X_k)$ for an indecomposable serial ring R with c(R) = n(R). The following example shows that $id(P_k) < id(X_k)$ happens.

Let ${}_RR = (8,9,10,11,11,9,10,11)$. Then $id(P_2) = 2 < 4 = id(X_2)$.

For an indecomposable serial ring R with typical left admissible sequence, we can calculate $id({}_RR)$ explicitly because we can easily calculate its right admissible sequence. We give here two typical examples.

Example 7.2. (Iwanaga [9]) Let R be an indecomposable serial ring with left admissible sequence $(c,c+1,\dots,c+(n-1))$. Then $id({}_RR) < \infty$ if and only if $c < n$ or $c \equiv 0$ (mod.n). If the conditions above are satisfied, then $id({}_RR) \leqq 2$.

Proof. By Theorem 2.4, we have only to calculate $id(P_k)$ for $c \leqq n$. So we assume $c \leqq n$. Let $P_1,\dots,P_n$ be the corresponding left Kupisch series. Then $P_n$ is the only one projective injective indecomposable module.

Case I. $c = n$. In this case, the river G(R) is the following by the above remark and Proposition 5.3.

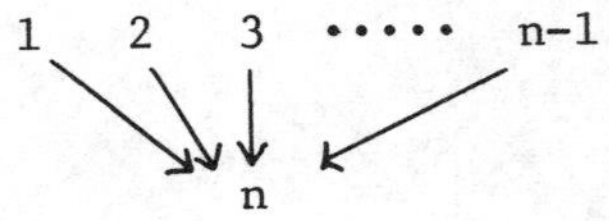


; $h(R) = 1$

By Theorem 6.3 we have $id(P_k) = 0$ or 2. Therefore $id(P_k) = 2$ if $1 \leqq k \leqq n - 1$, and $id(P_n) = 0$.

Case II. $c < n$. Since $Soc(P_1) \cong top(P_{[2-c]})$ and $E(P_1) \cong P_n$, we have $d([2-c]) = c + n - 1$. By [5,Theorem 2.2] and by the assumption $c < n$, we can obtain the following formulae easily.

$d([k+1]) = n - k$ if $k \leqq n - c$

$= 2n - k$ if $k > n - c$, for $1 \leqq k \leqq n$.

By Lemma 1.1, we have

$\omega_1(P_k) = 0$ if $k = n$

$\omega_1(P_k) = d([k+1-c(k)]) - c(k) = n - k > 0$ if $k < n$.

Let $1 \leqq k < n$. Then we have

$\omega_2(P_k) = d([k+1]) - \omega_1(P_k) = 0$ if $k \leqq n - c$

$= n$ if $k > n - c$.

Let $n - c < k < n$. Then we have

$\omega_3(P_k) = d([k+1+\omega_1(P_k)]) - \omega_2(P_k) = 0.$

Therefore $id(P_k) = 1$ if $1 \leqq k \leqq n - c$

$= 2$ if $n - c < k < n$

$= 0$ if $k = n$.

Conclusion If $c = 1$, then $id(_RR) = 1 = gl.dim\ R$ (Cf. Theorem 3.1). If $1 < c < n$, then $id(_RR) = 2$ and $id(P_1) = 1$. If $c = n$, then $id(_RR) = id(P_k) = 2$ for all k but n. This completes the proof.

Example 7.3. Let $H = H(c,n)$ be an indecomposable serial ring with left admissible sequence $(c,c+1,\ldots,c+1)$ and $n(H) = n$ ($c+1$ occurs $(n-1)$ times in the admissible sequence). Then the following statements hold.

I. If $c = 1$, then $gl.dim\ H = n - 1$.

II. Assume $c$ 1. Then the following conditions are equivalent.

(1) $id(_HH) < \infty$. (2) $id(H_H) < \infty$. (3) $(c+1,n) = 1$.

As such is the case, we have $id(_HH) = id(H_H) = 2h$ where h is the least integer such that $0 \leqq h \leqq n - 1$ and $1 + h(c+1) \equiv 0$ (mod.n).

Proof. The assertion I is immediate from Theorem 3.1. To prove the assertion II, we can assume $2 \leqq c \leqq n + 1$ by Theorem 2.4. In the following discussion, we shall let $P_1, \ldots, P_n$ be a corresponding left Kupisch series for H.

(i) The case $c = n + 1$. Let $P'_1, \ldots, P'_n$ be a left Kupisch series for H(1,n) with $|P'_1| = 1$ and $|P'_2| = \cdots = |P'_n| = 2$. By I, we have $\mathrm{id}(P'_1) = n - 1$. By Theorem 2.4, we have $\mathrm{id}(P_1) = n - 1$ if n is odd, and $\mathrm{id}(P_1) = \infty$ if n is even. As clearly seen, the equation $1 + h(c+1) \equiv 0$ (n) has a solution if and only if n is odd. If $n = 2m - 1$, then $h = m - 1$ is the desired solution for the equation. Then $2h = 2m - 2 = n - 1 = \mathrm{id}({}_H H)$.

(ii) The case $c = n$. Then we have $(c+1,n) = 1$. It is clear that the river G(H(c,n)) is the following.

$$1 \to 2 \to 3 \to \cdots \to n$$

Therefore we have $\mathrm{id}({}_H H) = 2(n-1)$ by Theorem 5.4. As is clearly seen, $h = n - 1$ is the desired solution.

(iii) The case $1 < c < n$. Let $X_k = P_k/N^c P_k$ and $X'_k = P_k/N^{c-1}P_k$. Then the set $\{P_2, \ldots, P_n, X_n\}$ is a complete set of non-isomorphic injective indecomposable modules and $\mathrm{Soc}(X_n) \cong \mathrm{top}(P_{[1-c]})$. We can show the next sublemma easily.

Sublemma. Let $X_k$ be non-injective. If $[k+1] \neq [1-c]$, then $\Omega^2(X_k)$ $X_{\rho(k)}$ where $\rho(k) = [k+(c+1)]$.

(iii)-1) The case $(c+1,n) = 1$. Let h be the least integer such that $0 \leq h \leq n-1$ and $1 + h(c+1) \equiv 0\ (n)$. Then we have $[\rho^q(1)+1] \neq [1-c]$ for any $0 \leq q < h$. By making use of Sublemma repeatedly, we have $\Omega^{2h}(X_1) \cong X_{\rho^h(1)} = X_n$. This shows $\mathrm{id}({}_H H) = 2h$.

(iii)-2) The case $(c+1,n) \neq 1$. If $[\rho^p(1)+1] \neq [1-c]$ for any $p \geq o$, then it follows from Sublemma that $\Omega^{2q}(X_1) \cong X_{\rho^q(1)}$ for any $q \geq 0$. Since $(c+1,n) \neq 1$, we see $\rho^q(1) \neq n$ for any $q \geq 0$. Hence $\mathrm{id}({}_H H) = \mathrm{id}(P_1) = \infty$ in this case. Finally assume that there exists an integer $p \geq 0$ such that $[\rho^p(1)+1] = [1-c]$. Let p be the least integer satisfying the above condition. Let $k^+ = [k+1]$ and $k^- = [k-1]$. Then we have the injective corepresentations;

$$0 \to X_1 \to P_2 \to P_{\rho(1)} \to X_{\rho(1)} \to 0$$

$$0 \to X_{\rho(1)} \to [P_{\rho(1)}+] \to P_{\rho^2(1)} \to X_{\rho^2(1)} \to 0$$

$$\cdots\cdots\cdots\cdots\cdots\cdots\cdots\cdots\cdots\cdots\cdots\cdots$$

$$0 \to X_{\rho^{p-1}(1)} \to [P_{\rho^{p-1}(1)}+] \to P_{\rho^p(1)} \to X_{\rho^p(1)} \to 0$$

$$0 \to X_{\rho^p(1)} \to [P_{\rho^p(1)}+] \to X_n \to X'_n \to 0$$

Let q be any integer such that $0 \leq q \leq p$. Suppose $E(X'_{\rho^q(1)}) \cong X_n$. Then we have $\rho^q(1) = n$, which contradicts the assumption $(c+1,n) \neq 1$. Therefore we have $E(X'_{\rho^q(1)}-) \cong P_{\rho^q(1)}+$. For any integer q, $0 \leq q < p$, we have $\rho^q(1) \neq [1-c]$ by the choice of p. Therefore we have $E^1(X'_{\rho^q(1)}-) \cong E(\mathrm{top}(P_{\rho^q(1)})) \cong P_{\rho^{q+1}(1)}-$. Also we have $E^1(X'_{\rho^p(1)}-) \cong E(\mathrm{top}(P_{\rho^q(1)})) \cong E(\mathrm{top}(P_{\rho^{q+1}(1)}-)) \cong P_n$. Therefore we have the injective corepresentations;

$$0 \to X'_n \to P_2 \to [P_{\rho(1)}-] \to [X'_{\rho(1)}-] \to 0$$

$$0 \to [X'_{\rho(1)}-] \to [P_{\rho(1)}+] \to [P_{\rho^2(1)}-] \to [X_{\rho^2(1)}-] \to 0$$

..........................................

$$0 \to [X'_{\rho}p{-}1_{(1)}-] \to [P_{\rho}p{-}1_{(1)}+] \to [P_{\rho}p_{(1)}-] \to [X'_{\rho}p_{(1)}-] \to 0$$

$$0 \to [X'_{\rho}p_{(1)}-] \to [P_{\rho}p_{(1)}+] \to P_n \to X'_n \to 0$$

This shows $id(P_1) = \infty$ and so $id({}_HH) = \infty$. This completes the proof of our assertion II because the admissible sequence of $H(c,n)$ is also $(c,c+1,\ldots,c+1)$.

## REFERENCES


[1] H. Bass, On the ubiquity of Gorenstein rings, Math. Z. 82 (1963), 8-28.

[2] S. U. Chase, A characterization of the triangular matrices, Nagoya Math. J. 18 (1961), 13-25.

[3] D. Eisenbud and P. Griffith, Serial rings, J. Algebra 17 (1971), 389-400.

[4] D. Eisenbud and P. Griffith, The structure of serial rings, Pacific J. Math. 36 (1971), 109-121.

[5] K. R. Fuller, Generalized uniserial rings and their Kupisch series, Math. Z. 106 (1968), 248-260.

[6] Y. Iwanaga, On the rings with self-injective dimension $\leq 1$, Osaka J. Math. 15 (1978), 33-46.

[7] Y. Iwanaga, On rings with finite self-injective dimension, Comm. Algebra 7 (1979), 394-414.

[8] Y. Iwanaga, On rings with finite self-injective dimension II, Tsukuba J. Math. 4 (1980), 107-113.

[9] Y. Iwanaga, On rings with finite self-injective dimension,

(Thesis), University of Tsukuba, 1979.

[10] H. Kupisch, Beitrage zur Theorie nichthalbeinfacher Ringe mit Minimal Bedingung, J. Reine Angew. Math. 201 (1959), 100-112.

[11] I. Murase, On the structure of generalized uniserial rings I, Coll. of Gen. Ed. University of Tokyo 13 (1963), 1-22.

[12] I. Murase, On the structure of generalized uniserial rings II, Coll. of Gen. Ed. University of Tokyo 13 (1963), 131-158.

[13] I. Murase, On the structure of generalized uniserial rings III, Coll. of Gen. Ed. University of Tokyo 14 (1964), 12-25.

[14] T. Nakayama, On Frobeniusean algebras II, Ann. of Math. 42 (1941), 1-21.

[15] H. Sato, Gorenstein rings with semigroup bases, to appear in J. Algebra.

# Smooth Affine PI Algebras

*William F.Schelter**
University of Texas
Austin, Texas 78712

## 1. Introduction

We wish to discuss a certain class of algebras, finitely generated as algebras over an algebraically closed field $k$,and satisfying the identities of n by n matrices. The free algebra on m generators satisfying these identities we denote by $k<X,m,n>$; it is usually called the algebra of generic matrices. It has many good properties which distinguish it from one of its random homomorphic images. Of course it has the freeness property. Another property is that all points of its *Spec* are accessible from $Spec_n$ by curves. In the commutative case, for n equal to 1, a homomorphic image is called smooth if locally it spectrum looks like $k^s$, that is to say like the *Spec* of the polynomial ring in 5 variables. More precisely, upon completion at a point it should become power series in $s$ variables. In the noncommutative case for $n = 2$, the dimension of the $2 \times 2$ algebra of generic matrices is 5. Thus it is impossible to say that a noncommutative homomorphic image of say dimension 1 or 2 should look like the *Spec* of the 'polynomials' in 1 or 2 variables.

Nevertheless such a 1 or 2 dimensional algebra might have some sort of freeness: for such an algebra $R$ we say it is ***smooth***, if for any

$$\begin{array}{ccc} R & \simeq & R \\ \tilde{f}\downarrow & & \downarrow f \\ \tilde{S} & \xrightarrow{g} & S \end{array} \qquad (1.1)$$

$f$ and for any $g$ with nilpotent kernel we can find an $\tilde{f}$ to complete the commutative diagram (1.1). We also restrict the $S$ and $\tilde{S}$ to have ***pidegree*** no greater than that of R. We need the restriction on ***pidegree*** or we would not have the polynomial ring $k[x,y]$ being smooth. The polynomial ring $k[x]$ would however be smooth in any event, since we can always lift maps from it.

*The author would like to thank the National Science Foundation for its support and also to thank M. Artin for many helpful discussions.



*F. van Oystaeyen (ed.), Methods in Ring Theory, 483–488.*

## 2. Some Elementary Properties of Smoothness

In the commutative case, the equivalence of smoothness and the above condition is due to Grothendieck (see [5], [3] or [7] Theorem 28M ). The nilpotent condition allows one to obtain infinitesimal information about the behaviour of $R$. In the noncommutative case clearly the generic matrix algebras are smooth. We wish to describe two instances when we can verify smoothness. These are in the two basic cases of $R$ being Azumaya or of it being one dimensional. The latter case will be dealt with in the next section.

If $R$ is Azumaya one would want it to be smooth precisely when the centre is. We first prove the following lemma.

**LEMMA (2.1).** *If $R$ is smooth and of pidegree $n$ and if $c$ is the evaluation of a central polynomial for $n \times n$ matrices then $R[c^{-1}]$ is smooth.*

*Proof.* Given a map $f$ from $R[c^{-1}]$ to $S$ we can extend the induced map from $R$ to a map $\tilde{f}$ from $R$ to $\tilde{S}$ since $R$ is smooth. The image of $c$ will be smooth under this map since it is the evaluation of a central polynomial and the *pidegree* of $\tilde{S}$ is not any larger than the that of $R$. Now since $c$ is central and has an inverse when mapped to $S$ it will have an inverse in $\tilde{S}$ by the nilpotence of the kernel. QED

We some time wish to use a slightly different criterion for smoothness.

**PROPOSITION (2.2).** *$R$ is smooth if and only if for any algebra $\tilde{R}$ of pidegree less than or equal to that of $R$ and any algebra surjection $\tilde{R} \to R$ with nilpotent kernel, the surjection must split.*

*Proof.* We first verify the second criterion implies our definition of smoothness. We have that $R$ is the homomorphic image of a generic matrix algebra $F$ so we suppose $R = F/I$. Now we have

$$\begin{array}{ccc} F/I^2 & \leftarrow & F/I \\ \downarrow & & \downarrow \\ \tilde{S} & \to & S \end{array}$$

where the arrows on the bottom and the right are the given, the top arrow exists by our assumption, and the left hand arrow exists since the kernel of the bottom arrow may be taken to be square zero. The splitting is given by composing the two arrows.

We now verify the remaining direction. We suppose that R is smooth and that we are given the map $\tilde{R} \to R$ with nilpotent kernel and we must split it. In the following diagram

$$\begin{array}{ccc} R & \simeq & R \\ \downarrow & & \downarrow \\ \tilde{R} & \to & R \end{array}$$

we see the existence of the left hand arrow by smoothness of R and this provides the splitting. QED

We are now ready to prove the desired result on smoothness of Azumaya algebras.

**PROPOSITION (2.3).** *If A is a prime Azumaya algebra with centre C then A is smooth if and only if C is smooth.*

*Proof.* We first assume that C is smooth. Let

$$0 \to N \to B \xrightarrow{f} A \to 0$$

be any exact sequence where $f$ is an algebra homomorphism and $N^2$ is zero. We know that $B$ must also be Azumaya since its *Spec* is the same set theoretically as that of $A$. By the Lemma it suffices to check that $f$ splits. But we know that

$$A \simeq B \otimes_C C/C \cap N$$

where $C$ is the centre of $B$, since N is centrally generated. The above isomorphism is a ring isomorphism. Now

$$C \to C/C \cap N \to 0$$

is split by a map $\alpha$ and so

$$B \otimes_C C \xleftarrow{1 \otimes \alpha} B \otimes_C C/C \cap N$$

is a splitting as required.

We now show the reverse direction. So assume that A is smooth. Suppose we are given a sequence

$$0 \to N \to D \to C \to 0$$

where $D$ is commutative and $N^2 = 0$. We can assume that C is local, since smoothness of C is a local question, and we are allowed to localise and retain smoothness by the Lemma on localisation. Thus $A$ is free of rank $m = n^2$, that is we have $A \simeq C^{(m)}$. Let $B = D^{(m)}$, so without loss of generality $B/NB = A$. B can be endowed with the structure of an Azumaya algebra so that the canonical maps

$$\begin{array}{ccc} D & \to & C \\ \downarrow & & \downarrow \\ B & \to & A \end{array}$$

are ring homomorphisms.

In order to do this one lifts the multiplication map $\lambda: A \otimes A \to A$ to $\lambda': B \otimes_D B \to B$ which we can do since B is free. Although $\lambda'$ may not be associative it can be made so. Let

$$h(a,b,c) = \lambda'(\lambda'(a \otimes b) \otimes c) - \lambda'(a \otimes \lambda'(b \otimes c))$$

be the obstruction to associativity. Then $h: B \otimes B \otimes B \to NB$ is a 3-cocycle, and so is a coboundary over A. This last statement follows from the fact that since R is Azumaya its Hochschild dimension as an algebra over its centre is 1.Thus

$$h(a,b,c) = ag(b \otimes c) - g(ab \otimes c) + g(a \otimes bc) - g(a \otimes b)c$$

where $a,b,c \in A$. Let $g'(a,b) = g(\bar{a},\bar{b})$ for $a,b \in B$. Then

$$g' + \lambda' : B \otimes B \rightarrow B$$

forms a new lifting of the multiplication, and this new multiplication is assosciative. $B$ will be Azumaya since $A$ is. Its centre is $D$ since $B$ is a $D$ algebra of rank $n^2$.

Now since A is smooth, the map $B \rightarrow A$ will split. This induces a splitting map on the centres. QED

We now wish to discuss some examples of smooth algebras.

**EXAMPLE** (2.4). The following is the simplest non Azumaya algebra of dimension 1. It is prime and geometrically integrally closed, that is there are no rings between it and its quotient ring such that the inclusion is geometric and integral (see the first paragraph of section 3).

$$\begin{bmatrix} k[u] & k[u] \\ uk[u] & k[u] \end{bmatrix}$$

We shall see that it is a particular case of the more general Theorem, and we shall present there the proof in this case.

**EXAMPLE** (2.5). This is a two-dimensional example which looks much like (2.4). It has only two points in $Spec_1$

$$\begin{bmatrix} k[u,v] & k[u,v] \\ (u,v) & k[u,v] \end{bmatrix}$$

This is part of a more general phenomenom which we will discuss in a future article. The above algebra if localised at the origin, has two simple modules, one of *proj dim* = 1 and one with *proj dim* = 2. Thus unlike the commutative affine case an algebra having all its simple modules of the same homological dimension is a different concept from smoothness. Indeed the above example could be altered to make it have the former property, by putting just $(u)$ in the lower left corner. But it would no longer be smooth.

**EXAMPLE** (2.6). This is similar to example (2.4) but it is not smooth.

$$\begin{bmatrix} k[u] & k[u] \\ u^2k[u] & k[u] \end{bmatrix}$$

The proof of this involves looking at a specific central deformation, and showing it does not split. Of course the above example is not geometrically integrally closed since it maps to geometrically to example (2.4).

### 3. Smooth Algebras of Dimension One

In dimension one we might expect that smoothness would be independent of pidegree. The commutative polynomial ring $k[x]$ satisfies the condition of (1.1) even for those $\tilde{S}$ which are not commutative. Of course in the one dimensional commutative case smoothness is the same as being a dedekind domain. Thus we might hope in view of examples that the 'normal' or integrally closed prime one dimensional rings be the smooth ones. This is much to strong a condition however, since it would eliminate examples like (2.4). Such a ring does have the property of being integrally closed with respect to geometric maps. Recall from[6] that geometric maps are the algebra homomorphisms such that the corresponding map on *Spec* sends $Spec_i$ to $Spec_i$. Although the full matrix algebra is integral over the above example, the above inclusion is not a geometric map since the point $u = 0$ in the full matrix ring maps to the two points of $Spec_1$ in the example. It is known that the one dimensional affine rings are finite modules over their centres, and so the result of Brumer[1] applies to this case. He gives the structure of hereditary orders over an order in a full matrix ring.

By using the structure theory we are able to deduce the following fact.

**THEOREM (3.1).** *A geometrically integrally closed (or hereditary) affine prime pi algebra R over an algebraically closed field is smooth. Indeed any algebra homomorphism onto R ,with nilpotent kernel, splits.*

We shall only indicate the proof in the case of example (2.4) although the general proof is similar but much more technical.

It suffices to show that

$$H^2(R,M) = 0$$

holds for every $R$–$R$ bimodule M. Here we are speaking of the Hochschild cohomology (see[4] ). It is important to note that we are speaking of bimodules where only $k$ is assumed central, or equivalently of left $R^e = R \otimes_k R^{op}$ modules. There is the exact sequence of bimodules

$$0 \to \Omega_R \to R^e \xrightarrow{\mu} R \to 0$$

where the map $R^e \xrightarrow{\mu} R$ is multiplication. Thus

$$Ext^1_{R^e}(\Omega_R, M) = Ext^2_{R^e}(R,M)$$

holds for all M. Now the right hand side is equal to $H2(R,M)$ so to verify its vanishing, we need only check that $\Omega$ is a projective $R^e$ module. If we think of

$$R = \begin{bmatrix} 1 & 1 \\ u & 1 \end{bmatrix} \quad R^e = \begin{bmatrix} 1 & v \\ 1 & 1 \end{bmatrix}$$

and so we have

$$R^e = \begin{bmatrix} 1 & v & 1 & v \\ 1 & 1 & 1 & 1 \\ u & uv & 1 & v \\ u & u & 1 & 1 \end{bmatrix}$$

where we allow entries from the ideal of $k[u,v]$ generated by the entry in our schematic description. If we denote the $i$th column of $R^e$ by $P_i$ then we have

$$\mu(P_2) = \mu(P_3) = 0$$

so

$$\Omega_A \simeq K \oplus P_2 \oplus P_3.$$

Let

$$\Sigma = P_1 + P_4 = \begin{bmatrix} & & 1 & \\ & & 1 & \\ & & (u,v) & \\ & & 1 & \end{bmatrix}$$

Observe the following exact sequences hold

$$0 \to K \to P_1 \oplus P_4 \to \Sigma/(u\text{-}v)P \to 0 \tag{3.2}$$

$$0 \to \Sigma/(u\text{-}v)P \to P/(u\text{-}v)P \to P/\Sigma \to 0 \tag{3.3}$$

$$0 \to L \to P_1 \oplus P_2 \to P \to P/\Sigma \to 0 \tag{3.4}$$

where the middle map in (3.2) is just the restriction of $\mu$ and so is just the sum of the columns taken modulo $(u\text{-}v)P$.

If we can show that $L$ is projective then we will have the projective dimension of $P/\Sigma$ is $\leq 2$. Thus by (3.3) the projective dimension of $P/\Sigma$ is no more than 1. So (3.2) tells us that $K$ is projective.

Now to see that L is projective we note that

$$L = \{(a,-a): a \in P_1 \text{ and } -a \in P_2\} = P_1 \cap P_4 \simeq P_2$$

and this completes the proof. QED

Is the converse of the above theorem true? Are smooth algebras hereditary. It is true, as M.Auslander mentioned to me, that algebras satisfying the stronger conclusion of the above theorem, namely that *all* nilpotent algebra extensions split, are hereditary. To see this one simply uses the identity

$$H^2_{A^e}(A, Hom_k(B,C)) \simeq Ext^2_A(B,C)$$

(see for example [2] ) to obtain global dimension of A equal to 1 from the vanishing of $H^2$.

## 4. References


1. A.Brumer, "Structure of hereditary orders," *Bull. Amer. Math. Soc.* **69** pp. 721-724 (1963).
2. H. Cartan and S. Eilenberg, *Homological Algebra,* Princeton Univ. Press, Princeton (1956).
3. M. Demazure and P. Gabriel, *Groupes Algebriques,* Masson, Paris (1970).
4. G.Hochschild, "On the cohomology theory for associative algebras," *Ann. of Math.* **46** pp. 58-67 (1945).
5. Alexander Grothendieck, "EGA IV," *I.H.E.S. Publications Mathematiques* **20** pp. 69-147 Presses Universitaires de France, (1964).
6. M.Artin and W.Schelter , "A version of Zariski's Main Theorem for polynomial identity rings," *Amer.J.Math.* **101** pp. 301-330 (1979).
7. Matsumura, *Commutative Algebra,* W.A.Benjamin, New York (1970).

# QUESTIONS ON SKEW FIELDS

A. H. Schofield

Trinity College, Cambridge, U.K.

There are certain skew fields that arise quite naturally and about which we know very little at present; they include the skew fields of fractions of the Weyl algebras henceforth called the Weyl skew fields and written as $D_n$, which are closely linked to the skew fields of factions of the enveloping algebras of finite dimensional Lie algebras, and the skew fields of fractions of group rings of torsion free polycyclic by finite groups; all these may be subsumed in the class of skew fields that are iterated Ore or finite extensions of their centres. For such skew fields it ought to be possible to find some useful notion of transcendence degree over the central subfield k which would give us some idea of the coarse skew subfield structure of such skew fields and ideally would be the maximal length of chains of infinite extensions starting at k and finishing at the skew field. Many of the questions in the following list are based on attempts to set up such a theory.



F. van Oystaeyen (ed.), Methods in Ring Theory, 489–495.

We begin with a notion of dubious value for such purposes, the Gelfand Kirillov transcendence degree, which was originally invented to distinguish between the various Weyl skew fields. In the case of the skew fields of fractions of nilpotent group rings it already fails to be the maximal length of chains of infinite extensions. Given a skew field D that is a k algebra, we define $\text{G.K.tr.deg.}(D) = \sup_{V} \inf_{b\in D} \overline{\lim_{n\to\infty}} \frac{\log[\Sigma(bV)^i : k]}{\log n}$,

$V\subseteq D, [V:k] < \infty$

Stafford raised the following three questions to give some ideas of the difficulties inherent in this notion.

$[D:E]_l$ is the left dimension of D over E and $[D:E]_r$ is the right dimension.

1. If $E<F$ are k algebras and skew fields and $[F:E]_l$ is finite, does G.K.tr.deg.(E) = G.K.tr.deg.(F)?

2. If $E<F$ are k algebras and skew fields and $[F:E]_l$ is finite, can it be shown that G.K.tr.deg.(F)$<\infty$ implies that G.K.tr.deg. (E) $<\infty$?

In the case of the skew field of fractions of a torsion free nilpotent group ring, Lorenz (1) has shown that G.K.tr. degree is equal to the rate of growth of the nilpotent group; from this comes the next question.

3. If G is a torsion free polycyclic by finite group that is not nilpotent by finite, is the G.K.tr.degree of the skew field of fractions of its group ring over k infinite?

The next notion of interest is a rather less obvious line

of attack on these matters. If $D^{o}\otimes_{k}D$ is a noetherian ring, all skew subfields of D must be finitely generated; this follows from the work in (4). This suggests that $D^{o}\otimes_{k}D$ may contain much useful information about the skew field D; in other words, in order to study the skew field D maybe we should look at the D bimodules. Therefore we shall consider for the time being the class of skew fields D such that $D^{o}\otimes_{k}D$ is noetherian; it includes all the skew fields mentioned so far. This class of skew fields has the merit of avoiding the Kurosh problem since it is easy to see that such a D has no transcendental elements over k if and only if it is finite dimensional over k. The various dimensions of this ring will certainly give interesting invariants, but the one that appears to have most promise is the Krull dimension of $D^{o}\otimes_{k}D$. Stafford (6) showed that the Krull and global dimension of $D^{o}\otimes_{k}D$ for D the skew field of fractions of the group ring of a torsion free polycyclic by finite group was equal to the Hirsch length of the group; if $D>E$, then K.dim.$(E^{o}\otimes_{k}E)$ is at most K.dim.$(D^{o}\otimes_{k}D)$ by faithful flatness, and if $[D:E]_{l}$ is finite, they are equal. K.dim. $(D^{o}\otimes_{k}D) = 0$ if and only if $[D:k]$ is finite; K.dim.$(D^{o}\otimes_{k}D) = 1$ if and only if D is finite dimensional over its centre which is finitely generated of transcendence degree 1 over k (5). What happens for skew fields such that K.dim.$(D^{o}\otimes_{k}D) = 2$ is not so clear; there are many examples of such skew fields such as the finitely generated commutative fields of transcendence degree 2, the first Weyl skew field and its multiplicative analogues, generated by x and y over k subject to the relation $xy = \lambda yx$ for $\lambda \epsilon k$ and not a root of unity. For all these examples, one can show that any skew subfield over which it is of infinite rank must be commutative of transcendence degree 1 over k (5); so, in these cases, K.dim.$(D^o \otimes_k D)$ is equal to the maximal length of chains of skew fields $D = D_o > D_1 > \ldots > D_n = k$ where $[D_i : D_{i+1}]_1$ is infinite.

4. If $D^o \otimes_k D$ is noetherian, does K.dim.$(D^o \otimes_k D)$ equal the maximal length of a chain of skew fields between D and k such that all steps are infinite dimensional on the left?

There is some apparent difficulty in the formulation of the last question since the left dimension of an extension of skew fields may be finite whilst the right is infinite; for the skew fields such that K.dim.$(D^o \otimes_k D) = 2$ discussed before, it is possible to show that this cannot happen (5). From this, we have the following question; a positive answer to it would tidy up the form of the last question.

5. If $D^o \otimes_k D$ is noetherian, and E is a skew subfield of D, does $[D:E]_1$ imply that $[D:E]_r$ is finite? If so, are they equal? In particular, if $D_1$ is the first Weyl skew field, and E is a skew subfield, does $[D_1:E]_1 = [D_1:E]_r$?

Another possible use for the Krull dimension of $D^o \otimes_k D$ is in the study of embeddings of skew fields. Results of Resco (3) on the transcendence degree of commutative subfields of a skew field may be regarded in this light.

6. If $D^{o} \otimes_k E$ is noetherian for all skew fields E and for the skew field F, K.dim.$(D^{o} \otimes_k F) \geqslant$ K.dim.$(D^{o} \otimes_k D)$, does D embed in $M_n(F)$ for some integer n?

In general, it is impossible to better the conclusion of question 6 by pulling the embedding of D in $M_n(F)$ down to an embedding of D in F; however, there are instances where there remains hope that this can be done.

7. If $M_n(D)$ contains a commutative subfield of transcendence degree t, does D?

8. If $M_n(D)$ contains the first Weyl skew field, does D?

Whilst on the subject of commutative subfields of skew fields, we should mention the following question of Resco.

9. If the transcendence degree of commutative subfields of $M_n(D)$ over k is at most t, is the transcendence degree of commutative subfields of $M_n(D(y)\ )$ over k(y) at most t?

There are a couple more questions about the skew subfield structure of special skew fields, which at present appear to be rather hard.

10. If $D_1 > E$, where $D_1$ is the first Weyl skew field, and E is not commutative, is E isomorphic to $D_1$?

11. Let F be a commutative field of characteristic 0, and let $\delta$ be a derivation on F with fixed field k; let D be $F(x:\delta)$, the

skew field of fractions of the ring of differential polynomials. There is a discrete valuation v defined on D by $v(F) = 0$, and $v(x) = -1$. Let $a \epsilon D$; then its centraliser $C_D(a)$ is commutative. If $c(a) \neq 0$, is $[C_D(a):k(a)]<\infty$ ?

One thing that assists greatly in the study of $D_1$ and similar skew fields is that there is a decent maximal order inside the skew field; unfortunately, this is not known to be the case for finite skew field extensions of $D_1$; in a contrary direction, however, Makar-Limanov (2) has shown that $D_1$ contains a free algebra. He raises the following question.

12. Does every skew field finitely generated but of infinite dimension over its centre contain a free algebra?

Since this theorem would solve the Kurosh problem we seem to be a long way from solving this problem.

He also raises the following question.

13. What is the outer automorphism group of $D_1$?

References

1. Lorenz — Division rings generated by nilpotent subgroups, preprint.

2. Makar-Limanov — Free subalgebras of division rings, preprint.

3. Resco — Dimension theory for division rings, Israel J. of Maths. 35 215 - 221.

4. Resco, Small, Wadsworth — Tensor products of skew fields and finite generation of subfields, Proc. AMS 77 7 - 10.

5. Schofield — Stratiform skew fields (to appear).

6. Stafford — Dimensions of division rings, preprint.

# TORSION UNITS IN GROUP RINGS

Sudarshan K. Sehgal

Department of Mathematics, University of Alberta
Edmonton, Alberta, Canada

Let $\mathcal{U}\,\mathbb{Z}G$ be the group of units of $\mathbb{Z}G$, the integral group ring of a finite group $G$. Obviously, $\pm g \in \mathcal{U}\,\mathbb{Z}G$ for $g \in G$; these are called trivial units. The augmentation of a unit has to be $\pm 1$. Thus we have, $\mathcal{U}\,\mathbb{Z}G = \pm\,\mathcal{U}_1\,\mathbb{Z}G$ with $\mathcal{U}_1\,\mathbb{Z}G$ denoting the group of units of augmentation one. It is a classical result of G. Higman [4] that every torsion unit of a commutative integral group ring is trivial. Hughes and Pearson [5] showed that there are two nonconjugate units of order 3 in $\mathcal{U}_1\,\mathbb{Z}S_3$, where $S_3$ is the symmetric group on three elements. These units can not be conjugate to trivial units. Accordingly, Zassenhaus made the following conjectures.

> Let $u$ be a torsion unit of $\mathcal{U}_1\,\mathbb{Z}G$ (written $u \in T\mathcal{U}_1\,\mathbb{Z}G$). Then there is a group element $g \in G$ and a unit $\alpha$ of the rational group algebra $\mathbb{Q}G$ such that $u = \alpha^{-1}g\alpha = g^{\alpha}$ (written $u \sim g$). (ZC1)

> $\mathbb{Z}G = \mathbb{Z}H \Rightarrow H = G^{\alpha}$ for some $\alpha \in \mathcal{U}\,\mathbb{Q}G$. (ZC2)

Since any finite subgroup of $\mathcal{U}_1\,\mathbb{Z}G$ is linearly independent, this conjecture says that any finite subgroup of $\mathcal{U}_1\,\mathbb{Z}G$ of maximal possible order is conjugate to $G$ by an element of $\mathcal{U}\,\mathbb{Q}G$. These two conjectures are special cases of another more general conjecture.



*F. van Oystaeyen (ed.), Methods in Ring Theory, 497–504.*

Any finite subgroup of $\mathcal{U}_1\ \mathbb{Z}G$ is conjugate in $\mathcal{U}\mathbb{Q}G$ to a subgroup of $G$. (ZC3)

It is useful to be able to strengthen (ZC2) to

$\mathbb{Z}G = \mathbb{Z}H \Rightarrow H = G^{\alpha}$, $\alpha \in \mathbb{Z}_{(\pi)}G$, where $\mathbb{Z}_{(\pi)}$ is the semilocalization of $\mathbb{Z}$ at a prescribed set of primes $\pi$. (ZC4)

There has been no work done in the noncommutative case on (ZC3). We shall report on (ZC1) in §2 and §3 and discuss (ZC2) and (ZC4) in §4.

## §2. METABELIAN GROUPS

Given $u \in T\mathcal{U}_1\ \mathbb{Z}G$, how should one find a $g \in G$ so that $u \sim g$? In general, one does not even know if $G$ has an element of the same order as $u$. However, if $G$ is metabelian, there is a useful argument of Whitcomb available to us. Let $A$ be a normal abelian subgroup of $G$ with $G/A$ abelian. Let $u \in T\mathcal{U}_1\ \mathbb{Z}G$. Then going mod $A$, $\bar{u}$ is a unit of finite order. We conclude by Higman's theorem that there is $g \in G$ with $\bar{u} = \bar{g}$. Let

$$\Delta(G,A) = \langle (1-a)/a \in A \rangle_{\mathbb{Z}G}$$

be the kernel of the homomorphism $\mathbb{Z}G \to \mathbb{Z}(G/A)$. Also, write $\Delta(G)$ for $\Delta(G,G)$. We have,

$$u = g + \sum_a \alpha_a(1-a), \quad \alpha_a \in \mathbb{Z}G, \quad a \in A,$$

$$= g + \sum n_a(1-a) (\mathrm{mod}\ \Delta(G)\Delta(A)), \quad n_a = \text{augmentation of } \alpha_a.$$

Writing $W_A$ for the Whitcomb ideal $\Delta(G)\Delta(A)$ we get

$$u \equiv g + \sum n_a(1-a) (\mathrm{mod}\ W_A), \quad n_a \in \mathbb{Z}.$$

By using the identity

$$(1-x) + (1-y) = (1-xy) + (1-x)(1-y)$$

we deduce that

$u \equiv ga_o \pmod{W_A}$ for some $a_o \in A$.

We write $u \equiv g_u \pmod{W_A}$ for some $g_u \in G$. Using the well known [2] fact:

$1 + W_A$ is torsion free,

we conclude that any torsion subgroup T of $\mathcal{U}_1\ \mathbb{Z}G$ is isomorphic to a subgroup of G by the map

$T \ni u \to g_u \in G.$

Thus, one might propose a strong Zassenhaus conjecture for metabelian groups.

Strong ZC: G metabelian, $u \in T\mathcal{U}_1\ \mathbb{Z}G$, $u \equiv g \bmod(W_A) \Rightarrow u \sim g$.

Theorem 1. Strong ZC is true if

(i) G is nilpotent class two (Ritter-Sehgal).
(ii) G is a metacyclic p-group (Sehgal-Weiss).

We shall indicate briefly how this is proved.

(i) First reduction: It is enough to find $\alpha$ in KG where K is any extension of $\mathbb{Q}$.
(ii) Second reduction: It is enough to prove

$\chi(u) = \chi(g)$ for all absolutely irreducible characters $\chi$ of G.

This is so because $u \equiv g \pmod{W_A} \Rightarrow u^k \equiv g^k \pmod{W_A}$ for all k.

By induction we may assume that $\chi$ is faithful. Let A be the centre of G.

$$g \text{ central} \Rightarrow ug^{-1} \equiv 1 \pmod{W_A} \Rightarrow ug^{-1} = 1 \Rightarrow u = g \Rightarrow \chi(u) = \chi(g).$$

g not central $\Rightarrow \chi(u) = \chi(g) = 0$ follows from (iii).

(iii) In case G is nilpotent class two or a metacyclic p-group, the faithful irreducible complex characters of G vanish outside the centre of G.

If G is a metacyclic p-group the result is due to

Sehgal-Weiss and will appear elsewhere. If $G$ is nilpotent class two the result is well known. Here is an easy argument of Marciniak for this.

Let $g$ be a non central element of $G$. Then there is an $x \in G$ such that $xgx^{-1} \neq g$. But since $G$ is nilpotent class two, $xgx^{-1}g^{-1} = z$ is central. We have

$$xgx^{-1} = gz, \quad \chi(g^x) = \chi(g)\xi$$

where $\xi$ is a root of unity $\neq 1$. Thus $\chi(g)(1-\xi) = 0$ implying $\chi(g) = 0$. □

After this theorem one gets hopeful about the strong ZC. Unfortunately, this strong conjecture is not always true. There is a counterexample due to Ritter-Sehgal. Let

$$D_{10} = \langle a^5 = 1 = x^2 \mid a^x = a^{-1} \rangle.$$

In $\mathbb{Z} D_{10}$ we have a unit

$$u = -a^2 + a^3 + a^4 + (a^3-1)x$$

of order 5 satisfying

$$u \equiv a^3 \mod(\Delta(A)\Delta(G) = {}_A W)$$
$$\equiv a^2 \mod(\Delta(G)\Delta(A) = W_A)$$

but $u \sim a \sim a^4$. Decompose $\mathbb{C}D_{10}$ as

$$\mathbb{C}\langle x \rangle \oplus C_{2\times 2} \oplus \mathbb{C}_{2\times 2} .$$

Write $T_1$ and $T_2$ for the two 2-dimensional irreducible representations:

$$T_1(a) = \begin{pmatrix} \xi & \\ & \xi^4 \end{pmatrix}, \quad T_2(a) = \begin{pmatrix} \xi^2 & \\ & \xi^3 \end{pmatrix}$$

$$T_1(x) = T_2(x) = \begin{pmatrix} 0 & 1 \\ 1 & 0 \end{pmatrix};$$

here $\xi$ is a primitive 5-th root of unity. Now one checks by direct computation that $T_1(u) \sim T_1(a)$ and $T_2(u) \sim T_2(a)$ which implies that $u \sim a$. □

## §3. METACYCLIC GROUPS

Even for metacyclic groups (ZC1) is not completely settled. The first result in this case is due to Bhandari-Luthar [1] who gave a proof for the group $G = C_p \rtimes C_q$ which is the semidirect product of a cyclic group, $C_p$, of order $p$ by a cyclic group, $C_q$, of order $q$, with $p$ and $q$ primes, $p$ dividing $(q-1)$. Of course, as stated in the last section, even the strong ZC is true for metacyclic p-groups. There is the following.

Theorem 2 (Ritter-Sehgal). (ZC1) is true in the following cases:

(1) $G = \langle a \rangle \rtimes \langle x \rangle$, $O(a) = p^m$, $O(x) = s$, $(s,p) = 1$, $p$ a prime;

(2) $G = \langle a \rangle \rtimes \langle x \rangle$, $O(a) = n$; $O(x) = q$, a prime; $q \nmid n$.

The special case of (1) when the action of $\langle x \rangle$ on $\langle a \rangle$ is faithful was proved by Polcino Milies and Sehgal [7]. The general case is reduced to this.

Idea of Proof of Theorem 2

We indicate briefly how (1) is proved in the faithful action case. We need to find the conjugating element $\alpha \in \mathbb{C}G$. For this it is enough to prove:

given $u \in T\mathcal{U}_1\ \mathbb{Z}G$, there is a $g \in G$ such that $\rho(u) \sim \rho(g)$ for every irreducible representation $\rho$ of $\mathbb{C}G$.

We observe that either $O(u)/p^m$ or $O(u)/s$. If $o(u)/s$ it is shown by "Hilbert 90" argument that there is a $g \in G$ with $u \sim g$. It remains to consider the case: $O(u)/p^m$. Then $\bar{u} = 1$ in the homomorphic image $\mathbb{Z}(G/\langle a \rangle)$. Thus $u \in 1 + \Delta(G, \langle a \rangle)$. It follows that $\rho(u) = 1$ for all linear representations $\rho$ as is the case for any $a^i$. We have only to show that there is an $i$ such that $\rho(u) \sim \rho(a^i)$ for all nonlinear $\rho$. It is easy to see that $\rho_i(u) = \rho_i(a^{n_i})$. It takes a bit of work to show that $n_i = n_j$. □

## §4. ($\mathbb{Z}$C2), (ZC4) AND THE ISOMORPHISM PROBLEM

(ZC2) asks the question:

$$\mathbb{Z}G = \mathbb{Z}H \Rightarrow H = G^{\alpha} \quad \text{for some} \quad \alpha \in \mathcal{U}QG?$$

This is known to be true for nilpotent class 2 (Sehgal [9]) and metacyclic and some metabelian groups (Peterson [6]). It is also known (Whitcomb [11]) that if $G$ is the dihedral group of order 8 then $\alpha$ can not be chosen in $\mathbb{Z}G$. It is more interesting to ask (as in (ZC4)) if $\alpha$ can be chosen in some semilocalization. More precisely, suppose $G$ has a normal abelian subgroup A. Let $R$ be the semilocalization at $|A|$, i.e.

$$R = \mathbb{Z}_{|A|} = \bigcap_{p \,|\, |A|} \mathbb{Z}_{(p)} ;$$

here, $\mathbb{Z}_{(p)}$ denotes the localization at the prime $p$. The question is:

$$\mathbb{Z}(G/A) = \mathbb{Z}(H/B) \Rightarrow H/B = (G/A)^{\alpha}, \quad \alpha \in \mathcal{U}R(G/A)?$$

A consequence of an affirmative answer to this question is the following theorem which is essentially due to Whitcomb.

Theorem 3. Suppose that $G$ has a normal abelian subgroup A. Suppose that every group basis of $\mathbb{Z}(G/A)$ is conjugate to $G/A$ in $R(G/A)$. Then

$$\mathbb{Z}G = \mathbb{Z}H \Rightarrow G \simeq H.$$

Proof. We know that $H$ has a normal abelian subgroup $B$ such that $\Delta(G,A) = \Delta(H,B)$, $\mathbb{Z}(G/A) = \mathbb{Z}(H/B)$. By hypothesis, there exists an $\bar{\alpha} \in \mathcal{U}R(G/A)$ such that $H/A = (G/A)^{\bar{\alpha}}$. Lift $\bar{\alpha}$ to $\alpha$, a unit of $RG$. For $h \in H$, we have

$$h \equiv \alpha^{-1} g \alpha \bmod \Delta_R(G,A).$$

Replacing $G^{\alpha}$ by $G$ we get

$$h \equiv g \bmod \Delta_R(G,A).$$

Since $R$ is a subring of the rationals with denominators relatively prime to the order of $A$ we can conclude by the Whitcomb argument mentioned earlier that

$$h \equiv g_o \bmod \Delta_R(G)\,\Delta_R(A).$$

It is easy to check that $h \to g_o$ is an isomorphism $H \simeq G$. □

When is the hypothesis of the theorem satisfied? The first two cases are known, the third is the exciting announcement made by Roggenkamp at this conference:

(1) G/A metacyclic (Endo-Miyata-Sekiguchi)

(2) G/A satisfying (ZC2) with $(|G/A|,|A|) = 1$ e.g. G/A nilpotent class two; (Sehgal-Sehgal-Zassenhaus).

(3) G/A nilpotent class two (Roggenkamp-Scott).

## REFERENCES

[1] Bhandari, A.K. and Luthar, I.S., Torsion units of integral group rings of metacyclic groups, J. Number Theory (to appear).

[2] Cliff, G., Sehgal, S.K. and Weiss, A. 1981, Units of integral group rings of metabelian groups, J. Algebra 73, pp. 167-185.

[3] Endo, S., Miyata, T. and Sekiguchi, K. 1982, Picard groups and automorphism groups of integral group rings of metacyclic groups, J. of Algebra 77, pp. 286-310.

[4] Higman, G. 1940, The units of group rings, Proc. London Math. Soc (2), 46, pp. 231-248.

[5] Hughes, I. and Pearson, K.R. 1972, The group of units of the integral group rings $\mathbb{Z} S_3$, Canad. Math. Bull. 15, pp. 529-534.

[6] Peterson, G. 1977, On the automorphism group of an integral group ring II, Illinois J. Math. 21, pp. 836-844.

[7] Polcino Milies, C. and Sehgal, S.K., Torsion units in integral group rings of metacyclic groups, J. Number Theory (to appear).

[8] Ritter, J. and Sehgal, S.K., On a conjecture of Zassenhaus on torsion units in integral group rings, Math. Ann. (to appear).

[9] Sehgal, S.K. 1969, On the isomorphism of integral group rings I, Canad. J. Math. 21, pp. 410-413.

[10] Sehgal, Sudarshan, Sehgal, Surinder and Zassenhaus, H., Isomorphism of integral group rings of abelian by nilpotent class two groups, Comm. Algebra (to appear).

[11] Whitcomb, A. 1968, The group ring problem, Ph.D. Thesis, University of Chicago.

ON THE LENGTH OF DECOMPOSITIONS OF CENTRAL SIMPLE ALGEBRAS IN TENSOR PRODUCTS OF SYMBOLS

J.-P. Tignol

Université Catholique de Louvain,
B-1348 Louvain-la-Neuve, Belgium

Abstract : We investigate the minimal number of symbols which are needed to decompose any central simple algebra according to the Merkurjev-Suslin theorem.

1. INTRODUCTION

Let e,d be integers which have the same prime factors and such that e divides d and let k be a field containing a primitive e-th root of unity $\xi$ (so that its characteristic does not divide e nor d). We denote by C(e,d/k) the class of all simple k-algebras which have degree d (i.e. whose dimension over their center is $d^2$) and exponent dividing e (so that their e-th tensor power over the center is a matrix algebra).

Merkurjev and Suslin recently proved [ 7 ] that every algebra $A \in C(e,d/k)$ is similar to a tensor product of symbols of degree e :

$$A \sim A_\xi(a_1,b_1) \otimes_F \dots \otimes_F A_\xi(a_r,b_r) \qquad (1)$$

where F denotes the center of A and, for $a,b \in F^\times$, the symbol $A_\xi(a,b)$ is the F-algebra of degree e generated by two elements $x,y$ subject to the relations :



F. van Oystaeyen (ed.), Methods in Ring Theory, 505–516.

$$x^e = a \qquad y^e = b \qquad yx = \xi\, x\, y$$

(compare [ 8,§15 ]).

The aim of this paper is to review some known facts about, and make some new observations on, the minimal number of symbols which are needed to decompose every algebra in $C(e,d/k)$. We denote this minimal number by $\ell(e,d/k)$. In other words, $\ell(e,d/k)$ is the least integer r such that for every algebra $A \in C(e,d/k)$ elements $a_1,b_1,\dots,a_r,b_r$ for which relation (1) holds can be found in the center of A. It is not clear *a priori* that such a least integer r exists (i.e. that $\ell(e,d/k) < \infty$), but this is easy to prove by generic methods : denote by $UD(e,d/k)$ a generic k-algebra of exponent e and degree d (i.e. a generic object for $C(e,d/k)$), whose existence has been shown by Saltman [ 14 ]. Using Procesi's arguments in [ 9, Corollario 8 ] with $UD(e,d/k)$ instead of the ring of generic d x d matrices, it can be seen that if $UD(e,d/k)$ is similar to a tensor product of r symbols of degree e, then every algebra in $C(e,d/k)$ has the same property. Therefore, $\ell(e,d/k)$ is also the least integer r such that $UD(e,d/k)$ is similar to a tensor product of r symbols of degree e.

Throughout this paper, when the notation $\ell(e,d/k)$ is used, it is always assumed that the integers e,d and the field k satisfy the hypotheses above, viz : e and d have the same prime factors, e divides d, and k contains a primitive e-th root of unity.

## 2. RESULTS

A) General case :

We start with a very simple observation :

2.1. Proposition : *If* d *divides* d', *then* $\ell(e,d/k) \leqslant \ell(e,d'/k)$.

This readily follows from the definition of $\ell(e,d/k)$, since every simple k-algebra A of degree d and exponent dividing e is similar to a simple k-algebra A' of degree d' and exponent dividing e (namely : $A' = M_{d'd^{-1}}(A)$).

Observe however that the corresponding statement with a divisibility relation on exponents instead of degrees : "*If* e *divides* e', *then* $\ell(e,d/k) \leqslant \ell(e',d/k)$" is *not* clear (and might be false, although $C(e,d/k) \subset C(e',d/k)$), because decomposing (up to similarity) simple k-algebras of degree d and exponent dividing e might require fewer symbols of degree e' than symbols of degree e.

B) Reduction to algebras of prime-power degree :

If $e = p_1^{\varepsilon(1)} \dots p_s^{\varepsilon(s)}$ and $d = p_1^{\delta(1)} \dots p_s^{\delta(s)}$, where $p_1,\dots,p_s$ are distinct prime numbers, then every simple algebra of exponent e and degree d decomposes as a tensor product of algebras with exponents $p_1^{\varepsilon(1)},\dots,p_s^{\varepsilon(s)}$ and degrees $p_1^{\delta(1)},\dots,p_s^{\delta(s)}$ (respectively). Therefore, $\ell(e,d/k)$ can be calculated from the various $\ell(p_i^{\varepsilon(i)},p_i^{\delta(i)}/k)$ for $i=1,\dots,s$ :

2.2. <u>Theorem</u> : $\ell(e,d/k) = \max\,\{\ell(p_i^{\varepsilon(i)},p_i^{\delta(i)}/k) \mid i=1,\dots,s\}$.

The proof lies in §3 below.

C) Algebras of degree $p^n$ :

Let p be a prime number and, for simplicity, denote :

$$\ell_p(m,n/k) = \ell(p^m,p^n/k).$$

The main result of this paper is the following :

2.3. <u>Theorem</u> : $\ell_p(m,n/k) \geqslant n$.

(For m=1, this result is improved in 2.10 below). The idea of the proof is to show that the generic algebra $UD(p^m,p^n/k)$ does not decompose (up to similarity) as a tensor product of less than n symbols of degree $p^m$. (Details are given in §4 below.)

In fact, we prove this result without restriction on the degrees of symbols (see lemma 4.1), and the following theorem can then be easily derived :

2.4. Theorem : *If* $m \geqslant 2$, *the algebra of* $p^{m-2} \times p^{m-2}$ *matrices over* $UD(p^m, p^n/k)$ *is not isomorphic to a product of symbols.*

By elementary arguments, an upper bound for $\ell_p(m,n/k)$ can be obtained from the decomposition of algebras of exponent less than $p^m$ :

2.5. Proposition : *For any integer* q *between* 1 *and* m - 1,

$$\ell_p(m,n/k) \leqslant \ell_p(m-q,\ n-q/k) + \ell_p(q,\ n+m\ \ell_p(m-q,n-q/k)/k).$$

Using the same arguments more ingeniously, so as to take advantage of the fact that algebras of degree 4 are elementary abelian crossed products, Snider has found a better bound for $\ell_2(2,2)$ (unpublished) :

$$\ell_2(2,2) \leqslant 5.$$

D) Algebras of prime exponent :

As above, let p be a prime number. For short, let

$$\ell_p(n/k) = \ell_p(1,n/k).$$

The exact value of $\ell_p(n/k)$ is known in a few cases only :

2.6. Theorem

a) $\ell_2(1/k) = 1$

b) (Wedderburn) $\ell_3(1/k) = 1$

c) (Albert) $\ell_2(2/k) = 2$

d) (Amitsur *et al.*, Elman *et al.*) *If* k *has characteristic zero, then* $\ell_2(3/k) = 4$.

(a) means that every simple algebra of exponent 2 and degree 2 is a quaternion algebra; for a proof, see [ 1,Theorem 9.26 ]. Similarly, (b) follows from the cyclicity of algebras of degree 3; see [ 1, Theorem 11.5 ] for a proof, and (c) asserts that every simple algebra of exponent 2 and degree 4 is a tensor product of quaternion algebras; an easy proof is given in [ 10 ]. Finally, (d) follows from the combination of two results : over any field k of characteristic not two, $\ell_2(3/k) \leqslant 4$ (see [ 15 ]) and if k is any field of characteristic zero, $\ell_2(3/k) > 3$ (see [ 3 ], [ 5,§5 ]) or, in other words, there exist indecomposable simple k-algebras of exponent 2 and degree 8.

If $p \geqslant 5$, the exact value of $\ell_p(1/k)$ is not known. (Of course, $\ell_p(1/k) = 1$ if and only if every simple k-algebra of degree p is cyclic, so the calculation of $\ell_p(1/k)$ involves the solution of the cyclicity problem for algebras of prime degree.) Nevertheless, an upper bound for $\ell_p(1/k)$ can be obtained by a corestriction argument :

2.7. Theorem (Rosset and Tate) $\ell_p(1/k) \leqslant (p-1)\,!$

Although this theorem is not explicitly stated in [ 11 ], it follows from the proof of the "splitting theorem" of [ 11 ] and from [ 12, Corollary 1 ].

In contrast to theorem 2.6(c), for odd p indecomposable algebras of exponent p and degree $p^2$ have been found recently by Rowen and by this author. We may thus conclude :

2.8. Theorem (Rowen) *If* k *has characteristic zero and* p *is odd, then* $\ell_p(2/k) > 2$.

If k contains a primitive p-th root of unity but not a primitive $p^2$-th root of unity, this follows from [ 13, theorem 4 ]; if k contains a primitive $p^2$-th root of unity, it follows from [ 16, théorème 2 ].

This theorem also yields some information on $\ell_p(n/k)$ for $n > 2$, because of the following result :

2.9. Proposition : *For any* $n \geqslant 1$, $\ell_p(n+1/k) \geqslant \ell_p(n/k) + 1$.

The idea of the proof is to show that if a central simple algebra A of degree $p^n$ and exponent p over a field F containing k does not decompose in less than r symbols of degree p, then the algebra $A \otimes_F A_\xi(x,y)$ over $F(x,y)$, where $\xi$ is a primitive p-th root of unity in F and $x,y$ are independent indeterminates over F, does not decompose in less than r + 1 symbols of degree p; see [ 16,Corollaire 2.13 ] for details.

Comparing theorem 2.6(d), theorem 2.8 and proposition 2.9, we readily get :

2.10 Corollary : *If* k *has characteristic zero, then* $\ell_p(n/k) \geqslant n+1$ *for* $n \geqslant 3$ *if* $p = 2$ *and for* $n \geqslant 2$ *if* p *is odd.*

Remark : During the conference, Saltman pointed out that the assumption on the characteristic of k can be removed in 2.6(d), 2.8 and 2.10. His methods are still unpublished.

## 3. PROOF OF THEOREM 2.2 :

3.1. Lemma : *Let* u *and* v *be relatively prime integers and let* $w = uv$. *Then every tensor product of a symbol of degree* u *and a symbol of degree* v *is a symbol of degree* w.

Proof : Let $\zeta$ be a primitive w-th root of unity in F. Denote $\xi = \zeta^v$ and $\eta = \zeta^u$, so that $\xi$(resp $\eta$) is a primitive u-th (resp. v - th) root of unity of F. Let also u' and v' be integers such that $uu' + vv' = 1$. Then a direct calculation, using lemma 15.5 of [ 8 ], shows that

$$A_\xi(a,b) \otimes A_\eta(c,d) \simeq A_\zeta(x,y)$$

where

$$x = a^{v} c^{u} \quad \text{and} \quad y = b^{vv'} d^{uu'}.$$

Q.E.D.

3.2. Let now $e = p_1^{\varepsilon(1)} \dots p_s^{\varepsilon(s)}$ and $d = p_1^{\delta(1)} \dots p_s^{\delta(s)}$, where $p_1,\dots,p_s$ are distinct prime numbers. The proof that

$$\ell(e,d/k) = \max \{\ell(p_i^{\varepsilon(i)}, p_i^{\delta(i)}/k) \mid i=1,\dots,s\}$$

falls in two parts :

*Claim 1* : $\ell(e,d/k) \geqslant \ell(p_i^{\varepsilon(i)}, p_i^{\delta(i)}/k)$ for $i=1,\dots,s$.

Let $A_i$ be a simple k-algebra of degree $p_i^{\delta(i)}$ whose exponent divides $p_i^{\varepsilon(i)}$ and let $A = M_n(A_i)$, where $n = d\, p_i^{-\delta(i)}$, so that A is a simple k-algebra of degree d and exponent dividing e. Assume

$$A \sim S_1 \otimes \dots \otimes S_r$$

where $S_1,\dots,S_r$ are symbols of degree e. Let $m = ep_i^{-\varepsilon(i)}$ and let m' be an integer such that $mm' \equiv 1 \bmod p_i^{\varepsilon(i)}$ ; since $A \sim A_i$, we have

$$A^{mm'} \sim A_i^{mm'} \sim A_i$$

whence

$$A_i \sim S_1^{mm'} \otimes \dots \otimes S_r^{mm'}. \tag{2}$$

By lemma 15.5 of [ 8 ], $S_i^m$ is similar to a symbol of degree $p_i^{\varepsilon(i)}$ and $S_i^{mm'}$ is similar to a symbol of degree $p_i^{\varepsilon(i)}$, since m' is relatively prime to $p_i^{\varepsilon(i)}$. Relation (2) thus shows that $A_i$ has a decomposition with r symbols of degree $p_i^{\varepsilon(i)}$ ; since we could choose $r = \ell(e,d/k)$, this proves the first claim.

*Claim 2* : $\ell(e,d/k) \leqslant \max \{\ell(p_i^{\varepsilon(i)}, p_i^{\delta(i)}/k) \mid i=1,\dots,s\}$.

Let A be a simple k-algebra of degree d whose exponent divides e. By [ 1,Theorem 5.18 ], there is a decomposition of A :

$$A \simeq A_1 \otimes \ldots \otimes A_s$$

where $A_i$ has degree $p_i^{\delta(i)}$ and exponent dividing $p_i^{\varepsilon(i)}$, for $i=1,\ldots,s$. Let $r = \max\{\ell(p_i^{\varepsilon(i)}, p_i^{\delta(i)}/k \mid i=1,\ldots,s\}$. Then each algebra $A_i$ is similar to a tensor product of r symbols of degree $p_i^{\varepsilon(i)}$ (some of which may be trivial) :

$$A_i \sim S_{i1} \otimes \ldots \otimes S_{ir},$$

Then $$A \sim (\otimes_{i=1}^{s} S_{i1}) \otimes \ldots \otimes (\otimes_{i=1}^{s} S_{ir})$$

and each factor $\otimes_{i=1}^{s} S_{ij}$ is a symbol of degree e, by lemma 3.1. This relation thus shows that A is similar to a tensor product of r symbols, whence $\ell(e,d/k) \leqslant r$ and the proof of theorem 2.2 is complete.

## 4. PROOF OF THEOREMS 2.3 and 2.4

In this §, we let $U = UD(p^m, p^n/k)$ and we denote by F the center of U.

4.1. Lemma : *The algebra* U *is not similar to a tensor product of (strictly) less than* n *symbols (of possibly different degrees).*

Proof : If U is similar to such a tensor product, then it is split by a Galois extension M of F whose Galois group G is abelian and generated by at most n-1 elements. By the same arguments as in the (unpublished) proof of [ 2,property b, p.15 ], it can be shown that every simple k-algebra A of degree $p^n$ whose exponent divides $p^m$ has a Galois splitting field M' with Galois group G' isomorphic to a subgroup of G; like G, the group G' is thus abelian and generated by at most n-1 elements.

Consider then $K = \overline{k}((x_1))((y_1))\ldots((x_n))((y_n))$, an iterated

Laurent series field in 2n indeterminates over the algebraic closure $\bar{k}$ of k and

$$A = A_\xi(x_1,y_1) \otimes_K \dots \otimes_K A_\xi(x_n,y_n),$$

where $\xi$ is a primitive p-th root of unity in K. The algebra A is a simple k-algebra of degree $p^n$ and exponent p; therefore it is split by a Galois extension M' of K whose Galois group G' is abelian and generated by at most n-1 elements. On the other hand, it follows from [ 4 ] that M' contains a maximal subfield of A; such a maximal subfield is Galois over K with elementary abelian Galois group (see for instance [ 6,p.102 ]), whence the elementary abelian group of order $p^n$ is a homomorphic image of G'. This is a contradiction, since G' is generated by at most n-1 elements.

Q.E.D.

4.2. *Proof of theorem 2.3* : Theorem 2.3 is a special case of lemma 4.1, since in theorem 2.3 we need only to consider symbols of degree $p^m$.

4.3. *Proof of theorem 2.4* : Let U' be the algebra of $p^{m-2} \times p^{m-2}$ matrices over U and assume U' is isomorphic to a tensor product of symbols :

$$U' \simeq S_1 \otimes_F \dots \otimes_F S_r \; ; \qquad (3)$$

then, letting $p^{s(i)}$ be the degree of $S_i$ and comparing degrees of both sides, we get :

$$n + m - 2 = s(1) + \dots + s(r). \qquad (4)$$

Of course, $s(i) \geqslant 1$ for i=1,...,r; moreover, at least one of s(1),...,s(r) is equal to (or greater than) m, since the exponent of U' is $p^m$. Therefore, equation (4) yields : $n + m - 2 \geqslant m+(r-1)$, whence

$$r \leqslant n - 1,$$

Then, relation (3) shows that U is similar to a tensor product of less than n-1 symbols : this contradicts lemma 4.1. Q.E.D.

## 5. PROOF OF PROPOSITION 2.5

Let A be a simple k-algebra of degree $p^n$ and exponent dividing $p^m$. Then, $A^{p^q}$ is similar to an algebra of degree at most $p^{n-q}$, by [1, theorem 5.16], and its exponent divides $p^{m-q}$. Therefore, we can find symbols $S_1,\ldots,S_r$ of degree $p^{m-q}$ such that

$$A^{p^q} \sim S_1 \otimes \ldots \otimes S_r \tag{5}$$

and

$$r \leqslant \ell_p(m-q, n-q/k). \tag{6}$$

Since the center contains a primitive $p^m$-th root of unity, each symbol of degree $p^{m-q}$ is the $p^q$-th power of a symbol of degree $p^m$, by lemma 15.5 of [8]. Let thus $S_i = S_i'^{p^q}$ for some symbol $S_i'$ of degree $p^m (i=1,\ldots,r)$. From relation (5), it follows that $A \otimes (S_1' \otimes \ldots \otimes S_r')^{-1}$ is a simple k-algebra of degree $p^{n+rm}$ and exponent dividing $p^q$, whence

$$A \otimes (S_1' \otimes \ldots \otimes S_r')^{-1} \sim T_1 \otimes \ldots \otimes T_s \tag{7}$$

for some symbols $T_1,\ldots,T_s$ of degree $p^q$ and for

$$s \leqslant \ell_p(q, n+rm/k). \tag{8}$$

Since $T_1,\ldots,T_s$ are $p^{m-q}$-th powers of symbols of degree $p^m$, relation (7) shows that A is similar to a tensor product of (r+s) symbols of degree $p^m$. Since, by (6) and (8),

$$r+s \leqslant \ell_p(m-q, n-q/k) + \ell_p(q, n + m\ell_p(m-q, n-q/k)/k),$$

the proof is complete.

REFERENCES

1. ALBERT, A.A. *Structure of Algebras*, Amer. Math. Soc. Coll. Pub. 24, Providence, R.I., 1961.

2. AMITSUR, S.A. Division algebras. A survey, Contemporary Math. 13(1982), pp. 3-26.

3. AMITSUR, S.A., ROWEN, L.H. and TIGNOL, J.-P. Division algebras of degree 4 and 8 with involution, Israel J. Math. 33 (1979) pp 133-148.

4. AMITSUR, S.A. and TIGNOL, J.-P. Totally ramified splitting fields of central simple algebras over Henselian fields, to appear.

5. ELMAN, R., LAM, T.-Y., TIGNOL, J.-P. and WADSWORTH, A.R. Witt rings and Brauer groups under multiquadratic extensions, I, to appear in : Amer. Math. J.

6. JACOBSON, N. *PI-Algebras. An introduction*, Lecture Notes in Math. 441, Springer, Berlin, 1975.

7. MERKURJEV, A.S. and SUSLIN, A.A. K-cohomology of Severi-Brauer varieties and norm residue homomorphism (preliminary version), preprint.

8. MILNOR, J. *Introduction to algebraic K-theory*, Ann. of Math. Studies 72, Princeton Univ. Press, Princeton, N.J. 1971.

9. PROCESI, C. Relazioni tra geometria algebrica ed algebra non commutativa. Algebre cicliche e problema di Luroth, Boll. U.M.I. 18-A(1981) pp 1-10.

10. RACINE, M.L. A simple proof of a theorem of Albert, Proc. Amer. Math. Soc. 43(1974)pp 487-488.

11. ROSSET, S. Abelian splitting of division algebras of prime degrees, Comm. Math. Helvetici 52(1977)pp 519-523.

12. ROSSET, S. and TATE, J. A reciprocity law for $K_2$-traces, Comm. Math. Helvetici 58(1983)pp 38-47.

13. ROWEN, L.H. Cyclic division algebras, Israel J. Math. 41(1982), pp 213-234; correction, ibid. 43(1982)pp 277-280.

14. SALTMAN, D.J. Indecomposable division algebras, Comm. Algebra 7(1979), pp 791-817.

15. TIGNOL, J.-P. Central simple algebras with involution, in : *Ring theory, proceedings of the 1978 Antwerp Conference* (Van Oystaeyen, ed.) M. Dekker, New York, 1979, pp 279-285.

16. TIGNOL, J.-P. Algèbres indécomposables d'exposant premier, to appear in : Advances in Math.

# A DUALITY THEOREM FOR HOPF ALGEBRAS

Michel Van den Bergh

University of Antwerp

## Introduction.

Let k be a commutative ring. All objects we consider are k objects. Let H be a finite Hopf algebra over k (i.e. is f. g. and projective over k). There is a natural action $H^{\star} \otimes (R \# H) \rightarrow R \# H$. In [3] corollary 12.7 S. Montgomery and R. Blatter proved that the natural map $(R \# H) \# H^{\star} \rightarrow \mathrm{End}_R(R \# H)$ is an algebra isomorphism. We consider R # H as a right as a right R module via the map $R \rightarrow R \# H$). We will give a different proof using non-commutative Galois theory.

Remark : We frequently use the SIGMA notation of Sweedler, usually without explicit mention.

Throughout S R is a ring extension and $\rho : S \rightarrow S \otimes_k H$ is a k map satisfying the following conditions.

(1) S is a H comodule

The author is supported by an NFWO grant.



*F. van Oystaeyen (ed.), Methods in Ring Theory, 517–522.*

(2) $\rho$ is a k algebra map (or equivalently the corresponding map $H^{\star} \otimes_R S \to S$ is measuring)

(3) $\forall\, r \in S \ : \rho\,(r) = r \otimes 1$

We consider $S \otimes H$ and $S \otimes S$ as right S modules in the following way

$s, s' \in S \qquad h \in H \qquad (s \otimes h)s' = ss' \otimes h$

$s, s, s'' \in S \qquad\qquad (s \otimes s')s'' = s \otimes s'\, s''$

We define a map

$$V_S : S \otimes_R S \to S \otimes_k H : s \otimes s' \to \sum_s s_{(1)}\, s' \otimes s_{(2)}.$$

This map is well defined since

$\forall\, s, s' \in S, r \in R$

$V_S\,(s \otimes rs') = \Sigma_s\, s_{(1)}\,(r s') \otimes_k s_{(2)}$ and

$V_S(s r \otimes s') = \sum_s (s_{(1)} r)\, s' \otimes s_{(2)}.$

It is also clear that $V_S$ is right S-linear.

Proposition.

The following diagram is commutative

$$\begin{array}{ccc} S \# H^{\star} & \xrightarrow{\ \varphi\ } & \mathrm{Hom}_R(S,S) \\ \Big\downarrow \alpha & & \Big\downarrow \beta \\ \mathrm{Hom}_S(S \otimes_h H, S) & \xrightarrow{\ V_S^{\star}\ } & \mathrm{Hom}_S(S \otimes_R S, S) \end{array} \qquad (1)$$

The maps $\alpha, \beta, \varphi$ are defined as follows.

$\alpha\,(s \# a^{\star})\,(t \otimes t') = s\, t\, a^{\star}(t')$

$\beta\,(f)(t \otimes t') = f(t)t'$

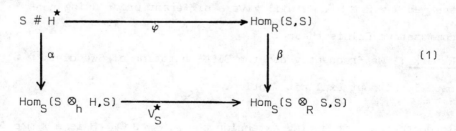

$\varphi\ (s \# a^{\star})s' = (s \# a^{\star}).s' = \Sigma_{s'}\ ss'_{(1)}\ a^{\star}(s'_{(2)})$.

Proof.

$V_S^{\star}\ (\alpha(s \# h^{\star}))(t \otimes t') = \alpha(s \# h^{\star})(V_S(t \otimes t')) =$

$\Sigma_t\ \alpha(s \# h^{\star})\ (t_{(1)}\ t' \otimes t_{(2)}) = \Sigma_t\ s\ t_{(1)}\ t'\ h^{\star}(t_{(2)})$

$\beta(\varphi(s \# h^{\star}))(t \otimes t') = \varphi(s \# h^{\star})(t).t' = \Sigma_t\ s\ t_{(1)} h^{\star}(t_{(2)})t' =$

$\Sigma_t\ s\ t_{(1)}\ t'\ h^{\star}(t_{(2)})$.

Remark : Diagram (1) is the same diagram as in [2 p. 67]. To make it commutative we used a map S that differs slightly from that of Sweedler.

Definition : S/R is H-Galois ⇔

(1) S is f.g. projective as right R-module

(2) $V_S$ is an isomorphism.

Proposition (2)

If S/R is H-Galois then $\varphi : S \# H^{\star} \to Hom_R(S,S)$ is an algebra isomorphism.

Proof.

(1) $\alpha$ is an isomorphism since $S \# H^{\star} \cong S \otimes H^{\star}$ as a k module and H/k is projective

(2) $\beta$ is always an isomorphism.

(3) $V_S^{\star}$ is an isomorphism since $V_S$ is an isomorphism.

Suppose from here on that R is an H module algebra. Let S be R #H

We will show that S/R is H-galois.

Define :

$$\rho : R \# H \to R \# H \otimes H : a \# h \to \sum_h a \# h_{(1)} \otimes h_{(2)}$$

Proposition (3) : P satisfies the 3 conditions mentioned in the beginning of this note.

Proof :

(1) R # H is a comodule for P. This is trivial from the fact that H is a H comodule.

(2) $\rho$ is an algebra map

$$\rho((a \# h)(b \# h')) = \rho(\sum_h a(h_{(1)}.b) \# h_{(2)}h')$$

$$\sum_{h'}\sum_h a(h_{(1)}.b) \# h_{(2)}\ h'_{(1)} \otimes h_{(3)}\ h'_{(2)}$$

$$\rho(a \# h)\,\rho(b \# h') = \sum_{h'}\sum_h (a \# h_{(1)} \otimes h_{(2)})(b \# h'_{(1)} \otimes h'_{(2)})$$

$$= \sum_{h'}\sum_h (a \# h_{(1)})(b \# h'_{(1)}) \otimes h_{(2)}\ h'_{(2)}$$

$$\sum_{h'}\sum_h a(h_{(1)}.b) \# h_{(2)}\ h'_{(1)} \otimes h_{(3)}\ h'_{(2)}$$

(3) $r \in R$

$$\rho(r) = \rho(r \# 1) = r \# 1 \otimes 1 = n \otimes 1$$

Theorem (4). R # H/R is H-galois.

Proof. Condition (1) of the definition is trivially satisfied. The only thing we have to check is that $V_S$ is an isomorphism.

Define

$$\xi : R \# H \otimes_k H \to R \# H \otimes_R R \# H$$

$$c \# k \otimes k' \to \sum_{k'} c \# k'_{(2)} \otimes S^{-1}(k'_{(1)})k$$

where S is the antipode. S is bijective (see the remark at the end).

$\xi(V_S(a \# h \otimes b \# h')) =$

$\xi(\Sigma_h (a \# h_{(1)})(b \# h') \otimes h_{(2)}) =$

$\xi(\Sigma_h a(h_{(1)}.b) \# h_{(2)} h' \otimes h_{(3)} =$

$\Sigma_h a(h_{(1)}.b) \# h_{(4)} \otimes S^{-1}(h_{(3)} h_{(2)} h' =$

$\Sigma_h a(h_{(1)}.b) \# h_{(4)} \otimes S^{-1}(S(h_{(2)} h_{(3)})h' =$

$\Sigma_h a(h_{(1)}.b) \# h_{(3)} \otimes S^{-1}(\varepsilon(h_{(2)}))h' =$

$\Sigma_h a(h_{(1)}.b) \# \varepsilon(h_{(2)})h_{(3)} \otimes$

$(a \# h)(b \# 1) \otimes h' =$

$(a \# h) \otimes (b \# h')$

and

$V_S(\xi(a \# h \otimes h')) \Sigma_{h'} V_S(a \# h'_{(2)} \otimes S^{-1}(h'_{(1)})h) =$

$\Sigma_{h'} (a \# h'_{(2)}) (S^{-1}(h'_{(1)})h) \otimes h'_{(b)}) =$

$\Sigma_{h'} a \# S^{-1}(h'_{(1)}S(h'_{(2)})) h \otimes h'_{(3)} =$

$\Sigma_{h'} a \# S^{-1}(\varepsilon(h'_{(1)})) h \otimes h'_{(2)} =$

$\Sigma_{h'} a \# h \otimes \varepsilon(h'_{(1)})h'_{(2)} =$

$a \# h \otimes h'$.

Corrolary.

$\varphi: (R \# H) \# H^{\star} \to \mathrm{Hom}_R (R \# H^{\star}, R \# H^{\star})$ is an algebra isomorphism. So $(R \# H) \# H^{\star}$ is Morita equivalent with R.

Remark :Let H/k be a finite Hopf algebra with antipode S. Then S is bijective. It clearly suffices to check this for k local with maximal ideal m. But then H/mH is a finite dimensional Hopf-algebra over a field.

So $\overline{S}$ is bijective [ 1 page 101 ] . This implies by Nakayama that S is bijective.

References

[1] M.E. Sweedler : Hopf Algebras, Benjamin, New York, 1969.

[2] S. Chase, M.E. Sweedler : Hopf Algebras and Galois theory Springer Verlag LNM 97, 1969.

[3] J. Blattner and S. Montgomery: A Duality Theorem for Hopf Module Algebras, to appear.

NOTE ON CENTRAL CLASS GROUPS OF ORDERS OVER KRULL DOMAINS

F. Van Oystaeyen
University of Antwerp, UIA, Belgium

Abstract

The relative Picard group with respect to the set of prime ideals of height one may be considered to be a noncommutative equivalent of the class group; indeed, if A is a maximal order over a Krull domain C then Picent(A,$\kappa_1$) is exactly the central class group CCl(A). The group CCl(A)/Cl(C) measures how A differs from being a reflexive Azumaya algebra; this follows from some results of L. Le Bruyn [ 11 ] but in the first part of these notes we present simplified proofs and correct some disturbing errors in [ 11 ]. In the second part we focus on maximal orders graded by an arbitrary group. Careful use of the multilinear Razmyslov polynomial allows to free earlier results from the rather restrictive conditions usually imposed on the group in order to assure that the center of the order is graded too. This note is only partly a survey, some of the new points of view presented await further application.



F. van Oystaeyen (ed.), Methods in Ring Theory, 523–540.

Introduction.

An extensive study of relative Picard groups for commutative rings has been undertaken in [26] . It is possible to extend this notion to noncommutative rings and in particular the "reflexive Picent" turns out to be an interesting group. If A is a maximal order in a central simple algebra Z with center K = Q(C), C = Z(A) then C is completely integrally closed. Moreover, if $X^{(1)}(-)$ denotes the set of prime ideals of height one of the ring (-), then $C = \cap \{C_p, p \in X^{(1)}(C)\}$. Define a kernel functor $\kappa_1$ by the filter $\mathcal{L}(\kappa_1) = \cap \{\mathcal{L}(\kappa_p), p \in X^{(1)}(C)\}$ = {ideals of C not in p, for every $p \in X^{(1)}(C)$}. The group Picent$(A,\kappa_1)$, defined in [26], may be called the central class group of A, actually if C is a Krull domain then this group may be described in terms of divisorial (fractional) ideals of A and we will then use the notation CCl(A), cf. [26 ],[ 9 ] . Reflexive Azumaya algebras in the sense of Yuan [30 ] , Orzech [18 ] or F. Van Oystaeyen, A. Verschoren [26 ] are a well understood kind of maximal orders over Krull domains. So the first step in studying maximal orders over Krull domains is to find a way of relating the orders to certain reflexive Azumaya algebras. Here CCl(A) comes in handy. In [11 ] L. Le Bruyn investigated necessary and sufficient conditions for the equivalence of the statements : 1° A is a reflexive Azumaya algebra; 2° CCl(A) = Cl(C). Two of the main results of [11 ] are reintroduced here in Section 1, one with a very simplified proof and the second with a correct proof which is not available in loc. cit. .

After having introduced the class group it will be necessary to calculate it sometimes. This is a difficult task in general, but if the (maximal) order A is graded by a group G then this extra information may facilitate calculation considerably. Earlier results in this vein were derived under the assumption that the center of A is graded by G too and this leads to severe restriction on G e.g. G has to be ordered, torsionfree abelian ... . In Section 2 we obtain in a very elementary way that these restrictions are unnecessary. The construction of generalized Rees rings $\check{A}(\Phi)$, cf. [ 15] [ 12], allows to reduce problems about Ccl(A) to problems about Ccl($\check{A}(\Phi)$) which is an epimorphic image of Ccl(A). Utilizing a suitable construction, $\check{A}(\Phi)$ will turn out to be a reflexive Azumaya algebra and then one may infer properties of A from properties of $\check{A}(\Phi)$ by using graded methods. Some applications of this method are in [ 12] , [ 19] ...

## 1. The Central Class Group and Reflexive Azumaya Algebras Revisited

There exist several different types of non-commutative Krull rings e.g. M. Chamarie, [ 3 ] ,H. Marubayashi, [ 13] , E. Jespers [ 9 ] , ..., but all these notions coincide if one considers rings with polynomial identities. In the latter case all the different notions of Krull rings reduce to the notion of a maximal order,A, over a Krull domain, C, in a central simple algebra, $\Sigma$. We use the terminology of M. Chamarie [ 3 ] and of F. Van Oystaeyen, A. Verschoren [26 ] . In particular, an ideal I of A will be reflexive as a left ideal if and only if it is reflexive as a right ideal, so we simple state that I is reflexive (c-ideal in M. Chamarie's terminology). The set of reflexive ideals of A with the product defined by taking the double dual of the internal product $I \star J = (IJ)^{\star\star}$, is an Abelian group D(A). A reflexive ideal is prime if and only if it is a maximal reflexive ideal and D(A) turns out to be the free abelian group generated by the reflexive prime ideals of A and these are exactly the prime ideals of A of height one. Note that all of this is already in Asano's papers. If A and $A_1$ are context-equivalent or Morita-equivalent C-orders, then D(A) and $D(A_1)$ are canonically isomorphic. A prime ideal P of A has height one if and only if $P \cap C \in X^{(1)}C$ and if this happens then P is the unique prime ideal of A lying over $p = P \cap C$. From [3 ] we recall :

1.1. Proposition. Consider a prime ideal P of A and put $p = P \cap C$. The following properties hold :

1. $A_P = C_P \otimes_C A$, but this "central localization" result need not hold for an arbitrary A-module.

2. The left (and right) Ore condition with respect to $\mathcal{C}(P)$ holds if and only if P is the unique prime ideal lying over p. In particular if $P \in X^{(1)}(A)$ then the Ore conditions hold at P.

1.2. Corollary a. The kernel functor $\kappa_1 = \inf\{\kappa_p, P \in X^{(1)}(A)\}$ is a central kernel functor.

b. If M is a divisorial C-lattice and also an A-module then $M_P \cong M_p$ holds at every prime ideal P of A, $p = P \cap C$.

It is well-known, cf. [ 26] , that a reflexive Azumaya algebra over a Krull domain is necessarily a maximal order, and for being an Azumaya algebra it is necessary and sufficient that A is flat as a C-module, cf. [ 18] . The central class group CCl(A) is defined to be $D(A)/P_C(A)$ where $P_C(A)$ is the subgroup of D(A) consisting of the ideals generalted by a single central element in K. In the terminology of [ 26] one may relate Picent$(A,\kappa_1)$ to CCl(A) and one has equality e.g. in case A is a reflexive Azumaya algebra. In the latter situation one easily derives that D(A) = D(C) and CCl(A) = Cl(C) . It is the converse of this implications we aim to investigate.

So let A be a maximal order over a Krull domain C in the K-central simple algebra Σ. In D(A) we may also consider the subgroup I(A) consisting of invertible ideals (with respect to the common product of fraction ideals). It is not hard to verify that A. Frölich's group Picent(A) equals $I(A)/P_C(A)$ in this case.

1.3. Lemma. For an intermediate ring $A \subset B \subset \Sigma$, the following statements are equivalent :

1. B is a (left) localization of A at a kernel functor $\kappa$.
2. There exists a $Y \subset X^{(1)}(A)$ such that $B = \cap \{A_P, P \in Y\}$
3. B is a central localization of A.

Proof. Follows from results of M. Chamarie [3], in particular Theorem 4.2.3. in loc. cit.

It is a consequence of the foregoing lemma that any left (or right) localization $Q_\kappa(A)$ is a subintersection of A and $A \to Q_\kappa(A)$ is a central extension which satisfies condition P.D.E. (no blowing up!) We obtain :

1.4. Proposition : Let $A \subset B \subset \Sigma$ be as in the lemma. The following sequence of abelian groups is exact :

$$1 \to H \to CCl(A) \to CCl(B) \to 1.$$

Moreover, H is generated by those $P \in X^{(1)}(A)$ such that $P \notin \mathcal{L}(\kappa)$.

1.5. Proposition. The following sequence is exact :

$$1 \to Cl(C) \to CCl(A) \to \bigoplus_P \mathbb{Z}/e_P\,\mathbb{Z} \to 1,$$

where $e_P$ is the ramification index in $\Sigma$ of the essential valuation of C associated to $C_p$, $p = P \cap C \in X^{(1)}(C)$. The exponent of the torsion group $\bigoplus_P \mathbb{Z}/e_P\,\mathbb{Z}$ is bounded by the p.i. degree of $\Sigma$.

Proof. cf. L. Le Bruyn, F. Van Oystaeyen, [12].

1.6. Corollary. The condition CCl(A) = Cl(C) may now be translated into a condition on the ramification of the essential valuations of C in $\Sigma$, in casu : all of these have to be unramified !

The proof of the consequent lemma is based uopn the idea of proof utilized in [11] but we did introduce a considerable simplification of the argument.

1.7. Lemma. Let A be a maximal order over a discrete valuation ring C. If CCl(A) = 1, or equivalently if C is unramified in $\Sigma$ (i.e. A is unramified) then either A is an Azumaya algebra over C, or else the center of A/mA is a purely inseparable extension of C/mC, where m is the maximal ideal of C.

Proof. If Z(A/mA) $=$ C/m then p.i. deg (mA) = p.i. deg(A) and by the Artin-Procesi theorem it then follows that A is an Azumaya algebra; the converse of this is obvious. The existence of non-trivial separable elements in Z(A/mA) over C/m would provide non-zero irreducible polynomials p',q' in Z(A/mA) [ t ] lying over a separable p in (C/mC) [ t ], hence this leads to prime ideals P,Q of A [ t ] such that $P \cap A = Q \cap A = mA$ and P and Q lie over the same prime ideal of C [ t ]. Since A [ t ] is a maximal order over the Krull domain C [ t ] we may apply Proposition 1.1.(1) to A [ t ] and derive that $A[t]_P P$ is the unique maximal ideal of $A[t]_P$, hence it is the Jacobson raical too. Moreover, we have :
$A[t]_P/Q_P(P) = A[t]/P = (A/mA)[t]/P'$, for some P' generated by an irreducible polynomial $p(t) \in Z(A/mA)[t]$. This implies that the elements of $\mathcal{C}(P)$ are invertible because they are invertible modulo the Jacobson radical of $A[t]_P$; it is then a classical fact (note : A [ t ] is Noetherian so results of [ 8 ], [10] apply) that A [ t ] satisfies the Ore conditions whith respect to $\mathcal{C}(P)$. By Proposition 1.1.2. it follows that P = Q, or p' = q', or Z(A/m A) is purely inseparable over C/mC. □

1.8. Lemma. Let A ba a maximal order over a discrete valuation ring C. The following properties are equivalent :

1. A is an Azumaya algebra.

2. CCl(A) = 1 and there exists a separable splitting field L of $\Sigma$ such that $A \otimes_C D$ is hereditary, where D is the integral closure of C in L.

<u>Proof.</u> The proof is based on the proof given in [11 ] but we did correct some errors. It is clear that only the implications 2 ⇒ 1 needs a proof.

Suppose that Z(A/mA) is purely inseparable over C/mC. The canonical map Br(C/mC) → Br(Z(A/mA)) is then surjective, so up to changing to a suitable matric ring over $\Sigma$ (and over A) we may assume that A/mA contains a C/mC-central simple algebra B such that $A/mA = B \otimes_{C/mC} Z(A/mA)$. The inverse image $A_1$ of B in A is a C-order contained in A and mA is a common ideal of A and $A_1$. Obviously mA is the unique nonzero prime ideal of $A_1$ and $Z(A_1/mA) = C/mC$. By definition D is a semilocal Dedekind domain and its nonzero prime ideals $m_1,\ldots,m_q$ over mD. Let $D_j$ be the discrete valuation ring $D_{m_j}$, $j = 1,\ldots,q_1$ and write $m'_j$ for the maximal ideal of $D_j$. Localizing the inclusions: $(A \otimes_C D)m \subset A_1 \otimes_C D \subset A \otimes D$, at $m_j$ say, yields : $(A \otimes_C D_j)m'_j \subset A_1 \otimes_C D_j \subset A \otimes_C D_j$. Since $(A_1 \otimes_C D_j)/(A \otimes_C D_j)m'_j = B \otimes_{C/mC} (D/m_jD)$ is central simple, it follows that $(A \otimes_C D_j)m'_j$ is the unique nonzero prime ideal of $A_1 \otimes_C D_j$. Therefore, ideals of $A \otimes_C D_j$ intersect $A_1 \otimes_C D_j$ in subsets of $(A \otimes_C D_j)m'_j$. Furthermore, the unique prime ideal of $Z(A/mA) \otimes_{C/mC} (D/m_j D)$ is the nilradical because $Z(A/mA)$ is a purely inseparable extension of $C/mC$ and $D/m_j D$ is algebraic over $C/m C$. Consequently there is only one prime ideal of $A \otimes_C D_j$ lying over $(A \otimes_C D_j)m'_j$. Harada's determination of the structure of hereditary orders in matrix rings (see also M. Artin, [ 1 ] )learns that the number of prime ideals of $A \otimes_C D_j$ lying over $m'_j$ is at least as big as the number of diagonal blocks of type $M_{n_j}(D_j)$ in the structure of $A \otimes_C D_j$ (which is hereditary since $A \otimes_C D$ was). This establishes that $A \otimes_C D_j = M_n(D_j)$ where n = p.i. deg(A) and therefore $A \otimes_C D$ is a reflexive Azumaya algebra over the Dedekind domain D hence it is an Azumaya algebra but then A is Azumaya over C , and then $Z(A/mA) = C/mC$. Consequently condition 2. excludes the alternative for being Azumaya given in Lemma 1.7.

1.9. <u>Remark 1.</u> Condition 2 in Lemma 1.8. may be replaced by the equivalent condition :

2'. $CCl(A) = 1$ and there exists a separable splittting field L of $\Sigma$ such that $A \otimes_C D$ is a tame D-order in $\Sigma \otimes_K L$. One easily verifies that the proof given remains valid if one starts from 2'.

2. The proof given above may easily be adapted to the situation where one consideres 2" : there exists a faithfully flat extension D of C such that D/m D is separable over C/mC which splits $\Sigma$, i.e. the field of fractions of D splits $\Sigma$.

Now we return to the general case. If A is a maximal order over a Krull domain C in $\Sigma$ and if L is a separable splitting field of

$\Sigma$ then we say that A is <u>L-tame</u> if $A \underset{C}{\otimes} D$ is a tame D-order in $\Sigma \underset{K}{\otimes} L$, where D is the integral closure of C in L. In the terminology of [11] we say that A is <u>Zariski-tamifiable</u> if for every $p \in X^{(1)}(C)$ there exists a separable splitting field L(p) of $\Sigma$ such that $A \underset{C}{\otimes} S(p)$ is a tame S(p)-order, where S(p) is the integral closure of $C_p$ in L.

1.10. <u>Theorem.</u> The following properties of a maximal order A over a Krull domain C are equivalent :

1. A is a reflexive Azumaya algebra over C.

2. CCl(A) = Cl(C) and A is Zariski-tamifiable

3. CCl(A) = Cl(C) and A is L-tame for some separable splitting field L of $\Sigma$.

<u>Proof.</u> The implications 1 ⇒ 3 ⇒ 2 are obvious, so let us prove 2 ⇒ 1. For every $p \in X^{(1)}(C)$, $A_p$ is a maximal $C_p$-order such that $CCl(A_p) = 1$. By Lemma 1.8. it follows that $A_p$ is an Azumaya algrbra over $C_p$. Since A is a maximal order it is $\kappa_1$-finitely presented over C, cf. [26 ] , thus the local conditions imply that A is a reflexive Azumaya algebra, cf. Proposition III.3.8. of [ 26] .

1.11. <u>Remark.</u> Let me point out that there do exist some splitting fields which do appear in a natural way as candidates for checking the condition 2. above. For any $p \in X^{(1)}(C)$ we first reduce to the complete local case at p and use the fact that $\hat{\Sigma}$ has an unramified splitting field, $\hat{L}_u$ say. If $\hat{S}_u$ is the integral closure of $\hat{C}_p$ in $\hat{L}_u$ then unramifiedness of both $\hat{A}$ and $\hat{S}_u$ entails that $\hat{A} \otimes_{\hat{C}_p} \hat{S}_u$ is a maximal order in $M_n(\hat{L}_u)$, hence an Azumaya algebra over $\hat{C}_p$. If $\hat{S}_u$ is separable or étale over $\hat{C}_p$ then the foregoing allows to deduce that A is an Azumaya algebra too by descent arguments. Note that separability of $\hat{S}_u$ contradicts the existence of purely inseparable elements in $Z(A_p/pA_p)$ over $C_p/pC_p = \overline{C}_p$ which are not contained in the latter. Indeed, if $\overline{S}_u$ (the residue of $\hat{S}_u$) is separable over $\overline{C}_p$, then $Z(A_p/p\,A_{p,\text{insep}}) \otimes_{\overline{C}_p} \overline{S}_u$ is a commutative field (by linear disjointness) of dimension larger than p.i.deg $(A_p/pA_p)$ (over $Z(A_p/pA_p)$) contained in $A_p/pA_p$.

## 2. The Central Class Group of a Graded Maximal Order.

In this section A is a maximal order over a Krull domain C and we assume that A is graded by an arbitrary group. Generalizations for just tame C-orders or gradations by cancellative semigroups are possible but outside the scope of this note. First we write down a slightly strenghtened version of a result which first appeared in **[23]**, but the proof remained the same.

2.1. Lemma. Let R' be a graded ring of type G such that R satisfies the identities of n x n - matrices. If every homogeneous element of R which happens to be central is invertible then R is an Azumaya algebra.

Proof. R has a multilinear central polynomial f (we will refer to f as the Razmyslov polynomial). Since f is not an identity for R it cannot vanish at every homogeneous substitution, say $c = f(\lambda_1,\ldots,\lambda_r) \neq 0$ for some $\lambda_i \in h(R)$. If c is invertible then the Formanek center of R equals Z(R) and R is an Azumaya algebra.□.

2.2. Corollary. A P.I. ring R which is gr-simple of type G is an Azumaya algebra.

Proof. If $c \in Z(R)$ is homogeneous in R then $Rc = R$ because R has no graded ideals and the lemma applies. □

Without any extra effort we have now obtained a generalization of known results in the sense that we do not assume the center Z(R) to be graded by G. In order to derive this from the fact that R is graded one has to impose conditions on G. e.g. G is abelian, or G is ordered.

2.3. Lemma. If R is a prime P.I. ring which is graded of type G for an arbitrary group G then every graded ideal I of R contains nonzero homogeneous elements in $I \cap Z(R)$.

Proof. If the Razmyzlov polynomial f vanishes for every substitution of the variables by homogeneous elements $i_{\sigma_1},\ldots,i_{\sigma_m}$ in a graded ideal I of R then also $f(i_1,\ldots,i_m = 0$ for every $i_1,\ldots,i_m \in I$. Pick $0 \neq c$ in $I \cap Z(R)$, then $c^N f(r_1,\ldots,r_m) = 0$ for all $r_1,\ldots,r_m \in R$ (N is the total degree of f). Since f is not an identity for R and since Z(R) is a domain, the latter is a contradiction. Therefore , $d = f(i_{\sigma_1},\ldots,i_{\sigma_m}) \neq 0$ for some homogeneous elements $i_{\sigma_1},\ldots,i_{\sigma_m} \in I$, and it is clear that

$d \in I \cap Z(R)$ is homogeneous in R. □

2.4. Corollary Let R be a maximal order over Krull domain C and suppose that R is graded by an arbitrary group G. Let $Q^h$ be the graded ring of fractions of R obtained by inverting the central homogeneous elements of R. Let $Q^g$ be the graded ring obtained by graded localization of R (cf. [ 21 ] [ 16 ] ) at the graded kernel functor $\tau$ with graded filter $\mathcal{L}^g(\tau)$ generated by the graded essential left ideals of R (i.e. $Q^g$ is the injective hull of R in the category R-gr). Then $Q^g = Q^h$ is an Azumaya algebra.

Proof. From M. Chamarie's definition of a Krull ring it follows that $\tau$ has finite type and therefore $Q^g = Q^g_\tau(R) = Q_{\underline{\tau}}(R)$, where $\underline{\tau}$ is the (ungraded) kernel functor corresponding to the filter of left ideals generated by the graded essential left ideals of R, (cf. [ 16 ] Section II.9, II.10 and in particular Proposition II.10.2.). By Lemma 1.3. : $Q^g = \cap\{R_P, P \in X^{(1)}(R)$, P contains no proper graded ideal of R $\}$ and $Q^h = \cap \{R_P, P \in X^{(1)}(R)$, P contains no homogeneous element of R$\}$.
By Lemma 2.3. $Q^g = Q^h$ follows. By Lemma 2.1. it follows that $Q^g = Q^h$ is an Azumaya algebra. □

2.5. Remark. In [ 16 ] is has been pointed out that a gr-Goldie ring need not always have a gr-simple gr-Artinian ring of fractions the above shows that this does hold in the case considered without any restriction on the gradation. Another consequence of $Q^h = Q^g$ is that one knows $Z(Q^g)$ well; i.e. $Z(Q^g)$ is

$(h^{\star}(R) \cap Z(R))^{-1} Z(R)$, where $h^{\star}(R)$ consists of the non-zero homogeneous elements of R.

2.6. Theorem. Let R be a maximal order over a Krull domain C and suppose that R is graded of type G. The following sequence of abelian groups is exact :

$1 \to H \to CCl(R) \to Cl(Z(Q^h)) \to 1$, where H is the subgroup generated by the classes of the $P \in X^{(1)}(R)$ such that $p = P \cap Z(R)$ contains a nonzero element of h(R).

Proof. Since $Q^h$ is an Azumaya algebra this follows immediately from Proposition 1.4. □

In the situation of the theorem we define the graded central class group $CCl^g(R)$ to be H. If G is torsion free abelian then $CCl^g(R)$ is generated by the classes of graded prime ideals of height one and then it is not hard to derive that $CCl^g(R) \cong CCl(R)$ and also $Picent^g(R) = Picent(R)$, where the group $Picent^g(R)$ is then defined to be the subgroup of Picent(R) generated by the classes graded representatives. If in this case R is an extension of $R_e$ then $R_e$ is also a maximal order over a Krull domain. At the other extreme we may consider the case where G is finite, in particular we may be interested in rings which are strongly graded by a finite group, as in [ 17] . In this case, for $P \in X^{(1)}(R)$ we have that $P_g$, the ideal generated by $P \cap h(R)$, is semiprime and $P_g = P_1 \cap \ldots \cap P_n$ with $P_1 = P$. If R is prime then $P_g$ is a nonzero graded ideal so it suffices to apply Lemma 2.3. and Theorem 2.6. to obtain :

2.7. Proposition. If R is a maximal order over a Krull domain C which is strongly graded by a finite group G then $CCl^g(R) = CCl(R)$ and $Z(Q^h)$ is factorial.

Another surprizing way of formulating a consequence of Theorem 2.6. is the following :

2.8. Proposition : If there exists a gradation of type G on a maximal order R over a Krull domain C then the essential valuations of C that ramify in $Q(R) = \Sigma$ correspond to $p \in X^{(1)}(C)$ such that p contains a nonzero homogeneous element of R.

Proof. Combining Proposition 1.5. and Theorem 2.6. in the following exact diagram (where $S_h = h^{\star}(R) \cap (R)$)

$$\begin{array}{ccccccccc} & & & & 1 & & & & \\ & & & & \uparrow & & & & \\ & & & & \oplus_P \mathbb{Z}/e_P\mathbb{Z} & & & & \\ & & & & \uparrow & & & & \\ 1 & \longrightarrow & H & \longrightarrow & CCl(R) & \xrightarrow{\ \pi\ } & Cl(S_h^{-1}C) & \longrightarrow & 1 \\ & & & & \uparrow j & & \| & & \\ 1 & \longrightarrow & j^{-1}H & \longrightarrow & Cl(C) & \longrightarrow & Cl(S_h^{-1}C) & \longrightarrow & 1 \\ & & & & \uparrow & & & & \\ & & & & 1 & & & & \end{array}$$

it is immediately clear that for a ramifying $P \in X^{(1)}(R)$, both P and $P^{e_P}$ have the same image under $\pi$ i.e. $c_\sigma P' \subset P^{e_P}$ for some $c_\sigma \in S_h$, P' a finitely generated ideal (we can take a centrally generated (left) ideal) such that $(P')^{\star\star} = P$. Then $c_\sigma P \subset (c_\sigma P')^{\star\star} \subset P^{e_P}$ and $c_\sigma \subset P$ follows if $e_P > 1$. □

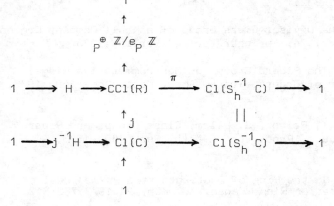

Remark. In other words if R allows a G-gradation then all ramification happens in the graded class group! Hence even if $CCl^g \neq CCl$ then the graded class group contains all the essential information on the order.

References :

[1] M. Artin, Left Ideals in Maximal Orders, LNM 917, 146-181 Springer Verlag, Berlin, 1982.

[2] M. Auslander, O. Goldman, Maximal Orders, Trans. Amer. Math. Soc. 97, 1960, 1-24.

[3] M. Chamarie, Anneaux de Krull non-commutatifs, Thèse, Univ. Claude-Bernard, Lyon I, 1981.

[4] R. Fossum, Maximal Orders over Krull Domains, J. of Algebra, 10, 1968, 321-332.

[5] R. Fossum, The Divisor Class Group of a Krull Domain, Ergebn. der Math. Wiss. 74, Springer Verlag, Berlin, 1973.

[6] A. Fröhlich, The Picard Group of Non-commutative Rings, in Particular of Orders, Trans. Amer. Math. Soc. 180, 1973, 1-45.

[7] A. Fröhlich, I. Reiner, S. Ullom, Class Groups and Picard Groups of Orders, Proc. London Math. Soc. 180, 1073, 405-434.

[8] A. Heinicke, On the Ring of Quotients at a Prime Ideal of a Right Noetherian Ring, Canad. J. Math, 24,1972,703-712.

[9] E. Jespers, P. Wauters, On Central Ω-Krull Rings and their Class Groups, to appear.

[10] J. Lambek, G. Michler, The Theory at a Prime Ideal of a Right Noetherian,

[11] L. Le Bruyn, On the Jespers-Van Oystaeyen Conjecture, J. of Algebra, 1983.

[12] L. Le Bruyn, F. Van Oystaeyen, Generalized Rees Rings and Relative Maximal Orders Satisfying Polynomial Identities, J. of Algebra, 83, No 2, 1983, 409-436.

[13] H. Marubayashi, Noncommutative Krull Rings, Osaka J. Math. 13, 1976, 491-500.

[14] H. Marubayashi; E. Nauwelaerts, F. Van Oystaeyen, Graded Rings over Arithmetical Orders, Comm. in Algebra.

[15] C. Năstăsescu, F. Van Oystaeyen, On Strongly Graded Rings and Crossed Products, Comm. in Algebra.

[16] C. Năstăsescu, F. Van Oystaeyen, Graded Ring Theorey, North Holland, Library of Math. 28, Amsterdam, 1982.

[17] E. Nauwelaerts, F. Van Oystaeyen, Orders strongly graded by finite groups, to appear.

[18] M. Orzech, Brauer Groups and Class Groups for a Krull Domain, LNM 917, 68-90, Springer Verlag, Berlin, 1982.

[19] R. Puystjens, J. Van Geel, Diagonalization of Matrices over Gr-PID, Linear Algebra and Applications, to appear?

[20] C. Procesi, Rings with Polynomial Identities, Marcel Dekker Monographs, New York, 1973.

[21] F. Van Oystaeyen, On Graded Rings and Modules of Quotients, Com. in Algebra 6, 1978, 1923-1959.

[22] F. Van Oystaeyen, Graded Prime Ideals and the Left Ore Conditions, Com. in Algebra 8, 1980, 861-868.

[23] F. Van Oystaeyen, Graded P.I. Rings, Bull. Soc. Math. Belg. XXXII, 1980, 21-28.

[24] F. Van Oystaeyen, Generalized Rees Rings and Arithmetically Graded Rings, J. of Algebra, Vol. 82, 1, 1983, 185-193.

[25] F. Van Oystaeyen, Some Constructions of Rings, in Proceedings of "The E. Noether Days 1982", J. P. Algebra, 1984.

[26] F. Van Oystaeyen, A. Verschoren, Relative Invariants of Rings. The Commutative Theory, Marcel Dekker Monographs in Pure and Applied Math. Vol. 79, New York, 1983.

[27] F. Van Oystaeyen, A.Verschoren, Relative Invariants of Rings, Orders, Monograph to appear.

[28] A. Verschoren, On the Picard Group of a Grothendieck Category, Com. in Algebra, 8, 1980, 1169-1194.

[ 29 ] A. Verschoren, Relative Picard and Brauer Groups, Thesis, UIA, 1982.

[ 30 ] S. Yuan, Reflexive Modules and Algebra Class Groups over Noetherian Integrally Closed Domains, J. of Algebra 32, 1974, 405-417.

# ON THE PICARD GROUP OF A QUASI-AFFINE SCHEME

A. Verschoren (★)

University of Antwerp, U.I.A.

## 0. Introduction.

In [6] the restriction map $\varphi_f$ : Pic(R) → Pic($R_f$) was studied for any noetherian integrally closed domain R and any prime element f ∈ R. In particular, it was proved in loc. cit that $\varphi_f$ is always injective and several sufficient conditions for the surjectivity of $\varphi_f$ were given. If should be noted that Pic(R) is just the Picard group of the affine scheme Spec(R), whereas Pic ($R_f$) is the Picard group of the open affine subscheme of Spec(R) determined by f. The obvious question in this context is of course : given X = Spec(R) and U ⊂ X, a not necessarily affine open subscheme of X, what can one say about the restriction map $\varphi_U$ : Pic(X) → Pic(U). In particular, one may ask when $\varphi_U$ is surjective or what Ker $\varphi_U$ is. It is the purpose of this note to provide answers to these questions.

## 1. Some torsion-theoretic background.

1.1. For simplicity's sake, let us assume R to be a noetherian domain. We denote by $X_R$ the affine scheme associated to R. The open subsets of $X_R$ for the Zariski topology are of the form $X_I$ for some ideal I of R. Here $X_I$ consists of the P ∈ $X_R$ with I ⊄ P.

An idempotent kernel functor σ in R-mod, the category of all R-modules, is a left exact subfunctor of the identity with the property that σ(M/σM)=0 for all M in R-mod. The definition and elementary properties of the associated localization functor $Q_\sigma$ may be found in [3,7,8]. Let us just mention that it is left exact and that there is for each M ∈ R-mod a canonical morphism $j_M$ : M → $Q_\sigma$(M) with the property that Ker($j_M$) and

(★) The author is supported by N.F.W.O.



*F. van Oystaeyen (ed.), Methods in Ring Theory, 541–549.*

Coker $(j_M)$ are $\sigma$-torsion. Let us call an R-module M $\sigma$-closed if $j_M$ is an isomorphism, i.e. $Q_\sigma(M) = M$. We denote by $\mathcal{L}(\sigma)$ the idempotent filter associated to $\sigma$, i.e. $\mathcal{L}(\sigma)$ consists of all ideals I of R such that $\sigma(R/I)=R/I$. It is well known that for any M $\in$ R-mod and any $m \in M$ we have $m \in \dot{\sigma} M$ if and only if there exists an ideal $I \in \mathcal{L}(\sigma)$ with $Im = 0$. In particular, for any $m \in Q_\sigma(M)$ we may find $I \in \mathcal{L}(\sigma)$ such that $Im \subset M/\sigma M \subset Q_\sigma(M)$.

1.2. Let us give some examples. If I is an ideal of R, then we may associate to it an idempotent kernel functor $\sigma_I$ with idempotent filter $\mathcal{L}(I)$, where $\mathcal{L}(I)$ consists of all ideals L of R containing a power $I^n$ of I. The associated localization is denoted by $Q_I(-)$. Let M $\in$ R-mod and let $\widetilde{M}$ denote the associated quasi-coherent sheaf on $X_R$, then one may prove, cf. [ 8,5], that $\Gamma(X_I,\widetilde{M}) = Q_I(M)$ for any ideal I of M.
Another example may be provided as follows. If $P \in X_R$, then we associate to it the idempotent kernel functor $\sigma_{R-P}$ with idempotent filter $\mathcal{L}(R-P)$ consisting of all ideals I of R with $I \not\subset P$. The associated localization functor $Q_{R-P}(-)$ is just localization at the prime ideal P, in the usual sense. The idempotent kernel functors $\sigma_{R-P}$ play a fundamental role in the theory of localization at idempotent kernel functors. Indeed, for any idempotent kernel functor $\sigma$, let us denote by $C(\sigma)$ the set of all ideals I of R, maximal with respect to the property $I \notin \mathcal{L}(\sigma)$. It is then easy to see that actually $C(\sigma) \subset X_R$. Moreover; one may show (e.g. [ 7,8 ] ) that $\sigma = \inf \{\sigma_{R-P};\ P \in C(\sigma)\}$, where inf of a family of idempotent kernel functor is defined in the obvious way.

1.3. Recall the following definition from [ 9 ] . Let $\sigma$ be an idempotent kernel functor in R-mod, then an R-module P is said to be $\sigma$-invertible if it is $\sigma$-closed and if there exists a $\sigma$-closed R-module Q such that $Q_\sigma(P \otimes_R Q) = Q_\sigma(R)$ (upto isomorphism in R-mod). The set of isomorphism classes [ P ] of $\sigma$-invertible R-modules P may be given the structure of a commutative group with multiplication induced by $[P_1].[P_2] = [Q_\sigma(P_1 \otimes_R P_2)]$ . We will denote this group by Pic(R,$\sigma$). The identity of Pic(R,$\sigma$) is the class $[Q_\sigma(R)]$ and for any $[P] \in \mathrm{Pic}(R,\sigma)$ the inverse $[P]^{-1}$ is the class of $P^\star = \mathrm{Hom}_R(P,R)$. One of the results of [ 9 ] may now be stated as

1.4. Theorem. Let R be a noetherian domain and I an ideal of R, then $\mathrm{Pic}(X_I) = \mathrm{Pic}(R,\sigma_I)$. □

1.5. Let U be a quasi-affine subscheme of $X_R$, then we may identify U with $X_I$ for some ideal I of R. In order to study the canonical restriction map $\mathrm{Pic}(X_R) \to \mathrm{Pic}(X_I)$, it clearly suffices by the foregoing to look at the map $\varphi_I : \mathrm{Pic}(R) \to \mathrm{Pic}(R,\sigma_I) : [P] \to [Q_I(P)]$ .
In this note we will restrict to the case $I \in X_R$.

## 2. The kernel of $\varphi_I$

2.1. From now on R will denote a noetherian integrally closed domain and I will denote a fixed (nonzero) prime ideal of R. We will write Z for the set of all minimal prime ideals of R. We say that M is reflexive if M is finitely generated and if $M = M^{\star\star}$, the double dual of M. Let $\sigma = \inf\{\sigma_{R-P};\ P \in Z\}$, then it has been pointed out in [ 10 ] that $C(\sigma) = Z$ and that for any finitely generated torsion-free R-module M we have $M^{\star\star} = Q_\sigma(M)$. If L is an ideal of R we will write $L^{(n)}$ for $(L^n)^{\star\star}$; similarly, let us write $L \perp M$ for $(LM)^{\star\star}$ for any ideal L of R and any R-module M. We then have

2.2. Lemma. For any reflexive R-module M the following statements are equivalent :

2.2.1. $Q_I(M) = Q_I(R)$ ;

2.2.2. There exists a positive integer $n \in \mathbb{N}$ such that $I^n \perp M = R$.

<u>Proof.</u> (1) $\Rightarrow$ (2) Since M is reflexive, it is torsionfree, so M embeds canonically into $Q_I(M)$. On the other hand M is finitely generated, so we may find a positive integer $n \in \mathbb{N}$ such that $I^nM \subset R$. Choose n minimal as such and let us prove that for all minimal prime ideals p of R this induces an isomorphism $(I^nM)_p = R_p$ (everything within $M \otimes K = K$, the field of fractions of R). Indeed, if $p \not\supset I$, then $Q_I(M) = Q_I(R)$ yields an isomorphism $Q_I(M)_p = Q_I(I^nM)_p = (I^nM)_p = Q_I(R)_p = R_p$. On the other hand, if $p \supset I$, then $I = p$ since p is minimal.
Assume $(I^nM)_I \hookrightarrow R_I$ is not an isomorphism, then since $R_I$ is local and the map is obviously injective, we have $I^nM \subset (I^nM)_I \subset I R_I$. Let us show that this implies $I^n M \not\subset I$. Indeed, otherwise $I_I^n M_I \subset I_I$, hence $I_I^{n-1} M_I \subset R_I$, since $I_I$ is invertible in $R_I$ as $R_I$ is a discrete valuation ring. Moreover, for all other minimal primes p we clearly have $I_p^{n-1}M_p = M_p \subset R_p$, so $I^{n-1}M = I^{n-1} \cap M_p \subset \cap (I^{n-1}M)_p \subset \cap R_p = R$, where the intersection is taken over all minimal prime ideals of R. However, n was chosen to be minimal with the property $I^n M \subset R$, so we derived a contradiction, i.e. $I^nM \not\subset I$. It follows that we may find $x \in I^nM - I \subset R - I$ such that $x \in IR_I$, i.e. there is $g \in R - I$ with $g x \in I$. This yields a contradiction, so indeed $(I^nM)_I = R_I$. Finally, we thus find that $R = \cap R_p = \cap (I^nM)_p = I^n \perp M$.

(2) $\Rightarrow$ (1) Let $I^n \perp M = R$, i.e. $(I^nM)^{**} = R$ and pick $P \in C(\sigma_I)$. Then $I^n \not\subset P$, hence $R_P = [(I^nM)^{**}]_P = [(I^nM)_P]^{**}$, where the dual is now taken in $R_P$-mod. But $(I^nM)_P = (I^n)_PM_P = M_P$, so $R_P = (M_P)^{**} = (M^{**})_P = M_P$. Since this holds for all $P \in C(\sigma_I)$, we find that $Q_I(M) = Q_I(R)$, indeed. □

2.3. Corollary. If M is a reflexive R-module and I is not minimal, then $Q_I(M) = Q_I(R)$ implies $M = R$.

Proof. If $I \not\in Z = C(\sigma)$, then clearly $I \in \mathcal{L}(\sigma)$, since for any prime ideal P of R and any idempotent kernel functor $\tau$ in R-mod, we have either $\tau(R/P) = R/P$ or $\tau(R/P) = 0$. It follows that $I^n \in \mathcal{L}(\sigma)$ for all positive integers $n \in \mathbb{N}$. Choose n as in (2.2.), then $\sigma(M/I^nM) = M/I^nM$, hence $Q_\sigma(I^nM) = Q_\sigma(M)$, i.e. $M = I^n \perp M$, proving the assertion. □

2.4. Corollary For any projective R-module M the following statements are equivalent :

2.4.1. $Q_I(M) = Q_I(R)$ ;

2.4.2. There exists a positive integer $n \in \mathbb{N}$ such that $I^{(n)}M = R$.

Proof. It suffices to note that for M projective, we have

$$I^n \perp M = Q_\sigma(I^nM) = Q_\sigma(I^n \otimes M) = \bigcap_{p\in Z} (I^n \otimes M)_p =$$

$$= \bigcap_{p\in Z} I^n \otimes R_p \otimes M = (\bigcap_{p\in Z} I^n_p) \otimes M = I^{(n)} \otimes M$$

$$= I^{(n)}M. \quad \square$$

Let < I > denote the group generated by the the isomorphism classes of invertible ideals amongst the $I^{(n)}$, $(n \in \mathbb{N})$, then

2.5. Theorem. With the above notations, the following sequence

$$0 \longrightarrow \langle I \rangle \longrightarrow \mathrm{Pic}(R) \xrightarrow{\varphi_I} \mathrm{Pic}(R,\sigma_I)$$

is exact.

Proof. First note that if $I^{(n)}$ is invertible, i.e. $[I^{(n)}] \in \mathrm{Pic}(R)$, then $Q_I(I^{(n)}) = Q_I(R)$ since for each $P \in C(\sigma_I)$, the canonical inclusion $I^{(n)} = (I^n)^{**} \to R$ induces an isomorphism $[I^{(n)}]_p = R_p$ as we have seen previously. It follows immediately that $\langle I \rangle \subset \mathrm{Ker}(\varphi_I)$. Conversely, if $[P] \in \mathrm{Ker}(\varphi_I)$, then $Q_I(P) = Q_I(R)$ (upto isomorphism), hence by (2.4.)

we may find a positive integer $n \in \mathbb{N}$ such that $I^{(n)}P = R$. Let $Q \in$ R-mod be such that $P \otimes Q = R$, then we obtain that :

$$Q = R \otimes Q = (I^{(n)}P) \otimes Q = (I^{(n)} \otimes P) \otimes Q = I^{(n)},$$

i.e. $I^{(n)}$ is invertible, i.e. $[I^{(n)}] \in \langle I \rangle$. This finishes the proof. □

2.6. Corollary. [ 6 ] . Let f be a prime element of R, then the canonical map $Pic(R) \to Pic(R_f)$ is injective.

Proof. Let I = Rf, then we may apply the foregoing, i.e. $Ker(\varphi_f) = \langle Rf \rangle$. Now $Rf^n$ is free, for all $n \in \mathbb{N}$, hence $\langle Rf \rangle = 0$, which proves the assertion. □

2.7. Corollary. Let I be an invertible prime ideal of R, then the kernel of the canonical restriction $Pic(R) \to Pic(Q_I(R))$ is generated by $[I] \in Pic(R)$. □

2.8. Note. It is clear that if $\sigma$ is an idempotent kernel functor in R-mod whose idempotent filter is generated by prime ideals , then (2.2.) may be reformulated (and proved) in the obvious way, in order to apply to $\sigma$. It may then be checked that the foregoing applies to more general open subsets of $X_R$, than those determined by $I \in X_R$. We leave details as an exercise to the reader.

## 3. The surjectivity of $\varphi_I$

3.1. Lemma. Let $\sigma$ be an arbitrary idempotent kernel functor in R-mod and let P be a $\sigma$-invertible module, then we may find a reflexive R-module $P_1$ such that $Q_\sigma(P_1) = P$.

Proof. If Q is a $\sigma$-closed R-module with the property that $Q_\sigma(P \otimes_R Q) = Q_\sigma(R)$, then we know from [ 9 ] that $P = Hom_R(Q, Q_\sigma(R))$. It follows that

P is torsion-free, since if $p \in P$ and $s \in R - \{0\}$, then viewing p as an R-linear homomorphism from Q to $Q_\sigma(R) \subset K$, we have that $sp = 0$ implies $s\,p(Q) = 0$ in K, hence $p(Q) = 0$ and $p = 0$. We may thus view all localizations as calculated within $P \otimes_R K$. As P is $\sigma$-invertible, it is in particular $\sigma$-finitely presented hence there is $P' \subset P$ which is finitely generated and such that $\sigma(P/P') = P/P'$. Let $P_1 = (P')^{**}$, then it suffices to verify that $Q_\sigma(P_1) = P$ in order to finish the proof. But $Q_\sigma(P_1) =$
$= \bigcap_{p\in C(\sigma)} P_{1,p} = \bigcap_{p\in C(\sigma)} (P')_p^{**} = \bigcap_{p\in C(\sigma)} P_p^{**}$, the double duals being taken in $R_p$-mod for each $p \in C(\sigma)$. But now, since $Q_\sigma(P \otimes_R Q) = Q_\sigma(R)$, we have $P_p \otimes Q_p = (P \otimes_R Q)_p = R_p$ for all $p \in C(\sigma)$, hence $P_p$ is invertible, hence free in $R_p$-mod, so $P_p^{**} = P_p$ within $P \otimes_R K$. We thus obtain $Q_\sigma(P_1) =$
$= \bigcap_{p\in C(\sigma)} P_p = Q_\sigma(P) = P$, which proves the assertion. □

<u>3.2. Corollary</u>. If for all maximal ideals m of R with $I \subset m$ we have $Pic(R_m, \sigma_I) = 0$, then de restriction morphism

$$Pic(R) \to Pic(R,\sigma_I)$$

is surjective.

<u>Proof.</u> Let P be a $\sigma_I$-invertible R-module and let Q be a reflexive R-module with $Q_I(Q) = P$. It suffices to prove that Q is invertible. Now, if m is a maximal ideal with $I \not\subset m$, then $Q_m = Q_I(Q)_m = P_m$ is invertible. On the other hand, if $I \subset m$, then by assumption $Q_I(Q_m) = Q_I(R_m)$, hence, by (2.2.) there exists a positive integer $n \in \mathbb{N}$ with $I^n \perp Q_m = R_m$. But then $R_m = (I^nQ_m)^{**} = (I_m^nQ_m)^{**} = Q_m^{**} = Q_m$, all double duals taken in $R_m$-mod. It follows that $Q_m$ is invertible for $I \subset m$ as well. Globalizing yields that Q is invertible, which finishes the proof. □

The foregoing criterion is closely linked to the following.

3.3. Proposition. With notations as before, the following statements are equivalent.

3.3.1. $\mathrm{Pic}(R_p, \sigma_I) = 0$ for all prime ideals $p \in \mathcal{L}(I)$;

3.3.2. $R_p$ is parafactorial for all $p \supsetneq I$.

Proof. (2) ⇒ (1). Let $P \in \mathrm{Pic}(R_p, \sigma_I)$, then by (3.1.) we may find a reflexive R-module Q such that $Q_I(Q) = P$. Let $Y \subset X_R$ consist of all prime ideals p such that $Q_p$ is not free, then Y is closed and contained in $X_R - X_I$. Moreover, $I \notin Y$, as $R_I$ is a discrete valuation ring, hence $Q_I$ is free being reflexive. Just as in [6] it follows that $(X_R, Y)$ is a parafactorial couple, hence there exists an invertible sheaf $\underline{P} = \tilde{P}_1$ on $X_R$ with $\underline{P}|X_R - Z = \tilde{Q}|X_R - Z$. But then $P = Q_I(Q) = Q_I(P_1)$, where $P_1$ is an invertible R-module.

(1) ⇒ (2). Let us mimick the proof of a similar statement in [6]. Pick $I \subsetneq p$, then depth $R_p \geq 2$, as $pR_p/IR_p$ contains a nonzero divisor. So, let $\underline{P}$ be an invertible sheaf on $U = \mathrm{Spec}(R_p) - \{pR_p\}$, then by [4] it suffices to prove that $\underline{P}$ is trivial. By (3.1.) we may find a reflexive $R_p$-module Q such that $\tilde{Q}|U = \underline{P}$. Now, $U \supset \mathrm{Spec}(R_p) - \{I\}$, so $Q_I(Q)$ is a $\sigma_I$-invertible $Q_I(R_p)$-module, cf. (1.4.). By assumption we then have $Q_I(Q) = Q_I(R_p)$ and from (2.1.) it follows that there is a positive integer n such that $I^n \perp Q = R_p$ in $R_p$-mod. But, as Q is an $R_p$-module and $I \in \mathcal{L}(R-p)$, it follows that $I^n \perp Q = Q^{**} = Q$, i.e. $Q = R_p$ and $\underline{P}$ is trivial as we asserted. □

## References

[1] Bourbaki, N., Algèbre commutative, Chap. VII, Herman, Paris, 1965.

[2] Fossum, R., The divisor class group of a Krull domain, Springer Verlag, Berlin, 1973.

[3] Golan, J. Localization in noncommutative rings, Marcel Dekker, New York, 1975.

[4] Grothendieck, A. Dieudonné, Eléments de géométrie algébrique IV, (1,4), Publ. Math. IHES 20, 32, 1964,1967.

[5] Hartshorne, R. , Algebraic Geometry, Springer Verlag, Berlin, 1977.

[6] Ischebeck, F. Uber die Abbildung Pic(A) → Pic($A_f$), Math. Ann. 243 (1979) 237-245.

[7] Stenstrom, B, Ring of Quotients, Springer Verlag, Berlin, 1975.

[8] Ven Oystaeyen, F. Prime Spectra in Non-commutative Algebra, LNM 444, Springer Verlag, Berlin, 1975.

[9] Verschoren, A., On the Picard Group of a Grothendieck Category, Comm. in Algebra 8 (1980) 1169-1194.

[10] Verschoren, A, On the reflexive class group, to appear.

[11] Verschoren, A., Exact sequences for relative Brauer Groups and Picard Groups, to appear.

DUALITY THEORY FOR QUASI-INJECTIVE MODULES

J.M. Zelmanowitz

Department of Mathematics
University of California, Santa Barbara,
California 93106 U.S.A.

ABSTRACT

The objective of this article is to characterize a quasi-injective module in terms of the duality it determines.

INTRODUCTION

Every Morita duality between full subcategories $\mathcal{C}$ of R-Mod and $\mathcal{D}$ of Mod-S is determined by an R-S-bimodule U which is a balanced injective cogenerator on both sides. (Recall that a Morita duality is a category equivalence between $\mathcal{C}$ and $\mathcal{D}$ where ${}_R R \in \mathcal{C}$, $S_S \in \mathcal{D}$, and $\mathcal{C}$ and $\mathcal{D}$ are closed under submodules and factor modules.) If U is assigned the discrete topology then every Morita duality extends to a duality between the closed submodules of direct products of copies of ${}_R U$ and Mod-S (see [4], [5], [7]).

Quite generally, if U is an R-S-bimodule with a structure of topological R-module such that the elements of S induce continuous R-endomorphisms of U then there is a duality between the category $\mathrm{Refl}_R U$ of U-reflexive topological R-modules and the category $\mathrm{Refl}\ U_S$ of U-reflexive right S-modules (Proposi-



*F. van Oystaeyen (ed.), Methods in Ring Theory, 551–566.*

tion 2.1). Our principal objective in this article is to provide a full description of this duality when ${}_RU$ is a (topologically) quasi-injective module (that is, every continuous R-homomorphism from a submodule of U to U can be extended to a continuous R-endomorphism of U). We are able to do this when the topology on U is either discrete (so that ${}_RU$ is quasi-injective in the usual algebraic sense) or is compact with no small submodules (that is, there exists an open neighborhood of 0 which contains no non-zero open submodule). This latter type of topology arises in the study of Pontryagin duality because the quotient topology on $\mathbb{R}/\mathbb{Z}$ is a topology of this sort.

The main result of this paper (Theorem 3.3) reads in part as follows. If a bimodule ${}_RU_S$ has a topology of one of the above types and S is naturally isomorphic to the group of continuous R-endomorphisms of U, then $\mathrm{Refl}_R U$ is the class of modules copresented by ${}_RU$, Refl $U_S$ is the class of modules cogenerated by $U_S$, and for every submodule K of ${}_RU$ there is a natural isomorphism between the U-double dual of K and the double annihilator of K. Conversely, the latter conditions imply that ${}_RU$ is quasi-injective. Theorem 3.3 also provides a lengthy list of other properties of such a module.

In Theorem 3.7 we apply the above result to give a description of the duality associated with a quasi-injective self-cogenerator. The only difference between this situation and the one described in the preceding paragraph is that here $\mathrm{Refl}_R U$ consists of the closed submodules of direct products of copies of ${}_RU$.

This kind of duality has been rather extensively investigated. Perhaps the first person to explicitly consider (discrete) quasi-injective modules was Sandomierski [6] who concentrated particularly on the situation of a discrete quasi-injective self-cogenerator with essential socle which is also linearly compact. This setting was later examined in [4] and [7]. Duality for discrete quasi-injectives was treated also in [3]. For quasi-injective self-cogenerators with topologies of the types treated in this paper, an extensive study and a characterization in terms of "good dualities" is given in [4; Proposition 4.6]. The descriptions in this paper are of a different sort, prompted by the identification of $\mathrm{Refl}_R U$ with the modules copresented by ${}_R U$. Our principal interest is in the discrete case, but a simultaneous treatment of compact modules with no small submodules is made possible by the results contained in [1] and [4].

§1. Throughout this paper, R will denote an arbitrary ring, S a ring with identity element, and U an R-S-bimodule endowed with a structure of a topological group such that the elements of S induce continuous R-endomorphisms of U. Homomorphisms will be written on the side of a module opposite to that of the scalars. Mod-S will stand for the category of right S-modules and $\mathrm{Hom}_S(L,L')$ will often be abbreviated $H_S(L,L')$.

For M and N topological R-modules $\mathrm{Cont}_R(M,N)$ or $C_R(M,N)$ will denote the group of continuous R-homomorphisms from M to N. By a topological R-monomorphism $f:M \to N$ we mean an R-monomorphism which is continuous and open onto Mf endowed with the subspace topology; and a topological R-isomorphism will indicate an R-isomorphism which is also a homeomorphism. R-Top will denote the category of Hausdorff topological R-modules with the ring R having the discrete topology. Note that to say that the topology is Hausdorff is equivalent to requiring that the intersection of the (basic) open neighborhoods of 0 is 0.

For any right S-module L the R-module $\mathrm{Hom}_S(L,U)$ will always be assumed to carry the topology induced by considering $\mathrm{Hom}_S(L,U)$ as an R-submodule of the product module $U^L$. That is, the basic open neighborhoods of 0 are of the form $\mathcal{O}(x_1,\ldots,x_t) = \{f \in \mathrm{Hom}_S(L,U) \mid fx_1,\ldots,fx_n \in \mathcal{O}\}$ for some $x_1,\ldots,x_t \in L$ and open neighborhood $\mathcal{O}$ of 0 in U. When U is discrete this is called the finite topology on $\mathrm{Hom}_S(L,U)$.

We will require the following elementary facts which were needed also in [7].

(1.1) <u>For any index set</u> $X$ <u>the canonical</u> R-<u>isomorphism</u> $\theta : \mathrm{Hom}_S(S^{(X)}, U) \to U^X$ <u>is a topological</u> R-<u>isomorphism</u>.

$\theta$ is defined by $(f)\theta = \{f(e_x)\}_{x\in X}$ where $f \in \mathrm{Hom}_S(S^{(X)}, U)$ and $e_x \in S^{(X)}$ is the element with a 1 in the x-coordinate and 0 in all other coordinates.

(1.2) <u>For</u> $L, L' \in$ Mod-S <u>each</u> $g \in \mathrm{Hom}_S(L,L')$ <u>induces an element</u> $g^* \in \mathrm{Cont}_R(\mathrm{Hom}_S(L',U), \mathrm{Hom}_S(L,U))$ <u>defined by</u> $(f)g^* = f \circ g$ <u>for</u> $f \in \mathrm{Hom}_S(L',U)$. <u>If</u> $g$ <u>is an epimorphism then</u> $g^*$ <u>is a topological R-monomorphism</u>.

(1.3) <u>Suppose that</u> $M' \xrightarrow{f} M \xrightarrow{g} M'' \to 0$ <u>is an exact sequence in</u> R-Top. <u>If</u> $g$ <u>is an open mapping then the induced sequence</u> $0 \to \mathrm{Cont}_R(M'',U) \xrightarrow{g^*} \mathrm{Cont}_R(M,U) \xrightarrow{f^*} \mathrm{Cont}_R(M',U)$ <u>is exact</u>.

While we are principally interested in the situation when $U$ is a discrete module, our main results are valid also for the case where $U$ is a compact module with <u>no small submodules</u> (that is, there is an open neighborhood of 0, called a <u>small neighborhood</u>, which contains no non-zero submodules of $U$). Of course a discrete module has no small submodules.

(1.4) <u>If</u> $U$ <u>has no small submodules and</u> $S = \mathrm{Cont}_R(U,U)$ <u>then, for any index set</u> $X$, <u>there is a natural isomorphism</u> $\mathrm{Cont}_R(U^X,U) \cong S^{(X)}$. [4; Corollary 3.4].

<u>Proof</u>.

A homomorphism $S^{(X)} \to \mathrm{Cont}_R(U^X,U)$ is obtained by assigning to $\{s_x\}_{x\in X} \in S^{(X)}$ the continuous R-homomorphism defined by $\{u_x\}_{x\in X} \to \sum_{x\in X} u_x s_x$. This homomorphism is an isomorphism because, with the product topology on $U^X$, every element of $\mathrm{Cont}_R(U^X,U)$ must factor through a finite direct sum of copies of U. □

§2. For $M \in R\text{-Top}$ and $L \in \text{Mod-}S$ we define natural homomorphisms $\mu_M : M \to \text{Hom}_S(\text{Cont}_R(M,U),U)$ and $\nu_L : L \to \text{Cont}_R(\text{Hom}_S(L,U),U)$ in in the usual way: $(m\mu_M)f = mf$ and $g(\nu_L x) = gx$ for $m \in M$, $x \in L$, $f \in \text{Cont}_R(M,U)$ and $g \in \text{Hom}_S(L,U)$. $\nu_L x$ is continuous for each $x \in L$ and $\mu_M$ is a continuous R-homomorphism.

We call $M \in R\text{-Top}$ U-*reflexive* if $\mu_M$ is a topological R-isomorphism; similarly $L \in \text{Mod-}S$ is called U-*reflexive* if $\nu_L$ is an S-isomorphism. We will say that there is a *U-duality* between full subcategories $\mathcal{C} \subseteq R\text{-Top}$ and $\mathcal{D} \subseteq \text{Mod-}S$ when $\text{Cont}_R(\_,U) : \mathcal{C} \to \mathcal{D}$, $\text{Hom}_S(-,U) : \mathcal{D} \to \mathcal{C}$, and every module in $\mathcal{C}$ and $\mathcal{D}$ is U-reflexive.

The following elementary observation shows that U-dualities always exist.

PROPOSITION 2.1 *There is a U-duality between* $\text{Refl}_R U = \{M \in R\text{-Top} \mid M$ *is* U-*reflexive*$\}$ *and* $\text{Refl } U_S = \{L \in \text{Mod-}S \mid L$ *is* U-*reflexive*$\}$.

*Proof*. It suffices to show that for each $M \in \text{Refl}_R U$ and each $L \in \text{Refl } U_S$, $M^* = \text{Cont}_R(M,U)$ and $L^* = \text{Hom}_S(L,U)$ are U-reflexive. This follows from the identities $\mu^*_M \circ \nu_{M^*} = 1_{M^*}$, $\mu_{L^*} \circ \nu^*_L = 1_{L^*}$ together with (1.2) and (1.3). □

A topological R-module M is *quasi-injective* if for every topological R-monomorphism $i : N \to M$ and every $f \in \text{Cont}_R(N,M)$ there exists $f' \in \text{Cont}_R(M,M)$ with $if' = f$. If $_RM$ is discrete then this agrees with the usual definition of a quasi-injective module. The following key result is due to Lambek [2; Proposition 5.2] in the discrete case and Menini and Orsatti [4; Proposition 3.9] in the compact case.

PROPOSITION 2.2 *Suppose that* U *is a quasi-injective module which is discrete or is compact with no small submodules. Then* $_RU$ *is* $U^X$-*injective; that is, given a topological* R-*monomorphism*

$i : M \to U^X$ and $f \in \mathrm{Cont}_R(M,U)$ there exists $f' \in \mathrm{Cont}_R(U^X,U)$ with $if' = f$.

This can readily be extended as follows.

COROLLARY 2.3 *Suppose that $U$ is a quasi-injective module which is discrete or is compact with no small submodules. Then given a topological R-monomorphism $i : M \to U^X$ and $f \in \mathrm{Cont}_R(M,U^Y)$ there exists $f' \in \mathrm{Cont}_R(U^X,U^Y)$ with $if' = f$.*

We also require the following result which appears in [4; Proposition 3.9]; see also [7] for the discrete case.

PROPOSITION 2.4 *Suppose that $U$ is discrete or is compact with no small submodules and that $S = \mathrm{Cont}_R(U,U)$. If ${}_RU$ is quasi-injective then $\nu_L$ is an S-epimorphism for every $L \in$ Mod-S.*

We will find it helpful to be able to refer to the isomorphisms given in the following pair of lemmas. Here $\ell_U(L) = \{u \in U \mid uL = 0\}$ and $r_S(K) = \{s \in S \mid Ks = 0\}$.

LEMMA 2.5 *For any left ideal $L$ of $S$ the R-isomorphism $\phi : \mathrm{Hom}_S(S/L,U) \to \ell_U(L)$ defined by $f \to f(\bar{1})$ is a topological R-isomorphism.*

*Proof.* It is clear that $\phi$ is an R-isomorphism. The basic open sets in $\mathrm{Hom}_S(S/L,U)$ are of the form $\mathcal{O}(\bar{s}_1,\ldots,\bar{s}_t) = \{f \in \mathrm{Hom}_S(L,U) \mid f\bar{s}_1,\ldots,f\bar{s}_t \in \mathcal{O}\}$ for some $\bar{s}_1,\ldots,\bar{s}_t \in S/L$ and $\mathcal{O}$ an open neighborhood of 0 in U. Now $(\mathcal{O}(\bar{s}_1,\ldots,\bar{s}_t))\phi = \{f(\bar{1}) \mid f(\bar{s}_1),\ldots,f(\bar{s}_t) \in \mathcal{O}\} = \{f(\bar{1}) \mid f(\bar{1})s_1,\ldots,f(\bar{1})s_t \in \mathcal{O}\} = \{u \in \ell_U(L) \mid u \in \bigcap_{i=1}^{t} \mathcal{O}s_i^{-1}\} = \ell_U(L) \cap \bigcap_{i=1}^{t} \mathcal{O}s_i^{-1}$ which is an open subset of $\ell_U(L)$. Hence $\phi$ is an open map. On the other hand, if $\mathcal{O}$ is open in U then $\mathcal{O}\phi^{-1} = \{f \in \mathrm{Hom}_S(L,U) \mid f(\bar{1}) \in \mathcal{O}\} = \mathcal{O}(\bar{1})$ is open, and so $\phi$ is continuous. □

COROLLARY 2.6 *If* ${}_RK \subseteq {}_RU$ *and* $S/r_S(K)$ *is U-reflexive then so is* $\ell_U r_S(K)$.

*Proof*. Take $L = r_S(K)$ in Lemma 2.5 and apply Proposition 2.1. □

LEMMA 2.7 *Suppose that* $S = \mathrm{Cont}_R(U,U)$, *let* $K$ *be a closed R-submodule of* $U$, *and assign* $U/K$ *the quotient topology. Then the S-homomorphism* $\psi : r_S(K) \to \mathrm{Cont}_R(U/K,U)$ *defined by* $\overline{u}(\psi(s)) = us$ *for* $u \in U$ *and* $s \in S$ *is an S-isomorphism*.

The proof is obvious.

REMARK 2.8 We will have use for the following observation. *If* $0 \to M \xrightarrow{f} U^X \xrightarrow{g} U^Y$ *is exact in R-Top with* $f$ *a topological R-monomorphism then* $Mf$ *is closed in* $U^X$ *and, with the quotient topology on* $U^X/Mf$, $\mu_{U^X/Mf}$ *is a monomorphism*.

The next result relates U-reflexivity to an injectivity property.

PROPOSITION 2.9 *Suppose that* $S = \mathrm{Cont}_R(U,U)$ *and that* $K = \ell_U(L)$ *for some* $L \subseteq S$. *Then* $S/r_S(K)$ *is U-reflexive if and only if every* $f \in \mathrm{Cont}_R(K,U)$ *extends to an element of* $S = \mathrm{Cont}_R(U,U)$.

*Proof*. Assume that $S/r_S(K)$ is U-reflexive and let $f \in \mathrm{Cont}_R(K,U)$ be given. We have to show that $i^* : \mathrm{Cont}_R(U,U) \to \mathrm{Cont}_R(K,U)$ is an epimorphism where $i$ is the inclusion map of $K$ into $U$.

Consider the diagram

$$\begin{array}{ccc} S = \mathrm{Cont}_R(U,U) & \xrightarrow{i^*} & \mathrm{Cont}_R(K,U) = \mathrm{Cont}_R(\ell_U r_S(K),U) \\ \downarrow p & & \downarrow \phi^* \\ S/r_S(K) & \xrightarrow{\nu_{S/r_S(K)}} & \mathrm{Cont}_R(\mathrm{Hom}_S(S/r_S(K),U),U) \end{array}$$

where $p$ is the canonical epimorphism and $\phi : \mathrm{Hom}_S(S/r_S(K),U) \to \ell_U r_S(K)$ is the topological R-isomorphism given by Lemma 2.5. This diagram commutes, $\phi^*$ is an S-isomorphism by (1.3), and $\nu_{S/r_S(K)}$ is an S-isomorphism by hypothesis. It therefore follows that $i^*$ is an epimorphism.

Conversely, assume that elements of $\mathrm{Cont}_R(K,U)$ extend to continuous endomorphisms of ${}_RU$. Let $i$ denote the inclusion map of $K$ into $U$ and consider the commutative diagram

$$\begin{array}{ccc} S & = & \mathrm{Cont}_R(U,U) \\ \downarrow q & & \downarrow i^* \\ S/L & \xrightarrow{\eta} & \mathrm{Cont}_R(K,U) \end{array}$$

where $\eta$ is defined by $x(\eta\bar{s}) = xs$ for all $s \in S$ and $x \in K$, and $q$ is the canonical epimorphism. $\eta$ is an S-epimorphism because $i^*$ is. Since $\ker \eta = r_S(K)/L$, $\eta$ induces an S-isomorphism of $S/r_S(K)$ with $\mathrm{Cont}_R(K,U)$. By Proposition 2.1, $S/r_S(K)$ will be U-reflexive provided that $K$ is U-reflexive.

Consider now the exact sequence $0 \to K \xrightarrow{i} U \xrightarrow{p} U/K \to 0$ where $p$ is the quotient map of $U$ onto $U/K$ with the quotient topology. Applying (1.3) and the hypothesis yields an exact sequence $0 \to \mathrm{Cont}_R(U/K,U) \xrightarrow{p^*} \mathrm{Cont}_R(U,U) \xrightarrow{i^*} \mathrm{Cont}_R(K,U) \to 0$. From (1.2) we get a commutative exact diagram

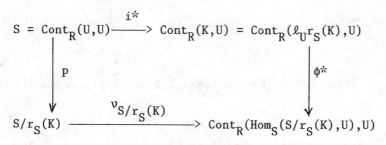

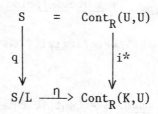

$$\begin{array}{ccccccccc} 0 & \longrightarrow & K & \xrightarrow{i} & U & \xrightarrow{p} & U/K & \longrightarrow & 0 \\ & & \downarrow \mu_K & & \downarrow \mu_U & & \downarrow \mu_{U/K} & & \\ 0 & \longrightarrow & H_S(C_R(K,U),U) & \xrightarrow{i^{**}} & H_S(C_R(U,U),U) & \xrightarrow{p^{**}} & H_S(C_R(U/K,U),U). & & \end{array}$$

$\mu_U$ is a topological R-isomorphism by (1.1), $\mu_K$ is a monomorphism because $K \subseteq U$, and $\mu_{U/K}$ is a monomorphism by the remark preceding this proposition ($K = \ell_U(L)$ induces an exact sequence $0 \to K \xrightarrow{i} U \xrightarrow{g} U^L$ where $ug = \{us\}_{s \in L}$). A diagram chase now establishes the fact that $\mu_K$ is onto, and $\mu_K$ is open because $i$ and $i^{**}$ are topological R-monomorphisms. Hence K is U-reflexive.

□

<u>§3</u>. We say that $M \in$ R-Top is <u>copresented</u> by ${}_RU$ if there exists an exact sequence $0 \to M \xrightarrow{f} U^X \xrightarrow{g} U^Y$ in R-Top with f a topological R-monomorphism; the exact sequence itself is called a <u>copresentation</u> of M. We let $\text{Copres}_R U$ denote the family of modules which are copresented by ${}_RU$. We indicate by $\text{Cogen}_R U$ (respectively, $\overline{\text{Cogen}_R U}$) the family of modules which are topologycally isomorphic to a (closed) submodule of a direct product of copies of ${}_RU$. We also write Cogen $U_S$ for the S-modules cogenerated by U. These families of modules are related to $\text{Refl}_R U$, the family of U-reflexive modules in R-Top, by the following observation.

PROPOSITION 3.1 $\text{Refl}_R U \subseteq \text{Copres}_R U \subseteq \overline{\text{Cogen}_R U}$.

<u>Proof</u>. Given any $L \in$ Mod-S, we can choose a presentation $S^{(Y)} \to S^{(X)} \to L \to 0$. Applying (1.2) and (1.1) yields a copresentation $0 \to \text{Hom}_S(L,U) \to U^X \to U^Y$ of $\text{Hom}_S(L,U)$. In particular, $\text{Refl}_R U \subseteq \text{Copres}_R U$. The second inclusion follows from the fact that kernels are closed in R-Top as was already noted in Remark 2.8.

□

We also require the next lemma, which is a consequence of Corollary 2.3.

LEMMA 3.2 Suppose that $U$ is a quasi-injective module which is discrete or is compact with no small submodules. For $M \in \text{Copres}_R U$ any topological $R$-monomorphism $g_1 : M \to U^X$ extends to a copresentation $0 \to M \xrightarrow{g_1} U^X \xrightarrow{g_2} U^Y$.

Proof. Fix a copresentation $0 \to M \xrightarrow{f_1} U^A \xrightarrow{f_2} U^B$ of $M$. By Corollary 2.3 there exist $\phi \in \text{Cont}_R(U^A, U^X)$ and $\psi \in \text{Cont}_R(U^X, U^A)$ with $f_1\phi = g_1$ and $g_1\psi = f_1$. Define $g_2 : U^X \to U^X \oplus U^B = U^{X \cup B}$ via $vg_2 = (v(\psi\phi - 1), v\psi f_2)$ for $v \in U^X$. $g_2$ is obviously in $\text{Cont}_R(U^X, U^{X \cup B})$, so it remains only to check that $0 \to M \xrightarrow{g_1} U^X \xrightarrow{g_2} U^X \oplus U^B$ is exact. If $m \in M$ then $mg_1g_2 = (mg_1(\psi\phi - 1), mg_1\psi f_2) = (mf_1\phi - mg_1, mf_1f_2) = (0,0)$, so $Mg_1 \subseteq \ker g_2$. On the other hand, if $vg_2 = (0,0)$ then $v\psi\phi = v$ and $v\psi f_2 = 0$. So $v\psi \subseteq \ker f_2 = Mf_1$. Write $v\psi = mf_1$ for some $m \in M$. Then $v = v\psi\phi = mf_1\phi = mg_1 \in Mg_1$. □

In order to describe the properties of the U-duality associated with a quasi-injective module we require the following concept. For $i : M \to M'$ an inclusion map with $M, M' \in R\text{-Top}$ then we say that $H_S(C_R(M,U),U)$ is isomorphic to $M'$ over $M$ provided that there exists a topological $R$-isomorphism $f : H_S(C_R(M,U),U) \to M'$ with $\mu_M f = i$. For $K$ an $R$-submodule of $U$ we define $\theta : S/r_S(K) \to C_R(K,U)$ by $x(\theta\bar{s}) = xs$ for any $s \in S$, $x \in K$. $\theta$ is clearly an $S$-monomorphism, and we will call it the canonical $S$-monomorphism from $S/r_S(K)$ to $C_R(K,U)$. In particular, $S/r_S(K)$ is in Cogen $U_S$ because $C_R(K,U) \subseteq U^K$.

THEOREM 3.3 Suppose that $U$ is discrete or is compact with no small submodules and that $S = \text{Cont}_R(U,U)$. Then the following conditions are equivalent.

(i) ${}_RU$ is quasi-injective.

(ii) *For every submodule* $K$ *of* ${}_RU$, $S/r_S(K)$ *is* U-*reflexive and the canonical* S-*monomorphism* $\theta : S/r_S(K) \to C_R(K,U)$ *is an isomorphism.*

(iii) Refl $U_S$ = Cogen $U_S$ *and for every submodule* $K$ *of* ${}_RU$, $H_S(C_R(K,U),U)$ *is isomorphic to* $\ell_U r_S(K)$ *over* $K$.

*When these conditions are satisfied then* $\mathrm{Refl}_R U = \mathrm{Copres}_R U$ *and the* U-*reflexive submodules of* ${}_RU$ *are the submodules of the form* $\ell_U(L)$ *for some* $L \subseteq S$.

Proof. (i) => (iii) From Proposition 2.4 we know that Refl $U_S$ = Cogen $U_S$. To show that $H_S(C_R(K,U),U)$ is isomorphic to $\ell_U r_S(K)$ over $K$, we may (by passing to the completion of $K$ if necessary) assume that $K$ is a closed submodule of ${}_RU$. Consider the exact sequence $0 \to K \xrightarrow{i} U \xrightarrow{p} U/K \to 0$ where $i$ is the inclusion map and $p$ is the quotient map. Applying (1.3) yields a commutative exact diagram

$$\begin{array}{ccccccccc} 0 & \longrightarrow & C_R(U/K,U) & \xrightarrow{p^*} & C_R(U,U) & \xrightarrow{i^*} & C_R(K,U) & \longrightarrow & 0 \\ & & \Big\downarrow \psi_K & & \Vert & & & & \\ 0 & \longrightarrow & r_S(K) & \hookrightarrow & S & & & & \end{array}$$

where $\psi_K$ is the S-isomorphism defined in Lemma 2.7. Hence $C_R(K,U) \cong S/r_S(K)$ under the canonical S-isomorphism $\theta : S/r_S(K) \to C_R(K,U)$. By (1.2) this yields a topological R-isomorphism $\theta^* : H_S(C_R(K,U),U) \to H_S(S/r_S(K),U)$. Composing $\theta^*$ with the topological R-isomorphism $\phi : H_S(S/r_S(K),U) \to \ell_U r_S(K)$ defined in Lemma 2.5 yields an isomorphism of $H_S(C_R(K,U),U)$ with $\ell_U r_S(K)$ over $K$.

(iii) => (ii) From the fact that $C_R(K,U) \in \mathrm{Cogen}\, U_S$ we conclude that $C_R(K,U)$ and $S/r_S(K)$ are U-reflexive. Next, let $\lambda : H_S(C_R(K,U),U) \to \ell_U r_S(K)$ be the topological R-isomorphism over $K$ given by hypothesis and let $\phi : H_S(S/r_S(K),U) \to \ell_U r_S(K)$

be the topological R-isomorphism given by Lemma 2.5. By (1.3), $\lambda^*$ and $(\phi^{-1})^*$ are S-isomorphisms. Since $\mu_K^* \circ \nu_{K^*} = 1_{K^*}$ where $K^* = C_R(K,U)$, $\mu_K^*$ is also an S-isomorphism.

Let $i : K \to \ell_U r_S(K)$ denote the inclusion map. $\mu_K \lambda = i$ so $\mu_K^* \lambda^* = i^*$, and therefore $i^*$ is an S-isomorphism. Hence the composition

$$S/r_S(K) \xrightarrow{\nu_{S/r_S(K)}} C_R(H_S(S/r_S(K),U),U) \xrightarrow{(\phi^{-1})^*} C_R(\ell_U r_S(K),U) \xrightarrow{i^*} C_R(K,U)$$

is an S-isomorphism, and a straightforward check establishes that this composition equals the canonical S-monomorphism $\theta : S/r_S(K) \to C_R(K,U)$.

(ii) => (i) From Proposition 2.9 it suffices to show that every $f \in \mathrm{Cont}_R(K,U)$ with $K$ a submodule of ${}_RU$ can be extended to $\ell_U r_S(K)$. Since $\theta : S/r_S(K) \to C_R(K,U)$ is an S-isomorphism, it follows from (1.2) that $H_S(C_R(K,U),U) \xrightarrow{\theta^*} H_S(S/r_S(K),U)$ is a topological R-isomorphism.

Next, observe that the diagrams

$$\begin{array}{ccc} U & \xrightarrow{\mu_U} & H_S(C_R(U,U),U) \\ \uparrow f & & \uparrow f^{**} \\ K & \xrightarrow{\mu_K} & H_S(C_R(K,U),U) \end{array} \qquad \text{and} \qquad \begin{array}{ccc} \ell_U r_S(K) & \xleftarrow{\phi} & H_S(S/r_S(K),U) \\ \uparrow & & \uparrow \theta^* \\ K & \xrightarrow{\mu_K} & H_S(C_R(K,U),U) \end{array}$$

commute, where $\phi$ is the topological R-isomorphism of Lemma 2.5 and $\mu_U$ is a topological R-isomorphism from (1.4) and (1.1). It follows that $\phi^{-1}(\theta^*)^{-1} f^{**} \mu_U^{-1}$ is the desired extension of $f$ to $\ell_U r_S(K)$.

Finally, we will show that conditions (i)-(iii) imply the last sentence of this theorem. To show that $\mathrm{Refl}_R U = \mathrm{Copres}_R U$, it suffices by Proposition 3.1 to demonstrate that if $0 \to M \xrightarrow{f} U^X \xrightarrow{g} U^Y$ is a copresentation of $M$ then $M$ is U-reflexive. We emulate the proof of Proposition 2.9. Applying (1.3) and Proposition 2.2 to the canonical exact sequence $0 \to M \xrightarrow{f} U^X \xrightarrow{p} U^X/Mf \to 0$ with the quotient topology on $U^X/Mf$, we get an exact sequence of S-modules

$$0 \to C_R(U^X/Mf,U) \xrightarrow{p^*} C_R(U^X,U) \xrightarrow{f^*} C_R(M,U) \to 0.$$

From Proposition 1.2 this yields a commutative exact diagram

$$\begin{array}{ccccccccc}
0 & \longrightarrow & M & \xrightarrow{\quad f \quad} & U^X & \xrightarrow{\quad p \quad} & U^X/Mf & \longrightarrow & 0 \\
 & & \uparrow{\mu_M} & & \uparrow{\mu_{U^X}} & & \uparrow{\mu_{U^X/Mf}} & & \\
0 & \longrightarrow & H_S(C_R(M,U),U) & \xrightarrow{f^{**}} & H_S(C_R(U^X,U),U) & \xrightarrow{p^{**}} & H_S(C_R(U^X/Mf,U),U). & &
\end{array}$$

Now $\mu_{U^X}$ is a topological R-isomorphism by (1.4) and (1.1), $\mu_M$ is a monomorphism because $M \in \text{Cogen}_R U$, and $\mu_{U^X/Mf}$ is a monomorphism by Remark 2.8. A diagram chase now establishes that $\mu_M$ is surjective, and it is open because $f$ and $f^{**}$ are topological R-monomorphisms. So $M$ is U-reflexive.

For any $L \subseteq S$, $\ell_U(L)$ is U-reflexive by Corollary 2.6. Now let $K$ be a U-reflexive submodule of ${}_RU$. Then, by what has just been shown, $K \in \text{Copres}_R U$. By applying Lemma 3.2 to the inclusion map $i : K \to U$ one concludes that $K = \ell_U(L)$ for some $L \subseteq S$. □

A module $M \in$ R-Top is said to be U-quotient-cogenerated if for every closed submodule $N$ of ${}_RM$ and every $m \in M \backslash N$ there exists $f \in \text{Cont}_R(M,U)$ with $Nf = 0$ and $mf \neq 0$. We let $\text{Cogen}^*_R U$ denote the class of U-quotient-cogenerated modules. When $U$ itself is U-quotient-cogenerated then $U$ is called a self-cogenerator; this is weaker than the definition of self-cogenerator used in [4] and [6]. We make the following elementary observation.

(3.4) When $S = \text{Cont}_R(U,U)$ then $U$ is a self-cogenerator if and only if $K = \ell_U r_S(K)$ for every closed submodule $K$ of ${}_RU$.

Recall that $M \in$ R-Top is linearly topologized if there is a neighborhood basis of 0 consisting of submodules of $M$. The equivalence of (i) and (ii) of the following proposition for the case when $U$ is discrete or compact appears in [4; Propositon 2.6].

PROPOSITION 3.5 For U linearly topologized the following conditions are equivalent.

(i) $U^{(t)}$ is a self-cogenerator for every positive integer t.

(ii) $\mathrm{Cogen}_R U \subseteq \mathrm{Cogen}_R^* U$.

(iii) Every topological R-monomorphism $f : M \to U^X$ with $Mf$ closed in $U^X$ extends to a copresentation $0 \to M \xrightarrow{f} U^X \xrightarrow{g} U^Y$ of M.

Proof. (i) => (ii) Suppose that N is a closed submodule of $M \in \mathrm{Cogen}_R U$ and let $i : M \to U^X$ be a topological R-monomorphism. We must show that given $m \in M\backslash N$ there exists $f \in \mathrm{Cont}_R(M,U)$ with $Nf = 0$ and $mf \neq 0$. For $x \in X$, let $\pi_x : U^X \to U$ denote the canonical projection onto the x-coordinate. The basic open neighborhoods of 0 in M are of the form $\bigcap_{k=1}^{t} V(i\pi_{x_k})^{-1}$ for V an open submodule of U.

Since N is closed in M there exists an open neighborhood $m + \bigcap_{k=1}^{t} V(i\pi_{x_k})^{-1}$ of m with $(m + \bigcap_{k=1}^{t} V(i\pi_{x_k})^{-1}) \cap N = \emptyset$. Set $p = \bigoplus_{k=1}^{t} i\pi_{x_k} \in \mathrm{Cont}_R(M,U^{(t)})$ and observe that $mp \notin Np + V^{(t)}$; for otherwise $(m + \bigcap_{k=1}^{t} V(i\pi_{x_k})^{-1}) \cap N = (m + V^{(t)}p^{-1}) \cap N$ would be non-empty.

$Np + V^{(t)} = \bigcup_{n \in N} np + V^{(t)}$ is open and hence closed in $U^{(t)}$. By hypothesis there exists $g_1 \in \mathrm{Cont}_R(U^{(t)},U)$ with $(Np + V^{(t)})g_1 = 0$ and $mpg_1 \neq 0$. $g_1$ induces $g \in \mathrm{Cont}_R(U^{(t)}/Np + V^{(t)},U)$ with $\overline{mp}g \neq 0$; here $U^{(t)}/Np + V^{(t)}$ carries the quotient topology. Let $\pi \in \mathrm{Cont}_R(U^{(t)}, U^{(t)}/Np + V^{(t)})$ denote the canonical epimorphism and set $f = p\pi g \in \mathrm{Cont}_R(M,U)$. Then $Nf = 0$ while $mf \neq 0$.

(ii) => (iii) We apply (ii) to choose for each $v \in U^X\backslash Mf$ an element $g_v \in \mathrm{Cont}_R(U^X,U)$ with $Mfg_v = 0$ and $vg_v \neq 0$. Set $Y = U^X\backslash Mf$ and define $g : U^X \to U^Y$ by $wg = \{wg_v\}_{v\in Y}$ for $w \in U^X$. Then $g \in \mathrm{Cont}_R(U^X,U^Y)$ and clearly $\ker g = Mf$, so $0 \to M \xrightarrow{f} U^X \xrightarrow{g} U^Y$ is exact in R-Top.

(iii) => (i) is clear. □

COROLLARY 3.6 *If* U *is linearly topologized and* $U^{(t)}$ *is a self-cogenerator for every positive integer* t *then* $\mathrm{Copres}_R U = \overline{\mathrm{Cogen}_R} U$.

We remark that if U is complete (in its canonical uniformity) and $U^{(t)}$ is a self-cogenerator for every positive integer t and $S = \mathrm{Cont}_R(U,U)$ then $\mathrm{Refl}_R U = \overline{\mathrm{Cogen}_R} U$. This can be found in [4; Corollary 2.5].

We conclude by giving a description of the U-duality associated with a quasi-injective self-cogenerator. This is the equivalence of conditions (i) and (v) in the following theorem. A different characterization of this U-duality appears in [4; Proposition 4.6].

THEOREM 3.7 *Suppose that* U *is discrete or is compact with no small submodules and that* $S = \mathrm{Cont}_R(U,U)$. *Then the following conditions are equivalent.*

(i) ${}_RU$ *is a quasi-injective self-cogenerator.*

(ii) ${}_RU$ *is quasi-injective and every closed submodule of* ${}_RU$ *is* U-*reflexive.*

(iii) $S/r_S(K)$ *is* U-*reflexive for every submodule* K *of* ${}_RU$ *and* $U^{(t)}$ *is a self-cogenerator for every positive integer* t.

(iv) *For every closed submodule* K *of* ${}_RU$, $S/r_S(K)$ *is* U-*reflexive and* $K = \ell_U r_S(K)$.

(v) $\mathrm{Refl}_R U = \overline{\mathrm{Cogen}_R} U$, $\mathrm{Refl}\ U_S = \mathrm{Cogen}\ U_S$, *and for every submodule* K *of* ${}_RU$, $H_S(C_R(K,U),U)$ *is isomorphic to* $\ell_U r_S(K)$ *over* K.

*Proof*. (i) => (iii) If U is a quasi-injective self-cogenerator and carries a discrete or compact topology then $U^{(t)}$ is a quasi-injective self-cogenerator for every positive integer t (see [1; Lemma 2.5]). Theorem 3.3 gives that $S/r_S(K)$ is U-reflexive for every submodule K of ${}_RU$.

(iii) => (iv) This follows from (3.4).

(iv) => (i) To show that ${}_RU$ is quasi-injective it suffices (by passing to the completion of $K$ if necessary) to show that if $f \in \mathrm{Cont}_R(K,U)$ with $K$ a closed submodule of $_RU$ then $f$ extends to an element of $\mathrm{Cont}_R(U,U)$. Since $K = \ell_U r_S(K)$ by hypothesis, we may apply Proposition 2.9 to get the desired extension.

(i) => (v) In view of Theorem 3.3, we need only show that $\mathrm{Copres}_R U = \overline{\mathrm{Cogen}_R U}$. For $U$ discrete this follows from (iii) and Corollary 3.6. For $U$ compact this was shown in [4; Corollary 2.5].

(v) => (ii) If $K$ is a closed submodule of $_RU$ then $K \in \overline{\mathrm{Cogen}_R U}$, and so is $U$-reflexive by hypothesis. $_RU$ is quasi-injective by Theorem 3.3.

(ii) => (iv) This follows from Theorem 3.3. □

## REFERENCES

(1) Bazzoni, S. "Pontryagin type dualities over commutative rings", 1979, Ann. di Mat. Pura ed Appl. 121, pp. 373-385.

(2) Lambek, J. "Localization at epimorphisms and quasi-injectives", 1976, J. Algebra 38, pp. 163-181.

(3) Lambek, J. and Rattray, B.A. "Localization and duality in additive categories", 1975, Houston J. Math. 1, 87-100.

(4) Menini, C. and Orsatti, A. "Good dualities and strongly quasi-injective modules", 1981, Ann. di Mat. Pura ed Appl. 127, pp. 187-230.

(5) Müller, B.J. "Duality Theory for linearly topologized modules", 1971, Math. Z., pp. 63-74.

(6) Sandomierski, F.L. "Linearly compact modules and local Morita Duality", 1972, Ring Theory, R. Gordon ed., Academic Press, New York, pp. 333-346.

(7) Zelmanowitz, J.M., and Jansen, W. "Duality and linear compactness", 1983, Research Report 83-6, McGill University.

LIST OF PARTICIPANTS

J. Alajbegovic, U.of Sarajevo, Sarajevo 71000, Yugoslavia

S.A. Amitsur, Hebrew U., Jerusalem, Israel

M. Artin, M.I.T., Cambridge, Mass. 02139, USA

M. Auslander, Brandeis U., Waltham, Mass. 02154, USA

B. Auslander, Brandeis U., Waltham, Mass. 02154, USA

H. Bass, Columbia U., New York, NY 10027, USA

J. Bergen, De Paul U., Chicago, IL.60614, USA

F. Bingen, Free U. of Brussels, 1050 Brussel, Belgium

J.L.Bueso, U. of Granada, Granada, Spain

S. Caenepeel, Free U. of Brussels, 1050 Brussel, Belgium

G. Cauchon, U. de Reims, 51062 Reims Cédex, France

M. Cohen, Ben Gurion U., BeerSheva, Israel

J. Hannah, U. College Galway, Galway, Ireland

M. Harada, Osaka City U., Osaka, Japan

I. Herstein, U. of Chicago, Chicago, IL 60637, USA

M. Hervé, Ecole Norm. Sup., 45 Rue d'Ulm, 75005 Paris, France

T. Hodges, U. of Utah, Salt Lake City, UT 84102, USA

R. Hoobler, City College, C.U.N.Y., New York, NY 10031, USA

A. Hudry, U. de Lyon I, 69621 Villeurbanne Cédex, France

R. Irving, U. of Washington, Seattle, Washington 98195, USA

T. Ishii, Kiuki U., 577 Higashi - Osaka - City, Japan

Y. Iwanaga, Shinsu U., Nagano 380, Japan

B. Fein, Oregon State U., Corvallis, OR 97331, USA

T. Ford, Florida Atlantic U., Boca Raton, Florida 33431, USA

E. Formanek, Philadelphia U., PA 16802, USA

E. Friedlander, Oxford U., Oxford, England

A. Giambruno, U. of Palermo, 90100 palermo, Italy

A. Goldie, University of Leeds, Leeds LS 29 JT, England

K. Goodearl, U. of Utah, Salt Lake City, UT 84112, USA

R. Gordon, Temple U., Philadelphia, Penn 19122, USA

W. Haboush, State U. of New York, Albany, NY 12222, USA

J. Haefner, U. of South Calif.,Los Angeles, CA 90089, USA

D.E. Haile, Indiana U., Bloomington, Indiana 47401, USA

W. Jansen, McGill U., Montreal, Canada

S. Jøndrup, Københavns U., 2100 København Ø, Danmark

T. Kanzaki, Osaka's Woman U., Sakai - City, Osaka 590, Japan

I. Kersten, U. Regensburg, 8400 Regensburg, Germany

W. Kimmerle, U. of Stuttgart, 7000 Stuttgart, Germany

A.A. Klein, Tel Aviv U., Ramat-Aviv, Tel-Aviv 69978, Israel

A. Kovacs, Cema 17, P.O.B.2250, Haifa, Israel

J. Krempa, U. of Warsaw, PKiN, 00-901, Warsaw, Poland

L. Le Bruyn, U. of Antwerp UIA, 2610 Antwerp, Belgium

D. Legrand, U. de Lille, Villeneuve d'Asque, Lille, France

S. Legrand, U. de Lille, Villeneuve d'Asque, Lille, France

T. Lenegan, U. of Edinburgh, Edinburgh EH93J2, Scotland, UK

A. Leroy, U.E. à Mons, 7000 Mons, Belgium

M. Lorenz, Max-Planck Inst. für Math., 5300 Bonn 3, Germany

L.G. Makar-Limanov, Wayne State U., Detroit MI 48202, USA

P. Malcolmson, Wayne State U., Detroit MI 48202, USA

M.P. Malliavin, 10 Rue St.Louis en l'Ile, 75004, Paris, France

P. Mamone, U.E. à Mons, 7000 Mons, Belgium

W.3rd. Martindale, U. of Massachusetts, Amherst, USA

Jara P. Martinez, U. de Granada, Granada, Spain

H. Marubayashi, College of Gen. Educ., Osaka U., Osaka, Japan

K. Masaike, Tokyo Gakugei U., Koganei, Tokyo, Japan

C.R. Miers, U. of Victoria, P.O.B. 1700, Victoria,BC.V8W242,CAN.

F.P. Milies, U. Sao Paulo, C.P.20570 Ag. Igratemi,Sao Paulo,Bras.

S. Montgomery, U. of South.Cal., Los Angeles, CA 90089, USA

H.G. Moore, Brigham Young U., Provo, Utah 84602, USA

B. Müller, McMaster U., Hamilton, Ontario L854K1, Canada

I. Musson, U. of Wisconsin, Madison, WI 53706, USA

C. Nastasescu, U. Bucuresti, Bucharest 1, Roumania

E. Nauwelaerts, LUC, Diepenbeek 3610, Belgium

J. Okninski, U. of Warsaw, PKiN 00-901, Warsaw,Poland

A. Ooms, LUC, Diepenbeek 3610, Belgium

A.J. Ornstein, Technion, Haifa, Israel

J. Osterburg, U. of Cincinnati, cincinnati, Ohio 45221, USA

D. Passman, U. of Wisconsin, Madison WI 53706, USA

E. Puczyłowski, U. of Warsaw, PKiN, 00-901 Warsaw, Poland

R. Puystjens, U. of Gent, 9000 Gent, Belgium

I. Reiten, U. of Trondheim, N7000 Trondheim, Norway

R. Resco, U. of Oklahoma, Norman, OK73019, USA

J.C. Robson, U. of Leeds, Leeds LS29JT, England

K. Roggenkamp, U. of Stuttgart, 7000 Stuttgart, Germany

H. Sato, Wakayama U., 1-1 Masago-cho, Wakayama, Japan

D.J. Saltman, U. of Texas, Austin, TX 78712, USA

M. Schacher, U. of California, Los Angeles, CA 90024, USA

W. Schelter, U. of Texas, Austin, TX 78712, USA

A. Schofield, Trinity College, Cambridge U., England

S. Sehgal, U. of Alberta, Edmonton T6G261, Canada

G. Sigurdsson, U. of Washington, Seattle, Washington 98495, USA

L.W. Small, U. of California, San Diego, Cal 92037, USA

P.F. Smith, U. of Glasgow, Glasgow, Scotland, UK

P. Smith, U. of South.Cal., Los Angeles, CA 90089-1113, USA
(now : U. of Warwick, England)

R. Snider, VPI & SU, Blacksburg, VA 24061, USA

J.T. Stafford, U. of Leeds, Leeds LS29JT, England

B. Stenström, Stockholms U., Box 6701, S-11385 Stockholm,Sweden

J.S. Strooker, U. of Utrecht, PB 80010, 3508 Utrecht, NL

J.P. Tignol, UCL, 1348 Louvain-la-Neuve, Belgium

K. Ulbrich, U. Regensburg, 8400 Regensburg, Germany

M. Van den Bergh, U. of Antwerp UIA, 2610 Antwerp, Belgium

J. Van Geel, U. of Gent, 9000 Gent

F. Van Oystaeyen, U. of Antwerp UIA, 2610 Antwerp, Belgium

A. Verschoren, U. of Antwerp UIA, 2610 Antwerp, Belgium

A. Wiedemann, U. of Stuttgart, 7000 Stuttgart, Germany

P. Wauters, KUL, 3030 Heverlee, Belgium

J. Zelmanowitz, U. of California, Santa Barbara, CA 93106, USA

A. Zollner, U. München, München, Germany

# Index

Admissible sequence corresponding to a Kupisch series 455

Almost maximal subfield 275

Approximation family of valuations 20

Ascending Loewy series 39

Atom 39

Bounded index 207

Bounded level 197

Brauer-Severi scheme induced over Y 135

Cancellable 404

Canonical S-monomorphism 560

Characteristic invariant 342

Circle operation 308

Clan 347

Compatible (approximation) family 21

Component regular module 410

Congruence subgroup 418

Conjugate actions 340

Connes spectrum 343

Contractible graph 438

Contractible to $T_1$ 439

Copresentation 559

Copresented module 559

Creating (left) E-limites 91

Cycle map 361

Cycles of maximal ideals 348

Dependent valuations 3

D-injective 317

Dimension formula 333

Disturbed additive function 441

Divisorially graded 293
Dominant indecomposable module 182

E-component functor 90
E-component functor for A in ⊂ 96
E-grading relative to z 105
Elementary automorphisms 416
Enough invertible ideals 365
E-pregrading 89
Equivalence invariants 332
Equivalent orders 185
Equivalent polarized sheaves 137
E-stabilizers 89
E-stable coproduct functor 98
E-stable coproduct system 98
E-stable endfunctor 98
Etale local structure 249
Exceptional Lie ideal 288
Extension property of simple modules 147

F-Artinian module 50,392
Final point 464
Finite length (of a lattice) 41
Finite topology 553
F-Noetherian module 50,392
Free subalgebras of division rings 271
F-stable 392

Generalized Bäkström order 438
Generalized Rees rings 313
Generic Azumaya algebra 136
Generic representation 135
Generic representation 141
Generic splitting variety 140
G-functors 107

GK-homogène 173
Gorenstein order 445
Graded completion 69
Graded Krull order 301
Graded localization 535
Graded maximal order 300
Graded vector fibres 255
Graded v-HC order 304
Graded v-H order 304
Gr-complete 68
Gr-Henselian rings 75,243
Gr-Henselization 79
Gr-I-complete 68
Grothendieck category 93
G-structure on a sheaf 115

H-Galois 519
Homogeneous bundle 116
Homogeneous sheaf 116

Ideal bon 172
Ideal level 417
Incomparable valuations 3
Infinite preprime 197
Infinite prime 197
Infinite torsion prime 197
Initial point 464
Integral ring extension 328

Köthe conjecture 207
Krull order 300
Krull ring 525
Kupisch series 455

Large Jacobson radical 16
Linearly topologized module 563

Link 347
Local signature 197
Loewy-length 39
L-tame order 532

Manis valuation 2
Manis valuation pair 3
Metabelian group 498
Metacyclic group 501
Mini quasi-injective module 147
Morite duality 551
Multilinear monic identity 207

Nakayama permutation 365
n-Gorenstein ring 456
n-Noetherian 393
Normal basis 201
Normal level conjecture 419
n-stable 393

Order 185

Partially ordered set of a Schurian order 446
P.I. ring (with bounded index) 209
Pointed module 409
Polynomial in a deviation 219
Prüfer domain 7
Prüfer domain 7
Pseudo-homothety 426

QF-3 ring 183
QF-2 ring 183
Quasi-Frobenius ring 323
Quasi-injective module 555
Quotient ring 185

Rationally irreducible action 271
Reduced length of a lattice 41
Reduced Loewy length 46
Rees ring 232
Reflexive Azumaya algebras 233
Reflexive ideal 526
Related $\tau$-critical modules 59
Relative Brauer groups 81
R-ideal (left) 185
Right colocal type 152
Right coserial 152
Right self-miniinjective 147
Right v-ideal 296
River 464
R-fractionary ideal 10
R-Priifer ring 8
R-regular element 1

$\sigma$-admissible coproduct system 100
Schurian orders 435
$\sigma$-closed (right) ideal 293
$\sigma$-closure 293
Self-cogenerator 563
Semi-artinian lattice "(
Semi-Artinian object 392
Semi-atomic 39
Semi-simple element 379
Semi-simple groups 126
Serial ring 455
$\sigma$-equivalent coproduct systems 101
Simple Noetherian rings 87
$\Sigma$-injective module 317
$\sigma$-invertible module 542
Small neighbourhood 554
Smooth algebra 483

Smooth maximal order 234
Socle 39
Space of polarized Azumaya algebras 136
Spectrum of a group action 343
Standard lie ideal 288
Strict Henselization 167
Strongly prime nonassociative ring 289
Strongly semiprime nonassociative ring 289
Strongly semiprime torsion theory 50
Subhereditary order 434
Subideal 288

τ-closure of a submodule 53
τ-critical module 55
τ-dense submodule 53,317
τ-finitely generated 318
Topological R-isomorphism 553
Topological R-monomorphism 553
τ-quasi-injective module 62
Trace invariants 342
Translation quiver 441
Trivial automorphism property (TAP) 451
τ-semicritical module 55
τ-torsion

U-duality 555
U-quotient-cogenerated 563
U-reflexive 555

V-closed ideal 3
V-ideal 297
V-idempotent 298
V-invertible 298

Weak Brauer-Severi scheme 262
Weil divisors 232

W-socle 59

z-compatible coproduct functor 105
Zariski local structure 245
Zariski tamifiable 532
Zassenhaus conjecture 451
Zassenhaus subgroup 282